The Encyclopedia of Global Warming Science and Technology

The Encyclopedia of Global Warming Science and Technology

Volume 1: A–H

Bruce E. Johansen

Greenwood Press
An Imprint of ABC-CLIO, LLC

Santa Barbara, California • Denver, Colorado • Oxford, England

Library of Congress Cataloging-in-Publication Data

Johansen, Bruce E. (Bruce Elliott), 1950–
The encyclopedia of global warming science and technology / Bruce E. Johansen.
v. cm.
Includes bibliographical references and index.
Contents: v. 1. A–H—v. 2. I–Z.
ISBN 978-0-313-37702-0 (hard copy : alk. paper) — ISBN 978-0-313-37703-7 (ebook)
1. Global warming—Encyclopedias. 2. Climatic changes—Encyclopedias.
I. Title.
QC981.8.G56J638 2009
577.27′603—dc22 2009005295

13 12 11 10 9 1 2 3 4 5

This book is also available on the World Wide Web as an eBook.
Visit www.abc-clio.com for details.

ABC-CLIO, LLC
130 Cremona Drive, P.O. Box 1911
Santa Barbara, California 93116-1911

This book is printed on acid-free paper ♾

Manufactured in the United States of America

Contents

List of Entries

Guide to Related Topics

Climate and Weather

Extremes of Heat and Cold around the World
Heat-Island Effect, Urban
Heat Waves
Jet Streams
Medieval Warm Period, Debate over Temperatures
Pliocene Paleoclimate
Solar Influences on Climate
Temperatures, Cold Spells
Temperatures, Recent Warmth
Temperatures and Carbon-Dioxide Levels
Thunderstorms and Tornadoes, Frequency and Severity of
"Urban Heat-Island Effect" and Contrarians
Wintertime Warming and Greenhouse-Gas Emissions
Worldwide Climate Linkages

Human Health

Asthma
Dengue Fever
Diseases and Climate Change
Hay Fever
Human Health, World Survey
Kidney Stones
Malaria in a Warmer World
West Nile Virus and Warming

Human Sources of Greenhouse Gases

Agriculture and Warming
Air Conditioning and Atmospheric Chemistry
Air Travel
Automobiles and Greenhouse-Gas Emissions
Coal and Climatic Consequences
Ecotourism
Fossil Fuels
Greenhouse-Gas Emissions, Worldwide
Hydrofluorocarbons (HFCs)
Jet Contrails, Role in Climate Change
Oceans, Carbon-Dioxide Levels
Tar Sands
War, Carbon Footprint

Hydrological Cycle

Deforestation
Desertification
Drought
Drought and Deluge: Anecdotal Observations
Drought and Deluge: Scientific Issues
Drought in Western North America
Hadley Cells
Hydrological Cycle
Jet Streams
Tropical Zones, Expansion of
Water Supplies in Western North America
Wildfires

Ice Melt

Glacial (Ice Age) Cycle, Prospective End of
Ice Melt, World Survey

Ice Melt—Antarctic

Antarctica and Climate Change
Antarctica and Debate over Inland Cooling
Antarctica and Speed of Ice Melt
Antarctic Oscillation
Antarctic Paleoclimatic Precedents
Antarctic Peninsula and Ice Shelf Collapse
Antarctic Peninsula and Warming

Ice Melt—Arctic

Ice Melt—Mountains

Oceans and Seas

Oceans and Seas—Plant and Animal Life

People

Plants and Animals

Plants and Animals—Aquatic

Fish Kills, New York Lakes
Flora and Fauna: Worldwide Survey
Food Web and Warming in Antarctica: Phytoplankton to Penguins
Great Barrier Reef, Australia
Ocean Food Web
Ocean Life: Whales, Dolphins, and Porpoises
Tropical Fish and Sharks off Maine
Tropical Fish and Warm-Climate Birds Migrate

Plants and Animals—Domesticated

Agriculture and Warming
Baseball Bats and Warming
Carbon Dioxide: Enhanced and Crop Production
Lobster Catches Decline in Warmer Water
Maple Syrup Wanes and Other Changes in New England
Salmon Decline in Warming Waters
Wine Grapes and Warming

Plants and Animals—Specific Species

Alligators Spread Northward
Anchovies Spread Northward
Armadillos Spread Northward
Bark Beetles Spread across Western North America
Coelacanth
Cuckoo Numbers Decline in Great Britain
Cyanobacterial Algal Blooms
Gelada Baboon
Giant Squid
Gray Whales and El Niño
Jellyfish Populations and Potency
Loggerhead Sea Turtles: Warmth Alters Gender Ratio
Palms in Southern Switzerland
Penguins, South African
Phytoplankton Depletion and Warming Seas
Pika Populations Plunge in the Rocky Mountains
Pine Beetles in Canada
Polar Bears under Pressure
Red Squirrels' Reproductive Cycle
Rice Yields Decrease with Rising Temperatures
Sea Birds Starve as Waters Warm
Spruce Bark Beetles in Alaska
Walruses and Melting Ice
Wildlife, Arctic: Musk Oxen, Reindeer, and Caribou
Wildlife, Arctic: Threats to Harp Seals in Canada

Politics, Economics, and Diplomacy

Contrarians (Skeptics)
Disaster Relief, Global Warming's Impact on
Economics of Addressing Global Warming, The
Gaia: The Eradication of Industrial Civilization?
Global Climate Coalition
Global Warming: Importance of the Issue
Global Warming: Origins as a Political Issue
Human Rights, Global Warming, and the Arctic
Intergovernmental Panel on Climate Change: The Politics of Climate Consensus
Kyoto Protocol
Legal Liability and Global Warming
National Security and Global Warming
Populations and Material Affluence
Poverty and Global Warming
Public Opinion
Public Protests Escalate over Energy-Policy Inertia
U.S. Climate Action Partnership (USCAP)
War, Carbon Footprint

Regional Effects (Places)

Alaska, Global Warming in
Amazon Valley, Drought and Warming
Australia: Heat, Drought, and Wildfires
Bangladesh, Sea-Level Rise in
China and Global Warming
Darfur
Greenland, Ice Melt
Island Nations and Sea-Level Rise
Land-Use Changes and the Amazon's Carbon Flux
Northwest and Northeast Passages
Sachs Harbour, Banks Island, Climate Change
Tanganyika, Lake: Warming Waters Choke Life
Venus

Regional Effects (Places)—Europe

Alps (and Elsewhere in Europe), Glacial Erosion
Great Britain, Weather Conditions and Leadership in Greenhouse Diplomacy
Ireland, Flora and Fauna in
Netherlands, Sea Levels
Russia, Glacial Collapse in
Saint Andrews Golf Course
Scandinavia and Global Warming
Scotland, Climate Changes in
Venice, Sea-Level Rise

Regional Effects (Places)—Mountains

Alps (and Elsewhere in Europe), Glacial Erosion
Andes, Glacial Retreat
Glacial Retreat: Comparative Photographs and Survey
Glacier National Park
Glaciers, Alaska
Glaciers, Andes
Glaciers, Rocky Mountains: Gone in 30 Years?
Himalayas, Glacial Retreat

Regional Effects (Places)—North America

Scientific Issues

Solutions

Preface

Building a Sustainable Future Is Not a Luxury

With the advent of a global financial crisis that may soon rival the Great Depression, I read a disturbing volley of reports asserting, with a flair for ironic punnery, that global warming is now on the "back burner." Can we "afford" such a "luxury," the reports ask, as if planning for a survivable future is a frill.

Building a sustainable future is not a luxury. The bad news is that we have no real choice. The really good news, however, is that creating a new energy infrastructure, done correctly, can function as an economic motor that will power our communities, and our world, out of a morass created by unchecked, short-sighted greed.

Just as our financial infrastructure needs to be reconstructed from its dangerous dependence on a surplus of borrowed money, so, too, our energy system must be recast from a fossil-fuel base that is living on borrowed environmental time.

The Need for Understanding of the Science

The Encyclopedia of Global Warming Science and Technology is offered to address a fundamental disconnect between concepts of global warming developed in scientific journals and much of the popular debate in the popular realm. This disconnect often vexes scientists, who realize that while they can propose a new course, it is the politicians, businesspeople, and other nonscientists who will ultimately decide how much, and how soon, the system by which we acquire and use energy (which drives greenhouse-gas production) will change.

One study of public opinion describes a mental landscape in which few people outside of scientific specialties in climate change (even many with training in other sciences) do not understand the delayed-effect nature of the problem. A large majority of people, asked to rate the most important threats to their lives, mention present-day, obvious problems (crime, loss of jobs and savings to financial turmoil, war, terrorism, and so on) over climate change and other environmental threats that are less obvious today but pose greater long-term risks. Knowledge of basic geophysical concepts having to do with thermal inertia and feedback loops is low, except among specialists. Even the statistical nature of accumulation (slowing the growth in greenhouse-gas levels, as opposed to actually reducing the total burden) escapes many people. Thus, many people think the specialists are overreacting, and instead favor a "go-slow" or "wait-and-see" approach (Sterman 2008, 532–533).

Many scientific concepts emerge only rarely on the op-ed pages, in the hands of political pundits whose audiences sometimes number in the millions. In their hands, the debate is phrased most often in political or moralistic terms. The more strident of these pundits dismiss global warming as a cult or theology, dismissing its scientific basis entirely.

In the scientific journals, the subject is studied with reference to the way in which the Earth system operates, invoking such concepts as thermal inertia, feedback loops, and various aspects of oceanic and atmospheric circulation in the context of paleoclimate (the Earth's climatic history). An understanding of such things is necessary at the popular level because the atmosphere gives us a reaction to today's greenhouse-gas emissions a half-century from now, demanding that popular opinion anticipate the future, and not just react to present conditions.

Scope of the Work

The scope of this work is at once global and local, involving all 6 billion–plus people of the Earth, and each individual on a personal basis. Readers may be surprised at the many ways in which a changing climate shapes the conduct of their daily lives, from sea level (for the many people who live near the coasts), to the survival of many plants and animals. Gigantic masses of ice erode at both poles, as a solitary Arctic trekker has his plans wrecked by unanticipated open water near the North Pole. Millions of marginal farmers cope with heat and drought as Arctic hunters fall to their deaths though rapidly thinning ice. The maple syrup harvest diminishes, and the quality of the wood used for major-league baseball bats declines. Jellyfish mass in unexpected places.

The format of this work is alphabetical, with a topical guide and detailed index that allows easier negotiation of the about 300 entries in two volumes. Each entry is followed by detailed lists of further readings; the work also includes a bibliography with more than 2,000 references cited, which may be the most detailed in the field. This bibliography indicates the wide range of source material that has gone into this set, from newspaper articles, to Internet sites, government reports, articles in scientific journals, and books from several fields. Climate change is an unusually multidisciplinary field with a range of knowledge that is growing unusually quickly, which, until recently, has lacked a shelf of encyclopedic references.

An Energy Revolution Is Overdue

In 100 years, students of history may remark at the nature of the fears that stalled responses to climate change early in the twenty-first century. Skeptics of global warming kept change at bay by appealing to most people's fear of change that might erode their comfort and employment security, all of which were wedded psychologically to the massive burning of fossil fuels. A necessary change in our energy base may have been stalled, these students may conclude, beyond the point at which climate change forced attention, comprehension, and action.

Technological change always generates fear of unemployment. Paradoxically, such changes also always generate economic activity. A change in our basic energy paradigm during the twenty-first century will not cause the ruination of our economic base, as some "skeptics" of climate change believe, any more than the coming of the railroads in the nineteenth century ruined an economy in which the horse was the major land-based vehicle of transportation. The advent of mass automobile ownership early in the twentieth century propelled economic growth, as did the transformation of information-gathering with computers in the recent past. The same developments also put out of work blacksmiths, keepers of hand-drawn accounting ledgers, and anyone who repaired manual typesetters.

We are overdue for an energy system paradigm shift. Limited supplies of oil and their location in the volatile Middle East provide arguments for new sources, along with accelerating climate change from greenhouse gases accumulating in the atmosphere. According to an editorial in *Business Week* on August 16, 2004,

> A national policy that cuts fossil-fuel consumption converges with a geopolitical policy of reducing energy dependence on Middle East oil. Reducing carbon dioxide emissions is no longer just a 'green" thing. It makes business and foreign policy sense, as well.... In the end, the only real solution may be new energy technologies. There has been little innovation in energy since the internal combustion engine was invented in the 1860s and Thomas Edison built his first commercial electric generating plant in 1882. (Carey and Shapiro 2004)

Before the end of this century, the urgency of global warming will become manifest to everyone. Solutions to our fossil fuel-dilemma—solar, wind, hydrogen, and others—will evolve during this century. Within our century, necessity will *compel* invention. Other technologies may develop that have not, as yet, even broached the realm of present-day science fiction, any more than digitized computers had in the days of the Wright Brothers a hundred years ago. We will take this journey because the changing climate, along with our own innate curiosity and creativity, will compel a changing energy paradigm.

Such change will not take place at once. A paradigm change in basic energy technology may require the better part of a century, or longer. Several technologies will evolve together. Oil-based fuels will continue to be used for purposes that require them. (Air transport comes to mind, although engineers already are working on ways to make jet engines more efficient.)

The coming energy revolution will engender economic growth and become an engine of wealth creation for those who realize the opportunities that it offers. Denmark, for example, is making every family a share-owner in a burgeoning wind-power industry. Solutions will combine scientific achievement and political change. We will end this century with a new energy system, one that acknowledges nature and works with its needs and cycles. Economic development will become congruent with the requirements of sustaining nature. Coming generations will be able to mitigate the effects of greenhouse gases without the increase in poverty so feared by "skeptics." Within decades, a new energy paradigm will be enriching us and securing a future that works with the requirements of nature, not against them.

Acknowledgments

What would I do without my support group? As an author of a large reference book, probably not much—or not enough to get the job done. No one cobbles together a 460,000-word reference set without a lot of behind-the-scene help: in this case, from James E. Hansen and the rest of the people at the Goddard Institute for Space Studies; Robert Hutchinson, my editor; Gail Baker, dean of the College of Communication, Fine Arts, and Media, University of Nebraska–Omaha (thanks for the new computer, Gail!); and Jeremy Lipschultz, director of the School of Communication (both also are important to freeing up time to write). Add my wife, Pat Keiffer, whose sage advice, warm companionship, and common sense contribute immeasurably to my work (the practical details of a good home, too). Shannon, Madison, and Samantha all contributed, as did my mother, without whose labor 59 years ago I would never have written anything.

FURTHER READING

Carey, John, and Sarah R. Shapiro. "Consensus Is Growing among Scientists, Governments, and Business That They Must Act Fast to Combat Climate Change." *Business Week*, August 16, 2004, n.p. (LEXIS).

Sterman, John D. "Risk Communication on Climate: Mental Models and Mass Balance." *Science* 322 (October 24, 2008):532–533.

Introduction

Endgame Approaching

During nearly two decades of study, writing, and commentary on global warming, I have been amazed at how quickly the endgame has chased us. Twenty years, ago the problem seemed comfortably in the future tense. Yes, the Arctic Ice Cap might melt, but not until the end of the century. Having lost nearly a quarter of its mass in one year alone (2007), projections of the ice cap's life in summer now range in the very low two figures. In 1995, 450 parts per million carbon dioxide in the atmosphere seemed a tolerable turning point before the Earth's ecosystems sustained major damage. The latest scientific assessments have lowered that limit to between 350 and 400, and we are now at 385, and rising rapidly.

This is the first encyclopedia of global warming to concentrate on science and technology. I have chosen this emphasis because a great need exists for public understanding of the scientific basis for what has become an enduring environmental issue and an acerbic public-policy debate. Nonscientists need to be conversant with the basics of an issue that is so vital to our future. The scientific academies of 13 countries on June 10, 2008, urged the world to act more forcefully to limit the threat posed by human-driven global warming. In a joint statement, the academies of the Group of 8 (G-8) industrial countries (Britain, Canada, France, Germany, Italy, Japan, Russia, and the United States) and of Brazil, China, India, Mexico, and South Africa called on the industrial countries to lead a transition to a low-carbon society and to move aggressively to limit the impacts from changes in climate that are already under way and impossible to stop.

At its basis, global warming is a scientific story. The most important concepts here bear on feedback loops, thermal inertia, and the compounding nature of greenhouse-gas emissions. The real news of global warming is not how warm it is today, because today's carbon emissions do not give us tomorrow's temperature. The real debate isn't over how much the oceans may rise from melting ice by the end of this century (one to three feet, perhaps), but how much melting will be "in the pipeline" by that time, for the next century and beyond. Because of thermal inertia, the wind we feel in our faces today carries the greenhouse forcing of roughly 50 years ago, when the amount of carbon dioxide, methane, and other heat-trapping gases that are being released into the air was one-fourth of today's combustion. Emissions of greenhouse gases worldwide have risen 70 percent since 1970 and could rise an additional 90 percent by 2030 under a "business-as-usual" scenario.

Across Alaska, northern Canada, and Siberia, scientists already are finding telltale signs that permafrost is melting more quickly. As permafrost melts, additional carbon dioxide and methane convert from solid form, stored in the Earth, to gas, in the atmosphere, retaining more heat. Once again, human contributions of greenhouse gases are provoking a natural process, like the trigger of a gun. This process compounds itself, accelerating over time. Thus, in 50 years, when our children are grandparents, the planetary emergency in which we are now tasting the first course will be a dominant theme in everyone's life, unless we act now. Within a decade or two, thermal inertia will take off on its own, portending a hot, miserable future for coming generations. Thermal inertia explains why so many scientists find the problem so urgent now.

Global warming can be tricky, not only because of its delayed effects, but also because it is not simplistically linear. Weather is variable, and many other forcings, or influences, come in to play. The association between rising carbon-dioxide levels and temperature is not seamless—a mistaken assumption made by contrarians who seem to jump on every cold wave as evidence of a new ice age, or at least proof that global warming's back has been broken. Other contrarians take such variations as proof that carbon dioxide has nearly nothing to do with warming because its rise and temperatures do not match exactly, year by year.

Worldwide temperatures hit an exceptional peak in 1998 with a very strong El Niño, for example, and then spent the next 10 years backing and filling, as cooler La Niña conditions set in. "Too many think global warming means monotonic relentless warming everywhere year after year," said Kevin Trenberth, a climate scientist at the National Center for Atmospheric Research in Boulder, Colorado. "It does not happen that way" (Revkin 2008d). "We're learning that internal climate variability is important and can mask the effects of human-induced global change," said the lead author of a paper on the subject in *Nature* Noel Keenlyside of the Leibniz Institute of Marine Sciences in Kiel, Germany (Keenlyside et al. 2008, 84–88). "In the end this gives more confidence in the long-term projections" (Revkin 2008d; Keenlyside et al. 2008, 84–88).

Kofi Annan, secretary general of the United Nations until 2007, said the following during his last speech in that position:

> The scientific consensus, already clear and incontrovertible, is moving toward the more alarmed end of the spectrum. Many scientists long known for their caution are now saying that warming has reached dire levels, generating feedback loops that will take us perilously close to a point of no return. A similar shift may be taking place among economists, with some formerly circumspect analysts saying it would cost far less to cut emissions now than to adapt to the consequences later. Insurers, meanwhile, have been paying out more and more each year to compensate for extreme weather events. And growing numbers of corporate and industry leaders have been voicing concern about climate change as a business risk. The few skeptics who continue to try to sow doubt should be seen for what they are: out of step, out of arguments, and just about out of time. (Annan 2006a, A-27; 2006b)

The Earth's rising temperature is gradually raising sea level both through thermal expansion of the oceans and the melting of glaciers and ice sheets. Scientists are particularly concerned by the melting of the Greenland ice sheet, which has accelerated sharply in recent years. If this ice sheet, a mile thick in some places, were to melt entirely, it would raise sea level by 23 feet, or seven meters. Even a one-meter (39 inch) sea-level rise would inundate vast areas of low-lying coastal land, including many of the rice-growing river deltas and floodplains of India, Thailand, Vietnam, Indonesia, and China (Brown 2006).

A one-meter rise in sea level could inundate half of Bangladesh's rice-growing land. About 30 million Bangladeshis would be forced to migrate. Lester Brown has written that

> Several hundred cities, including some of the world's largest, would be at least partly inundated by a one-meter, rise in sea level, including London, Alexandria, and Bangkok. More

> than a third of Shanghai, a city of 15 million people, would be under water. A one-meter rise combined with a 50-year storm surge would leave large portions of Lower Manhattan and the National Mall in Washington, D.C., flooded with seawater. (Brown 2006)

Even as we approach climatic endgame, carbon dioxide continued to race upward in 2007 at a rate of 2.9 percent, faster than any of the experts' worst predictions, to 8.47 gigatons (billions of metric tons) according to the Australia-based Global Carbon Project, an international consortium of scientists that tracks emissions. This output is at the very high end of scenarios outlined by the Intergovernmental Panel on Climate Change (IPCC) and could translate into a global temperature rise of more than 11 degrees Fahrenheit by the end of the century, according to the panel's estimates (Eilperin 2008c). Major contributors included China, India, and Brazil, which have doubled their carbon emissions in less the 20 years. Total carbon emissions from industrial nations as a whole have risen only slightly since 1990.

During September 2008, two scientists with the Scripps Institution of Oceanography at the University of California at San Diego published research in the *Proceedings of the National Academy of Sciences,* indicating that if greenhouse gas emissions had stopped completely as of 2005, The world's average temperature still would increase by 2.4 degrees C. (4.3 degrees F.) by the end of the twenty-first century. Richard Moss, vice president and managing director for climate change at the World Wildlife Fund, said the new carbon figures and research show that "we're already locked into more warming than we thought" (Eilperin 2008c; Ramanathan and Feng 2008, 14,245–14,250).

Even as greenhouse-gas concentrations race upward, an energy revolution is under way, using alternative sources such as solar, wind, and geothermal. William Moomaw, a lead author of a chapter in the IPCC's 2007 assessment on energy options and a professor of international environmental policy at Tufts University, said that

> Here in the early years of the 21st century, we're looking for an energy revolution that's as comprehensive as the one that occurred at the beginning of the 20th century when we went from gaslight and horse-drawn carriages to light bulbs and automobiles. In 1905, only 3 percent of homes had electricity. Right now, 3 percent is about the same range as the amount of renewable energy we have today. None of us can predict the future any more than we could in 1905, but that suggests to me it may not be impossible to make that kind of revolution again. (Revkin 2007b)

So much is true, perhaps—but will it be too little, too late?

Thomas Friedman, a columnist for the New York *Times*, visited Greenland and remarked:

> My trip with Denmark's minister of climate and energy, Connie Hedegaard, to see the effects of climate change on Greenland's ice sheet leaves me with a very strong opinion: Our kids are going to be so angry with us one day. We've charged their future on our Visa cards. We've added so many greenhouse gases to the atmosphere, for our generation's growth, that our kids are likely going to spend a good part of their adulthood, maybe all of it, just dealing with the climate implications of our profligacy. (Friedman 2008b)

Greenhouse gases have no morals, loyalty, nor party affiliation. Carbon dioxide is not having a debate with us. It merely retains heat. Humans are now determining the course of climate, and as we pass tipping points, human beings may lose any ability to influence the climatic future. Thus, we approach the endgame in the environment the human race has known since its origins.

FURTHER READING

Annan, Kofi. "As Climate Changes, Can We?" *Washington Post*, November 8, 2006a, A-27. http://www.washingtonpost.com/wp-dyn/content/article/2006/11/07/AR2006110701229_pf.html.

Annan, Kofi. "Global Warming an All-Encompassing Threat." Address to United Nations Conference on Climate Change, Nairobi, Kenya. Environment News Service, November 15, 2006b. http://www.ens-newswire.com/ens/nov2006/2006-11-15-insann.asp.

Brown, Lester R. "The Earth Is Shrinking." Environment News Service, November 20, 2006. http://www.ens-newswire.com.

Eilperin, Juliet. "Carbon Is Building Up in Atmosphere Faster Than Predicted." *Washington Post*, September 26, 2008c, A-2. http://www.washingtonpost.com/wp-dyn/content/article/2008/09/25/AR2008092503989_pf.html.

Friedman, Thomas. "Learning to Speak Climate." *New York Times*, August 6, 2008b. http://www.nytimes.com/2008/08/06/opinion/06friedman.html.

Keenlyside, N. S., M. Latif, J. Jungclaus, L. Kornblueh, and E. Roeckner. "Advancing Decadal-scale Climate Prediction in the North Atlantic Sector." *Nature* 453 (May 1, 2008):84–88.

Ramanathan, V., and Y. Feng. "On Avoiding Dangerous Anthropogenic Interference with the Climate System: Formidable Challenges Ahead." *Proceedings of the National Academy of Sciences* 105, no. 38 (September 23, 2008):14,245–14,250.

Revkin, Andrew C. "Climate Panel Reaches Consensus on the Need to Reduce Harmful Emissions." *New York Times*, May 4, 2007b. http://www.nytimes.com/2007/05/04/science/04climate.html.

Revkin, Andrew C. "In a New Climate Model, Short-term Cooling in a Warmer World." *New York Times*, May 1, 2008d. http://www.nytimes.com/2008/05/01/science/earth/01climate.html.

The Encyclopedia of Global Warming Science and Technology

A

Acidity and Rising Carbon-Dioxide Levels in the Oceans

By the early years of the twenty-first century, carbon dioxide levels were rising in the oceans more rapidly than any time since the age of the dinosaurs, according to work published by Ken Caldeira and Michael E. Wickett. They wrote:

> We find that oceanic absorption of CO_2 from fossil fuels may result in larger pH changes over the next several centuries than any inferred in the geological record of the possible 300 million years, with the possible exception of those resulting from rare, extreme events such as bolide impacts or catastrophic methane hydrate degassing. (Caldeira and Wickett 2003, 365)

A "bolide" is a large extraterrestrial body (such as a large asteroid), usually at least a half-mile in diameter and sometimes much larger, that impacts the Earth at a speed roughly equal to that of a bullet in flight. "Methane hydrate degassing" involves the rapid conversion of solid methane deposits on ocean floors to a gaseous form in the atmosphere by warming temperatures.

Already, by the year 2008, scientists measured levels of ocean-surface acidity 30 percent above preindustrial levels. Under a business-as-usual scenario, acidity levels could be 100 to 150 percent higher in a century, imperiling shelled animal life throughout much of the world's oceans. This danger is most notable in the colder waters of the Arctic and Antarctic, which hold more carbon dioxide than warmer oceans (Holland 2001, 110–111).

Scientists have begun to investigate what continued ocean acidification might do to other animals with calcium shells. Gretchen Hofmann of the University of California–Santa Barbara reported that rising ocean temperatures and acidification could be fatal to the purple sea urchin (*Stronylocentrotus purpuratus*). At a pH level of 7.8 the larvae of the purple sea urchin build skeletons with great difficulty. Warming the water in which the sea urchins live compounds the effect (Kintisch and Stokstad 2008, 1029).

Hofmann and Victoria Fabry of California State University–San Marcos studied the effects of temperature rise and decreased pH on the pteropod *Limacina helicina*, a swimming snail that is important to the food web in the oceans of the Southern Hemisphere. In their evolution, many pteropods have never experienced acid levels as high as those that will occur under business-as-usual carbon emissions in the next century.

Computer models forecast that polar waters will no longer sustain viable populations of pteropods at the rate that acidity has been increasing. Once acidity reaches levels that dissolve calcium shells in the tropics, "It's a doomsday scenario for coral reefs," said Caldeira—that is, for corals not already killed by rising water temperatures. He anticipates that coral reefs will survive only in walled-off areas where acidity has been controlled by humankind in open ocean enclosures. "Our emissions are huge compared with natural fluxes," said Caldeira. "If you could stop emissions and wait 10,000 years, natural processes would probably take care of most of it" (Holland 2001, 111). Emissions, however, are not being curtailed.

Jason M. Hall-Spencer and colleagues studied the effects of acidification in ecosystems at shallow coastal sites where volcanic carbon-dioxide vents lower the pH of the water and found that calcareous organisms, such as corals and sea urchins, were adversely affected. This is probably the first time that the effects of elevated acidity have been tested in ocean water. Most other studies have been done under laboratory conditions. The species populating the vent sites include a suite of organisms that are resilient to naturally high concentrations of carbon dioxide and indicate that ocean acidification may benefit highly invasive non-native algal species" (Hall-Spencer et al. 2008, 96).

As scientists learn more about the acidification of the oceans, the reality of the threat becomes more evident. The date at which increasing carbon-dioxide levels in the oceans are expected to change acidity enough to dissolve the calcium-carbonate shells of corals, planktons, and other marine animals has now advanced to the next few decades, sooner than previously projected. A team of scientists writing in *Nature* said that "In our projections, Southern Ocean surface waters will begin to become under-saturated with respect to aragonite, a metastable form of calcium carbonate, by the year 2050. By 2100, this under-saturation could extend throughout the entire Southern Ocean and into the subarctic Pacific Ocean" (Orr et al. 2005, 681).

Ocean Acidification in the Present Tense

Ocean acidification to a degree that damages coral calcification is no longer only a theory. Scientists investigated 328 colonies of massive *Porites* corals, which grow to more than six meters tall over decades to centuries, on the Great Barrier Reef off Australia. Results from 69 sections of the reef found that calcification had declined 14.2 percent between 1990 and 2005, impeding the reefs' growth by 13.3 percent (De'ath et al. 2009, 116). Such a sudden, massive decline in the reef's calcification had no precedent in recorded history, about 400 years. Increasing temperature stress and a rising carbon-dioxide level in the water around the reef are the probable causes. "This study has provided the first really vigorous snapshot of how calcification might be changing [worldwide]," said marine biologist Ove Hoegh-Guldberg of Australia's University of Queensland. "The results are extremely worrying" (Pennisi 2009, 27).

By 2008, scientists surveying waters near the west coast of North America found rising levels of acidified ocean water within 20 miles of the shoreline, raising concern for marine ecosystems from Canada to Mexico. Researchers on the *Wecoma*, an Oregon State University research vessel, discovered that the acidified upwelling from the deeper ocean is probably 50 years old. Future ocean acidification levels probably will rise as atmospheric levels of carbon dioxide increase (Feely et al. 2008, 1490; "Pacific Coast" 2008).

"When the upwelled water was last at the surface, it was exposed to an atmosphere with much lower CO_2 (carbon dioxide) levels than today's," said Burke Hales, an associate professor in the College of Oceanic and Atmospheric Sciences at Oregon State University, a co-author of the study. "The water that will upwell off the coast in future years already is making its undersea trek toward us, with ever-increasing levels of carbon dioxide and acidity." According to Hales, the researchers found that the 50-year-old upwelled water had carbon-dioxide levels of 900 to 1,000 parts per million (ppm), placing it "right on the edge of solubility" for calcium carbonate-shelled aragonites ("Pacific Coast" 2008). Continued carbon-dioxide overload in the oceans could make ocean water more acidic than it has been "for tens of millions of years and, critically, at a rate of change 100 times greater than at any time over this period" (Riebesell et al. 2007, 545).

Carbon dioxide is being injected into the oceans much more quickly than nature can neutralize it. Seawater is usually alkaline, about 8.2 pH. The pH scale is logarithmic, so a 0.1 decrease in pH, the change since the beginning of the Industrial Revolution, indicates a 30 percent increase in the concentration of hydrogen ions. Under a business-as-usual scenario, the pH will fall 0.5 by the year 2100, increasing the level of hydrogen ions to three times the preindustrial "baseline" concentration (Henderson 2006, 30).

Since the Industrial Revolution began, human beings have infused roughly 120 billion tons of carbon dioxide into the oceans. By 2006, the seas were absorbing an additional two billion tons of carbon dioxide per year. Every day, each citizen of the United States adds, on average, 40 pounds of carbon dioxide to the world ocean (Kolbert 2006b, 68–69). In 1800, the carbon-dioxide level of the atmosphere was 280 ppm, and the oceans' pH averaged 8.16. Today, atmospheric carbon dioxide is 380 ppm, and ocean pH averages 8.05. Some estimates suggest a fall of pH to 7.9 by the year 2100 (Ruttimann 2006, 978).

A report on ocean acidification by Britain's Royal Society said that "without significant action to reduce CO_2 emissions" there may be "no place in the future oceans for many of the species and ecosystems we know today" (Kolbert 2006b, 74). Stated more simply, increasing acidification of the oceans because of rising levels of carbon dioxide may threaten a large number of ocean species with extinction.

The effects of acidity in the oceans will continue long after burning of fossil fuels peaks on land. Ken Caldeira modeled ocean acidification for fossil-fuel burning that peaks in the year 2100 and found that the oceans will continue to become more acidic for centuries after that. At the surface, acidity will peak at about 2750 C.E. A kilometer deep in the ocean, acidification will rise for a thousand years. "People would know that the consequences of what we're doing in the next decade will last for thousands of years," said Caldeira (Ruttimann 2006, 979–980).

Corals at Risk

By the end of the twenty-first century, according to Caldeira, surface acidity around Antarctica will be roughly double preindustrial levels (a 0.2 decrease in pH), threatening life forms ability to maintain their shells (Kolbert 2006b, 70). Such a level would put about two-thirds of cold-water corals in corrosive waters (Kintisch and Stokstad 2008, 1029).

Acidification will affect corals acutely, dissolving their shells at a time when warming temperatures are already threatening their survival. "While bleaching … is an acute stress that's killing them off … acidification is a chronic stress that's preventing them from recovering," said Joanie Kleypas, a coral-reef scientist at the National Center for Atmospheric Research in Boulder, Colorado (Kolbert 2006b, 72).

Ilsa B. Kuffner of the U.S. Geological Survey in St. Petersburg, Florida, reported in *Nature Geoscience* that an increase in ocean acidity harms crustose coralline algae, the builders of coral reefs, as well as the reefs themselves. Kuffner and colleagues manipulated the pH of ocean water near reefs in Hawaii for conditions expected in the year 2100 and found, after seven weeks, that "[t]here was far less algae encrusted on clear plastic cylinders inside the more acidic tanks" (Fountain 2008, D-3). Instead, the space was taken by "soft" algae that do not secrete calcium carbonate. The secretion of calcium by crustose coralline algae acts as a mortar to maintain the structure of reefs.

Corals are among the richest biological areas of the oceans, and thus, acidification is a major, long-term threat to aquatic life. Thomas Lovejoy, who coined the term "biological diversity" in 1980, compared the effects of ocean acidification to "running the course of evolution in reverse." According to Lovejoy, the two most important biological factors for organisms in the ocean are temperature and acidity. The effects of changes in both provoked by human-induced carbon-dioxide emissions reach to the base of the oceanic food chain, with profound long-term implications favoring "lower" forms of life such as jellyfish and other invertebrates. In the very long run, some scientists fear human intervention in the oceanic system may be favoring the return to slime as a predominant life form there, according to German marine biologist Ulf Riebesell (Kolbert 2006b, 75).

Deep-water corals are at risk from increasing ocean acidification for several reasons. First, they are composed of aragonite, a carbonate material that is more soluble than the calcite used by corals closer to the surface. Carbonates' vulnerability to dissolution also increases in colder water at greater pressure. By the end of this century, two-thirds of deep-water corals (compared with none today) could be exposed to seawater that is corrosive to aragonite.

The death of corals (coccolithophores) near the surface of the oceans could amplify global warming because of albedo. When they bloom, these organisms lighten the surface, reflecting sunlight. They also produce dimethylsulfide, which accounts for much of the aerosolized sulfate in the air above the oceans, which "seed" cloud droplets. Without them, oceanic cloud cover could decline, allowing more sunlight and heat to reach the surface (Ruttimann 2006, 980).

Caldeira says that "[i]f you look at the business-as-usual scenario for emissions and its impact with respect to aragonite in surface waters, by the end of the century there is no place left with the kind of chemistry where corals grow today" (Henderson 2006, 31). Corals could become rare because temperatures and acidity levels are rising at the same time. In addition, pteropods, which form an important part of the food chain for cod, salmon, and whales in colder water, could find their shells dissolving at the lower pH levels anticipated by the year 2050 (Henderson 2006, 31).

An experiment conducted between 1996 and 2003 at Columbia University's Biosphere 2 lab in Tucson, Arizona, concluded that corals' growth in the lab was reduced by half compared with growth in aquariums, where the corals were exposed to a level of carbon dioxide projected to for the year 2050. Coupled with the warmer sea temperatures that climate change produces, Langdon said, corals may not survive by the end of the century. "It's going to be on a global scale and it's also chronic," Langdon said of ocean acidification. "Twenty-four/seven, it's going to be stressing these organisms.... These organisms probably don't have the adaptive ability to respond to this new onslaught" (Eilperin 2006d, A-1). Stanford University marine biologist Robert B. Dunbar has studied the effect of increased carbon dioxide on coral reefs in Israel and Australia's Great Barrier Reef. "What we found in Israel was [that] the community is dissolving," Dunbar said (Eilperin 2006d, A-1).

Caldeira has mapped areas where corals grow today and the pH levels of the water in which they live. He maintains that by the end of the twenty-first century, no seawater will be as alkaline as corals' present-day habitats. If carbon dioxide emissions continue at their current levels, "It's say goodbye to coral reefs" (Eilperin 2006d, A-1).

In addition to corals, rising carbon-dioxide levels in the oceans could threaten the health of many marine organisms, beginning with plankton at the base of the food chain. Regarding the acidification of the oceans, "We're taking a huge risk," said Ulf Riebesell, a marine biologist at the Leibniz Institute of Marine Sciences in Kiel, Germany. "Chemical ocean conditions 100 years from, now will probably have no equivalent in the geological past, and key organisms may have no mechanisms to adapt to the change (Schiermeier 2004b, 820).

Although the fate of plankton and marine snails may not seem as compelling as vibrantly colored coral reefs, they are critical to sustaining marine species such as salmon, redfish, mackerel, and baleen whales. "These are groups everyone depends on, and if their numbers go down there are going to be reverberations throughout the food chain," said John Guinotte, a marine biologist at the Marine Conservation Biology Institute. "When I see marine snails' shells dissolving while they're alive, that's spooky to me" (Eilperin 2006d, A-1).

Oceans No Longer Filter Carbon Dioxide

Before scientists detected the toll taken on ocean life by rising acidity, some climate experts had asserted that the oceans would help to control the rise in carbon dioxide by acting as a filter. Caldeira and Wickett said, however, that as carbon dioxide enters the oceans as carbonic acid, gradually raising the acidity of ocean water, it inhibits oceans' ability to absorb future emissions. According to their studies, the rate of change during the last century already matches that of 10,000 years before the industrial age. Caldeira pointed to acid rain from industrial emissions as a possible precursor of changes in the oceans: "Most ocean life resides near the surface, where the greatest change would be expected to come, but deep ocean life may prove to be even more sensitive to changes" (Toner 2003c).

Caldeira said that the only way to save the oceans is to aim for zero human-generated carbon-dioxide emissions. "People laugh at this," he said, but the oceans naturally absorb only 0.1 gigatons more carbon dioxide per year than they release. Now they are soaking up an extra 2 gigatons a year, more than 20 times the natural rate. "Even if we halve emissions," he said. "That will merely double the time until we kill off your favorite plant or animal" (Henderson 2006, 32).

The oceans have reached only one-third of their capacity for absorption of humankind's excess carbon dioxide, but even at this level, the rising level of carbon in ocean water (and therefore, its acidity) is impeding sea animals' ability to grow protective shells. Scientists who project this trend into the future find ample reason to worry about the sustainability of ocean life (Feely et al. 2004, 362; Sabine et al. 2004, 367; "Report Says" 2004).

Ocean Acidification in the Past

Warming-provoked acidification of the oceans has precedent in the Earth's natural history, notably during the Paleocene-Eocene Thermal Maximum, between 55 and 56 million years ago, when massive amounts of carbon dioxide, oxidized from methane clathrates, surged into the atmosphere from the oceans, raising sea-surface temperatures by 5°C in the tropics to about 9°C at high latitudes. An initial, rapid rise in temperatures over 1,000 years was followed by a slower increase during the next 30,000 years (Zachos et al. 2005, 1612).

Scientists have been studying the acidification of the oceans during this long-ago epoch as an analogue to similar conditions anticipated in response to human-provoked increases in carbon dioxide. One study of the problem, published in *Nature*, pointedly concluded the following:

> What, if any, implications might this have for the future? If combustion of the entire fossil-fuel reservoir (about 4,500 gigatons of carbon) is assumed, the impacts on deep-sea pH [acidity] and biota will likely be similar to those in the Paleocene-Eocene Thermal Maximum. However, because the anthropogenic carbon input will occur within just 300 years, which is less than the mixing time of the ocean, the impacts on ocean surface pH and biota will probably be more severe. (Zachos et al. 2005, 1614)

Some Organisms Benefit from Acidification

In an ocean of concern that acidification will harm calcium-shelled organisms, a few researchers have come up with test results indicating the opposite. Coccolithophores, single-celled, carbon-shelled algae that play an important role in the ocean food chain, may benefit from lower pH levels that seem to augment their ability to photosynthesize, according to results published in *Science* by M. Debora Iglesias-Rodriguez of the National Oceanography Center at the University of Southampton, England, who has been working with several colleagues (Iglesias-Rodriguez et al. 2008). These results contradict earlier tests, probably because of differences in how acid was added to water in test plots. In the earlier tests, acid was added (and pH lowered) directly. In the new tests, the acid was added indirectly, with the aid of carbon-dioxide bubbles, which Iglesias-Rodriguez and colleagues believe more closely resembles the way the process occurs in the oceans. In on-site tests, the mass of coccolithophores has increased 40 percent over the last 220 years as ocean pH has declined (Chang 2008a, A-11; Iglesias-Rodriguez et al. 2008, 336–340). "You cannot look at calcification in isolation," said Iglesias-Rodriguez. "You have to look at photosynthesis as well" (Chang 2008a, A-11). This study poses a question for future research: how low would pH have to go before the coccolithophores have no shells in which to photosynthesize?

Coccolithophores account for about one-third of oceans' calcium carbonate mass. "Our findings show that coccolithophores are already responding and will probably continue to respond to rising atmospheric CO_2 partial pressures, which has important implications for biogeochemical modeling of future oceans and climate," said the report (Iglesias-Rodriguez et al. 2008, 336). *See also:* Carbon-Dioxide Levels Worldwide; Carbon-Dioxide Levels and Paleoclimate; Oceans, Carbon-Dioxide Levels

FURTHER READING

Caldeira, Ken, and Michael E. Wickett. "Oceanography: Anthropogenic Carbon and Ocean pH." *Nature* 425 (September 25, 2003):365.

Chang, Kenneth. "Climate Shift May Aid Algae Species." *New York Times*, April 18, 2008a, A-11.

De'ath, Glenn, Janice M. Louygh, and Katharina E. Fabricius. "Declining Coral Calcification on the Great Barrier Reef." *Science* 323 (January 2, 2009):116–119.

Eilperin, Juliet. "Growing Acidity of Oceans May Kill Corals." *Washington Post*, July 5, 2006d, A-1, http://www.washingtonpost.com/wp-dyn/content/article/2006/07/04/AR2006070400772_pf.html.

Feely, Richard A., Christopher L. Sabine, J. Martin Hernandez-Ayon, Debby Ianson, and Burke Hales. "Evidence for Upwelling of Corrosive 'Acidified' Water onto the Continental Shelf." *Science* 320 (June 13, 2008):1490–1492.

Feely, Richard A., Christopher L. Sabine, Kitack Lee, Will Berelson, Joanie Kleypas, Victoria J. Fabry, and Frank J. Millero. "Impact of Anthropogenic CO_2 on the $CaCO_3$ System in the Oceans." *Science* 305 (July 16, 2004):362–366.

Fountain, Henry. "More Acidic Ocean Hurts Reef Algae as Well as Corals." *New York Times*, January 8, 2008, D-3.

Hall-Spencer, Jason M., Riccardo Rodolfo-Metalpa, Sophie Martin, Emma Ransome, Maoz Fine, Suzanne M. Turner, Sonia J. Rowley, Dario Tedesco, and Maria-Cristina Buia. "Volcanic Carbon Dioxide Vents Show Ecosystem Effects of Ocean Acidification." *Nature* 454 (July 3, 2008):96–99.

Henderson, Casper. "The Other CO_2 Problem." *New Scientist*, August 5, 2006, 28–33.

Holland, Jennifer S. "The Acid Threat: As CO_2 Rises, Shelled Animals May Perish." *National Geographic*, November 2001, 110–111.

Iglesias-Rodriguez, M. Debora, Paul R. Halloran, Rosalind E. M. Rickaby, Ian R. Hall, Elena Colmenero-Hidalgo, John R. Gittins, Darryl R. H. Green, Toby Tyrrell, Samantha J. Gibbs, Peter von Dassow, Eric Rehm, E. Virginia Armbrust, and Karin P. Boessenkool. "Phytoplankton Calcification in a High-CO_2 World." *Science* 320 (April 18, 2008):336–340.

Kintisch, Eli, and Erik Stokstad. "Ocean CO_2 Studies Look Beyond Coral." *Science* 319 (February 22, 2008):1029.

Kolbert, Elizabeth. "The Darkening Sea: What Carbon Emissions Are Doing to the Oceans." *The New Yorker*, November 20, 2006b, 66–75.

Orr, James C., Victoria J. Fabry, Olivier Aumont, Laurent Bopp, Scott C. Doney, Richard A, Feely, Anand Gnanadesikan, Nicolas Gruber, Akio Ishida,

Fortunat Joos, Robert M. Key, Keith Lindsay, Ernst Maier-Reimer, Richard Matear, Patrick Monfray, Anne Mouchet, Raymond G. Najjar, Gian-Kasper Plattner, Keith B. Rodgers, Christopher L. Sabine, Jorge L. Sarmiento, Reiner Schlitzer, Richard D. Slater, Ian J. Totterdell, Marie-France Weirig, Yasuhiro Yamanaka, and Andrew Yooi. "Anthropogenic Ocean Acidification over the Twenty-first Century and Its Impact on Calcifying Organisms." *Nature* 437 (September 29, 2005):681–686.

"Pacific Coast Turning More Acidic." Earth Observatory. Media Alerts Stories Archive. May 22, 2008, http://earthobservatory.nasa.gov/Newsroom/MediaAlerts/2008/2008052226903.html.

Pennisi, Elizabeth. "Calcification Rates Drop in Australian Reefs." *Science* 323 (January 2, 2009):27.

"Report Says Oceans Hit by Carbon Dioxide Use." *Boston Globe* in *Omaha World-Herald*, July 17, 2004, 5-A.

Riebesell, U., K. G. Schulz, R. G. J. Bellerby, M. Botros, P. Fritsche, M. Meyerhöfer, C. Neill, G. Nondal, A. Oschlies, J. Wohlers, and E. Zöllner. "Enhanced Biological Carbon Consumption in a High CO_2 Ocean." *Nature* 450 (November 22, 2007):545–548.

Ruttimann, Jacqueline. "Oceanography: Sick Seas." *Nature* 442 (August 31, 2006):978–980.

Sabine, Christopher L., Richard A. Feely, Nicolas Gruber, Robert M. Key, Kitack Lee, John L. Bullister, Rik Wanninkhof, C. S. Wong, Douglas W. R. Wallace, Bronte Tilbrook, Frank J. Millero, Tsung-Hung Peng, Alexander Kozyr, Tsueno Ono, and Aida F. Rios. "The Oceanic Sink for Anthropogenic CO_2." *Science* 305 (July 16, 2004):367–371.

Schiermeier, Quirin. "Researchers Seek to Turn the Tide on Problem of Acid Seas." *Nature* 430 (August 19, 2004b):820.

Toner, Mike. "Oceans' Acidity Worries Experts; Report: Carbon Dioxide on Rise, Marine Life at Risk." *Atlanta Journal and Constitution*, September 25, 2003c, n.p. (LEXIS)

Zachos, James C., Ursula Röhl, Stephen A. Schellenberg, Appy Sluijs, David A. Hodell, Daniel C. Kelly, Ellen Thomas, Micah Nicolo, Isabella Raffi, Lucas J. Lourens, Heather McCarren, and Dick Kroon. "Rapid Acidification of the Ocean During the Paleocene-Eocene Thermal Maximum." *Science* 308 (June 10, 2005):1611–1615.

Adirondacks and Warming

During the twentieth century, the average temperature in the Adirondack Mountains of Upstate New York increased more quickly than in other parts of New York State, confirming other trends that anticipate the most rapid warming at high altitudes and latitudes. From 1895 to 1999, the annual temperature in the Adirondacks rose 1.8°F, while New York State as a whole warmed by 1°F, said climate scientist Barrett Rock. Rock, of the University of New Hampshire's Complex Systems Research Center, asserted that the primary reason temperatures were rising so quickly was extensive logging, which removed much of the forest that had helped cool the region. "This is surprising. The Adirondacks have warmed significantly more than the rest of the state," Rock said (Capiello 2002).

The Adirondacks are considered particularly vulnerable because of their elevation and wide variety of habitats that support species that live in narrow temperature ranges. Very small changes in the area's climate could threaten their survival. When the climate changes, the forests of the Adirondacks could be susceptible to exotic pests and pathogens, threatening the region's timber industry and tourism economy. "Climate change is obviously affecting the Adirondacks," said Brian Houseal, the Adirondack Council's executive director. "It's going to change this entire region" (Capiello 2002). Rock, who used data from more than 300 federal monitoring stations nationwide to compute regional temperature rises, saw an even bigger difference between the Adirondacks and the rest of the state during winter.

Downslope to the east, weather records indicate that between 1815 and 1950 Lake Champlain failed to freeze completely only six times. Between 1950 and 2003, however, this lake failed to freeze more than 25 times, yet another signal of a warming climate in this area (Lowy 2004).

A local resident observed the following:

> To give you some idea of regional changes here ... normally we've had the highest snowfall east of the Rockies . . 537" on level one year at Sears Pond on Tug Hill Plateau ... usually from late November thru mid January ... then sub zero on and off from January thru late February. This year there was hardly any snow, and all of January the temperatures varied from mid-30s F up to near 50. Then in mid-February, we had a wind storm blow through here at 100 mph (we've had wind storms before, but never that strong or that early). Barn roofs were ripped off, trees downed, cows injured, three people killed, cars crushed, power out for days in some places, and those state road-number signs (usually very strong in most normal winds), some went down like wet tacos. Any fences, decorative or functional, were a joke ... all flattened. My property looks like a war zone, and I'm not alone. Any fool

can see the climate is in trouble, but George W. Bush and cronies want to hide the truth like they do with everything else. (Einhorn 2006)

FURTHER READING

Capiello, Dina. "Adirondacks Climate Growing Hotter Faster." *Albany Times-Union*, September 21, 2002.
Einhorn, Arthur. Personal communication. March 23, 2006.
Lowy, Joan. "Effects of Climate Warming Are Here and Now." Scripps-Howard News Service, May 5, 2004. (LEXIS)

Aerosols and Climate Change

The same industrial combustion of fossil fuels that generates carbon dioxide and methane also may have produced the haze that has shielded the Earth from full-strength sunlight, which in turn provokes global warming. As efforts to reduce visual pollution have accelerated, however, the shield of haze has partially dispersed, increasing warming. Not all aerosols produce a cooling effect, however. Darker particles, widely known as soot, may increase warming because of their albedo—that is, their ability to absorb heat. Aerosol pollution patterns may affect rainfall patterns as well as temperatures. In some parts of Asia, most notably India and China, rapid industrialization with coal-fired power has thickened the haze, known among some scientists as the "Asian Brown Cloud," which recent research indicates may increase warming significantly.

The Issue's Complexity

The complexity of aerosols' effects on the atmosphere is a major problem in climate modeling. Writing in *Nature*, Meinrat O. Andreae and colleges sketched the complexity of the problem:

> All aerosol types (sulfates, organics, mineral dust, sea salts, and so on) intercept incoming sunlight, and reduce the energy flux arriving at the Earth's surface, thus producing a cooling. Some aerosols (for example, soot) absorb light and thereby warm the atmosphere, but also cool the surface. This warming of atmospheric levels also may reduce cloudiness, yielding another warming effect. In addition to these "direct" radiative effects, there are several "indirect," cloud-mediated effects of aerosols, which all result in cooling: more aerosols produce more, but smaller, droplets in a given cloud, making it more reflective. Smaller droplets are less likely to coalesce into raindrops, and thus the lifetime of clouds is extended, again increasing the Earth's albedo. Finally, modifications in rainfall generation change the thermodynamic processes in clouds, and consequently the dynamics of the atmospheric "heat engine" that drives all of weather and climate. (Andreae, Jones, and Cox 2005, 1187)

This complex mixture of effects enhances warming as aerosols are removed from the atmosphere by pollution reduction. The amount of consequent warming is open to intense debate, however. The industrial aerosol haze that cools much of the planet seems to have thinned over the past decade or so, according to remote-sensing specialists reported in the journal *Science*. "If real, the thinning might explain the unexpectedly strong global warming of late, the accelerating loss of glacial ice, and much of rising sea levels" (Kerr 2007b, 1480).

The industrial pollution that increases greenhouse-gas concentrations in the atmosphere also may be retarding their effects, at least for a time. A workshop of top atmospheric scientists in Berlin, during June 2003, addressed the "parasol effect," in which industrial pollution's aerosols have been shielding the surface of the Earth from even greater warming. Without this pollution, the workshop concluded that the Earth might have warmed by 2.5°C during the twentieth century, instead of 0.6°C. Assuming that industrial pollution is stopped and the skies return to the pristine state of the preindustrial world, the scientists estimated that temperatures could rise 7° to 10°C by the year 2100, which "could be devastating for the Earth and all the life upon it, leading to mass extinctions of animal and plant species, desperate problems in food production and water supply, the collapse of many economies, and drastic changes in every aspect of our lives" (McGuire 2005, 55–56).

Nobel Laureate Paul Crutzen and Swedish meteorologist Brent Bolin, former chairman of the Intergovernmental Panel on Climate Change (IPCC) said during the workshop in Berlin that a diminishing aerosol "parasol effect" in the atmosphere during the twenty-first century could contribute to warming that may exceed IPCC estimates. "It looks like the warming today may have been only about a quarter of what we would have ... without aerosols," Crutzen said ("Global Warming's Sooty Smokescreen" 2003). Scientists at the Berlin workshop speculated that a growing load of aerosols in the atmosphere reduced warming by about 1.8°C during the twentieth century, two to three times as much as previously believed, indicating that the lower atmosphere is more sensitive to greenhouse gases than most models suggest.

In 1996, Jonathan T. Overpeck, working with the Paleoclimatology Program at the National Oceanic and Atmospheric Administration's (NOAA's) National Geophysical Data Center in Boulder, Colorado, led a team of scientists who conducted global climate model simulations to examine the potential role of tropospheric dust in glacial climates. Comparing "modern dust" with "glacial dust" conditions, they found patterns of regional warming that increased at progressively higher latitudes. The warming was greatest (up to 4.4°C) in regions with dust over areas covered with snow and ice. Under some circumstances, they wrote, "aerosols can reduce cloud cover and thus significantly offset aerosol-induced radiative cooling at the top of the atmosphere on a regional scale" (Ackerman et al. 2000, 1042).

The "Asian Brown Cloud"

Human activities now pump nearly as many aerosols into the atmosphere as natural processes. On the Indian subcontinent, these aerosols form a brownish haze that collects on the southern shores of Mount Everest, as well as a shield of dust that travels from heavily populated areas of Asia across the Pacific Ocean to North America. Urban haze also has been observed flowing over the Indian Ocean from cities in Asia. This blanket of pollution has been called the "Asian Brown Cloud." According to Veerabhadran Ramanathan and colleagues, "[a]nthropogenic sources contribute almost as much as natural sources to the global AOD [aerosol optical depth]" (Ramanathan et al. 2001, 2119).

On August 10, 2002, a team of international climatologists led by Professor Paul Crutzen, whose work on stratospheric ozone depletion won the 1995 Nobel Prize, said that the "Brown Cloud" comprises a 10-million-square-mile, three-kilometer-deep haze of man-made pollutants (mainly from burning wood and dung, as well as fuel for vehicles and power generation) spreading across the most thickly populated parts of the Asian continent, blocking as much as 15 percent of incoming sunlight. When it reaches the stratosphere, this haze can spread around the world in a matter of days.

Ramanathan, of the Scripps Institution of Oceanography, has researched the phenomenon for several years. He explained that it was not solely an Asian problem:

> We used to think that the human impact on climate was just global warming. Now we know it is more complex. The brown cloud shows that man's activities are making climate more unpredictable everywhere. Greenhouse gases like carbon dioxide are distributed uniformly, but the particulates in the brown cloud add to unpredictability worldwide. (Vidal 2002c)

More than 200 scientists have taken part in the Indian Ocean Experiment to study this haze.

The "Brown Cloud" has been detected obscuring the sky around the highest peaks in the Himalayas, and as far as 1,000 kilometers downwind from major Indian urban areas. The particles absorb heat, which tends to intensify global warming. At the same time, however, warming is mitigated because the haze partially obscures the sun, by about 10 percent, causing some loss in agricultural productivity, notably of Asian rice crops. The scientists are concerned that pollutants in the "Brown Cloud" may disrupt India's life-giving annual monsoon.

The "Brown Cloud" has been described as a "dynamic soup" of vehicle and industrial pollutants, carbon monoxide, and minute soot particles or fly ash generated by the regular burning of forests and wood used for cooking in millions of rural homes. As described by John Vidal in the *London Guardian*,

> At its seasonal peak, usually in January, the soot in the cloud bounces back sunlight into the upper atmosphere, and prevents evaporation from the sea, leading to less rainfall. This, in turn, is thought to be affecting the monsoon rains which determine agriculture, and adversely affecting the health and livelihoods of up to three billion people throughout Asia. (Vidal 2002c)

According to Ramanathan, "Some places will see more drying, others more rainfall. Greenhouse gases and aerosols may be acting in the same direction or may be opposing each other" (Vidal 2002c). "It is now undisputed that air pollutants and their chemical products can be transported over many thousands of kilometers. We urgently need data on the sources of the pollution, especially for China and India since they are contributing the bulk of the emissions," said a United Nations report (Vidal 2002c).

These aerosols, according to Ramanathan and colleagues,

> produce brighter clouds which are less efficient at releasing precipitation. These in turn lead to large reductions in the amount of solar irradiance reaching Earth's surface, a corresponding increase in solar heating of the atmosphere, changes in the atmospheric temperature structure, suppression of rainfall, and less efficient removal of pollutants. (Ramanathan et al. 2001, 2119)

Increasing density of aerosols also may weaken the hydrological cycle, "which connects directly to availability and quality of fresh water, a major environmental issue of the 21st century" (Ramanathan et al. 2001, 2119).

Ramanathan and colleagues have continued to study the role of the "Brown Cloud" in global warming. Research published in 2007 surprised many experts by attributing as much warming to this dry-season haze as emissions of greenhouse gases themselves. The contribution of this pollution to warming is especially dramatic on the Himalayan plateau, where it is contributing to glacial erosion (Ramanathan et al. 2007, 575; Pilewskie 2007, 541–542).

A study released during 2007 by scientists at Scripps Institution of Oceanography at the University of California–San Diego indicated that brown clouds over South Asia (which contain soot, sulfates, nitrates, hundreds of organic compounds, and fly ash) multiplied solar heating of the lower atmosphere by about 50 percent, a factor in the retreat of Himalayan glaciers. Ramanathan led this study, which refutes widespread assumptions that the brown clouds have reduced solar heating by blocking the sun, or "global dimming" (Ramanathan et al. 2007, 575).

"While this is true globally," Ramanathan explained, "this study reveals that over southern and eastern Asia, the soot particles in the brown clouds are intensifying the atmospheric warming trend caused by greenhouse gases by as much as 50 percent" ("Asian Brown Clouds" 2007). The brown-cloud effect also helps to explain why south Asia's warming trend is more pronounced at higher altitudes than closer to sea level.

Air pollution from Asian cooking fires, industry, and automobiles is drifting into the Himalayas, a source of drinking water for a billion people (mainly in China, Pakistan, and India), and accelerating the melting of glaciers, according to a United Nations report, "Atmospheric Brown Clouds: Regional Assessment with Focus on Asia," issued late in 2008. In Bangkok, New Delhi, Seoul, Tehran, Shanghai, and areas between, this haze has dimmed their sunlight by 20 percent or more.

"We used to think of this brown cloud as a regional problem, but now we realize its impact is much greater," said Ramanathan, who led the United Nations scientific panel. "When we see the smog one day and not the next, it just means it's blown somewhere else" (Jacobs 2008).

Henning Rodhe, a professor of chemical meteorology at Stockholm University, estimates that 340,000 people in China and India die each year from cardiovascular and respiratory diseases that can be traced to the emissions from coal-burning factories, diesel trucks, and kitchen stoves fueled by twigs (Jacobs 2008).

The INDOEX Experiment

An experiment headquartered on the Maldives Islands in the Indian Ocean took six weeks, cost $25 million, and utilized scientists from 15 countries, a research ship, and a C-130 military transport aircraft crammed with instruments (Fialka 2003, A-6). The INDOEX experiment used instruments on land and on aircraft together with measurements made by the National Aeronautics and Space Administration's (NASA's) Clouds and Earth's Radiant Energy System (CERES) sensor as it flew overhead aboard the Tropical Rainfall Measuring Mission (T.R.M.M.) satellite. The experiment's objective was to help scientists understand to what extent human-produced aerosols may offset global warming.

The Indian subcontinent offered the architects of the INDOEX campaign an ideal setting for their field experiment. The region was chosen for its unique combination of meteorology, landscape (relatively flat plains framed by the towering Himalayan Mountains to the north and open ocean to the south), and large southern Asian population (roughly 1.5 billion) with a growing economy. "Together, these features maximize the effects of aerosol pollution," Ramanathan explained ("New NASA" 2001). Because of human industry, automobiles, factories, and burning vegetation, particles build up in the atmosphere where they are blown southward over most of the tropical Indian Ocean. The Indo-Asian haze covered an area larger than that of the United States. Although the INDOEX team found atmospheric particles of natural origin, such as trace amounts of sea salts and desert dust, they also found that 75 percent of the aerosols over the region resulted from human activities, including sulfates, nitrates, black carbon, and fly ash. Most natural aerosols scatter and reflect sunlight back to space, thereby making our planet brighter. However, human-produced black carbon aerosol absorbs more light than it reflects, thereby making our planet darker.

"Ultimately, we want to determine if our planet as a whole is getting brighter or darker," Ramanathan stated. "We could not answer that question until we could measure the sunlight reflected at the top of the atmosphere with an absolute accuracy of 1 percent. The CERES sensors provide that accuracy for the first time ever from a space-based sensor" ("New NASA" 2001).

"A large reduction of sunlight at the surface has implications for the hydrological cycle because of the close tie between heat and evaporation," Ramanathan added. "It could change the heating structure of the atmosphere and perturb the climate system in ways we don't understand now. We don't know, for example, how this might affect the monsoon season" ("New NASA" 2001). *See also:* Soot, A "Wild Card" in Global Warming

FURTHER READING

Ackerman, A. S., O. B. Toon, D. E. Stevens, A. J. Heymsfield, V. Ramanathan, and E. J. Welton. "Reduction of Tropical Cloudiness by Soot." *Science* 288 (May 12, 2000):1042–1047.

Andreae, Meinrat O., Chris D. Jones, and Peter M. Cox. "Strong Present-day Aerosol Cooling Implies a Hot Future." *Nature* 435 (June 30, 2005):1187–1190.

"Asian Brown Clouds Intensify Global Warming." Environment News Service, August 1, 2007. http://www.ens-newswire.com/ens/aug2007/2007-08-01-02.asp.

Fialka, John J. "Soot Storm: A Dirty Discovery Over Indian Ocean Sets off a Fight." *Wall Street Journal*, May 6, 2003, A-1, A-6.

"Global Warming's Sooty Smokescreen Revealed." New Scientist.com, June 3, 2003.

Jacobs, Andrew. "Report Sees New Pollution Threat." *New York Times*, November 14, 2008. http://www.nytimes.com/2008/11/14/world/14cloud.html.

Kerr, Richard A. "Is a Thinning Haze Unveiling the Real Global Warming?" *Science* 315 (March 16, 2007b):1480.

McGuire, Bill. *Surviving Armageddon: Solutions for a Threatened Planet.* New York: Oxford University Press, 2005.

"New NASA Satellite Sensor and Field Experiment Shows Aerosols Cool the Surface but Warm the Atmosphere." National Aeronautics and Space Administration Public Information Release, August 15, 2001. http://earthobservatory.nasa.gov/Newsroom/MediaResources/Indian_Ocean_Experiment/indoex_release.html.

Pilewskie, Peter. "Aerosols Heat Up." *Nature* 448 (August 3, 2007):541–542.

Ramanathan, V., P. J. Crutzen, J. T. Kiehl, and D. Rosenfeld. "Aerosols, Climate, and the Hydrological Cycle." *Science* 294 (December 7, 2001):2119–2124.

Ramanathan, Veerabhadran, Muvva V. Ramana, Gregory Roberts, Dohyeong Kim, Craig Corrigan, Chul Chung, and David Winker. "Warming Trends in Asia Amplified by Brown Cloud Solar Absorption." *Nature* 448 (August 2, 2007):575–578.

Vidal, John. "You Thought It Was Wet? Wait until the Asian Brown Cloud Hits Town: Extreme Weather Set to Worsen through Pollution and El Niño: Cloud with No Silver Lining." *London Guardian*, August 12, 2002c, 3.

Agriculture and Warming

Temperatures in the twenty-first century will rise to such a degree that Europe's extreme heat of August 2003, which killed tens of thousands of people, will become average, devastating world agriculture and provoking a "perpetual food crisis" including crop failures in many regions, according to a study conducted by scientists at the University of Washington and Stanford University and published in *Science* on January 9, 2009. Yields of staples such as wheat, corn, and rice may be reduced 20 to 40 percent. The lead author of the study, University of Washington climate researcher David Battisti, said that effects will be most intense in the tropics and subtropics, where many people already live at the margin of survival (Mittelstaedt 2009). The scientists used data from 23 global climate models to show, with a high probability of more than 90 percent, that growing-season temperatures in the tropics and subtropics by the end of the twenty-first century will exceed the most extreme seasonal temperatures recorded from 1900 to 2006 (Battisti and Naylor 2009, 240).

How will agriculture fare in a warmer world? The MINK (Missouri-Iowa-Nebraska-Kansas) Study surveyed potential climate change in the central United States, North America's agricultural heartland. Under certain circumstances, the researchers found, higher levels of carbon dioxide might enhance growth of some crops, but as a whole, "[u]nder the best of these scenarios ... the productivity of the region's agriculture would be significantly diminished" (Rosenburg 1992, 151). Agriculture in present-day farming regions will be severely affected not only by heat stress, but also by reduced surface-water supplies, because most global climate models predict that as the atmosphere warms the interiors of continents would become hotter and drier, especially during the growing season. An additional problem facing farmers in Nebraska and Kansas is depletion and salinization of aquifers that already support a large part of agricultural production in both states, especially their drier western areas.

By 2050, Earth's population is expected to increase by half, or about three billion people, and roughly 75 percent of poor people will depend on agriculture. Hotter, drier weather combined with explosive bursts of precipitation may shorten growing seasons and threaten production in some areas (notably in Africa and India) where agricultural production is limited by the availability of water rather than the onset of cold weather. Hundreds of millions of people who already live at the margin may find their survival threatened.

Various modes of adaptation, such as breeding crops to resist more heat, flood, drought, or insect infestations may provide some help in the short range, but their ability to mitigate the destructive nature of warming probably will decline as temperatures rise. Accelerating climate change will provide farmers with an ever-changing environment, bereft of old environmental benchmarks.

A research report by the Consultative Group on International Agricultural Research (CGIAR) in Washington, D.C., a worldwide network of agricultural research centers, is already developing new crop strains that can withstand rising temperatures, drier climates, and increasing soil salinity. The CGIAR is also researching measures to reduce the carbon footprint of farming (Zeller 2006).

CGIAR studies have projected trends indicating that temperature increases and shifts in rainfall patterns probably will reduce growing periods in Sub-Saharan Africa by more than 20 percent, with some of the world's poorest nations in East and Central Africa at greatest risk. CGIAR also cited new research indicating that a generally warming climate will reduce wheat production in India's breadbasket. Production may decline about 50 percent by 2050, a decrease that could put as many as 200 million people at greater risk of chronic hunger.

While some studies anticipate rising food production in some areas during the early stages of global warming, others anticipate "negative surprises" in world agriculture, especially during the last half of the twenty-first century. Poor, developing nations may lose 334 million acres of prime farm land during the next half-century as temperatures rise and storms intensify, according to three studies in the *Proceedings of the National Academy of Sciences* (Howden et al. 2007, 19,691; Schmidhuber and Tubiello 2007, 19,703; Tubiello, Soussana, and Howden 2007, 19,686). By the last half of the twenty-first century even cooler regions that may benefit from earlier temperature rises could experience declines in productivity. The authors of these studies argue that extreme heat will join with other factors, such as the spread of weeds and diseases, to compound agricultural problems. These problems will inhibit increases in food production necessary to feed rising populations. These studies project that as many as 170 million people may be "at risk of hunger" by 2080 (Schmidhuber and Tubiello 2007, 19,703).

"Many people assume that we will never have a problem with food production on a global scale. But there is a strong potential for negative surprises," said Francesco Tubiello, a physicist and agricultural expert at the National Aeronautics and Space Administration's (NASA's) Goddard Institute for Space Studies, who coauthored three National Academy of Science reports. The authors of these studies said that much previous research work is oversimplified, and as a consequence, the potential for bigger, more rapid problems remains unexplored. Heat waves and extreme storms could have their greatest effects on crops at crucial germination or flowering times. Tubiello says this is already happening on smaller scales.

Some people believe that global warming will be ameliorated by adaptations such as forecasting systems that may advise farmers to switch crops or change the timing of planting. Crops are already being modified to survive heat and drought, as well. These may buy only a few decades, as warming continues. "After that," said Tubiello, "all bets are off" ("Warming Climate" 2007).

During 2008, the U.S. Department of Agriculture issued a revised climatic-zone map for gardeners. This map assigns zones to various plants for winter survival. In the 18 years since the map had been last revised (in 1990), growing zones for many plants have moved northward. Southern magnolias, for example, once restricted to coastal Virginia southward, now thrive into Pennsylvania. In 1990, kiwis died north of Oklahoma. In 2008, they fruited in St. Louis, Missouri.

The Arbor Day Foundation has made similar adjustments in its maps, last revised in 2005. New drafts of U.S. Department of Agriculture (USDA) maps may be difficult to find, however, because they were delayed by political bickering within the agency. Climate scientists and skeptics disagreed, as well as nursery owners (who feared they would lose money if they sold plants with money-back guarantees that died in the new zones) and other sellers who want to expand their marketing northward.

Agriculture contributes about 20 percent of human greenhouse-gas emissions. Low-till or zero-till farming can help keep carbon in the soil, and more selective use of nitrogen fertilizer may inhibit emission of nitrous oxide, a greenhouse gas 310 times more potent than the main greenhouse gas, carbon dioxide. In 2008, the National Farmers Union paid farmers in the United States $5.8 million to adopt environmentally constructive practices in a "carbon credit program," many of which (such as no-till farming and rotational grazing) capture carbon dioxide. Agencies that aid farmers already are developing new varieties of corn, wheat, rice, and sorghum, as well as programs to encourage more efficient use of water and soil resources and to develop new practices to reduce greenhouse-gas emissions from farming.

Warming and Canadian Grain Production

Climate-change skeptics often advance a simplistic solution to agricultural problems caused by warming temperatures and a more explosive hydrological cycle: move it all northward. Following their rather simplistic climatic logic, one can almost imagine corn sprouting along the shores of Hudson Bay. Only if the soils are right, however, will crops grow. Even today, wheat, soy, and canola are grown almost to the 60th parallel in the Peace River Valley in Northern Alberta and British Columbia. Some varieties of grain, rye, flax, and canola mature in 120 days. Given enough rain (a problem in some northern latitudes), warmer weather could benefit Canadian agriculture. Grain-growing areas in Russia also may move northward to areas where soil is suitable.

Alberta's and Saskatchewan's northern areas already are producing hardy varieties of wheat. In northern Ontario and part of western Quebec, the clay belts are being farmed. The Canadian Shield's soils do not support agriculture, mainly because the soil is poor, but this is not the dominant landform in the area.

Warmer temperatures, especially in August and September, would allow for increased farming in Canada. The fertile eastern townships of Quebec in the St. Lawrence Lowlands have been farmed since at least the early 1700s. Most of these areas are former lake or sea bottoms, with deep, rich soils. The limiting factor has been the growing season. The boreal forests of northern Canada include tens of millions of acres with abundant water, in which trees have grown, died, rotted, and re-grown for millennia. Add warmth, and some of these areas might become productive farmlands.

Eastern Canada north of the Great Lakes and the St. Lawrence River Valley is not generally suitable for grain. The western plains, from western Lake Superior to the Rocky Mountains, contain excellent grain-growing soils, however. Grain from Manitoba, Saskatchewan, and Alberta fed Great Britain from 1939 to 1941 during the early days of World War II.

The Industrial Scale of Agriculture

Many people who do not farm for a living share a stereotype of agriculture as a family affair, a builder of character, and a style of employment that evokes, for tillers of the soil, a basic sense of enjoyment from communion with nature. During the twentieth century, however, agriculture became progressively more mechanized on a massive scale suited to large, worldwide markets. Agriculture, like other modes of production in our machine culture, has come to demand less on human labor and an increasing amount more on fossil-fuel energy. Industrial-scale agriculture also requires copious amounts of synthetic fertilizers. For all except a few remaining (and often struggling) family farmers, agriculture has become as industrial as factory work.

Ecologist Barry Commoner described how the American farm has changed:

> Between 1950 and 1970, the total U.S. crop output increased by 38 per cent, although the acreage decreased by 4 per cent and the labor [number of people employed] fell by 58 per cent. This sharp increase in productivity was accomplished by an 18 per cent increase in the use of machinery and a 295 per cent increase in the application of synthetic pesticides and fertilizer. (Commoner 1990, 49)

Craig Benjamin, writing in *Native Americas*, described how large-scale monocultural farms make themselves vulnerable to a growing risk of pest attack in a warmer, more humid world:

> This impressive vulnerability of industrial agriculture is key to understanding how climate change will likely have an impact on global agriculture and on the relationship between industrial agriculture and indigenous farming communities. Faced with rapid and dramatic climate change, the impressively vulnerable industrial farm can conceivably continue to use large-scale irrigation and artificial fertilizers to counter the effects of changing temperature and precipitation. (Benjamin 1999, 80)

"Climate-Ready" Crops

Large multinationals were racing for dominance of "climate ready" crops. By 2008, three companies (Germany's BASF, the Swiss multinational Syngenta, and U.S. agribusiness giant Monsanto) had filed applications that would allow them to control two-thirds of gene families in worldwide patent office filings, according to ETC Group of Ottawa, Canada, which advocates causes that benefit subsistence farmers. These new crops are being bred to resist not only heat and drought, but also saltwater inundation, flooding, and increasing ultraviolet radiation (associated with depletion of stratospheric ozone).

The ETC Group maintained that the companies are engaged in "an intellectual-property grab," while the companies asserted that "gene-altered plants will be crucial to solving world hunger but will never be developed without patent protections" (Weiss 2008). Patenting genes will prevent farmers in poor countries from saving seeds for future harvests and would require them to purchase new seeds from the companies. The rush to patent new seeds also may prevent distribution by public-sector agencies affiliated with the United Nations and World Bank. "When a market is dominated by a handful of large multinational companies, the research agenda gets biased toward proprietary products," said Hope Shand, ETC's research director. "Monopoly control of plant genes is a bad idea under any circumstance. During a global food crisis, it is unacceptable and has to be challenged" (Weiss 2008).

Ranjana Smetacek, speaking for Monsanto, said these companies should be appreciated for developing crops that will survive adverse environmental conditions. "I think everyone recognizes that the old traditional ways just aren't able to address these new challenges. The problems in Africa are pretty severe" (Weiss 2008). Monsanto maintains that not all of its work is profit oriented. It has joined with BASF, for example, to support the Bill and Melinda Gates Foundation's development of drought-resistant corn provided royalty free to farmers in four southern African countries, "We aim to be at once generous and also cognizant of our obligation to shareholders who have paid for our research," Smetacek said (Weiss 2008).

The patents may be applied to a number of crops, including maize, wheat, rye, oat, triticale, rice, barley, soybean, peanut, cotton, rapeseed, canola, manihot, pepper, sunflower, tagetes, solanaceous plants, potato, tobacco, eggplant, tomato, *Vicia* species, pea, alfalfa, coffee, cacao, tea, Salix species, oil palm, coconut, perennial grass and forage crop plants (Weiss 2008).

The Response of Plants

While some climate contrarians argue that global warming will benefit agriculture by providing plants higher levels of carbon dioxide, Pim Martens, director of the International Centre for Integrated Assessment in the Netherlands, contends that increased growth may be counterbalanced by the growth of molds and other parasites that thrive in hot, humid weather. "It is generally believed that a climate change will have negative effects for global food production," Martens wrote (1999, 540). To cite one of many examples, the Mediterranean Fruit Fly could expand into Northern Europe during the next century with the degree of global warming projected by the IPCC.

Research by Fakhri A. Bazzaz and Eric D. Fajer cast doubt on the contrarians' assertions that a carbon-dioxide-enriched atmosphere will lead to more plant growth and greater agricultural yields:

> Studies have shown that an isolated case of a plant's positive response to increased CO_2 levels does not necessarily translate into increased growth for entire plant communities.... [P]hotosynthetic rates are not always greatest in CO_2-enriched environments. Often plants growing under such conditions initially show increased photosynthesis, but over time this rate falls and approaches that of plants growing under today's carbon-dioxide levels.... When nutrient, water, or light levels are low, many plants show only a slight CO_2 fertilization effect. (Bazzaz and Fajer 1992, 68–71)

Bazzaz and Fajer further asserted: "[W]e do not expect that agricultural yields will necessarily improve in a CO_2-rich future" (Bazzaz and Fajer 1992, 68). William R. Cline added that scarcity of water (which is forecast for many continental interiors as the atmosphere warms) also may reduce agricultural yields (Cline 1992, 91).

Studies by Martin Parry (1990, 1991) estimated that the European corn borer could move 165 to 200 kilometers northward (in the Northern Hemisphere) with each 1°C rise in temperature. The potato leaf-hopper, a major pest for soybeans, presently spends its winters along the Gulf Coast. Global warming could move this range northward. The range of the hornfly, which caused about $700 million a year in damage to beef and diary cattle across the United States during the late 1980s, could be similarly affected.

Horticulturalist N. C. Bhattacharya has found that while enriched carbon dioxide causes accelerated growth in most plants, others respond negatively. Increasing the growth rate of plants tends to accelerate depletion of soils, stunting later growth. Heat also may be detrimental to some plants even as their growth is being stimulated by rising carbon-dioxide levels in the atmosphere (Bhattacharya 1993).

Climatologist Y. A. Izrael summarized the rough road of agriculture in a warmer world:

> Estimates of the impact of doubled CO_2 on crop potential have shown that in the northern mid-latitudes summer droughts will reduce potential production by 10 to 30 per cent. The impact of climate change on agriculture in all, or most, food-exporting regions will entail an average cost of world agricultural production [of] no less than 10 per cent. (Izrael 1991, 83)

Parry elaborated: "While global levels of food production can probably be maintained in the face of climate change, the cost of this could be substantial" (1990, 279). Gains in production at higher latitudes are unlikely to balance reductions in the hotter midlatitudes, which are major grain exporters today (Parry 1990, 279).

Harold W. Bernard, in *Global Warming: Signs to Watch For* (1993), described global warming's anticipated role in the spread of wheat rust, a fungus that thrives in dry heat. Wheat rust destroyed millions of tons of wheat in North America during the dust-bowl decade of the 1930s. At the same time, the pale western cutworm, another pest that favors hot and dry conditions, damaged thousands of acres of wheat in Canada and Montana (Bernard 1993, 47).

William R. Cline, an economist who specializes in global warming, modeled global warming three centuries into the future. By that time, he expects that agricultural production in many contemporary breadbaskets will have been devastated. Indicating an average July maximum for Iowa between 100° and 108°F, Cline scoffs at the idea that higher carbon-dioxide levels in the atmosphere will enhance agricultural yields. By the year 2275, Cline anticipates that temperatures may become so hot that most staple grain crops in present-day Iowa (and surrounding states) may die of heat stress. Wheat, barley, oats, and rye simply will not grow in the climate that Cline projects for the U.S. Midwest roughly three centuries from today. In Eurasia, rice, corn, and sorghum will be pressed to their limits by such temperatures as far north as Moscow. By 2275, according to Cline, the atmosphere's carbon-dioxide load may be eight times the levels of the 1990s (Cline 1992).

According to Thomas R. Karl and colleagues (1997), increases in minimum temperatures are important because of their effect on agriculture. Observations over land areas during the latter half of the twentieth century indicate that average minimum temperatures have increased at a rate more than 50 percent greater than that of maximums.

The rise in minimum temperatures has lengthened the growing (frost-free) season in many parts of the United States; in the northeast, for example, Karl and colleagues wrote that the frost-free season began an average of 11 days earlier during the 1990s than during the 1950s. The compression of daily high and low temperatures may be related to increasing cloud cover and evaporative cooling in many areas, Karl and associates proposed. Clouds depress daytime temperatures because they reflect sunlight, as they also warm night-time temperatures by inhibiting loss of heat from the surface. "Greater amounts of moisture in the soil from additional precipitation and cloudiness inhibit daytime temperature increases because part of the solar energy goes into evaporating this moisture," they wrote (Karl, Nicholls, and Gregory 1997, 79).

In a warmer world, erosion caused by deluges *and* persistent drought probably will pose greater dangers to agriculture. This is not the paradox that it seems, because, according to Karl and colleagues, "not only will a warmer world be likely to have more precipitation, but the average precipitation event is likely to be heavier" (Karl, Nicholls, and Gregory 1997, 79). Karl cited data indicating that heavy precipitation events increased roughly 25 percent from the beginning to the end of the twentieth century. Karl described how a generally wetter world also will be a place in which drought may threaten flora and fauna more often:

> As incredible as it may seem with all this precipitation, the soil in North America, southern Europe and in several other places is actually expected to become drier in the coming decades. Dry soil is of particular concern because of its far-reaching effects, for instance, on crop yields, groundwater resources, lake and river ecosystems.... Several models now project significant increases in the severity of drought. (Karl, Nicholls, and Gregory 1997)

Karl and colleagues temper this statement by citing studies indicating that increased cloud cover may reduce evaporation in some of the areas that are expected to become drier. All in all, however, most of the world's flora and fauna will suffer significantly in a notably warmer world.

FURTHER READING

Battisti, David. S., and Rosamond L. Naylor. "Historical Warnings of Future Food Insecurity with Unprecedented Seasonal Heat." *Science* 323 (January 9, 2009):240–244.

Bazzaz, Fakhri A., and Eric D. Fajer. "Plant Life in a CO_2-rich World." *Scientific American*, January 1992, 68–74.

Benjamin, Craig. "The Machu Picchu Model: Climate Change and Agricultural Diversity." *Native Americas* 16, no. 3/4 (Summer/Fall 1999):76–81.

Bernard, Harold W., Jr. *Global Warming: Signs to Watch For.* Bloomington: Indiana University Press, 1993.

Bhattacharya, N. C. "Prospects of Agriculture in a Carbon-Dioxide-Enriched Environment." In *A Global Warming Forum: Scientific, Economic, and Legal Overview,* ed. Richard A. Geyer, 487–505. Boca Raton, FL: CRC Press, 1993.

Cline, William R. *The Economics of Global Warming.* Washington, DC: Institute for International Economics, 1992.

Commoner, Barry. *Making Peace with the Planet.* New York: Pantheon, 1990.

Howden, S. Mark, Jean Francois Soussana, Francesco Tubiello, Netra Chhetri, Michael Dunlop, and Holger Meinke. "Adapting Agriculture to Climate Change." *Proceedings of the National Academy of Sciences* 104 (December 11, 2007):19,691–19,696.

Izrael, Yu. A. "Climate Change Impact Studies: The IPCC Working Group II Report." In *Climate Change: Science, Impacts, and Policy,* eds. J. Jager and H. L. Ferguson, 83–86. Proceedings of the Second World Climate Conference. Cambridge: Cambridge University Press, 1991.

Karl, Thomas R., Neville Nicholls, and Jonathan Gregory. "The Coming Climate: Meteorological Records and Computer Models Permit Insights into Some of the Broad Weather Patterns of a Warmer World." *Scientific American* 276 (1997):79–83. http://www.scientificamerican.com/0597issue/0597karl.htm.

Martens, Pim. "How Will Climate Change Affect Human Health?" *American Scientist* 87, no. 6 (November/December 1999):534–541.

Mittelstaedt, Martin. "World Faces Perpetual Food Crisis: Study." *Globe and Mail* (Toronto), January 8, 2009. http://www.theglobeandmail.com/servlet/story/RTGAM.20090108.wclimate0108/BNStory/International/home.

Parry, Martin. *Climate Change and World Agriculture.* London: Earthscan, 1990.

Parry, Martin, and Zhang Jiachen. "The Potential Effect of Climate Changes on Agriculture." In *Climate Change: Science, Impacts, and Policy,* ed. J. Jager and H. L. Ferguson, 279–289. Proceedings of the Second World Climate Conference. Cambridge: Cambridge University Press, 1991.

Pegg, J. R. "Climate Change Increases Food Security Concerns." Environment News Service, December 5, 2006. http://www.ens-newswire.com/ens/dec2006/2006-12-05-01.asp.

Schmidhuber, Josef, and Francesco Tubiello. "Global Food Security under Climate Change." *Proceedings of the National Academy of Sciences* 104 (December 11, 2007):19,703–19,708.

Tubiello, Francesco, Jean Francois Soussana, and S. Mark Howden. "Crop and Pasture Response to Climate Change." *Proceedings of the National Academy of Sciences* 104 (December 11, 2007):19,686–19,690.

"Warming Climate Undermines World Food Supply." Environment News Service, December 3, 2007. http://www.ens-newswire.com/ens/dec2007/2007-12-03-05.asp.

Weiss, Rick. "Firms Seek Patents on 'Climate Ready' Altered Crops." *Washington Post*, May 13, 2008. http://www.washingtonpost.com/wp-dyn/content/article/2008/05/12/AR2008051202919_pf.html.
Zeller, Tom, Jr. "America's Breadbasket Moves to Canada." *New York Times*, December 5, 2006. http://thelede.blogs.nytimes.com/2006/12/05/americas-breadbasket-moves-to-canada.

Air Conditioning and Atmospheric Chemistry

Air conditioning (especially the window-mounted variety) is a large contributor to global warming based on its electricity use, as well as emissions from cooling agents. The most popular refrigerant for air conditioners, hydrochlorofluorocarbons (specifically HCFC-22), also harms the Earth's stratospheric ozone layer as it contributes to global warming by retaining heat. International pressure has grown rapidly for quick action to regulate HCFC-22. "We scientifically have proof: if we accelerate the phase-out of HCFC, we are going to make a great contribution to climate change," said Romina Picolotti, the chief of Argentina's environmental secretariat (Bradsher 2007b).

On a molecular level, a derivative of HCFC-22 is about 11,700 times as potent as carbon dioxide at absorbing heart in the atmosphere. Phasing it out could cut emissions of global-warming gases substantially. The 1987 Montreal Protocol that restricted stratospheric ozone-depleting gases contained an exception for HCFC-22 production by developing countries. China and India, two of these countries, have since become leading manufacturers of air-conditioning systems that use it. In addition to manufacturing units for export, air-conditioning use has been doubling every three years in China and India as their economies boom. The same is true in other countries. Use of HCFC-22 increased 100-fold in the African country of Mauritius between 2000 and 2005 due in large part to luxury hotel construction and rapid expansion of a fishing industry (with large exports) that uses refrigeration (Bradsher 2007b).

The Kyoto Protocol exempted HCFC-22 and other ozone-depleting substances on the grounds that the Montreal agreement had addressed the issue (Bradsher 2007b). The 1987 protocol allows developing countries to increase production of HCFC-22 until 2016, and then freezes production at that level until 2040, with a phaseout following. That timetable was drawn up in the early 1990s, when HCFC-22 was used mainly in industrial nations; developing countries were then believed to be too poor to afford it. Industrial countries (as defined in 1987) were required to phase out HCFC-22 manufacture by 2020. The European Union banned them completely in 2004, and the United States plans a ban in 2010. Many European and American air conditioners are now imported from India and China, using more modern (and expensive) refrigerating agents.

In a twist on its usual lack of environmental priorities, the George W. Bush administration by 2007 strongly supported a worldwide phaseout of HCFC-22, with lobbying muscle from DuPont. The giant chemical company stands to profit handsomely from the manufacture and sale of a replacement—hydrofluorocarbons, which contain no chlorine—and thus minimize impact on both ozone depletion and global warming. The United States, while resisting international diplomatic efforts to reduce greenhouse-gas emissions on the grounds that they will harm its economy, was beating the diplomatic drum to advance the phaseout date for HCFC-22 to 2020 from 2030 for industrial nations, and 2030 from 2040 for developing countries (Fialka 2007, A-8).

FURTHER READING

Bradsher, Keith. "Push to Fix Ozone Layer and Slow Global Warming." *New York Times*, March 15, 2007b. http://www.nytimes.com/2007/03/15/business/worldbusiness/15warming.html.
Fialka, John J. "U.S. Plots New Climate Tactic." *Wall Street Journal*, September 7, 2007, A-8.

Air Travel

One journey across the Atlantic Ocean from the United States to Europe on a jet aircraft will emit as much greenhouse gases *per person* as an average automobile commuter creates in an entire year. U.S. citizens, whose aviation mileage per person increased 400 percent between 1970 and 2006 (Hillman, Fawcett, and Rajan 2007, 55) contribute an enormous and growing carbon overload. The rest of the world is not far behind. Addressing air travel's impact on world climate

Jet and contrails. (Courtesy of Shutterstock)

is most necessary in the United States, the point of origin for a third of the world's commercial aviation. By 2004, air travel in the United States consumed 10 percent of all fossil-fuel energy, with passenger loads expected to double between 1997 and 2017, making air travel the fastest growing source of carbon dioxide and nitrous oxides in the economy (Flannery 2005, 282). Airline travel contributed an estimated 5 percent to the total of global human-generated greenhouse gases by 1990, an amount that was increasing at a much faster rate than overall fossil-fuel usage.

By 2008, several U.S. states and environmental groups were urging the U.S. Environmental Protection Agency (EPA) to regulate emissions of greenhouse gases by flights that use airports in the United States. "We want the EPA to take [its] head out of the sand and actively promulgate rules to reduce greenhouse-gas emissions," said California Attorney General Jerry Brown. "The EPA has taken a very passive and unimaginative approach to combating global warming" (Chea 2007, 6-B). California, Connecticut, New Jersey, New Mexico, Pennsylvania, and the District of Columbia said in a petition to the EPA that the Federal Aviation Administration expects aircraft emissions over the United States to increase 60 percent by 2025. The petition urges the EPA to be more aggressive in directing airlines to raise fuel efficiency, build lighter and more aerodynamic vehicles, and develop fuels that emit fewer pollutants, The Air Transport Association, the airlines' U.S. lobby, replied that fuel efficiency already has increased 103 percent since 1978, and will rise another 30 percent by 2025 (Chea 2007, 6-B).

During the 1990s, aviation was the fastest-growing mode of travel in the United States. To handle this growth, 32 of the nation's 50 busiest airports expanded their facilities. Sixty of the 100 largest airports built new runways. During the late 1990s, passenger and freight air transport mileage was doubling roughly every 10 years. The number of passengers was increasing by 8 percent a year, on average, as the volume of cargo rose by 13 percent annually.

According to a report published by the Institute of Public Policy Research, "Flying by jet plane is the least environmentally sustainable way to travel and transport goods" (Lean 2001c, 23). One European author wrote,

> So as we are flying into the Alps for our ski holiday we are contributing to their destruction. Our honeymoon flight to the Maldives is slowly sinking it under rising sea levels and destroying coral through bleaching associated with global warming; and finally our safari flight to Africa is contributing to drought, famine and disease. It's not an appetizing thought, is it? (Francis 2006)

Britain's Debate over Aviation's Carbon Footprint

A large headline on the front page of London's *Independent*, the most aggressive of Britain's newspapers on global warming, upbraided Prince Charles for emphasizing the worldwide risks of global warming as he jetted off to India on a private Airbus 319. The *Independent* described, in great, relishing detail, a roundtrip flight by Prince Charles from London to India via the Arab Republic of Egypt and Saudi Arabia that covered 9,272 miles and was responsible for 42 tons of carbon-dioxide emissions (Hickman 2006, 1). At the same time, Prince Charles was arguing that global warming was "the greatest challenge" facing humankind. The newspaper urged him to restrict his flying schedule.

In 2007, Prince Charles disclosed that his personal carbon footprint had shrunk 9 percent in a year to 3,775 tons. He has been taking private jets fewer times (substituting trains when available) and gassing up the Royal Jaguar with used cooking oil. Otherwise, Charles has gone "carbon neutral" with offsets worth $600,000 a year. His three mansions still have a carbon footprint the size of 500 average British homes, however ("Carbon Neutral Chic" 2007, A-14).

As newspapers argued over Prince Charles' carbon footprint, flight traffic was exploding in England, with a third runway planned at London's Heathrow, and similar extensions at London's Stansted, as well as Birmingham, Edinburgh, and Glasgow. Twelve other British airports also had announced expansion plans. According to the House of Commons' Environmental Audit Committee, the growth the government foresaw would require the equivalent of another Heathrow-size airport every five years (Monbiot 2006b). London's Heathrow itself was planning a sixth terminal while its fifth was under construction.

London's Heathrow, with 67 million passengers a year (200,000 a day) by 2007, is the world's busiest airport. One reason for the huge passenger load is the fuel-squandering nature of the world's hub-and-spoke airline routing system. A passenger in Equatorial Guinea, in West Africa, who wants to get to Johannesburg, in South Africa, connects in London—in at Gatwick, out at Heathrow.

One-fifth of the world's international airline passengers flew to or from an airport in the United Kingdom during 2006. The number of passengers in this hub increased fivefold in the 30 years ending in 2005, as the government envisaged that they would more than double by 2030, to 476 million a year (Monbiot 2006b). Carbon-dioxide emissions from air traffic originating in the United Kingdom rose 85 percent between 1990 and 2000, according to the U.K. Department of Trade and Industry (Houlder 2002b, 2).

The phenomenal success of budget airlines such as Ryanair and Easyjet contributed to a 76 percent increase in traffic through Britain's airports in a decade, to 215 million passengers in 2004 (Clark 2006). Even as they talked about reining in greenhouse-gas emissions, political leaders in the United Kingdom were promoting rapid increases in air travel and expansion of airports, much as automobile manufacturers in the United States talked up greenhouse-gas emissions caps while opposing higher gas-mileage standards. Both were "green" in theory and anything but in practice. Prime Minister Tony Blair said, "climate change is, without doubt, the major long-term threat facing our planet," just as British air mileage continued to explode (Lynas 2006, 12).

The Jetsons Come to Life

As scientists worry about the atmosphere's overload of greenhouse gases, commuting by air in the image of the twentieth-century cartoon family "The Jetsons" has taken on a life of its own. During the 1990s, a rising number of people in the United States were commuting to work via the airlines, sometimes thousands of miles per week. The *New York Times* has carried accounts of Manhattan jobholders who fly into the city from Rochester, New York. Silicon Valley, where the price of the average house was

$400,000 by the year 2000, was drawing weekly commuters, according to one *Times* account, from "Arizona, Idaho, Nevada, Oregon, and Utah" (Johnston 2000, 1-G). Some people who work at New York's Lincoln Center commute from Florida. The typical schedule includes long workdays Tuesday through Thursday, travel Monday and Friday, and a weekend at home hundreds, sometimes thousands, of miles from the office.

For those with a penchant for technology and no concern about rising greenhouse-gas levels in the atmosphere, the status vehicle in elite circles of the future may not be a sports-utility vehicle. According to one observer, it will be "the family plane" (Fallows 1999, 84). Such an aircraft, the Cirrus SR20, was being produced by the late 1990s, as the National Aeronautics and Space Administration (NASA) was "quietly advocating an Interstate Skyway Network" (Fallows 1999, 88). Between 2000 and 2005, the number of private jets in the United States grew by 40 percent.

The "flying car," to some people, is the next step beyond the private jet. Features in the *Wall Street Journal* have promoted "flying cars," with no mention of their impact on the atmosphere. Environmentally, "air cars" make Hummers look like bicycles. By 2006, about 100 people had paid $25,000 each to reserve a Moller International "Skycar," which was at the test stage. These Skycars are expected to have an initial sticker price of about $500,000 each (Stoll 2006, R-8). The Skycar is designed to taxi from home to a designated takeoff area (a "vertiport"), laid out on a parking lot or field. The Skycar, which its builders say will sell for about $60,000 once it is in mass production, will run on methanol, ethanol, diesel, or gasoline.

Air Travel's Effect on the Atmosphere

Aircraft emissions are especially damaging because much of their pollution takes place in the upper atmosphere, at jet-stream level. Carbon-dioxide emissions from air traffic originating in the United Kingdom rose 85 percent between 1990 and 2000, according to the U.K. Department of Trade and Industry. By 2004, air travel in the United States consumed 10 percent of all fossil-fuel energy, with passenger loads expected to double between 1997 and 2017, making air travel the fastest-growing source of carbon dioxide and nitrous oxides in the economy.

Air transport emits a toxic cocktail of gases. Atmospheric emissions from aircraft have three times the global-warming potential of their carbon-dioxide content alone. In addition to carbon dioxide, combustion of jet fuel injects water vapor and sulfur dioxide into the stratosphere, both of which enhance ozone depletion at that level. Nitrous oxides enhance ozone at lower levels, where it is a pollutant, and depletes it in the stratosphere, where it helps guard against ultraviolet radiation. The chemicals emitted into the atmosphere by combusting aviation fuel "cause the formation of polar stratospheric clouds, affect markedly the aerosol composition of the atmosphere, and intensify the greenhouse effect" (Kondratyev, Krapivin, and Varotsos 2004, 249).

"As far as climate change is concerned," wrote George Monbiot in the *London Guardian*,

> [t]his is an utter, unparalleled disaster. It's not just that aviation represents the world's fastest-growing source of carbon-dioxide emissions. The burning of aircraft fuel has a "radiative forcing ratio" of around 2.7. What this means is that the total warming effect of aircraft emissions is 2.7 times as great as the effect of the carbon dioxide alone. The water vapor they produce forms ice crystals in the upper troposphere (vapor trails and cirrus clouds) that trap the earth's heat. According to calculations by the Tyndall Centre for Climate Change Research, if you added the two effects together (it urges some caution as they are not directly comparable), aviation's emissions alone would exceed the government's target for the country's entire output of greenhouse gases in 2050 by around 134 per cent. (Monbiot 2006b)

The British government excludes international aircraft emissions from the target.

By 1999, in some air-traffic corridors above Europe, jet contrails covered as much as 4 percent of the sky at any given time. Vapor trails form cirrus clouds that contribute to global warming in addition to emissions from aircraft engines. Brian Hoskins, a research professor at the Royal Society and former head of meteorology at the United Kingdom's Reading University, said that if growth of air travel continued unrestricted, airline vapor trails would cover 10 percent of the sky over Britain by 2050 (Clover and Millward 2002, 1, 4). In the U.S. Northeast, jet contrails sometimes covered as much as 6 percent of the sky. Contrail coverage over Asia was increasing so quickly that contrails could multiply by a factor of 10 in 50 to 60 years. Jet contrails have a net atmospheric warming effect similar to that of high thin ice clouds, trapping outgoing long-wave radiation. The contrails also reflect some incoming solar radiation. Given this balance, night-time flights (December though February) at a site in southeast England "were responsible for most of the contrail radiative forcing" (Stuber et al. 2006, 864). Night flights "account for only 25 percent of daily air traffic, but contribute 60 to 80 per cent of the contrail forcing" (Stuber et al. 2006, 864). Winter flights account for only 22 percent of annual air traffic, but contribute half of the annual mean forcing. According to Stuber and colleagues, "These results suggest that flight rescheduling could help to minimize the climate impact of aviation" (2006, 864).

Flying above the Law

Because airline travel often crosses international borders, its greenhouse-gas emissions have been excluded from many tallies that add them up on a nation-by-nation basis. The airline industry routinely uses its international status to avoid national constraints such as energy-efficiency targets that go along with plans to reduce greenhouse-gas emissions. The 1944 Chicago Convention, now supported by 4,000 bilateral treaties, rules that no government may levy tax on aviation fuel. "The airlines," wrote British author George Monbiot, "have been bottle-fed throughout their lives" (Monbiot 2006b).

Despite its growing role in global warming, aviation is unregulated, unrecognized by international climate treaties such as the Kyoto Protocol, even as it grows faster than any other form of transport worldwide, averaging 12 percent a year. Taxes on airline fuel are banned by international treaty (the 1944 Chicago Convention on taxing fuel), giving the industry a unique incentive to avoid researching and enforcing efficiency exactly where our transportation system needs it most.

Jet fuel is not only one of the very few untaxed fuels in the world, but also it is zero-value-added tax rated. The airline industry argues that it is effectively taxed as U.K. airlines pay a duty of £5 per passenger for a European flight. This fee was halved from £10 per flight in 2000, which raised £900 million per year. However, airlines also benefit from a £9 billion tax subsidy per year in the United Kingdom (Francis 2006).

Air-pollution emissions from international flights have been routinely excluded from treaties to counter global warming and ozone depletion. Airports in Britain are exempt from pollution control, from planning permission requirements for many developments, and—in most cases—from statutory noise regulation. And, to top it all, air travel receives billions of dollars every year in subsidies from taxpayers of industrial countries (Lean 2001c, 23).

In June 2008, the European Union agreed that airlines arriving or departing from its soil will be required to buy carbon credits under the Union's cap-and-trade system. U.S. airline officials said that the move might be illegal under international conventions governing civil aviation, which ban taxation of international carriers.

Jet Fuel Efficiency

The U.S. aviation industry increased its fuel efficiency 23 percent from 2000 to 2006, flying 18 percent more miles, carrying 12 percent more passengers, all on roughly 5 percent less fuel, according to the U.S. Department of Transportation. According to the International Air Transport Association, jet engines by 2006 were 40 percent more fuel efficient than they were in the 1960s. The air-travel industry may have reduced the amount of fuel it burns to transport each of its passengers by about half since the mid-1970s, but the growth of the number of people taking to the air has more than extinguished these savings in fuel efficiency. The same relationship applies to nitrogen oxides and hydrocarbons.

Fuel consumption fell to 19.6 billion gallons in 2006, compared with 204 billion in 2000, as airliners added navigation technology to allow more direct routing, modified wings to make them more aerodynamic, and decreased weight with lighter seats and lower inventories of food and water, among other things. However, "[t]here is little left you can do other than renewing the fleet or flying less," said John Heimlich, chief economist of the Air Transport Association (Frank 2008, 1-B). American and Delta Air Lines jets, for example, now sometimes taxi using one engine to reduce fuel use. Both carriers promoted their efforts to modify wings so that they reduce drag and boost efficiency, as well as finding ways to reduce weight of aircraft by removing ovens, galleys, and water (Wilber 2007, D-1).

Discussions of airline fuel efficiency offer some promise of incremental improvements—flying 2,000 meters lower may save 6 percent of fuel; better planning to avoid long waits on take-off and more direct routing may save 10 percent. The Boeing Company has pitched its 787 Dreamliner as a marvel of efficiency that will stretch fuel supplies by 20 percent. There has been some talk of jet fuel from soya, which may lower greenhouse-gas emissions (but how many square miles of soybean fields would be required to get a Dreamliner from New York City to London and back?). At the end of the day, however, air travel remains a bad dream in any global-warming calculus. Every passenger on a long-haul flight is responsible for 124 kilograms (about 270 pounds) of carbon dioxide per hour (McGuire 2005, 188–189).

The long life of jets restricts the ability to raise an airline fleet's fuel efficiency. The Boeing 747, for example, is still flying 36 years after it was introduced. The Tyndall Centre predicts that the Airbus A380, new in 2006, still will be flying (although in an incrementally modified form) in 2070. "Switching to more efficient models," wrote Monbiot, "would mean scrapping the existing fleet" (Monbiot 2006b).

Future efficiency gains may be meager, however, due to the mature nature of jet-engine technology. With commercial aviation mileage expected to double by 2050, the search is on for solutions to an air transport system in which one airplane flying from New York City to Stockholm emits as much carbon dioxide as an average automobile commuter in 50 years. (Daviss 2007, 33). Various systems have been tried and abandoned that might make the aerodynamics of jet aircraft more efficient. Usually, systems meant to improve laminar flow do not pay for themselves over the life of an aircraft.

In September 2006, Richard Branson, owner of Virgin airlines, revealed plans to invest $3 billion to develop ecologically friendly plant-based jet fuel. At present, the use of hydrogen fuel or ethanol in place of the usual kerosene jet fuel faces formidable obstacles. Hydrogen fuel provides only 25 percent as much energy per volume as jet fuel, meaning that a hydrogen-powered aircraft would need huge fuel tanks and have to fly with a heavier load, reducing mileage. Because of the volume, fuel could not be carried in the wings, but in the body of the aircraft, increasing drag. Hydrogen would produce no carbon dioxide, but its output of water vapor at high altitudes would increase the size of contrails, which aggravate global warming. Plant-based fuel weighs two-thirds more by volume than kerosene for the same amount of thrust. It also freezes easily at high altitudes (Daviss 2007, 35).

The Association of British Travel Agents (ABTA) in late 2006 joined with two other British travel-industry organizations in a carbon-offset program. The program allowed agents to offer customers an opportunity to compensate for the climate-warming impact of their travel (most notably airline trips) by contributing financially toward environmental projects around the world. The ABTA joined with the Association of Independent Tour Operators and the Federation of Tour Operators to develop a plan to begin early in 2007. British Airways has introduced a "CO_2 Emissions Calculator" on its Web site that allows passengers to pay an offset for their flights (Michael and Carey 2006, A-6).

Carbon offsets may include helping to pay for a wind farm, developing a more energy-efficient technology, paying to plant trees, or contributing to other projects. Critics of offsetting contend that greenhouse-gas emissions should be reduced in reality rather than offset in theory. Offsets allow guilt-free carbon-dioxide indulgences (such as flying long distances) and parading of green corporate credentials. Writer George Monbiot has said the most destructive effect of the carbon offset trade is that it allows people to believe they can continue polluting with complacency by paying for a tree-planting company ("British Travel" 2006).

An International Panel on Climate Change (IPCC) report on air travel and global warming stimulated calls for environmental taxes and stricter emissions targets for airlines. The report was produced jointly by the United Nations and the World Meteorological Organization. "Transportation is the area of greatest growth (in terms of impact on the climate), and aviation is growing more rapidly than any other sector," said John Houghton of Britain, one of the report's co-authors ("Aircraft Pollution" 1999). "Policy options include more stringent aircraft engine emissions regulations, removal of subsidies and incentives [and] market-based options such as environmental levies," the report said ("Aircraft Pollution" 1999).

Proposals to lighten the carbon footprint of air travel include replacing all the metal in aircraft with lightweight composites, which could reduce fuel use about 25 percent, according to Ian Poll, an aeronautical engineer at Cranfield University. The Boeing 787 Dreamliner is half plastic reinforced with carbon fiber, by weight (Sanderson 2008, 264). High-fat-content algae someday may be cultivated as biofuel to blend with jet fuel. Other biofuels may use prairie grass (switchgrass) or jatropha, mixed with jet fuel.

Small planes may be powered by lithium batteries. A small solar-powered aircraft, the Solar Impulse, plans a round-the-world flight in 2009.

Eco-Tourism's Hidden Pollution

Steve McCrea, editor of the *Eco-Tourist Journal*, called air travel "eco-tourism's hidden pollution" (McCrea 1996). Tourists who take the utmost ecological care when they visit exotic locales rarely give a second thought to the greenhouse gases that they generate while reaching their destinations. According to McCrea,

> One ton of carbon dioxide enters the atmosphere for every 4,000 miles that the typical eco-tourist flies. A round trip from New York to San Jose, Costa Rica (the world's leading eco-tourist destination) is 4,200 miles, so the typical eco-tourist generates roughly 2,100 pounds of carbon dioxide by traveling to a week of sleeping in the rainforest. (McCrea 1996)

To balance carbon dioxide generated by their air travel, McCrea suggested that eco-tourists plant three trees for every 4,000 miles, to compensate not only for the carbon dioxide, but also for other greenhouse gases created by the combustion of jet fuel.

Even as the glaciers melt, travel agents flock to the banner of eco-tourism, advising clients to hurry up and see ancient ice before it's gone. In 2007, Betchart Expeditions, Inc., of Cupertino, California, offered a 12-day tour to "Warming Island," off Greenland, which had recently emerged from melting ice. This expedition company cited the island as "a compelling indicator of the rapid speed of global warming" (Naik 2007, A-12). The cost of the full tour was $5,000 to $7,000, plus airfare. In the Greenland coastal village of Illuissat (population 5,000), about 35,000 tourists arrived, most of them on cruise ships, during 2007, up from 10,000 about five years previously. Many came to witness the receding Jacobshavin glacier, which had lost nine miles in five years. Once upon a time, winter temperatures routinely fell to −40°F there; by 2007, −15°F was the usual winter's low. The harbor, which used to freeze solid, now remains liquid all year, allowing fisherman to pull halibut out of the water at all seasons, depleting stocks (Naik 2007, A-12).

Reducing Aviation's Carbon Footprint

Could airliners use hydrogen or biofuels? Mark Lynas argued probably not, for several reasons. Hydrogen is too bulky to work as a fuel, and in any case, its combustion output of water, when injected high in the stratosphere, would contribute to global warming rather than reducing it. As for biofuels, they do not have the energy density of kerosene (the major constituent of standard jet fuel), and the business of producing them in large quantities is already endangering food security and boosting deforestation across the tropics. There is simply no possibility that they could be produced in the volume needed to quench the thirst of jet aircraft in the long term (Lynas 2006, 14). Lynas added that the land Heathrow plowed up for runways once grew some of the best apples in England. Today, these apples are being replaced by airlifted produce with an enormous carbon footprint (Lynas 2006, 15).

Tim O'Riordan, of the Zuckerman Institute for Connective Environmental Research at the University of East Anglia, offered this explanation:

> Everyone who uses a car or flies regularly should consider what positives they can put back into the environmental battle by way of compensation. Maybe, for example, you already cycle to work, or take public transport, as a chosen alternative to driving. The next time you fly somewhere, either by choice or for business, the carbon dioxide and nitrous oxide production from the plane you use will cancel out a year of contributions from cycling. (Urquhart and Gilchrist 2002, 9)

During late November 2002, two official British studies called for an end to cheap flights and a ban on new airport runways. The Royal Commission on Environmental Pollution urged Britain's government to halt airport growth, raise fares, and place financial pressure on short-haul and no-frills carriers. At the same time, the British government's body of environmental advisers, the Sustainable Development Commission, said that proposals for new runways at Stansted, Heathrow, Luton, Rugby, and a potential new airport at Cliffe in Kent required a "fundamental rethink" (Clover and Millward 2002, 1, 4). The commission, a body of academics, businessmen, and people from public life, recommended that the price of a one-way ticket should rise by £40 to have any chance of mitigating climate change. Paul Ekins, an economist and a member of the commission, said, "We believe a stable climate is a good thing and worth modifying human behavior for" (Clover and Millward 2002, 1, 4). The commission called for the rapid growth of high-speed trains to replace short-haul services at transport hubs such as Amsterdam's Schipol airport (Clover and Millward 2002, 1, 4).

A "No-Flying Movement"

So what can fliers who are concerned about ruining the atmosphere do? The elegantly simple—and perhaps, for the time being, only—effective answer is don't fly as much. While airline manufacturers have gained some efficiency in recent decades, no nonfossil fuel can provide the thrust that airliners need to gain and maintain altitude and speed. No one is seriously entertaining the idea of solar-, wind-, or (thankfully) nuclear-driven aircraft. Like no other form of transport, the modern aircraft is a hostage of the fossil-fuel age and a penultimate producer of greenhouse gases.

"In researching my book about how we might achieve a 90 percent cut in carbon emissions by 2030," wrote Monbiot, "I have been discovering, greatly to my surprise, that every other source of global warming can be reduced or replaced to that degree without a serious reduction in our freedoms. But there is no means of sustaining long-distance, high-speed travel" (Monbiot 2006b).

In Britain, a "no-flying movement" has begun to take shape, as many people voluntarily are committing to avoid aviation at all for nonessential trips.

Newspaper travel supplements began to provide information on train or shipping alternatives. Vacationers have been advised to

> [t]ravel to the Alps by train and ... get a real sense of geography, of evolving culture and changing climatic zones. Arrive by air and all you see is identical airport terminals and thousands of other culture-shocked, aggravated travelers. Slow travel, like slow food, is about clawing back quality of life. (Lynas 2006, 14–15)

Environmental protesters in the United Kingdom tried to halt expansion at several British airports, including Heathrow and Stansted. On February 19, 2006, a convoy of more than 100 cars toured some of the villages to be affected if proposals to build a second runway at Stansted were approved (Clark 2006).

FURTHER READING

"Aircraft Pollution Linked to Global Warming; Himalayan Glaciers are Melting, with Possibly Disastrous Consequences." Reuters in *Baltimore Sun*, June 13, 1999, 13-A.

"British Travel Agents Launch Carbon Offset Scheme." Environment News Service, November 28, 2006. http://www.ens-newswire.com/ens/nov2006/2006-11-28-05.asp.

"Carbon Neutral Chic." Editorial, *Wall Street Journal*, July 9, 2007, A-14.

Cars and Climate Change. Paris, France: International Energy Agency, 1993.

Chea, Terence. "Environmental Groups Call for Planes to Reduce Emissions." *USA Today*, December 6, 2007, 6-B.

Clark, Andrew. "'Open Skies' Air Treaty Threat." *London Guardian*, February 20, 2006. http://www.guardian.co.uk/frontpage/story/0,,1713677,00.html.

Clover, Charles. "Air Travel Is a Threat to Climate." *London Daily Telegraph*, June 5, 1999.

Clover, Charles, and David Millward. "Future of Cheap Flights in Doubt; Ban New Runways and Raise Fares, Say Pollution Experts." *London Daily Telegraph*, November 30, 2002, 1, 4.

Daviss, Bennett. "Green Sky Thinking: Could Maverick Technologies Turn Aviation into an Eco-success Story? Yes, but Time Is Running Out." *New Scientist*, February 24, 2007, 32–38.

Fallows, James. "Turn Left at Cloud 109." *New York Times Sunday Magazine*, November 21, 1999, 84–89.

Flannery, Tim. *The Weather Markers: How Man Is Changing the Climate and What It Means for Life on Earth.* New York: Atlantic Monthly Press, 2005.

Francis, Justin. Responsibletravel.com. United Kingdom. Accessed April 30, 2006. http://www.responsibletravel.com/Copy/Copy101993.htm.

Frank, Thomas. "Planes Fly More, Emit Less Greenhouse Gas." *USA Today*, May 9, 2008, 1-B.

Hickman, Martin. "The Prince of Emissions." *London Independent*, April 1, 2006, 1.

Hillman, Mayer, Tina Fawcett, and Sudhir Chella Rajan. *The Suicidal Planet: How to Prevent Global Climate Catastrophe.* New York: St. Martin's Press/Thomas Dunne Books, 2007.

Houlder, Vanessa. "Rise Predicted in Aviation Carbon Dioxide Emissions." *London Financial Times*, December 16, 2002b, 2.

Johnston, David Cay. "Some Need Hours to Start Another Day at the Office." *New York Times* in *Omaha World-Herald*, February 6, 2000, 1-G.

Kondratyev, Kirill, Vladimir F. Krapivin, and Costas A. Varotsos. *Global Carbon Cycle and Climate Change.* Berlin, Germany: Springer/Praxis, 2004.

Lean, Geoffrey. "We Regret to Inform You that the Flight to Malaga Is Destroying the Planet; Air Travel Is Fast Becoming One of the Biggest Causes of Global Warming." *London Independent*, August 26, 2001c, 23.

Lynas, Mark. "Fly and Be Damned." *New Statesman* (London), April 3, 2006, 12–15. www.newstatesman.com/200604030006.

McCrea, Steve. "Air Travel: Eco-tourism's Hidden Pollution." *San Diego Earth Times*, August 1996. http://www.sdearthtimes.com/et0896/et0896s13.html.

McGuire, Bill. *Surviving Armageddon: Solutions for a Threatened Planet.* New York: Oxford University Press, 2005.

Michael, Daniel, and Susan Carey. "Airlines Feel Pressure as Pollution Fight Takes Off." *Wall Street Journal*, December 12, 2006, A-6.

Monbiot, George. "We Are All Killers: Until We Stop Flying." *London Guardian*, February 28, 2006b. http://www.monbiot.com/archives/2006/02/28/we-are-all-killers/.

Naik, Gautam. "Arctic Becomes Tourism Hot Spot, but is it Cool?" *Wall Street Journal*, September 24, 2007, A-1, A-12.

Rosenthal, Elisabeth. "Air Travel and Carbon on Increase in Europe." *New York Times*, June 22, 2008f. http://www.nytimes.com/2008/06/22/world/europe/22fly.html.

Sanderson, Katherine. "Flights of Green Fancy." *Nature* 453 (May 15, 2008):264–265.

Stoll, John D. "Visions of the Future: What Will the Car of Tomorrow Look Like? Perhaps Nothing Like the Car of Today." *Wall Street Journal*, April 17, 2006, R-8.

Stuber, Nicola, Piers Forster, Gaby Radel, and Keith Shine. "The Importance of the Diurnal and Annual Cycle of Air Traffic for Contrail Radiative Forcing." *Nature* 441 (June 15, 2006):864–867.

Urquhart, Frank, and Jim Gilchrist. "Air Travel to Blame as Well." *The Scotsman*, October 8, 2002, n.p. (LEXIS)

Webster, Ben. "Boeing Admits Its New Aircraft Will Guzzle Fuel." *London Times*, June 19, 2001, n.p. (LEXIS)

Wilber, Del Quentin. "U.S. Airlines under Pressure to Fly Greener; Carriers Already Trying to Save Fuel as Europe Proposes Plan." *Washington Post*, July 28, 2007, D-1. http://www.washingtonpost.com/wp-dyn/content/article/2007/07/27/AR2007072702256_pf.html.

Alaska, Global Warming in

Gunter Weller, director of the Center for Global Change and Arctic System Research at the University of Alaska in Fairbanks, said mean temperatures in Alaska had increased by 5°F in the summer and 10°F in the winter during 30 years. This makes Alaska one of the most rapidly warming places on Earth. Moreover, the Arctic ice field has shrunk by 40 to 50 percent over the last few decades and lost 10 percent of its thickness, studies show. "These are pretty large signals, and they've had an effect on the entire physical environment," Weller said (Murphy 2001, A-1).

Glaciers were melting so quickly in Alaska by mid-2005 that their anticipated demise was prompting some tourists to visit. The Travel Section of the *New York Times* headlined: "The Race to Alaska before It Melts" (Egan 2005). Cities and towns across the entire state (including Anchorage, Fairbanks, Juneau, and Nome) reported record-high temperatures during the summer of 2004. At Portage Lake, 50 miles south of Anchorage, "people came by the thousands to see Portage Glacier, one of the most accessible of Alaska's frozen attractions. Except, you can no longer see Portage Glacier from the visitor center. It has disappeared" (Egan 2005).

Barrow, in northernmost Alaska, experienced a high of 70°F, the highest in memory, during the summer of 2004, as salmon and porpoises were sighted offshore during Alaska's warmest summer on record. Previously, salmon had been so rare off Barrow that indigenous people had to ask their southern relatives how to cure and dry them. Some people swam in Prince William Sound, most of them for the first time. Baby walruses, which had been abandoned by their mothers, were observed swimming in open water, their usual ice-floe homes nowhere to be seen. Unable to survive on their own, most of the walrus pups died within a month (Petit 2004, 66–69).

For Native peoples—17 percent of Alaskans—global warming is a particular threat. The natural world "is our classroom," said Sterling Gologergen, an environmental specialist with the Nome-based Norton Sound Health Corporation (Rosen 2003, 1). In her home region of Alaska, traditional bowhead whaling migrations have been disrupted by rising temperatures. Walrus hunters must travel farther, at greater risk, to find animals at the edges of receding ice. Beavers, previously unknown in the region, are showing up in local streams, and their dams could interfere with water quality and fish runs.

Rapid warming in Alaska affects the economy in some unusual ways. For example, on Alaska's North Slope, most oil exploration takes place during the season that the tundra is frozen, the only time that heavy equipment may be easily moved without damage to the land. In 1970, the frozen season was 200 days. By 2004, it was about half that (Braasch 2007, 66).

Melting Permafrost and "Drunken Forests"

Some Alaskan forests have been drowning and turning gray as thawing ground sinks under

Nathan Weyiouanna's abandoned house at the west end of Shishmaref, Alaska, December 8, 2006, sits on the beach after sliding off during a fall storm in 2005. (AP/Wide World Photos)

them. Trees and roadside utility poles, destabilized by thawing, lean at crazy angles. The warming has contributed a new phrase to the English language: "the drunken forest" (Johansen 2001, 20). In Barrow, home of Pepe's, the world's northernmost Mexican restaurant, mosquitoes, another southern import, have become a problem. Barrow also has experienced its first thunderstorm on record. Temperatures in Barrow began to rise rapidly at about the same time that the first snowmobile arrived, in 1971. By the summer of 2002, bulldozers were pushing sand against the invading sea in Barrow.

In Kotzebue, Alaska, according to a report in the *London Guardian*, "[t]undra has turned from spongy to dry" (Campbell 2001, 11). In Fairbanks, portions of some golf courses are collapsing as permafrost begins to melt. North of Fairbanks, houses were lifted on hydraulic jacks to prevent sinking due to melting permafrost, "which is no longer permanent" (Egan 2002a, A-1). By 2002, the trans-Alaskan oil pipeline was being inspected for damage resulting from melting permafrost. The pipeline, built during the 1970s, was designed on assumptions that the permafrost would never melt.

Mark Lynas, extracting from his book *High Tide: News from a* Warming *World* (Flamingo 2004), wrote in the *London Guardian*:

> Roads all around Fairbanks are affected by thawing permafrost; driving over the gentle undulations is like being at sea in a gentle swell. In some places the damage is more dramatic—crash barriers have bent into weird contortions, and wide cracks fracture the dark tarmac. Permafrost damage now costs a total of $35 million every year, mostly spent on road repairs. Some areas of once-flat land look like bombsites, pockmarked with craters where permafrost ice underneath them has melted and drained away. These uneven landscapes cause "drunken forests" right across Alaska. In one spot near Fairbanks, a long gash had been torn through the tall spruce trees, leaving them toppling over towards each other. (Lynas 2004b, 22)

Across large parts of the Arctic, permafrost has melted because of record-high temperatures during recent decades. "It's really happening almost everywhere," said Vladimir Romanovsky, a geophysicist at the University of Alaska. The accelerating pace of permafrost melting has shocked researchers, according to one observer (Stokstad 2004a, 1618). A major reason that researchers are so concerned has been the potential that melting permafrost will release additional carbon dioxide into the atmosphere. For many thousands of years, the arctic has been locking carbon away in permafrost; estimates range from 350 to 450 gigatons, about a quarter of the Earth's soil carbon. In parts of Siberia, peat deposits that are hundreds of meters thick extend several thousand miles. Permafrost covers about a third of the land in the Northern Hemisphere (Stokstad 2004a, 1619).

During the last half of the twentieth century, researchers began to detect signs that carbon dioxide and methane from the Arctic's permafrost has been reentering the atmosphere because of rapidly warming temperatures. The area was converting from a "sink" to a net source of greenhouse gases. Walter Oechel, an ecologist at San Diego State University, led research at Toolik Lake and Barrow, Alaska, where they found that the Alaskan tundra was releasing more carbon dioxide than it was absorbing—a surprise that was, according to Oechel, "contrary to all that was known about the Arctic system" (Stokstad 2004a, 1620). As the melting of the Arctic's carbon stores continues, the region is "likely to be a huge positive feedback on global warming" (Stokstad 2004a, 1620). The warmer and drier the soil becomes, the more carbon dioxide it releases; warmer, wetter soil releases more methane. As larger volumes of green plants invade the arctic with warmer weather, however, they may absorb some of the excess carbon dioxide emitted from melting permafrost.

Raining on the Iditarod

During January and February of 2001, snow was in short supply across much of Alaska. For several weeks, unusual rain soaked areas around Anchorage. The tundra along Norton Sound, near the end of the 1,760-kilometer Iditarod race trail, was so bare in December that parts of it caught fire ("Late Snow" 2001). During training for the race in the Matanuska-Susitna Valley, 40 miles north of Anchorage, much of the ground was bare or ice covered. Some trails were so icy that dogs risked injury. Sufficient snow fell before the race's scheduled start in March, however.

In 2003, the Iditarod was postponed, for the first time, for lack of snow. Snow was shipped in for the race's start March 1, 2003, and the race route was revised to avoid some areas. As in 2001 and 2002, unusual rain soaked parts of the race route, as Alaska experienced its mildest winters in more than a century of recordkeeping. At the same time, the eastern United States was experiencing intense cold and much above-average snowfall. Vermont was having its coldest winter in 25 years, and many of the big cities on the U.S. Eastern Seaboard experienced repeated heavy snow and ice storms. During 2004, however, Alaska had abundant snow at race time—not something that previously would be cause for comment.

Shrubs Advance, Glaciers Retreat

The retreat of Alaskan glaciers is illustrated most graphically by comparing photographs taken recently with those of a century or more ago in the same locations. Bruce Molnia, a geologist with the U.S. Geological Survey, has gathered more than 200 glacier photos taken from the 1890s to the late 1970s. "Where masses of ice were once surging down wide mountain passes into the sea, or were hanging from high and perilously steep faces," wrote David Perlman of the *San Francisco Chronicle*, the surfaces in Molnia's images now stand bare. What remains from many of the retreating glaciers are stretches of open water or broad, snow-free layers of sediment" (Perlman 2004, A-18). "And as the glaciers disappear," Molnia said, "you get the amazing appearance of vegetation." On the tundra north of Alaska's Brooks Range, according to Molnia's observations (as cited by Perlman),

> [e]xplosive bursts of vegetation—willows, alders, birch and many shrubs—are thriving where permafrost once kept the tundra surface frozen in winter. The growth of shrubs across the tundra has increased by 40 percent in less than 60 years, [University of Washington geophysicist Ken] Tape said, and that perturbation is certainly due to the changing climate. (Perlman 2004, A-18)

"The abundance of Arctic shrubs is ... increasing, apparently driven by a warming climate," according to the authors of an article in the January 2005 edition of *Bioscience*. The paper was written by Matthew Sturm of the U.S. Army

Cold Regions Research and Engineering Laboratory in Fort Wainwright, Alaska, and seven other researchers. "The evidence for increasing shrub abundance is most comprehensive for northern Alaska," Sturm and colleagues wrote. "An extensive comparison of old [1940s] and modern photographs has shown that shrubs there are increasing in size and are colonizing previously shrub-free tundra" (Whipple 2005). Shrubs also are increasing in the western Canadian Arctic, the research team stated,

> but there the change is inferred from the recollections of long-time residents. In central Russia, a transect along the Pechore River has shown a decrease in tundra and a corresponding increase in shrubland, but for the vast tundra regions of Siberia, there are currently no data on which to make an assessment. (Whipple 2005)

The Sturm group added that despite the lack of specific data for those areas, "satellite remote sensing studies ... greatly strengthen the case for pan-Arctic expansion of shrubs" (Whipple 2005).

Changes in Alaskan Native Life

Native Alaskans are watching their natural world change rapidly. Mike Williams, a Yupiaq Eskimo and leader of the Alaska Inter-Tribal Council, which represents 229 native Alaskan tribes, has spent most of his life on the Kushokwim River in the western region of subarctic Alaska. Williams has watched as increasing erosion has forced Bethel and other Alaskan towns to repeatedly reinforce sea walls. Millions of dollars have been spent on erosion control for Bethel alone. Nipaciak and other smaller villages have been moved from their seacoast locations. "This used to be a village here and because of the erosion, it had wiped it out and people are moving way back," Williams said (Padden 2006). Seventy-seven-year-old James Willie, who is 76 years of age according to traditional calendars, said that the snow has changed. "It was a different cold. Snow wasn't, you know, it's just like feather. When it got a little bit warm it melted away fast" (Padden 2006). Katie Kernak said that forest fires have been the greatest change from rapid warming: "When [I] was growing up [I] never used to hear about any fires at all. But now in the summer it is smoky and there are all kinds of fires" (Padden 2006).

FURTHER READING

Braasch, Gary. *Earth Under Fire: How Global Warming Is Changing the World.* Berkeley: University of California Press, 2007.

Campbell, Duncan. "Greenhouse Melts Alaska's Tribal Ways; As Climate Talks Get Under Way In Bonn Today, Some Americans Are Ruing the Warming Their President Chooses to Ignore." *London Guardian,* July 16, 2001, 11.

Egan, Timothy. "Alaska, No Longer so Frigid, Starts to Crack, Burn, and Sag." *New York Times,* June 16, 2002a, A-1.

Egan, Timothy. "On Hot Trail of Tiny Killer in Alaska." *New York Times,* June 25, 2002b, F-1.

Egan, Timothy. "The Race to Alaska before It Melts." *New York Times.* Travel Section, June 26, 2005.

Johansen, Bruce E. "Arctic Heat Wave." *The Progressive,* October 2001, 18–20.

"Late Snow Smoothes the Way for Iditarod Sled-dog Race." Reuters in *Ottawa Citizen,* March 2, 2001, B-8.

Lynas, Mark. "Meltdown: Alaska Is a Huge Oil Producer and Has Become Rich on the Proceeds. But It Has Suffered the Consequences; Global Warming, Faster and More Terrifyingly Than Anyone Could Have Predicted." *London Guardian* (Weekend Magazine), February 14, 2004b, 22.

Murphy, Kim. "Front-row Exposure to Global Warming; Climate: Engineers Say Alaskan Village Could Be Lost as Sea Encroaches." *Los Angeles Times,* July 8, 2001, A-1.

Padden, Brian. "Native Alaskans Feel the Heat of Global Warming." Voice of America. August 11, 2006.

Perlman, David. "Shrinking Glaciers Evidence of Global Warming; Differences Seen by Looking at Photos from 100 Years Ago." *San Francisco Chronicle,* December 17, 2004, A-18.

Petit, Charles W. "Arctic Thaw." *U.S. News and World Report,* November 8, 2004, 66–69.

Rosen, Yereth. "Alaska's Not-so-Permanent Frost." *Christian Science Monitor,* October 7, 2003, 1.

Stokstad, Eric. "Defrosting the Carbon Freezer of the North." *Science* 304 (June 11, 2004a):1618–1620.

Whipple, Dan. "Climate: The Arctic Goes Bush." United Press International, January 10, 2005. (LEXIS)

Albedo

Albedo measures a surface's ability to reflect light and heat. The concept is important in global warming because changes in albedo have an important role in changing temperatures in any given location and, thus, the amount of heat absorbed. In the Arctic, albedo influences the speed at which ice or permafrost melts. Changes in albedo (Latin for "whiteness") are among the factors contributing to a rate of warming in the Arctic during the last 20 years that has been eight times the rate of warming during the previous 100 years ("Recent Warming" 2003). Recent increases in the number and extent of boreal forest fires also have been increasing the amount of soot in the atmosphere, which also changes albedo.

The melting of ocean-borne ice in polar regions can accelerate overall warming as it changes surface albedo. The darker a surface, the more solar energy it absorbs. Seawater absorbs 90 to 95 percent of incoming solar radiation, whereas snow-free sea ice absorbs only 60 to 70 percent of solar energy. If the sea ice is snow covered, the amount of absorbed solar energy decreases substantially, to only 10 to 20 percent. Therefore, as the oceans warm and snow and ice melt, more solar energy is absorbed, leading to even more melting. "It is feeding on itself now, and this feedback mechanism is actually accelerating the decrease in sea ice," said Mark Serreze of the University of Colorado (Toner 2003d, 1-A).

As high latitudes warm and the coverage of sea ice declines, thawing Arctic soils also may release significant amounts of carbon dioxide and methane now trapped in permafrost. Warmer ocean waters also eventually could release formerly solid methane and carbon dioxide from the seafloor. According to David Rind, of NASA's Goddard Institute for Space Studies in New York,

> [t]hese feedbacks are complex and we are working to understand them. Global warming is usually viewed as something that's 50 or 100 years in the future, but we have evidence that the climate of the Arctic is changing right now, and changing rapidly. Whatever is causing it, we are going to have to start adapting to it. (Toner 2003d, 1-A)

The warming of the Arctic "[w]ill definitely impact our weather in the United States," said Rind. "Those outbreaks of Arctic cold we get each winter might seem like something we could do without, but if we don't have them, we're going to receive a lot less winter precipitation," he said. "Computer models show that Kansas would be 10 degrees warmer during the winter—and get 40 percent less snow, which could make it very difficult to grow winter wheat there" (Toner 2003d, 1-A).

Albedo seems a simple process, but the gradual darkening of Arctic surfaces can produce significant changes in the amount of solar energy that a given area absorbs. Scientists have been trying to model how large this effect may be, and by 2005, their estimates were quite substantial: three watts per square meter *per decade*, or roughly equal to a doubling of atmospheric carbon dioxide in areas where albedo has changed to a marked degree. "The continuation of current trends in shrub and tree expansion could further amplify this atmospheric heating by two to seven times," the same study anticipated (Chapin et al. 2005, 657). Forest is generally darker in color than the tundra that it replaces. Warming recently has been shortening the snow-covered season by roughly 2.5 days per decade in the Arctic, accelerating the change in albedo (Foley 2005, 627).

James Hansen and colleagues, writing in the *Proceedings of the Royal Society* (Great Britain), said that the Earth's climate is remarkably sensitive to albedo-driven influences on climate (i.e., "forcings"). A small forcing can produce a large effect. They wrote:

> This allows the entire planet to be whipsawed between climate states. One feedback, the "albedo flip" ... provides a powerful trigger mechanism. A climate forcing that "flips" the albedo of a sufficient portion of an ice sheet can spark a cataclysm. Ice sheet and ocean inertia provides only moderate delay to ice sheet disintegration and a burst of added global warming. Recent greenhouse gas (GHG) emissions place the Earth perilously close to dramatic climate change that could run out of our control, with great dangers for humans and other creatures. Carbon dioxide (CO_2) is the largest human-made climate forcing, but other trace constituents are important. (Hansen et al. 2007)

"Only intense simultaneous efforts to slow CO_2 emissions and reduce non-CO_2 forcings can keep climate within or near the range of the past million years," Hansen and colleagues continued.

> The most important of the non-CO_2 forcings is methane (CH_4), as it causes the second-largest human-made greenhouse-gas (GHG) climate forcing and is the principal cause of increased tropospheric ozone (O_3), which is the third-largest GHG forcing. Nitrous oxide (N_2O) should also be a focus of climate mitigation efforts. Black carbon ("black soot") has a high global warming potential (about 2000, 500, and 200 for 20, 100, and 500 years, respectively) and deserves greater attention. Some

forcings are especially effective at high latitudes, so concerted efforts to reduce their emissions could still "save the Arctic," while also having major benefits for human health, agricultural productivity, and the global environment. (Hansen et al. 2007)

FURTHER READING

Chapin, F. S., III., M. Sturm, M. C. Serreze, J. P. McFadden, J. R. Key, A. H. Lloyd, A. D. McGuire, T. S. Rupp, A. H. Lynch, J. P. Schimel, J. Beringer, W. L. Chapman, H. E. Epstein, E. S. Euskirchen, L. D. Hinzman, G. Jia, C.-L. Ping, K. D. Tape, C. D. C. Thompson, D. A. Walker, and J. M. Welker. "Role of Land-Surface Changes in Arctic Summer Warming." *Science* 310 (October 28, 2005):657–660.

Foley, Jonathan A. "Tipping Points in the Tundra." *Science* 310 (October 28, 2005):627–628.

Hansen J., M. Sato P., Kharecha, G. Russell, D.W. Lea, and M. Siddall. "Climate Change and Trace Gases." *Philosophical Transactions of the Royal Society* A365 (2007):1925–1954.

"Recent Warming of Arctic May Affect World-wide Climate." National Aeronautics and Space Administration Press Release, October 23, 2003. http://www.gsfc.nasa.gov/topstory/2003/1023esuice.html.

Toner, Mike. "Arctic Ice Thins Dramatically, NASA Satellite Images Show." *Atlanta Journal and Constitution*, October 24, 2003d, 1-A.

Alligators Spread Northward

In mid-May, 2006, alligators were sighted in the backwaters of the Mississippi River as far north as Memphis. "It's possible that alligators have had a northern range expansion due to the mild winters we've had in the past 10 years," wildlife agent Gary Cook said ("Gators Spotted" 2006). The Tennessee Wildlife Resources Agency received reports of alligator sightings on McKellar Lake, a backwater of the Mississippi just south of Memphis, and at T. O. Fuller State Park, north of the city. Up to five alligators may have been seen, including one said to be close to seven feet long that was reportedly spotted on a bank beside McKellar Lake. "It was just laying out in the sun," said Kay Vescovi, a chemical plant manager at a nearby industrial park. "Nobody was concerned, just kind of shocked that they're this far north" ("Gators Spotted" 2006).

Stan Trauth, a zoology professor at Arkansas State University, said alligators spotted in the Memphis region could have been pets that were released into the wilds. "Moving north is not what they would want. They would want to stay in the more moderate climate," said Trauth, who has tracked alligators released in Arkansas by wildlife agents. Trauth said that the region is along the northern edge of animals' survival range. "They're reaching their physiological tolerance in the winter in this area," he said ("Gators Spotted" 2006).

FURTHER READING

"Gators Spotted in Mississippi River Backwaters at Memphis." Associated Press in *Daytona Beach News-Journal*, May 13, 2006, n.p.

Alps (and Elsewhere in Europe), Glacial Erosion

Rising temperatures have been melting ancient glaciers on the high Alps, causing devastating summer rockslides that have endangered the lives of many climbers, including 70 in one day (July 14, 2003), one of the largest mass rescues in the area's history. Most were plucked from the Matterhorn, which was racked by two major landslides that day. According to one observer, "Those climbing its slopes could have been forgiven for thinking the crown jewel of the Alps had started falling apart under their feet" (McKie 2003, 18).

According to Robin McKie, writing in *London's Observer*, "The great mountain range's icy crust of permafrost, which holds its stone pillars and rock faces together, and into which its cable car stations and pylons are rooted, is disappearing" (McKie 2003, 18). Several recent Alpine disasters, including the avalanches that killed more than 50 people at the Austrian resort of Galtur during 1999, have been blamed on melting permafrost. During August 2003, the freezing line in the Alps rose to 13,860 feet (4,200 meters), almost 4,000 feet above its usual summer maximum of 3,000 meters (9,900 feet) (Capella 2003).

More Danger Ahead

Scientists attending the 2003 International Permafrost Association conference in Zurich, Switzerland, said that conditions in the high Alps could get more dangerous in coming years, assuming continued warming. "I am quite sure what happened on the Matterhorn … was the result of the Alps losing its permafrost. We have found that the ground temperature in the Alps around the Matterhorn has risen considerably over the past decade. The ice that holds mountain slopes and rock faces together is simply disappearing. At this rate, it will vanish completely—with profound consequences," said conference organizer and civil engineer Michael Davies, a professor at Dundee University (McKie 2003, 18).

Air-temperature increases in the Alps are being amplified fivefold underground. A borehole dug at Murtel in southern Switzerland has revealed that frozen subsurface soils have warmed by more than 1°C since 1990. In addition to general air-temperature rises that are heating the ground, increased evaporation caused by warmer summers has caused heavier snows that insulate the soil and keep it warmer in winter. Ice also becomes more unstable as it warms, raising the danger of devastating landslides.

The melting permafrost in the Alps and other European mountain ranges does much more than spoil mountain climbers' treks. Landslides sometimes threaten alpine villages and ski resorts. Fear has been expressed that some villages may have to be abandoned. Rivers also may be blocked by debris, causing flash floods when these unstable mounds of earth subsequently collapse. According to a report in the *London Guardian*, Charles Harris of the earth-sciences department at Cardiff University, who coordinates research for the European Union, said that the main areas at risk are the Alps in Switzerland, Austria, France, Germany, and Italy, where the mountains are densely populated and the slopes are very steep. According to this account, among the places being monitored is the Murtel-Corvatsch Mountain above fashionable St. Moritz, and the Schilthorn, which towers above the Muran and Gandeg resorts, near Zermatt (Brown 2001, 3). Harris said that the Swiss Alps had warmed by 0.5°C to 1°C during the past 15 years (Clover 2001, 12).

An organization called Permafrost and Climate in Europe (PACE) monitors the effects of climate change on the stability of mountains. PACE literature contends that "[t]he combination of ground temperatures only slightly below the freezing point, [along with] high ice contents and steep slopes, makes mountain permafrost particularly vulnerable to even small climate changes" (Brown 2001, 3).

Flood Threats below the Alps

Melting glacial ice has been increasing the volume of the Rhine, Rhone, and Po rivers. Since 1850, the volume of Europe's glaciers has shrunk by 50 percent (Toner 2002b, 4-A). The great rivers of Europe that rise in the Alps may decline from swollen summer torrents to trickles as global warming melts ancient mountain ice fields during the twenty-first century. David Collins of Salford University presented findings to a Royal Geographical Society conference in Belfast indicating that the ice that is now melting in the Alps accumulated during the "Little Ice Age" between the fifteenth and eighteenth centuries.

"The [present] combination of warmer summers and drier winters, meaning less snow to feed the glaciers, has meant that the vast bank of ice on the mountain tops is disappearing," said Collins. "The ice is like money in the bank, if you keep drawing more than you put in, eventually it runs out" (Brown 2002a, 9). Before the rivers dry up, warming will supply a final torrent as the glaciers melt. The summer flows of the rivers fed from the Alps, including the Rhine, Rhone, Po, and Inn (which feeds the Danube) recently have been higher than they had been for centuries. The excessive water flow has been good news for those living on riverbanks in southern and eastern Europe who drew off the excess water for irrigation and domestic use. In France, the river water was used for cooling nuclear power stations.

Forecasts by Collins and his team showed that these boom times for water supply may soon end. When all the ice goes, the summer flow of the rivers will almost entirely depend on rainfall. Under some climate models, rainfall in southern Europe probably will decline even further under warmer conditions. "We can see serious potential problems but it is very hard to be precise because weather patterns could change again," he said (Brown 2002a, 9).

"Some of the glaciers—for example, there are a number of small ones at Gornergrat, near Zermatt—are now below the snow line in summer. This means they are doomed. The ice they are made of was laid down in snowfall two or three centuries ago and is melting away faster each year," Collins reported (Brown 2002a, 9). Collins said the reduction in the glaciers in the Alps had been matched by an increase in glacier size in the Jotunheimen range in Norway because increased precipitation in northern Europe, in this case falling as snow, had blanketed those glaciers and protected them from any temperature increase. This led to a net increase in the size of glaciers over the same period as those in the Alps were retreating.

The people of Macugnaga (pronounced maa-COON-yaga), an Alpine resort village in Italy, long ago learned to cope with the floods that sometimes accompany the melting snow in the spring, "But nothing," according to on account, "prepared them for the catastrophic flood threat they now face—a glacier rapidly melting from unusually warm temperatures" (Konviser 2002, C-1).

During July 2002, as many as 300 officials and volunteers struggled under a state of emergency "to prevent a gigantic glacier-fed lake from breaking through the giant ice wall that confines it" (Konviser 2002, C-1). If they failed, a devastating wall of water carrying chunks of glacier and mountainside could surge through this verdant valley. Known technically as a "glacier lake outburst flood," or GLOF, "it's an event previously seen only in the Himalayas where the slopes of the mountains are steeper. But scientists say the threat is both real, and a warning of things to come if the global-warming trend continues" (Konviser 2002, C-1).

"It's a dangerous situation because the border of the lake is ice, which isn't stable," said Claudia Smiraglia, a professor of physical geography at Milan University. "The glacier is always in motion" (Konviser 2002, C-1). Bruce I. Konviser, reporting for the *Boston Globe*, described the potential scope of the threat:

> If the water escapes, the 650 residents of Macugnaga and as many as 7,000 vacationers, depending on the time of year—would have approximately 40 minutes to gather their belongings and get to higher ground before the wave of water and mountain wipes out much, if not all, of the manmade structures, according to Luka Spoletini, a spokesman for the Italian government's Department of Civil Protection. (Konviser 2002, C-1)

FURTHER READING

Brown, Paul. "Melting Permafrost Threatens Alps; Communities Face Devastating Landslides from Unstable Mountain Ranges." *London Guardian*, January 4, 2001, 3.

Brown, Paul. "Geographers' Conference: Ice Field Loss Puts Alpine Rivers at Risk: Global Warming Warning to Europe." *London Guardian*, January 5, 2002a, 9.

Capella, Peter. "Europe's Alps Crumbling; Glaciers Melting in Heat Wave." Agence France Presse, August 7, 2003. (LEXIS)

Clover, Charles. "Geographers' Conference: Alps May Crumble as Permafrost Melts." *London Telegraph*, January 4, 2001, 12.

Konviser, Bruce I. "Glacier Lake Puts Global Warming on the Map." *Boston Globe*, July 16, 2002, C-1.

McKie, Robin. "Decades of Devastation Ahead as Global Warming Melts the Alps: A Mountain of Trouble as Matterhorn Is Rocked by Avalanches." *London Observer*, July 20, 2003, 18.

Toner, Mike. "Meltdown in Montana; Scientists Fear Park's Glaciers May Disappear within 30 Years." *Atlanta Journal and Constitution*, June 30, 2002b, 4-A.

Amazon Valley, Drought and Warming

Rising temperatures increase the probability of drought in areas of the Amazon, a risk that is made worse by deforestation that interrupts natural recycling of rainfall by lush forest. Areas already experiencing intense deforestation (that is to say, usually the ones with the most human development) face the greatest risk; areas in the northwest region of the valley, with the least human development, seem least likely to experience debilitating drought.

The potential for drought seems to operate on several levels. Some, such as inhibition of transpiration, are local. Others, including changes in the El Niño/La Niña cycle, occur outside the valley, but influence weather and climate there (Malhi et al. 2008, 169–172). Drought also tends to feed upon itself: a series of dry seasons desiccates trees, which recycle less rainfall, and intensifies the cycle. Human activities (clearing of land for farming and pasture, especially using fire) often makes things even worse. Deforestation leaves behind combustible material that makes future fires more intense and quicker spreading. "A tipping point may be reached," according to one study, "[w]hen grasses can establish in the forest understory, providing a renewable source of fuel for repeated burns" (Malhi et al. 2008, 171).

Human activity is pushing the Amazon Valley toward drought in many ways, of which fossil-fuel effluvia is only one. Rising sea-surface temperatures in the Atlantic Ocean change circulation patterns in the atmosphere in ways that can induce drier weather in parts of the valley, while changes in climate patterns from more frequent El Niño episodes—which also may be provoked by rising temperatures—do the same. Land-use changes associated with the spread of logging, ranching, and farming also cause parts of the shrinking rainforest to dry out, as world demand surges for tropical timber, soybeans, and

Amazon rainforest. (Courtesy of Getty Images/PhotoDisc)

free-range beef, among other products. Roads expand to link production centers in the Amazon Valley to ports on South America's Pacific coasts, allowing exports to the rapidly developing economies of China, India, and other parts of Asia. Meanwhile, demand for "green" ethanol causes more Amazon forest to be cleared to make way for sugarcane fields. In an area where 25 to 59 percent of rainfall is "recycled" by the forest, once roughly 30 to 40 percent of the rainforest goes, the area as a whole could pass a threshold (a "tipping point") into a drier climate on a long-range basis (Malhi et al. 2008, 169).

Severe Drought Plagues Amazon Valley

During 2005, a severe drought spread through the Amazon Valley at the same time that new evidence was being assembled that indicated damage from logging had been 60 to 123 percent more than previously reported. "We think this [additional logging] adds 25 percent more carbon dioxide to the atmosphere" from the Amazon than previously estimated, said Michael Keller, an ecologist with the U.S. Department of Agriculture's Forest Service, and co-author of an Amazon logging inventory published in *Science* (Asner et al. 2005, 480–481; Naik and Samor 2005, A-12).

This study differed from others that measured only the clear-cutting of large forest areas. The study by Asner and colleagues included these measures of deforestation and added trees cut selectively, while much of surrounding forest was left standing in the five Brazilian states (Mato Grosso, Para, Rondonia, Roraima, and Acre) that account for more than 90 percent of deforestation in the Brazilian Amazon (Asner et al. 2005, 480). In addition, the Amazon Valley's worst drought in about 40 years was causing several tributaries to evaporate, probably contributing even more carbon dioxide via wildfires.

A report of the United Nations Intergovernmental Panel on Climate Change, issued during April 2007, projected continued drying in the Amazon Valley. "By mid-century, increases in temperature and associated decreases in soil water are projected to lead to gradual replacement of tropical forest by savanna in eastern Amazonia," it anticipated, warning that "crop productivity is projected to decrease for even small local temperature increases" in tropical areas, "which would increase risk of hunger" (Rohter 2007).

By 2008, the drought that in some areas was the worst since recordkeeping began a century before may have been only an early indication of a new weather regime in the Amazon Valley. The valley holds nearly a quarter of the world's fresh water (Giles 2006a, 726) and could be caught in a double vise as the world warms, and as warmer Atlantic Ocean temperatures combine with El Niño events to provoke more frequent droughts. El Niño events tend to reverse the air circulation over the Amazon from east-west to west-east, setting up drying, downslope winds.

Spreading Deserts

According to one observer,

> Already the Amazon desert is emerging in Rondonia. As far as the eye can see, there is only red, cracked ground. The red dust fills the air and sky itself turns red. Ten years ago this area was jungle, then it was slashed and burned to form farmland. But the jungle soil is notoriously poor, so it was used for pasture. Once overgrazed it became desert. Now even the landless poor who created it have moved on. They have moved west, to begin the process deeper in the land that used to be the Amazon rainforest. (Brandenburg and Paxson 1999, 225)

"The Amazon is a kind of canary-in-a-coal-mine situation," said Daniel C. Nepstad, a senior scientist at the Woods Hole Research Center in Massachusetts and the Amazon Institute of Ecological Research in Belem (Rohter 2005b). "A warmer Atlantic not only helps give more energy to hurricanes, it also aids in evaporating air," said Luiz Gylvan Meira, a climate specialist at the Institute for Advanced Studies at the University of São Paulo.

> But when that air rises over the oceans in one region, it eventually has to come down somewhere else, thousands of miles away. In this case, it came down in the western Amazon, blocking the formation of clouds that would bring rain to the headwaters of the rivers that feed the Amazon. (Rohter 2005b)

Per Cox of the Center for Ecology and Hydrology in Winfrith, United Kingdom, anticipated that increased drought frequency could devastate about 65 percent of the Amazon's forest cover by the end of the twenty-first century (Cox et al. 2004, 137). Nepstad has been simulating drought in the Amazon Valley since 1998. In a 2006 paper for *Ecology*, Nepstad estimated that that reduced wood formation and tree mass killed largely by drought in the Amazon Valley added half a billion tons of carbon dioxide to the atmosphere during the 2005 drought—yet another example of cascading positive feedbacks that intensify carbon-related warming (Giles 2006a, 727). For comparative purposes, note that the Kyoto Protocol's goal as negotiated in 1997 was to remove twice that amount, one billion tons, from human-generated emissions.

Nepstad's fear is that a drier Amazon Valley "will be invaded by highly flammable grasses," adding still more carbon dioxide to the atmosphere (Giles 2006a, 727). In Acre state of western Brazil, parched trees turned to tinder during the 2005 drought, and the number of forest fires recorded during 2005 tripled to nearly 1,500 by September compared with a year earlier. The resulting smoke, which may itself have intensified the drought by impeding the formation of storm clouds, was so thick on some days that residents took to wearing masks when they went outdoors (Rohter 2005b). "Because droughts remain registered in the soil for up to four years, the situation is still very critical and precarious, and will remain so," Nepstad said. Where there are "forests already teetering on the edge," he added, the prospect of "massive tree mortality and greater susceptibility to fire" must be considered (Rohter 2005b).

FURTHER READING

Asner, Gregory P., David E. Knapp, Eben N. Broadbent, Paulo J. C. Oliveira, Michael Keller, and Jose N. Silva. "Selective Logging in the Brazilian Amazon." *Science* 310 (October 21, 2005):480–481.

Brandenburg, John E., and Monica Rix Paxson. *Dead Mars, Dying Earth.* Freedom, CA: The Crossing Press, 1999.

Cox, P. M., R. A. Betts, M. Collins, P. P. Harris, Huntingford, and C. D. Jones. "Amazonian Forest Dieback Under Climate-carbon Cycle Projections for the 21st Century." *Theoretical and Applied Climatology* 78 (2004):137–156.

Giles, Jim. "The Outlook for Amazonia Is Dry." *Nature* 442 (August 17, 2006a):726–727.

Malhi, Yadvinder, J. Timmons Roberts, Richard A. Betts, Timothy J. Killeen, Wenhong Li, and Carlos A. Nobre. "Climate Change, Deforestation, and the

Fate of the Amazon." *Science* 319 (January 11, 2008):169–172.

Naik, Gautam, and Geraldo Samor. "Drought Spotlights Extent of Damage in Amazon Basin." *Wall Street Journal*, October 21, 2005, A-12.

Rohter, Larry. "A Record Amazon Drought, and Fear of Wider Ills." *New York Times*, December 11, 2005b. http://www.nytimes.com/2005/12/11/international/americas/11amazon.html.

Rohter, Larry. "Brazil, Alarmed, Reconsiders Policy on Climate Change." *New York Times*, July 31, 2007. http://www.nytimes.com/2007/07/31/world/americas/31amazon.html.

Amphibians, Extinctions and Warming

By 2006, an increasing number of scientific studies pointed to warming temperatures as a crucial (but not sole) reason why several species of amphibians (frogs and toads) have been going extinct or face sharp population declines around the world.

These include studies of the Monteverde harlequin frog (*Atelopus sp.*), which vanished along with the golden toad (*Bufo periglenes*) during the late 1980s from the mountains of Costa Rica. According to analysis by J. Alan Pounds, resident scientist at the Tropical Science Center's Monteverde Cloud Forest Preserve in Costa Rica, and colleagues in *Nature* (2006, 161–167), an estimated 67 percent of roughly 110 species of *Atelopus*, which are endemic to the American tropics, are now extinct, having been afflicted by a pathogenic chytrid fungus (*Batrachochytrium dendrobatidis*). Pounds and colleagues concluded with more than 99 percent confidence that large-scale environmental warming is a key factor in the disappearances. They found that rising temperatures at many highland localities have encouraged growth of *Batrachochytrium*, encouraging outbreaks. "With climate change promoting infectious disease and eroding biodiversity, the urgency of reducing greenhouse-gas concentrations is now undeniable," they concluded (Pounds et al. 2006, 161).

"Disease is the bullet killing frogs, but climate change is pulling the trigger," Pounds said. "Global warming is wreaking havoc on amphibians and will cause staggering losses of biodiversity if we don't do something first" (Eilperin 2006a, A-1). The chytrid fungus kills frogs by growing on their skin, then attacking their epidermis and teeth, as well as by releasing a toxin. Higher temperatures allow more water vapor into the air, which forms a cloud cover that leads to cooler days and warmer nights. These conditions favor the fungus, which grows and reproduces best at temperatures between 63° and 77°F. "There's a coherent pattern of disappearances, all the way from Costa Rica to Peru," Pounds said. "Here's a case where we can show that global warming is affecting outbreaks of this disease" (Eilperin 2006a, A-1).

Amphibians Threatened Worldwide

In 2004, the first worldwide survey of 5,743 amphibians species (frogs, toads, and salamanders) indicated that one in every three species was in danger of extinction, many of them likely victims of an infectious fungus possibly aggravated by drought and warming (Stokstad 2004b, 391). Findings were contributed by more than 500 researchers from more than 60 countries. "The fact that one-third of amphibians are in a precipitous decline tells us that we are rapidly moving toward a potentially epidemic number of extinctions," said Achim Steiner, director-general of the World Conservation Union based in Geneva (Seabrook 2004, 1-C).

"Amphibians are one of nature's best indicators of the overall health of our environment," said Whitfield Gibbons, a herpetologist at the University of Georgia's Savannah River Ecology Laboratory (Seabrook 2004, 1-C; Stokstad 2004b, 391). Amphibians are more threatened and are declining more rapidly than birds or mammals: "The lack of conservation remedies for these poorly understood declines means that hundreds of amphibians species now face extinction," wrote the scientists who conducted the worldwide survey (Stuart et al. 2004, 1783–1786).

In North and South America, the Caribbean, and Australia, a major culprit appears to be the highly infectious fungal disease *chytridiomycosis*. New research shows that prolonged drought may cause outbreaks of the disease in some regions, although some scientists attribute the disease spread to global warming. Other threats include loss of habitat, acid rain, pesticides and herbicides, fertilizers, consumer demand for frog legs, and a depletion of the ozone layer that leaves the frogs' skin exposed to radiation (Seabrook 2004, 1-C). Gibbons said that the loss of wetlands and other habitat to development, agriculture, and other reasons may be the leading cause of amphibian declines in Georgia. Pollution also may be playing a significant role. "It's hard to find a pristine stream anymore," he said (Seabrook 2004, 1-C).

Warming and the Decline of Oregon's Western Toad

Global warming could be playing a role in the decline of the western toad in Oregon, according to a report that was among the first to link climatic change with amphibian die-offs in North America (Pounds 2001, 639). Toad population declines may be part of ecological chain reaction in which rapid global warming plays a role. Pounds, who has researched the decline of the golden toad in Costa Rica, described the situation in Oregon: "In the crystal-clear waters surrounded by snow-capped peaks in the Cascade range, the jet-black [western toad] embryos are suffering devastating mortality. They develop normally for a few days, then turn white and die by the hundreds of thousands" (Pounds 2001, 639).

Research by Joseph M. Kiesecker of Pennsylvania State University and colleagues indicated that the toads' fatal infection results from a complex sequence of events provoked by warming temperatures. Kiesecker and colleagues wrote that "[e]levated sea-surface in this region [the western United States] since the mid-1970s, which [has] affected the climate over much of the world, could be the precursor for pathogen-mediated amphibian declines in many regions" (Kiesecker, Blaustein, and Belden 2001, 681). "Reductions in water depth due to altered precipitation patterns expose the embryos to damaging ultraviolet B (UV-B) radiation, thereby opening the door to lethal infection by a fungus, *Saprolegnia ferax*" (Pounds 2001, 639).

Depth of water influences the amount of UV-B radiation that reaches the toads' eggs. Reduced water depth is related to the El Niño/La Niña cycle, which Kiesecker and colleagues theorized is related to climate change. Similar patterns have been detected in cases of massive frog and toad mortality from other parts of the world. In some areas, fungus-borne diseases afflict lizards as well as frogs and toads (Pounds 2001, 639–640). Given what Kiesecker and others have found, Pounds asserted that "[t]here is clearly a need for a rapid transition to cleaner energy sources if we are to avoid staggering losses of biodiversity" (2001, 640). Kiesecker's results support those of Pounds who, in 1999, reported that warmer and drier periods in the cloud forest atop the Continental Divide at Monte Verde were tied to El Niño events and had caused massive population reductions in more than 20 frog species, including the disappearance of the golden toad (Souder 2001).

The role of warming involves several factors causing the decline of amphibians. For example, Ross A. Alford and colleagues wrote in *Nature*:

> Is global warming contributing to amphibian declines and extinctions by promoting outbreaks of the chytrid fungus *Batrachochytrium dendrobatidis*? Analyzing patterns from the American tropics, Pounds et al. envisage a process in which a single warm year triggers die-offs in a particular area (for instance, 1987 in the case of Monteverde, Costa Rica). However, we show … that populations of two frog species in the Australian tropics experienced increasing developmental instability, which is evidence of stress, at least two years before they showed chytrid-related declines. Because the working model of Pounds et al. is incomplete, their test of the climate-linked epidemic hypothesis could be inconclusive. (Alford, Bradfield, and Richards 2007)

Amphibian Deformations and Ultraviolet Radiation

Three studies published in the July 1, 2002, edition of *Environmental Science and Technology* indicated that ultraviolet radiation could be responsible for increasing numbers of malformed frogs and other amphibians. "We really wanted to fill the gap between the findings of other laboratory research and what might happen in natural environments," said Dr. Steve Diamond, an environmental toxicologist at the U.S. Environmental Protection Agency, and author of all three papers ("UV Radiation" 2002).

In one of the studies, Diamond and his colleagues kept frog eggs in small outdoor containers while exposing them to varying degrees of UV radiation—from 25 to 100 percent of natural sunlight. As the eggs developed, the researchers observed hatching success, tadpole survival, and the presence of malformations. The scientists found that the frequency of malformations increased with increasing UV radiation, with half of the frogs experiencing malformations at 63.5 percent of the intensity of natural sunlight.

According to an Environment News Service report,

> In the second study, the researchers measured levels of dissolved organic carbon (DOC) in water in

wetlands in Wisconsin and Minnesota. The top five to 20 centimeters of wetlands absorb as much as 99 percent of UV radiation, the researchers found. The third study involved a survey of 26 wetlands in the same region to estimate the specific level of risk of frogs living in these environments. Using a combination of computer models, historical weather records and DOC measurements, they concluded that UV radiation posed a risk to amphibians living in three of the 26 wetlands. ("UV Radiation" 2002)

FURTHER READING

Alford, Ross A., Kay S. Bradfield, and Stephen J. Richards. "Ecology: Global Warming and Amphibian Losses." *Nature* 447 (May 31, 2007): E-3, E-4.

Eilperin, Juliet. "Warming Tied to Extinction of Frog Species." *Washington Post,* January 12, 2006a, A-1. http://www.washingtonpost.com/wp-dyn/content/article/2006/01/11/AR2006011102121_pf.html.

Kiesecker, Joseph M., Andrew R. Blaustein, and Lisa K. Belden. "Complex Causes of Amphibian Population Declines." *Nature* 410 (April 5, 2001):681–684.

Pounds, J. Alan. "Climate and Amphibian Decline." *Nature* 410 (April 5, 2001):639–640.

Pounds, J. Alan, Martin R. Bustamante, Luis A. Coloma, Jamie A. Consuegra, Michael P. L. Fogden, Pru N. Foster, Enrique La Marca, Karen L. Masters, Andres Merino-Viteri, Robert Puschendorf, Santiago R. Ron, G. Arturo Sanchez-Azofeifa, Christopher J. Still, and Bruce E. Young. "Widespread Amphibian Extinctions from Epidemic Disease Driven by Global Warming." *Nature* 439 (January 12, 2006):161–167.

Seabrook, Charles. "Amphibian Populations Drop." *Atlanta Journal-Constitution,* October 15, 2004, 1-C.

Souder, William. "Global Warming and a Toad Species' Decline." *Washington Post,* April 9, 2001, n.p. (LEXIS)

Stokstad, Erik. "Global Survey Documents Puzzling Decline of Amphibians." *Science* 306 (October 15, 2004b):391.

Stuart, Simon N., Janice S. Cranson, Neil A. Cox, Bruce E. Young, Ana S. L. Rodrigues, Debra L. Fischman, and Robert W. Walker. "Status and Trends of Amphibian Declines and Extinctions Worldwide." *Science* 306 (December 3, 2004):1783–1786.

"UV Radiation Linked to Deformed Amphibians." Environment News Service." June 21, 2002. http://www.ens-newswire.com/ens/jun2002/2002-06-21-09.asp#anchor4.

Anchovies Spread Northward

The anchovy, usually found in the Mediterranean Sea, has been caught by several fishing vessels as far north as Donegal Bay in Ireland. Four vessels working on mackerel and herring during December 2001 reported catches of anchovies, with one of the most substantial hauls made in Donegal Bay. John Molloy, a scientist with Ireland's Marine Institute, told the *Irish Times* that one vessel working out of Greencastle County also caught some off Malin Head. "It is very unusual, particularly so far north," Molloy said. "Occasionally you'd get individual specimens in trawl hauls, but nothing of any note" (Siggins 2001, 1).

Kevin Flannery, a sea-fishery officer with the Department of the Marine, said that anchovies had also been caught off the Old Head of Kinsale. "It would seem that the shoals are staying well offshore and may be swimming within a warm movement of water that had been carried up north of Biscay and is holding this area of high pressure with it," Flannery said (Siggins 2001, 1).

FURTHER READING

Siggins, Lorna. "Warm-water Anchovies Landed by Trawlers in Donegal Bay." *Irish Times,* December 12, 2001, 1.

Andes, Glacial Retreat

Hundreds of Andean glaciers are retreating, and scientists say that their erosion is a direct result of rising temperatures. During three decades (1970–2000), Peru's glaciers lost almost a quarter of their 1,225-square-mile surface (Wilson 2001, A-1). An analysis of 268 mountain stations between 1 degree North and 23 degrees South latitude (1°N and 23°S) along the tropical Andes indicates a temperature rise of 0.11°C per decade, compared with a global average of 0.6°C between 1939 and 1998 (Bradley et al. 2006, 1755). Glacial retreat, much of it rapid, has been observed all along the spine of the Andes and on adjacent high plateaus.

Benjamin Morales, the dean of Peru's glaciologists, calls the glaciers of Peru (80 percent of the world's tropical ice pack) "the world's most sensitive thermometers," because they react to the smallest changes in temperatures (Wilson 2001, A-1). Many of the Peruvian glaciers that are now melting formed more than a million years ago. Today, some of these glaciers are melting so quickly that—

> residents have watched a usually painstakingly slow geological process with their own eyes. Since 1967, when Peru began monitoring its glaciers, scientists

French glaciologist Edouard Perroy (right) and Rolando Fuertes of the French Institut de Recherche pour le Developpement set a Global Positioning System device at the Huayna Potosi glacier in the outskirts of El Alto, Bolivia, Wednesday, October 31, 2007. (AP/Wide World Photos)

> estimate, the ice caps have lost 22 percent of their volume, enough to fill more than 5.6 million Olympic-size swimming pools. (Wilson 2001, A-1)

Glaciologist Lonnie Thompson has estimated that many of Peru's glaciers could disappear during the next 15 years (Wilson 2001, A-1). In Bolivia, the mass of glaciers and mountain snowcaps has shrunk 60 percent since 1978, raising a specter of water shortage for La Paz, home of 1.5 million people, as well as nearby El Alto (Forero 2002, A-3).

The 18,700-foot-high Quelccaya Ice Cap in the Andes of southeastern Peru has been steadily shrinking at an accelerating rate, losing 10 to 12 feet a year between 1978 and 1990, up to 90 feet a year between 1990 and 1995, and 150 feet a year between 1995 and 1998. The glacier retreated between 100 and 500 feet, depending on location, between 1999 and 2004. The Peruvian National Commission on Climate Change forecast in 2005 that Peru will lose all its glaciers below 18,000 feet in 10 years. Within 40 years, the commission said that all of Peru's glaciers will be gone (Regaldo 2005, A-1).

For 25 years, Thompson has been tracking a particular Peruvian glacier, Qori Kalis, where the pace of shrinkage accelerated enormously during the late 1990s. From 1998 to 2000, the glacier receded an average of 508 feet a year, according to Thompson. "That's thirty-three times faster than the rate in the first measurement period," he said, referring to a previous study of the glacier that covered the years 1963 to 1978 (Revkin 2001a, A-1). The Qori Kalis outlet glacier in Peru has been retreating about 660 feet (200 meters) a year (as of 2004–2005), 40 times the rate when measurement began in the 1970s (Braasch 2007, 37).

The Quelccaya Ice Cap in the Peruvian Andes is retreating more than 500 feet a year, or from 22 to 17 square miles between 1974 and 1998. Water from hundreds of glaciers in a section of the Andes known as the Cordillera Blanca ("White Range") drives the rural economy of Peru. The water runoff moistens wheat and potatoes along the mountain slopes. It also lights residents' houses with electricity generated by a hydroelectric plant on the river (Wilson 2001, A-1).

As hydropower generated from glacier runoff becomes undependable, these areas may come to rely more on fossil fuels for power. Bradley and colleagues pointed out that similar conditions apply in areas of East Africa and New Guinea with some reliance on glaciers that (as in the Andes) may disappear during the next 20 years (2006, 1756).

Andes' Water Supplies

Lima, a city of eight million people situated in the Atacama, one of the driest deserts on Earth, receives nearly all of its water during a six-month dry season from glacial ice melt. Within a few decades, at present melting rates, Lima's people will experience severe water shortages. The same water is used to generate much of Lima's electricity. Even as its wells go dry and mountain glaciers shrink, Lima has been adding 200,000 residents a year. Bolivia's capital La Paz and Ecuador's capital Quito face similar problems (Lynas 2004a, 236–237). Quito receives part of its drinking water from a rapidly retreating glacier on Volcano Antizana. Within a few years, if freezing levels continue to rise, the life-giving waters supplying all three cities may diminish to a trickle for the last time.

Many Peruvians who face drought in the long term are being lulled into a sense of plenty in the short term by increasing glacial runoff. According to Scott Wilson, writing in the *Washington Post*, the short term glacial runoff "has made possible plans to electrify remote mountain villages, turn deserts into orchards and deliver potable water to poor communities. In some mud-brick villages scattered across the valley, new schools will open and factories will crank up as the glacier-fed river increases electricity production" (Wilson 2001, A-1).

"In the long run … these long-frozen sources of water will run dry," said Cesar Portocarrero, a Peruvian engineer who worked for Electroperu, the government-owned power company, who has monitored Peru's water supply for 25 years (Revkin 2002b, A-10). In the meantime, evidence of changing climate has appeared in Portocarrero's hometown, Huaraz, a small city at 10,000 feet in the Andes. "I was doing work in my house the other day [in 2002] and saw mosquitoes," Portocarrero said. "Mosquitoes at more than 3,000 meters [almost 10,000 feet]. I never saw that before. It means really we have here the evidence and consequences of global warming" (Revkin 2002b, A-10).

About 99 percent of the Chacaltaya glacier in Bolivia has disappeared since 1940, said World Bank engineer Walter Vergara, in his report, "The Impacts of Climate Change in Latin America." One of the highest glaciers in South America, Chacaltaya is one of the first glaciers to melt because of climate change. Although the glacier is more than 18,000 years old, it is expected to vanish by 2010. Since 1970, glaciers in the Andes have lost 20 percent of their volume, according to a report by Peru's National Meteorology and Hydrology Service. Eventual melting could curtail water supplies to 30 million people, the report said ("Melting Andean Glaciers" 2008).

El Niño conditions in the equatorial Pacific, which may become more frequent as temperatures warm, cause drought in the Andes, eroding glaciers and providing little new snow to replace melting. The albedo of the glaciers also diminishes (they become grayer, flecked with dust). Three El Niños hit Bolivia during the 1990s.

Ice melt in Peru also may make some areas more vulnerable to the frequent earthquakes that afflict the area.

> Glaciers usually melt into the rock, filling in fissures with water that expands and freezes when the temperatures drop. What scientists fear is that, with increased melting, more water and larger ice masses are pulling apart the rock and making the ice cap above more susceptible to the frequent seismic tremors that rock the area. (Wilson 2001, A-1)

FURTHER READING

Braasch, Gary. *Earth Under Fire: How Global Warming Is Changing the World.* Berkeley: University of California Press, 2007.

Bradley, Raymond S., Mathias Vuille, Henry F. Diaz, and Walter Vergara. "Threats to Water Supplies in the Tropical Andes." *Science* 312 (June 23, 2006):1755–1756.

Forero, Juan. "As Andean Glaciers Shrink, Water Worries Grow." *New York Times*, November 24, 2002, A-3.

Lynas, Mark. *High Tide: The Truth about Our Climate Crisis.* New York: Picador/St. Martins, 2004a.

"Melting Andean Glaciers Could Leave 30 Million High and Dry." Environment News Service, April 28, 2008. http://www.ens-newswire.com.

Regaldo, Antonio. "The Ukukus Wonder Why a Sacred Glacier melts in Peru's Andes." *Wall Street Journal*, June 17, 2005, A-1, A-10.

Revkin, Andrew C. "A Message in Eroding Glacial Ice: Humans Are Turning Up the Heat." *New York Times*, February 19, 2001a, A-1.
Revkin, Andrew C. "Forecast for a Warmer World: Deluge and Drought." *New York Times*, August 28, 2002b, A-10.
Wilson, Scott. "Warming Shrinks Peruvian Glaciers; Retreat of Andean Snow Caps Threatens Future for Valleys." *Washington Post*, July 9, 2001, A-1.

Antarctica and Climate Change

Antarctica, which contains 90 percent of Earth's fresh water, has become a land of climatic paradoxes. Ice sheets have been melting around the edges of the continent more quickly than anticipated, as warming temperatures accelerate glaciers' movement into the surrounding oceans. The Antarctic Peninsula is among the most rapidly warming areas on Earth, where large ice shelves have been crumbling into the surrounding seas for several years. Temperatures on the West Antarctic Peninsula have risen 8.8°F in winter since 1950, and 4.5°F in summer. At the same time, sections of Antarctica's interior have experienced a pronounced cooling trend, whereas most other areas of Earth have warmed. Some Antarctic ice shelves are even disintegrating in the winter.

Are Antarctic ice sheets thickening or thinning? Is sea ice expanding or contracting? Both questions are open to debate. These debates are of great interest for the rest of the world because significant melting of land-based Antarctic ice could raise sea levels and inundate coastal residences of many hundreds of millions of people. Whatever the outcome of these debates, many observations indicate that a climate change–provoked breakdown in the Antarctic food chain, which begins with krill and ends with penguins and whales. These problems also may be intensified by human-caused declines in stratospheric ozone levels.

Melting Ice in West Antarctica

Scientists watch West Antarctica for signs that its glaciers could erode into the ocean and raise world sea levels. The western ice sheet is much smaller than the main body of ice east of the South Pole.

A study published in the online journal *Nature Geoscience* (January 13, 2007) estimated that West Antarctica's ice sheets lost 132 billion tons of ice between 1996 and 2006. (These are estimates that the authors of the study express with a rather large potential margin of error; the figure of 132 billion tons is listed as plus or minus 60 billion tons). "Without doubt, Antarctica as a whole is now losing ice yearly, and each year it's losing more," said Eric Rignot, professor of Earth Science at the University of California–Irvine, a senior scientist with NASA's Jet Propulsion Laboratory, and lead author of the paper (Kaufman 2008a; Rignot et al. 2008).

The cause of the melting, Rignot said, may involve changes in the Antarctic Circumpolar Current that circles much of the continent. As wind patterns change, this current's relatively warm water is brushing the continent and melting edges of glaciers deep underwater. This current, which flows about 200 yards below the frigid surface water, began to warm sharply during the 1980s. "Something must be changing the ocean to trigger such changes," said Rignot. "We believe it is related to global climate forcing" (Kaufman 2008a).

A few days before this report, the head of the Intergovernmental Panel on Climate Change (IPCC) said that the panel's next periodic report, due in 2012, should closely examine the "frightening" possibility that the ice of Greenland and Antarctica could melt rapidly at the same time. "Both Greenland and the West Antarctic ice sheet are huge bodies of ice and snow, which are sitting on land," said Rajendra Pachauri, chief of the IPCC, the United Nations' scientific advisory group. "If, through a process of melting, they collapse and are submerged in the sea, then we really are talking about sea-level rises of several meters" (Kaufman 2008a).

This study is based on a satellite mapping that covered 85 percent of Antarctica. The study found that the larger ice stores of East Antarctica have not yet begun to erode, except along the coasts. Snowfall and ice loss along the coats are still in balance in that area. In West Antarctica, however, ice loss has increased by 59 percent over the past decade, while the yearly loss along the Antarctic Peninsula has increased by 140 percent, to 60 billion metric tons (Kaufman 2008a).

Rapid Sea-Level Surges

In our time, with an assist from rising temperatures prompted by human greenhouse-gas emissions, several reports indicate that Antarctic glaciers are accelerating toward the sea, especially around the fringes of the West Antarctic ice sheet—a harbinger, perhaps, of an accelerating sea-level rise worldwide in coming decades.

Even without human intervention, however, sea levels have changed rapidly during glacial cycles in the past. P. U. Clark and colleagues investigated sea-level changes roughly 14,200 years ago that resulted in sea-level surges of about 40 millimeters a year over 500 years, much more rapid than the one- to two-millimeter sea-level rise of the twentieth century (Clark et al. 2002, 2438; Sabadini 2002, 2376). These "melt-water pulses" probably originated in Antarctica, mostly from ice-sheet disintegration.

The Earth's history reveals periods during which cataclysmic collapse of the West Antarctic ice sheet caused the sea level to rise at an average rate of about one meter every 20 years for centuries. Within such periods, there must have been times when the rate of sea-level change was even faster. Because hundreds of large cities today are located on coastlines around the world, such abrupt ice-sheet collapse would have devastating consequences. It is impossible to specify the exact level of global warming necessary to cause such an ice sheet collapse, but, as argued above, it is nearly certain that West Antarctica and Greenland would disintegrate at some point if global warming approaches 3°C, as it would under a business-as-usual greenhouse-gas scenario (Hansen 2006b, 32–33).

Wilkins Ice Shelf Breaks Apart in Winter

Satellite photos from NASA during mid-2008 showed that Antarctica's Wilkins Ice Shelf is even closer to breaking from the Antarctica Peninsula, and experts said that the effects of warming there now look irreversible. In an unusual twist, the ice shelf has been disintegrating during the *winter*, in the absence of summertime surface warmth and sunshine that can lead to disintegration. The winter melting is probably due to relatively warm upwelling of ocean water under the ice.

By the end of May 2008, long, thin blocks had broken off the Wilkins Ice Shelf and were moving away toward the southwest. By June 28, 2008, the portion of the ice shelf connecting to Charcot Island had narrowed, assuming an almost hourglass shape. Immediately northeast of this skinny stretch of shelf, the darker parts of the ice mélange appear to be melting. Farther northeast, the large blocks of ice have begun to drift apart.

An image taken from space on July 17, 2008, showed the continued breakup of the ice shelf. Large, relatively intact plates of ice drifted toward the northeast from the thin piece of shelf that still stretched toward the nearby island. Large blocks of ice in the northeast continued their northward drift, some separated by areas of open water.

Continued Antarctic Melting

During 2007, NASA researchers using 20 years of data from space-based sensors (1987 through 2006), found that the Antarctic Ice Cap has been melting over time farther inland from the coast. Ice and snow also are melting at higher altitudes. Melting is increasing on Antarctica's largest (eastern) ice shelf. Snow and ice in Antarctica has been melting as far inland as 500 miles away from the coast and as high as 1.2 miles above sea level in the trans-Antarctic mountains. The same study also found that melting has been increasing on the Ross Ice Shelf, both in geographic area and duration. The study was published September 22, 2007, in *Geophysical Research Letters.*

As in Greenland, satellite sensors found that melting snow and ice on the surface was forming ponds, with meltwater filling small cracks, which cause larger fractures in the ice shelf. "Persistent melting on the Ross Ice Shelf is something we should not lose sight of because of the ice shelf's role as a 'brake system' for glaciers," said the study's lead author, Marco Tedesco, a research scientist at the Joint Center for Earth Systems Technology, which is cooperatively managed by NASA's Goddard Space Flight Center, Greenbelt, Maryland, and the University of Maryland.

> Ice shelves are thick ice masses covering coastal land with extended areas that float on the sea, keeping warmer marine air at a distance from glaciers and preventing a greater acceleration of melting. The Ross Ice Shelf acts like a freezer door, separating ice on the inside from warmer air on the outside. So the smaller that door becomes, the less effective it will be at protecting the ice inside from melting and escaping. ("NASA Researchers" 2007)

A team of NASA and university scientists found clear evidence that extensive areas of snow melted in west Antarctica in January 2005 in response to temperatures that were much above average. Combined, the affected regions comprise an area as large as California. The area of widespread melting reached to within about 300 miles of the

South Pole. This is one of a number of exceptional warm episodes noted in polar regions during recent years; others have been described by scientists and residents on Baffin Island, Northern Alaska, Western Greenland, and northern Sweden.

Andrew Revkin wrote of this report in the *New York Times*:

> Balmy air, with a temperature of up to 41°F in some places, persisted across three broad swathes of West Antarctica long enough to leave a distinctive signature of melting, a layer of ice in the snow that cloaks the vast ice sheets of the frozen continent. The layer formed the same way a crust of ice can form in a yard in winter when a warm day and then a freezing night follow a snowfall, the scientists said. (Revkin 2007d)

Increasing Snowfall over Interior Antarctica

Global warming can work in contradictory ways, however, full of counterindications. Witness the fact that sea-level rise may be slowed by increasing snowfall over Antarctica, provoked by rising temperatures. The eastern half of Antarctica is gaining weight, more than 45 billion tons a year, according to one scientific study. Data from satellites bouncing radar signals off the ground show that the surface of eastern Antarctica appears to be slowly growing higher, by about 1.8 centimeters a year, as snow and ice pile up (Chang 2005, A-22). As temperatures rise, so does the amount of moisture in the air, causing snowfall to increase in cold areas such as Antarctica. "It's been long predicted by climate models," said Dr. Curt H. Davis, a professor of electrical and computer engineering at the University of Missouri. In the meantime, however, another study found that changes in snowfall had been insignificant since the 1950s (Monaghan et al. 2006, 827–831).

Satellite radar altimetry measurements suggest that the East Antarctic ice sheet interior north of 81.6°S increased in mass by 45 ± 7 billion tons per year from 1992 to 2003. Comparisons with contemporaneous meteorological model snowfall estimates suggest that the gain in mass was associated with increased precipitation. A gain of this magnitude is enough to slow sea-level rise by 0.12 ± 0.02 millimeters per year (Davis et al. 2005). The accumulation occurring across 2.75 million square miles of eastern Antarctica corresponds to a gain of 45 billion tons of water a year or, equivalently, the removal of the top 0.12 millimeter of the world's oceans. According to Davis, Antarctica "is the only large terrestrial ice body that is likely gaining mass rather than losing it" (Chang 2005, A-22).

The data from two European Space Agency satellites cover the period from 1992 to 2003; however, because the satellites did not pass directly over the South Pole, they did not provide any information for a 1,150-mile-wide circular area around the pole. Assuming that snow was falling there at the same rate seen in the rest of Antarctica, the total gain in snowfall would correspond to a 0.18-millimeter-a-year drop in sea levels (Chang 2005, A-22). *See also:* West Antarctic Ice Sheet

FURTHER READING

Chang, Kenneth. "Warming Is Blamed for Antarctica's Weight Gain." *New York Times*, May 20, 2005, A-22.

Clark, P. U., J. X. Mitrovica, G. A. Milne, and M. E. Tamisiea. "Sea-level Fingerprinting as a Direct Test for the Source of Global Meltwater Pulse." *Science* 295 (March 29, 2002):2438–2441.

Davis, Curt H., Yonghong Li, Joseph R. McConnell, Markus M. Frey, and Edward Hanna. "Snowfall-Driven Growth in East Antarctic Ice Sheet Mitigates Recent Sea-Level Rise." *Science* 308 (June 24, 2005):1898–1901.

Hansen, James E. "Declaration of James E. Hansen." *Green Mountain Chrysler-Plymouth-Dodge-Jeep, et al., Plaintiffs v. Thomas W. Torti, Secretary of the Vermont Agency of Natural Resources, et al., Defendants.* Case Nos. 2:05-CV-302 and 2:05-CV-304, Consolidated. United States District Court for the District of Vermont. August 14, 2006b. http://www.giss.nasa.gov/~dcain/recent_papers_proofs/vermont_14aug20061_textwfigs.pdf.

Kaufman, Marc. "Escalating Ice Loss Found in Antarctica; Sheets Melting in an Area Once Thought to Be Unaffected by Global Warming." *Washington Post*, January 14, 2008a. http://www.washingtonpost.com/wp-dyn/content/article/2008/01/13/AR2008011302753_pf.html.

Monaghan, Andrew J., David H. Bromwich, Ryan L. Fogt, Sheng-Hung Wang, Paul A. Mayewski, Daniel A. Dixon, Alexey Ekaykin, Massimo Frezzotti, Ian Goodwin, Elisabeth Isaksson, Susan D. Kaspari, Vin I. Morgan, Hans Oerter, Tas D. Van Ommen, Cornelius J. Van der Veen, and Jiahong Wen. "Insignificant Change in Antarctic Snowfall Since the International Geophysical Year." *Science* 313 (August 11, 2006):827–831.

"NASA Researchers Find Snowmelt in Antarctica Creeping Inland." NASA Earth Observatory, September 20, 2007. http://earthobservatory.nasa.gov/Newsroom/NasaNews/2007/2007092025613.html.

Revkin, Andrew C. "Analysis Finds Large Antarctic Area Has Melted." *New York Times*, May 16, 2007d. http://www.nytimes.com/2007/05/16/science/earth/16melt.html.

Rignot, Eric, Jonathan L. Bamber, Michiel R. van den Broeke, Curt Davis, Yonghong Li, Willem Jan van de Berg, and Erik van Meijgaard. "Recent Antarctic Ice Mass Loss from Radar Interferometry and Regional Climate Modelling." *Nature Geoscience* 1 (January 13, 2008):106–110. http://www.nature.com/ngeo/journal/vaop/ncurrent/abs/ngeo102.html.

Sabadini, Roberto. "Ice Sheet Collapse and Sea Level Change." *Science* 295 (March 29, 2002):2376–2377.

"Wintertime Disintegration of Wilkins Ice Shelf." NASA Earth Observatory. July 22, 2008. http://earthobservatory.nasa.gov/Newsroom/NewImages/images.php3?img_id=18095.

Antarctica and Debate over Inland Cooling

In an apparent contradiction of temperature trends across most of the world, some areas of interior Antarctica have cooled steadily for more than two decades as areas on the fringes of the continent have warmed rapidly. Data assembled by Peter Doran, an associate professor of earth and environmental sciences at the University of Illinois at Chicago, found that temperatures in the McMurdo Dry Valleys of East Antarctica have declined at a rate of 1.2°F per decade since 1986. Similar trends have been observed across the continent's interior since 1978.

The apparent cooling of inland Antarctica has been used by climate skeptics to refute global warming as an idea, much to the consternation of scientists involved in research. Doran and other members of the National Science Foundation's Long-term Ecological Research Team assembled temperature data in the Dry Valleys near McMurdo Sound, a snow-free mountainous desert of chill, arid soils, bleak, bedrock outcroppings and ice-covered lakes which is home to many microscopic invertebrates, mostly nematodes (Gugliotta 2002, A-2).

Doran stressed that although scientists could not explain the falling temperatures, his research "does not change the fact that the planet has warmed up on the whole. The findings simply point out that Antarctica is not responding as expected" (Gugliotta 2002, A-2). Doran also warned that "[y]ou don't want to overstate the effects" of the cooling trend, because any rise in sea level caused by global warming this century is expected to come from thermal expansion of existing oceans and not from any theoretical melting of the southern ice cap (Gugliotta 2002, A-2).

By 2009, a detailed analysis of Antarctic temperature records for half a century indicated that cooling there has been limited. This study maintains that assertions of inland warming have been "substantially incomplete owing to the sparseness and short duration of the observations" (Steig et al. 2009, 459). Instead, Eric J. Steig and colleagues, writing in *Nature*, found that "significant warming extends well beyond the Antarctic Peninsula to cover most of West Antarctica, an area of warming much larger than previously reported. West Antarctic warming exceeds 0.1°C per decade over the past 50 years, and is strongest in winter and spring" (Steig et al. 2009, 459).

The scientists included satellite measurements to interpolate temperatures in the vast areas between Antarctic weather stations. "We now see warming is taking place on all seven of the earth's continents in accord with what models predict as a response to greenhouse gases," said Steig, a professor of earth and space sciences at the University of Washington in Seattle (Chang 2009). What cooling has been recorded in parts of Antarctica cannot be traced directly to increases in the strength of the westerlies in the upper atmosphere, this study said. "Instead, regional changes in atmospheric circulation and associated changes in sea surface temperature and sea ice are required to explain the enhanced warming in West Antarctica" (Steig et al., 2009, 459).

Measuring Temperatures in the Dry Valleys

The fragile ecosystem of the Dry Valleys requires four to six weeks of above-freezing temperatures during the southern summer, Doran said. This period of relative warmth causes meltwater from hillside glaciers to cascade downward in seasonal arroyos that feed life in local lakes. Researchers have found that temperatures had been dropping, not rising, since 1986, with the most pronounced declines in summer and autumn. Glacial ice has not been melting, and so the streams are not flowing, lakes are shrinking, and microorganisms are disappearing (Doran et al. 2002, 517).

Doran said that his team also studied data collected since 1966 from permanent installations throughout the Antarctic. Previous studies had indicated overall warming, but the researchers found that these calculations relied disproportionately on readings from the Antarctic Peninsula. When the researchers corrected for this distortion, they found that Antarctica as a whole had become considerably colder. "Temperatures were rising between 1966 and 1978," Doran said, but then they started to fall and have continued falling ever since (Doran et al. 2002, 517).

> Our spatial analysis of Antarctic meteorological data demonstrates a net cooling on the Antarctic continent between 1966 and 2000, particularly during summer and autumn. The McMurdo Dry Valleys have cooled by 0.7°C per decade between 1986 and 2000, with similar pronounced seasonal trends. Summer cooling is particularly important to Antarctic terrestrial ecosystems that are poised on the interface of ice and water.... Continental Antarctic cooling, especially the seasonality of cooling, poses challenges to models of climate and ecosystem change. (Doran et al. 2002, 517)

Doran and colleagues did not speculate on the causes of the temperature decline. They do know that temperatures in the Dry Valleys rise when the wind blows and clouds cover the sky. Doran explained that as winds roll downhill off the Antarctic plateau into the Dry Valleys, the air compresses and heats up as a result, an effect similar to the Chinook winds of the western United States or the dry, warm Santa Ana winds of Southern California (Gugliotta 2002, A-2). At the same time, relatively warm summer winds gathering speed over the ocean bring warmer air in from the coast to promote the thaw, he said. Wind generally brings clouds that add to the warming, Doran added (Gugliotta 2002, A-2). Recently, however, "We're getting a decrease in winds from both directions," Doran said, and, perhaps as a consequence, temperatures in the Dry Valleys are dropping. "It's clearly connected to the winds, but what's controlling the decrease in the winds is not clear" (Gugliotta 2002, A-2).

Once Doran and colleagues' results were made public, several newspaper reports rushed to simplify them into a worldwide cooling trend, taking this news as a reason to refute global warming. The authors of the studies expressed caution. One of the scientists involved in studies indicating that the Ross Ice Shelf was thickening was Slawek Tulaczyk of the University of California–Santa Cruz. Tulaczyk said that press misinterpretations left him increasingly frustrated by sometimes-careless media coverage of the global warming issue.

Data Distortion of Contrarians

When Tulaczyk and Ian Joughin of the Jet Propulsion Laboratory in Pasadena reported in *Science* that the movement of glacial ice streams appeared to be slowing on the Ross Ice Shelf, allowing the ice to thicken, a headline over an editorial in the *San Diego Union-Tribune* minced no words: "Scientific Findings Run Counter to Theory of Global Warming." The editorial sarcastically asked, "Oh dear. What will the doomsayers say now? How will they explain away yet two more scientific studies that clearly contradict the global warming orthodoxy?" (Davidson 2002, A-8) A headline in the *National Post*, a generally right-wing Canadian newspaper, declared: "Antarctic Ice Sheet has Stopped Melting, Study Finds" (Davidson 2002, A-8). "Is Another Ice-Age On the Way?" asked an editorial in the *Rocky Mountain News* (Davidson 2002, A-8). Analyzing these reports in the *San Francisco Chronicle*, Keay Davidson commented that "Some media mistakenly equated the phenomenon studied by Joughin and Tulaczyk—a change in ice flow rates—with ice melting rates. The mistake contributed to the erroneous belief that the studies constituted, as it were, scientific 'tests' of the global warming theory" (Davidson 2002, A-8).

Contrary to some news reports, "the ice-sheet growth that we have documented in our study area has absolutely nothing to do with any recent climate trends," Tulaczyk said, emphasizing those words in an e-mail to the *San Francisco Chronicle* (Davidson 2002, A-8). The thickening of Antarctic ice in certain regions—especially "Ice Stream C" of the Whillans Ice Stream, adjacent to the Ross Ice Shelf—results from the complex internal dynamics of the ice itself. These particular ice-flow changes were unrelated to global warming caused by combustion of fossil fuels; such changes occurred for many millennia before the Industrial Revolution boosted atmospheric levels of heat-trapping gases. The area with the greatest ice thickening is on an ice stream that stopped flowing about 150 years ago (Davidson 2002, A-8).

"I keep repeating to journalists that climate science is much like economics. Both deal with complex systems," Tulaczyk observed. "Just as a single stock going up or down cannot be interpreted as a reliable indicator of economic recovery or collapse, we have to accept the occurrence of contradictory trends in the global climate" (Davidson 2002, A-8). Contrary to some reports attributed to his research, "Global warming is real and happening right now," asserted Doran. He said that the cooling trend in Antarctica appears to be a surprising, regional exception to the overall planetary warming (Davidson 2002, A-8).

Antarctic Inland Cooling in a Global Context

Doran emphasized:

> Our paper does not change the global [temperature] average in any significant way.... Although we have said that more area of the continent is cooling than warming, one just has to look at the paper itself ... to see that it is a close call.... Our analysis suggests that about two-thirds of the main continent has been cooling in the last 35 years, but there is one-third of the continent that has been warming if you remove the [Antarctic] Peninsula. And with the Peninsula included, it shrinks to 58 percent cooling. (Davidson 2002, A-8)

Doran bluntly advised the public: "If you want the facts, you have to go to the original scientific peer-reviewed literature, and avoid the broken-telephone effect of the popular press (Davidson 2002, A-8).

Doran also replied on the op-ed page of the *New York Times*:

> [M]isinterpretation had already become legend, and in the four and half years since, it has only grown. Our results have been misused as "evidence" against global warming by Michael Crichton in his novel *State of Fear* and by Ann Coulter in her latest book, *Godless: The Church of Liberalism*. Search my name on the Web, and you will find pages of links to everything from climate discussion groups to Senate policy committee documents—all citing my 2002 study as reason to doubt that the earth is warming. One recent Web column even put words in my mouth. I have never said that "the unexpected colder climate in Antarctica may possibly be signaling a lessening of the current global warming cycle." I have never thought such a thing either. (Doran 2006)

Doran said his data did find that 58 percent of Antarctica cooled from 1966 to 2000. During that period, however, the rest of the continent was warming. Climate models created after Doran's study was released

> have suggested a link between the lack of significant warming in Antarctica and the ozone hole over that continent. These models, conspicuously missing from the warming-skeptic literature, suggest that as the ozone hole heals—thanks to worldwide bans on ozone-destroying chemicals—all of Antarctica is likely to warm with the rest of the planet. An inconvenient truth? (Doran 2006)

FURTHER READING

Chang, Kenneth. "Study Finds New Evidence of Warming in Antarctica." *New York Times*, January 22, 2009. http://www.nytimes.com/2009/01/22/science/earth/22climate.html.

Davidson, Keay. "Media Goofed on Antarctic Data; Global Warming Interpretation Irks Scientists." *San Francisco Chronicle*, February 4, 2002, A-8.

Doran, Peter. "Cold, Hard Facts." *New York Times*, July 27, 2006. http://www.nytimes.com/2006/07/27/opinion/27doran.html.

Doran, Peter T., John C. Priscu, W. Berry Lyons, John E. Walsh, Andrew G. Fountain, Diane M. McKnight, Daryl L. Moorhead, Ross A. Virginia, Diana H. Wall, Gary D. Clow, Christian H. Fritsen, Christopher P. McKay, and Andrew N. Parsons. "Antarctic Climate Cooling and Terrestrial Ecosystem Response." *Nature* 415 (January 30, 2002):517–520.

Gugliotta, Guy. "In Antarctica, No Warming Trend; Scientists Find Temperatures Have Gotten Colder in Past Two Decades." *Washington Post*, January 14, 2002, A-2.

Steig, Eric J., David P. Schneider, Scott D. Rutherford, Michael E. Mann, Josefino C. Comiso, and Drew T. Shindell. "Warming of the Antarctic Ice-Sheet Surface since the 1957 International Geophysical Year." *Nature* 457 (January 22, 2009):459–462.

"Study Confirms Antarctica Warming." Environment News Service, September 6, 2006.

Antarctica and Speed of Ice Melt

Ice loss in Antarctica increased by 75 percent during the decade ending in 2007 as glacier movement to the sea accelerated. By 2007, glacier discharge into the sea had caught up with the rate in Greenland. Scientists led by Eric Rignot of NASA's Jet Propulsion Laboratory, Pasadena, California, and the University of California–Irvine, detected a rapid increase, doubling the contribution to world sea-level rise from 0.3 millimeters (0.01 inches) a year in 1996, to 0.5 millimeters (0.02 inches) a year in 2006 ("Antarctic Ice Loss" 2008; Rignot et al. 2008, 106).

Rignot said the losses, which were primarily concentrated in West Antarctica's Pine Island

Bay sector and the northern tip of the Antarctic Peninsula, "are caused by ongoing and past acceleration of glaciers into the sea." This is mostly a result of warmer ocean waters, which bathe the buttressing floating sections of glaciers, causing them to thin or collapse. "Changes in Antarctic glacier flow are having a significant, if not dominant, impact on the mass balance of the Antarctic ice sheet," he said ("Antarctic Ice Loss" 2008; Rignot et al. 2008, 106). The team found that the net loss of ice mass from Antarctica increased from 112 (plus or minus 91) gigatons a year in 1996 to 196 (plus or minus 92) gigatons a year in 2006. A gigaton is one billion metric tons, or more than 2.2 trillion pounds ("Antarctic Ice Loss" 2008).

The collapse of ice shelves does not, by itself, raise sea levels, but their demise could speed the melting of land-based ice that could accelerate sea-level rises in the future. "Loss of ice shelves surrounding the continent could have a major effect on the rate of ice flow off the continent," said Scambos (Toner 2002a, A-1). At least five of the ice shelves on the Antarctic Peninsula now are receding faster than at any time in recent history.

Ice Moves toward the Sea

After 2002, breakup of the Larsen Ice Shelf increased glacial movement toward the sea, according to data from two satellites. Some theorists and climate modelers anticipated beginning in the 1970s that global warming could provoke ice shelves in the oceans around Antarctica to melt and release glaciers, increasing the amount of ice being pushed into the sea. "People thought they were wrong, because they thought those models were simplistic," said glaciologist and remote-sensing specialist Eric Rignot of the Jet Propulsion Laboratory in Pasadena, California. "What we are seeing now is that they were not wrong," he said. "They were right" (Nesmith 2004, 9-A).

Rignot and Scambos, of the National Snow and Ice Data Center at the University of Colorado, were lead authors of two articles published in *Geophysical Research Letters.* Their work indicates a two- to sixfold increase in "centerline speed" for glaciers that were left exposed by collapse of the Larsen B Ice Shelf. One glacier, Hektoria, lost about 38 meters of height within six months (Rignot et al. 2008; Scambos et al. 2004). These observations provide a glimpse of what could happen on a larger scale if other large ice shelves in Antarctica—for example, the huge Ross Ice Shelf—break up. "These glaciers are themselves too small to contribute any significant amount of ice to the sea-level problem," Rignot said, "but the Ross Ice Shelf buttresses much larger glaciers. If these are released, the impact would be much greater" (Nesmith 2004, 9-A).

Like a Cork in a Bottle

Some scientists compare the "tongue" of a glacier (where the body of ice meets the sea) to a cork in a bottle. "The tongue of the glacier or the cork in the bottle do not represent that much," said Claudio Teitelboim, director of the Center for Scientific Studies, a private Chilean institution that cooperates with NASA to survey the ice fields of Antarctica and Patagonia. "But once the cork is dislodged, the contents of the bottle flow out, and that can generate tremendous instability" (Rohter 2005a).

Glaciers flowing into Antarctica's Amundsen Sea, which help drain the West Antarctic ice sheet, were thinning twice as fast near the coast by 2004 as they had during the 1990s. Warmer seawater erodes the bond between coastal ice and the bedrock below, "like weakening the cork in a bottle," said Robert H. Thomas, a glacier expert for NASA in Wallops Island, Virginia. "You start to let stuff out" (Revkin 2004c, A-24). Thomas and several co-authors wrote in *Science* during 2004 that "[r]ecent aircraft and satellite laser altimeter surveys of the Amundsen Sea sector of West Antarctica show that local glaciers are discharging about 250 cubic kilometers of ice per year to the ocean, about 60 percent more than is accumulated within their catchment basins" (Thomas et al. 2004, 255). At present, such discharge could raise world sea levels about 0.2 millimeters per year—not a startling amount. However, the long-term implications of such ice flow may be more ominous: "Most of these glaciers flow into floating ice shelves over bedrock up to hundreds of meters deeper than previous estimates, providing exit routes for ice from further inland if ice-sheet collapse is underway" (Thomas et al. 2004, 255).

The anticipated thinning of the West Antarctic ice sheet is being intensely examined by scientists. Andrew Shepherd and colleagues, writing in *Geophysical Research Letters*, have used satellite altimetry to associate thinning ice in the area to melting caused by the motion of warming ocean currents. In a related study, Anthony J. Payne and colleagues used an ice-flow model to describe how thinning of the Pine Island Glacier affects the ice sheet further inland (Payne et al. 2004).

Measuring Antarctic Ice Loss

Measurements of time-variable gravity from the Gravity Recovery and Climate Experiment (GRACE) satellites determined mass variations of the Antarctic ice sheet during 2002–2005. Researchers found that the ice sheet mass decreased significantly, at a rate of 152 ± 80 cubic kilometers per year of ice, equivalent to 0.4 ± 0.2 millimeters per year of global sea-level rise. Most of this mass loss came from the West Antarctic ice sheet. By 2006, the Antarctic ice sheet was losing about 36 cubic miles of ice a year. Increasing snowfalls in the interior (a product of warmer temperatures) had ceased to balance accelerating melting along the coasts of the continent. These findings imply that global sea levels will rise more rapidly than earlier thought.

These measurements found that the amount of water pouring annually from the ice sheet into the ocean (which is equal to the amount of water used by U.S. citizens in three months) is causing global sea level to rise by 0.4 millimeters a year (Eilperin 2006b, A-1). "The ice sheet is losing mass at a significant rate," said Isabella Velicogna, the study's lead author and a research scientist at University of Colorado at Boulder's Cooperative Institute for Research in Environmental Sciences. "It's a good indicator of how the climate is changing. It tells us we have to pay attention" (Eilperin 2006b, A-1).

Richard Alley, a Pennsylvania State University glaciologist who has studied the Antarctic ice sheet but was not involved in the GRACE project, said that more research is needed to determine whether the shrinkage is a long-term trend, because the new report is based on just three years of data. "One person's trend is another person's fluctuation," he said. However, said Alley, the switch in balance from accumulating ice to net melting in Antarctica had not been anticipated this soon by many researchers. "It looks like the ice sheets are ahead of schedule" in terms of melting, Alley said. "That's a wake-up call. We better figure out what's going on" (Eilperin 2006b, A-1).

FURTHER READING

"Antarctic Ice Loss Speeds Up, Nearly Matches Greenland Loss." NASA Earth Observatory. January 23, 2008. http://www.jpl.nasa.gov/news/news.cfm?release=2008-010.

Eilperin, Juliet. "Antarctic Ice Sheet Is Melting Rapidly; New Study Warns of Rising Sea Levels." *Washington Post*, March 3, 2006b, A-1.

Monaghan, Andrew J., David H. Bromwich, Ryan L. Fogt, Sheng-Hung Wang, Paul A. Mayewski, Daniel A. Dixon, Alexey Ekaykin, Massimo Frezzotti, Ian Goodwin, Elisabeth Isaksson, Susan D. Kaspari, Vin I. Morgan, Hans Oerter, Tas D. Van Ommen, Cornelius J. Van der Veen, and Jiahong Wen "Insignificant Change in Antarctic Snowfall Since the International Geophysical Year." *Science* 313 (August 11, 2006):827–831.

Nesmith, Jeff. "Antarctic Glacier Melt Increases Dramatically." *Atlanta Journal-Constitution*, September 22, 2004, 9-A.

Payne, A. J., A. Vieli, A. P. Shepherd, D. J. Wingham, and E. Rignot. "Recent Dramatic Thinning of Largest West Antarctic Ice Stream Triggered by Oceans." *Geophysical Research Letters* 31, no. 23 (December 9, 2004), L23401, doi:10.1029/2004GL021284.

Revkin, Andrew. "Antarctic Glaciers Quicken Pace to Sea; Warming Is Cited." *New York Times*, September 24, 2004c, A-24.

Rignot, Eric, Jonathan L. Bamber, Michiel R. van den Broeke, Curt Davis, Yonghong Li, Willem Jan van de Berg, and Erik van Meijgaard. "Recent Antarctic Ice Mass Loss from Radar Interferometry and Regional Climate Modelling." *Nature Geoscience* 1 (2008):106–110.

Rohter, Larry. "Antarctica, Warming, Looks Ever More Vulnerable." *New York Times*, January 25, 2005a. http://www.nytimes.com/2005/01/25/science/earth/25ice.html.

Scambos, T. A., J. A. Bohlander, C. A. Shuman, and P. Skvarca. "Glacier Acceleration and Thinning after Ice Shelf Collapse in the Larsen B Embayment, Antarctica." *Geophysical Research Letters* 31, no. 18 (September 22, 2004), doi:10.1029/2004GL020670.

Thomas, R., E. Rignot, G. Casassa, P. Kanagaratnam, C. Acuòa, T. Akins, H. Brecher, E. Frederick, P. Gogineni, W. Krabill, S. Manizade, H. Ramamoorthy, A. Rivera, R. Russell, J. Sonntag, R. Swift, J. Yungel, and J. Zwally. "Accelerated Sea-Level Rise from West Antarctica." *Science* 306 (October 8, 2004):255–258.

Toner, Mike. "Huge Ice Chunk Breaks off Antarctica." *Atlanta Journal and Constitution*, March 20, 2002a, A-1.

Velicogna, Isabella, and John Wahr. "Measurements of Time-Variable Gravity Show Mass Loss in Antarctica." *Science* online, March 2, 2006a, doi:10.1126/science.1123785.

Antarctic Oscillation

The Antarctic Peninsula has warmed several degrees Celsius as some of its ice shelves have disintegrated into the ocean, even as parts of interior Antarctica are cooling, and as some of its glaciers thicken. An explanation of this apparent paradox may lie in a climatic "master switch" over the high southern latitudes, a circular wind pattern (the Antarctic Oscillation, or Southern Annular Mode) that has been driven faster by the depletion of stratospheric ozone. According to one analysis, the cooling of inland Antarctica may reverse as the rupture in stratospheric ozone heals (Shindell and Schmidt 2004).

The work of David W. J. Thompson and Susan Solomon may be "the strongest evidence yet" that a shift in the Antarctic Oscillation "could explain a number of different components of [Antarctic] climate trends," according to David Karoly, a meteorologist at Monash University in Clayton, Australia (Kerr 2002a, 825). The researchers linked cooling in the stratosphere induced by depletion of ozone levels to acceleration of the winds. "During the summer-fall season," Thompson and Solomon wrote, "[t]he trend toward stronger circumpolar flow has contributed substantially to the observed warming over the Antarctic Peninsula and Patagonia and to the cooling over eastern Antarctica and the Antarctic plateau" (Thompson and Solomon 2002, 895).

Writing in the May 3, 2002, edition of *Science*, Thompson, a professor of atmospheric science at Colorado State University, and Solomon, a senior scientist at the National Oceanic and Atmospheric Administration in Boulder, Colorado, asserted that ozone depletion over the Antarctic may help explain both contradictory trends. "Ozone seems to be capable of tickling the Southern Hemisphere patterns," Thompson said (Chang 2002, A-16). Thompson and Solomon assert that a vortex of winds blowing around Antarctica that traps cold air at the South Pole has strengthened in the past few decades, keeping the cold air even more confined. The Antarctic Peninsula lies outside the wind vortex and thus escapes the cooling effect. Ozone depletion may be a key causal factor in strengthening the wind pattern, according to Thompson and Solomon. "That's where we speculate," Dr. Thompson said, "and the emphasis is on the word 'may'" (Chang 2002, A-16).

Scientists already knew that ozone depletion has cooled the upper atmosphere. Thompson and Solomon's research indicates that over parts of Antarctica the troposphere, the lowest six miles of the atmosphere, also has cooled. "It's a lot of food for thought in there," said Dr. John E. Walsh, a professor of atmospheric science at the University of Illinois (Chang 2002, A-16). Walsh said the data tying the cooling to stronger winds was convincing. "My one reservation," he said, "is the link to the ozone" (Chang 2002, A-16). He noted that the ozone hole was usually largest in November or December, but that the greatest cooling had been about six months later. Thompson agreed that ozone depletion could not explain the whole climactic picture, and said other influences like ocean currents probably played important roles, too. "I seriously doubt it's the only player," he said. "I think it's one of many" (Chang 2002, A-16).

The idea that stratospheric ozone depletion has been a factor in driving a stronger circumpolar vortex (and resulting cooling inside the vortex, with warming outside) has been gaining support. In 2003, Nathan P. Gillett and David W. J. Thompson published results of a modeling study supporting this effect during the spring and summer. "The results," they wrote in *Science*, "provide evidence that anthropogenic emissions of ozone-depleting gases have had a distinct impact on climate not only at stratospheric levels but at Earth's surface as well" (Gillett and Thompson 2003, 273; Karoly 2003, 236–237).

Mark P. Baldwin and colleagues wrote in *Science*:

> The resulting ozone "hole" leads to a relative reduction in solar heating and a stronger vortex. Observations and recent model simulations show that the strengthening of the polar vortex during spring leads to lower surface temperatures over Antarctica and higher temperatures in the mid-latitudes of the Southern Hemisphere that persist into summer. (Baldwin et al. 2003, 317)

FURTHER READING

Baldwin, Mark P., David W. J. Thompson, Emily F. Shuckburgh, Warwick A. Norton, and Nathan P. Gillett. "Weather from the Stratosphere?" *Science* 301 (July 18, 2003):317–318.

Chang, Kenneth. "Ozone Hole Is Now Seen as a Cause for Antarctic Cooling." *New York Times*, May 3, 2002, A-16.

Gillett, Nathan P., and David W. J. Thompson. "Simulation of Recent Southern Hemisphere Climate Change." *Science* 302 (October 10, 2003):273–275.

Karoly, David J. "Ozone and Climate Change." *Science* 302 (October 10, 2003): 236–237.
Kerr, Richard A. "A Single Climate Mover for Antarctica." *Science* 296 (May 3, 2002a):825–826.
Shindell, D. T., and G. A. Schmidt. "Southern Hemisphere Climate Response to Ozone Changes and Greenhouse Gas Increases." *Geophysical Research Letters* 31 (2004), L18209, doi:10.1029/2004GL020724.
Thompson, David W. J., and Susan Solomon. "Interpretation of Recent Southern Hemisphere Climate Change." *Science* 296 (May 3, 2002): 895–899.

Antarctic Paleoclimatic Precedents

Scientists have studied the historic climate (paleoclimate) of Antarctica in search of conditions similar to those forecast for the end of the twenty-first century, as temperatures and atmospheric carbon-dioxide levels rise. A study of Antarctica 24 million years ago, when carbon-dioxide levels were high compared with long-term averages, has revealed major expansions and contractions of Antarctica's ice, leading to worldwide sea-level changes of up to 200 feet.

An Analogous Period 24 Million Years Ago?

T. R. Naish and colleagues have presented sediment data from shallow marine cores in the Western Ross Sea that exhibit well-dated cyclic variations that link the extent of the East Antarctic ice sheet directly to orbital cycles during the Oligocene-Miocene transition (24.1 to 23.7 million years ago) (Naish et al. 2001, 719). During this transition, planetary temperatures averaged 3° to 4°C warmer than today, and the atmospheric carbon-dioxide level was twice as high as today. These studies, they said, "[s]hould help to provide realistic analogues for their future behavior following the increased levels of atmospheric CO_2 and temperature projected for the end of this century" (Naish et al. 2001, 723).

Another international research team studied sediment and fossil samples from 3,000 feet below the surface in the Cape Roberts area of the East Antarctic ice sheet, which covers most of the continent. The scientists focused on a 400,000-year period roughly 24 million years ago when carbon-dioxide levels were twice as high as today and temperatures were up to 4°C higher. These conditions are similar to those anticipated for the Earth in the next 50 to 100 years because of human-induced rises in greenhouse-gas levels.

The team, whose study was published October 18, 2001, in *Nature*, stated, "Studies of Antarctic ice sheets during that time should help to provide realistic analogues for their future behavior following the increased levels of atmospheric CO_2 and temperature projected for the end of this century" (Dalton 2001, 7). Peter Webb, professor of geological sciences at Northern Illinois University, commented:

> To find out what we are in for, we had to go back to a time when there were similar conditions. We wanted to find a comparable period in the past when carbon-dioxide levels were as high as they are expected to reach. There is every indication that this will be a model for what will happen in the future. We found the ice sheets 24 million years ago were smaller and more dynamic than today. There is good evidence for a lot of instability at the margins of the sheets. (Dalton 2001, 7)

Webb continued:

> We know there are glacial cycles, with the ice altering more when the climate is warmer, but it is now a question of working out how the speed and frequency of these cycles will change. They were probably faster in the past and will probably be faster in the future. (Dalton 2001, 7)

Researchers have identified the West Antarctic ice sheet as the main contributor of a massive and relatively rapid increase in global sea levels at the end of the last ice age. This event provoked a surge of fresh water from melting glacial ice into the world ocean that caused a rise of 20 meters in 500 years. Writing in *Science*, hydrologist Victor Baker of the University of Arizona said that during the last 40 years evidence has emerged for "repeated, catastrophic failures of ice dams" (O'Neill 2002). Baker said that "the southern edge of the Laurentide Ice Sheet, which covered the central and eastern regions of northern North America, was repeatedly drained by a super-flood that had a major influence on global climate during the last glacial period" (O'Neill 2002).

Ice Sheets during a Super Greenhouse?

The paleoclimate story has considerable nuance—in at least one instance about 90 million years ago, for example, very large glaciers appear to have formed for a time when alligators had swum in the Arctic roughly 200,000 years earlier. This finding, described in *Science*, January 11,

2008, calls into question a linear conviction that a "super-greenhouse" effect will, without fail, melt all glaciers (Bornemann et al. 2008, 189). That, or glacial cycles can occur with what, in geological time, could only be regarded as awesome speed, if the scientists' proxies are not misrepresenting what actually happened.

In this study, which was based on proxies—organic molecules in ocean sediments and chemicals in ancient fossil shells, scientists wrote that the "super greenhouse climate" was "not a barrier to the formation of large ice sheets, calling into question the common assumption that the poles were always ice-free during past periods of intense global warming" ("Study Says Glaciers" 2008).

The glaciers that formed during a 200,000-year period probably contained ice sheets that were about two-thirds the size of today's Antarctic Ice Cap. At the same time, surface temperatures in the west Atlantic ocean probably were more than 95° to 99°F. The Turonian period was noted for large climatic variations because of changes in the Earth's orbit.

Lest climate contrarians jump on this study as a reason not to worry about global warming, Thomas Wagner, a German scientist at Newcastle University in England who was among its authors, said, "It's difficult to draw a direct relationship between our findings and the current discussion on the climate. The results, however, show that even in a very warm world it is possible, at least temporarily, to build up larger ice caps in cooler regions" ("Study Says Glaciers" 2008). That, and some scientists say that the proxies are not iron-clad: "We're right at the limit of what our proxies tell us, or a bit beyond," said Timothy Bralower, a paleoceanographer at Pennsylvania State University. Even if the proxies are correct, the research team asserted that the ice formed only "under certain rare conditions" (Bornemann et al. 2008, 191).

FURTHER READING

Bornemann, Andrè, Richard D. Norris, Oliver Friedrich, Britta Beckmann, Stefan Schouten, Jaap S. Sinninghe Damstè, Jennifer Vogel, Peter Hofmann, and Thomas Wagner. "Isotopic Evidence for Glaciation during the Cretaceous Super-greenhouse." *Science* 319 (January 11, 2008):189–192.

Dalton, Alastair. "Ice Pack Clue to Climate-change Effects." *The Scotsman,* October 18, 2001, 7.

Kerr, Richard A. "More Climate Wackiness in the Cretaceous Super-greenhouse?" *Science* 319 (January 11, 2008a):145.

Naish, T. R., K. J. Woolfe, P. J. Barnett, G. S. Wilson, C. Atkins, S. M. Bohaty, C. J. Backer, M. Claps, F. J. Davey, G. B. Dunbar, A. G. Dunn, C. R. Fielding, F. Florindo, M. J. Hannah, D. M. Harwood, S. A. Henrys, L. A. Krissek, M. Lavelle, J. van der Meer, W. C. McIntosh, F. Niessen, S. Passchier, R. D. Powell, A. P. Roberts, L. Sagnotti, R. P. Scherer, C. P. Strong, F. Talarico, K. L. Verosub, G. Villa, D. K. Watkins, P.-N. Webb, and T. Wonik. "Orbitally Induced Oscillations in the East Antarctic Ice Sheet at the Oligocene/Miocene Boundary." *Nature* 413 (October 18, 2001):719–723.

O'Neill, Graeme. "The Heat Is On." *Sunday Herald-Sun* (Sydney, Australia), March 31, 2002.

"Study Says Glaciers Formed During a Very Warm Period." Reuters in *New York Times,* January 11, 2008. http://www.nytimes.com/2008/01/11/world/europe/11glacier.html.

Antarctic Peninsula and Ice Shelf Collapse

During March 2008, a chunk of ice about seven times the size of Manhattan Island collapsed into the sea near the base of the Antarctic Peninsula. The 160-square-mile piece of the Wilkins Ice Shelf was holding back a larger mass of ice (about the size of Connecticut), which then began to move toward the ocean. This particular chunk of ice was only one of several ice shelves adjacent to the Antarctic Peninsula that have collapsed into the ocean in recent years, becoming spectacular poster images for global-warming advocates. Kevin Krajick, writing in *Science,* remarked that glaciologists in Antarctica "are keeping an eye on an alarming trend: sudden, explosive calving [of icebergs] in parts of Antarctica. The fear is that if this continues, it may hasten the death of glaciers at an unanticipated rate" (Krajick 2001b, 2245).

After February 2008, when a large section of the Wilkins Ice Shelf on the western side of the Antarctic Peninsula fell apart, what was left has been held in place by a thin ice bridge connecting Charcot Island on the north to Latady Island on the south. Initially, the ice bridge was about 6 kilometers wide, but within five months, during Antarctic fall and winter, the bridge's width ebbed by half. As the summer of 2008–2009 arrived in November 2008, the European Space Agency's Advanced Synthetic Aperture Radar detected new cracks on the ice shelf's seaward edge.

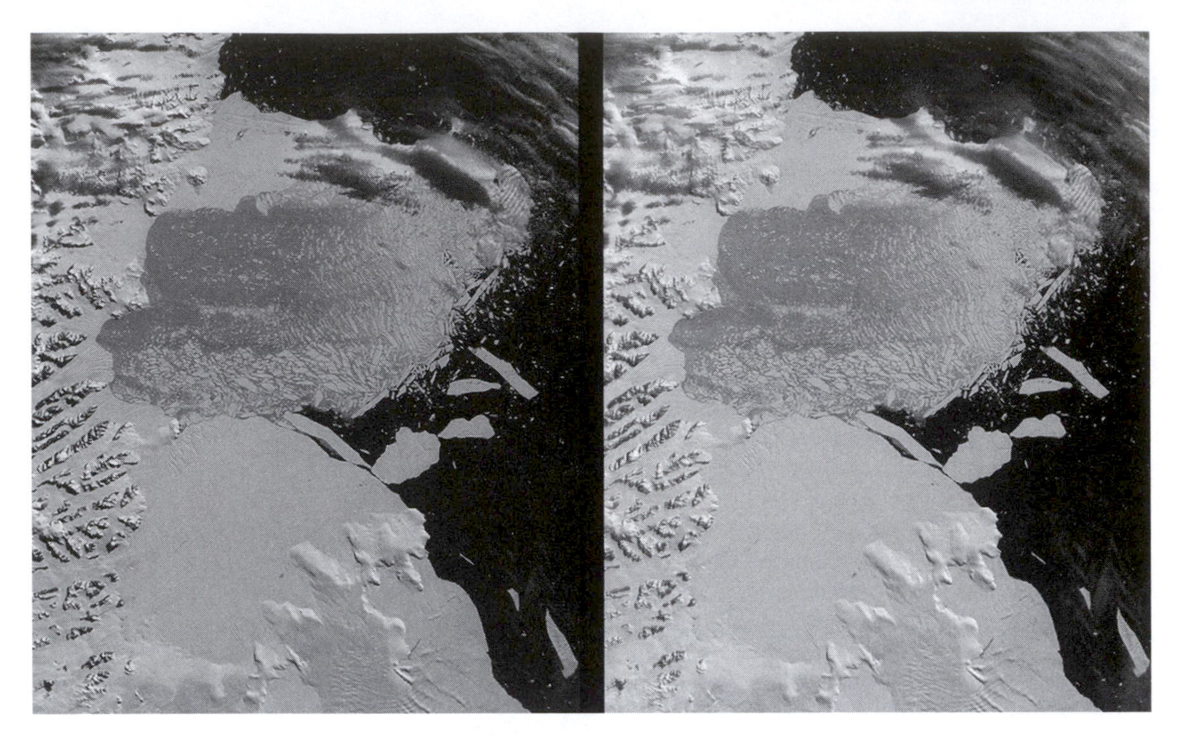

Single and multiangle views of the breakup of the northern section of the Larsen B Ice Shelf are shown in this image pair from the Multi-angle Imaging SpectroRadiometer. The Larsen B Ice Shelf collapsed and broke away from the Antarctic Peninsula during February and March 2002. (NASA)

Ted Scambos, a glacier expert at the National Snow and Ice Data Center, a joint operation of the Commerce Department and the University of Colorado, was quoted as saying that the rapid fracturing is too rapid to be explained by temperature rises alone. He surmised that "summer temperatures are now high enough to form melt pools on the glacial surfaces, which percolate rapidly into small weaknesses to form crevasses. Once a complex of crevasses hits sea level, sea water rushes in, re-freezes, and the mass blows apart" (Krajick 2001b, 2245). Scambos then said that such fracturing might spread to the Ross Ice Sheet (which is much closer to the South Pole than the Antarctic Peninsula) within about 50 years.

The Scope of Ice Loss

The ice shelves of Antarctica lost 3,000 square miles of surface area during 1998 alone. During March 2000, one of the largest icebergs ever observed broke off the Ross Ice Shelf near Roosevelt Island. Designated B-15, its initial 4,250-square-mile (11,007-square-kilometer) area was almost as large as the state of Connecticut. In mid-May 2002, another massive iceberg broke off the Ross Ice Shelf, according to the National Ice Center in Suitland, Maryland.

The new iceberg, named C-19 to indicate its location in the Western Ross Sea, was the second to break from the Ross Ice Shelf in two weeks. On May 5, researchers spotted a new floating ice mass named C-18, measuring about 41 nautical miles long and four nautical miles wide. An iceberg 200 kilometers (120 miles) long, 32 kilometers (20 miles) wide, and about 200 meters (660 feet) thick calved from the Ross Ice Shelf during late October 2002. The iceberg, called C-19, is one of the biggest observed in recent years, said David Vaughan of the British Antarctic Survey ("Monster Iceberg" 2002).

Vaughan said he believed that the emergence of several very large icebergs from the Ross Ice Shelf in so short a time should not be a cause for worry about global warming.

> That's the normal, natural cycle of the ice shelf. There are areas where we've seen ice shelves retreating over the last 50 years, and we think that is a

> response to climate change, but this is not one of those areas. The Antarctic Peninsula is where climate has been changing most rapidly and where the ice shelves have been retreating. But they don't retreat like this, producing one big iceberg every now and again. They retreat by year-on-year production of lots of little icebergs, a kind of constant retreat. ("Monster Iceberg" 2002)

The Antarctic Peninsula is a different story. Disintegration of ice shelves there seems to be bona fide evidence of rapid climate change. Near the northernmost tip of the peninsula, the Larsen Ice Shelves (known by scientists as, from north to south, "A," "B," and "C") have been disintegrating since 1995, when the "A" shelf fell apart; "B" followed in 1998, losing 1,000 square miles over four years. A 1,250-square-mile section of the Larsen B Ice Shelf disintegrated in just 35 days, setting thousands of icebergs adrift in the Weddell Sea.

"We knew what was left of the Larsen B Ice Shelf would collapse eventually, but this is staggering," said David Vaughan, a British glaciologist. "It's just broken apart. It fell over like a wall and has broken as if into hundreds of thousands of bricks" (Vidal 2002b, 3). The collapse of the Larsen B Ice Shelf "is unprecedented during the Holocene," that is, during the last 10,000 years, according to a scientific team writing in *Nature:*

> This is the largest single event in a series of retreats by ice shelves in the peninsula over the last 30 years. Satellite images indicated that during 2002 another massive iceberg, larger in area than Delaware, broke away from the Thwaites ice tongue, a sheet of glacial ice that extends into the Amundsen Sea nearly a thousand miles from the Larsen Ice Sheet. (Domack et al. 2005, 681)

Melting Linked to Human Activity

Scientists during 2006 reported direct evidence associating the collapse of an Antarctic ice shelf during 2002 with global warming. The researchers found that stronger westerly winds in the northern Antarctic Peninsula, provoked mainly by rising levels of human-induced greenhouse gases, were a major contributor to dramatic summer warming that led to the retreat and collapse of the Larson B Ice Shelf. "This is the first time that anyone has been able to demonstrate a physical process directly linking the breakup of the Larsen Ice Shelf to human activity," said lead author Gareth Marshall from the British Antarctic Survey (BAS)" ("Antarctic Ice Shelf" 2006). The study said that global warming as well as human-induced stratospheric ozone loss has changed Antarctic weather, forcing warm winds eastward and over the natural barrier created over a mile-high mountain chain on the Antarctic Peninsula. The warm winds sometimes raise temperatures by 5° to 10°C ("Antarctic Ice Shelf" 2006).

Warming riding these warm winds created conditions that allowed meltwater to drain into crevasses on the Larsen Ice Shelf, a key process that led to its breakup in 2002. "Climate change does not impact our planet evenly—it changes weather patterns in a complex way that takes detailed research and computer modeling techniques to unravel," Marshall said. "What we've observed at one of the planet's more remote regions is a regional amplifying mechanism that led to the dramatic climate change we see over the Antarctic Peninsula" ("Antarctic Ice Shelf" 2006).

FURTHER READING

"Antarctic Ice Shelf Collapse Tied to Global Warming." Environment News Service, October 16, 2006. http://www.ens-newswire.com/.

Domack, Eugene, Diana Duran, Amy Leventer, Scott Ishman, Sarah Doane, Scott McCallum, David Amblas, Jim Ring, Robert Gilbert, and Michael Prentice. "Stability of the Larsen B Ice Shelf on the Antarctic Peninsula during the Holocene Epoch." *Nature* 436 (August 4, 2005):681–685.

Krajick, Kevin. "Tracing Icebergs for Clues to Climate Change." *Science* 292 (June 22, 2001b):2244–2245.

"Monster Iceberg Heads into Antarctic Waters." Agence France Presse, October 22, 2002. (LEXIS)

"New Cracks in the Wilkins Ice Shelf." NASA Earth Observatory, December 9, 2008. http://earthobservatory.nasa.gov/IOTD/view.php?id=36060.

Rignot, E., G. Casassa, P. Gogineni, W. Krabill, A. Rivera, A., and R. Thomas. "Accelerated Ice Discharge from the Antarctic Peninsula Following the Collapse of Larsen B Ice Shelf." *Geophysical Research Letters* 31, no. 18 (September 22, 2004), doi:10.1029/2004GL020697.

Vidal, John. "Antarctica Sends Warning of the Effects of Global Warming: Scientists Stunned as Ice Shelf Falls Apart in a Month." *London Guardian*, March 20, 2002b, 3.

Antarctic Peninsula and Warming

Great Britain's Rothera Station on the western shore of the Antarctic Peninsula may be the most rapidly warming locale on Earth. Average temperatures there rose 20°F during the last quarter of the twentieth century (Bowen 2005, 33).

In 2001, Tom Spears, in the *Ottawa Citizen*, described

> [a] gentle Antarctic winter [that] is bringing mild breezes along the seashore, weather for pleasant walks in a light jacket, snow that's mushy enough to make snowballs. This is the depth of winter in the southern hemisphere, yet part of Antarctica remains balmy, even above freezing, as shown by many of the scientific research stations that report daily weather. (Spears 2001, A-1)

Scientists from the British Antarctic Survey reported in December 2004 that grass was growing on some parts of the Antarctic Peninsula in places that until recently had been covered by ice year-round (Rohter 2005a).

On August 7, 2001, mid-winter in the Southern Hemisphere, temperatures rose to a balmy −1°C at Argentina's Orcadas base, on an island near the mainland. On the continent itself, the same day brought highs of −3°C in the Jubany and Great Wall research stations, and −4°C at the Russian Bellingshausen station. "A breeze at Bellingshausen dropped the wind chill to −13°C, but that's still pretty mild for Antarctica," Spears wrote. "Yet parts of the Antarctic continue to bask in relative warmth. Tomorrow's forecast is for a high of 5°C at Palmer Station, and in recent years it has risen as high as 9°C in winter" (Spears 2001, A-1).

The Antarctic Peninsula is usually the continent's warmest region, with average temperatures of about −10°C in mid-winter. In recent years, however, the mid-winter average has risen from −4°C to −3°C, according to climatologist Henry Hengeveld, Environment Canada's global-warming expert (Spears 2001, A-1). Warming on the Antarctic Peninsula has been much more rapid than over the rest of Antarctica; climate proxies indicate the warming is "unprecedented over the last two millennia" (Vaughan et al. 2003, 243).

Furthermore, measurements of 244 marine glaciers fronts on the Antarctic Peninsula that have been draining into the sea (with records dating to 1945) indicated that 87 percent of them are retreating and that "a clear boundary between mean advance and retreat has migrated progressively southward" (Cook et al. 2005, 544). According to the authors, the study indicated that "increased drainage of the Antarctic Peninsula is more widespread than previously thought" (Cook et al. 2005, 544).

FURTHER READING

Bowen, Mark. *Thin Ice: Unlocking the Secrets of Climate in the World's Highest Mountains.* New York: Henry Holt, 2005.

Cook, A. J., A. J. Fox, D. G. Vaughan, J. G. Ferrigno. "Retreating Glacier fronts on the Antarctic Peninsula over the past Half-Century." *Science* 308 (April 22, 2005):541–544.

Rohter, Larry. "Antarctica, Warming, Looks Ever More Vulnerable." *New York Times*, January 25, 2005a. http://www.nytimes.com/2005/01/25/science/earth/25ice.html.

Spears, Tom. "Antarctica Rides Global 'Heatwave': Continent's Warm Coast Causes Concern." *Ottawa Citizen*, August 8, 2001, A-1.

Vaughan, D. G., G. J. Marshall, W. M. Connolley, C. L. Parkinson, R. Mulvaney, D. A. Hodgson, J. C. King, C. J. Pudsey, and J. Turner. "Recent Rapid Regional Climate Warming on the Antarctic Peninsula." *Climatic Change* 60, no. 3 (October 2003):243–274.

Anthropogenic Warming in the Twentieth Century

Anthropogenic (human-generated) warming has raised Earth's surface temperature significantly during the last 130 years, according to work by Robert Kaufmann of Boston University's Center for Energy and Environmental Studies and David Stern of the Australian National University's Centre for Resource and Environmental Study. A team led by Kaufmann analyzed historical data for greenhouse-gas levels, human sulfur emissions, and variations in solar activity between 1865 and 1990. The greenhouse gases studied included carbon dioxide, methane, nitrous oxide, and chlorofluorocarbons 11 and 12. Using the statistical technique of "cointegration," Kaufmann and Stern compared these factors over time with global surface temperature in both the Northern and Southern hemispheres. According to its authors, this was the first study to make a statistically meaningful link between human activity and temperature independent of climate models, Kaufmann said ("Century of Human Impact" 2002).

This study found that the impact of human activity has been different in the Earth's two hemispheres. In the Northern Hemisphere, the warming effect of greenhouse gases was offset to some degree by the cooling effect of industrial sulfur emissions, making the effect on temperature more difficult to detect. In the Southern Hemisphere, with its overwhelming coverage of oceans, where human sulfur emissions are lower, the effects were easier to detect. "The countervailing effects of greenhouse gases and sulfur emissions undercut comments by climate change skeptics, who argue that the rapid increase in atmospheric concentrations of greenhouse gases between the end of World War II and the early 1970s had little effect on temperature," said Kaufmann ("Century of Human Impact" 2002). During this period, Kaufmann said, "The warming effect of greenhouse gases was hidden by a simultaneous increase in sulfur emissions. But, since then, sulfur emissions have slowed, due to laws aimed at reducing acid rain, and this has allowed the warming effects of greenhouse gases to become more apparent" ("Century of Human Impact" 2002).

Peter A. Stott and colleagues used a "comparison of observations with simulations of a coupled ocean-atmosphere model" to estimate the roles of various natural and anthropogenic "forcings" (i.e., albedo-driven influences on climate) to account for rising temperatures during the twentieth century. Stott and colleagues found both natural and human-induced factors in the warming trend. "Natural forcings were relatively more important in the early-century warming and anthropogenic forcings have played a dominant role in warming observed in recent decades," this researchers observed (Stott et al. 2000, 2136). Stott and colleagues reported "the most comprehensive simulation of twentieth-century climate to date" (Zwiers 2000, 2081). However, some of the data used in the model (such as those for the role of the sun) are available for only the last half of the century.

Removing the masking effects of volcanic eruptions and El Niño events from the global mean temperature record has revealed a more gradual but stronger global-warming trend over the last century, according to an analysis by Tom Wigley, a climate expert at the National Center for Atmospheric Research. This analysis supports scientists' assertions that human activity is influencing the Earth's climate. These findings are published in the December 15, 2000, issue of *Geophysical Research Letters* (Wigley 2000, 4101). "Once the volcanic and El Niño influences have been removed," said Wigley, "the overall record is more consistent with our current knowledge, which suggests that both natural and anthropogenic influences on climate are important and that anthropogenic influences have become more substantial in recent decades" ("New Study" 2000).

FURTHER READING

"Century of Human Impact Warms Earth's Surface." Environment News Service. January 24, 2002. http://ens-news.com/ens/jan2002/2002L-01-24-09.html.

"New Study Shows Global Warming Trend Greater without El Nino and Volcanic Influences." Environmental Journalists' Bulletin Board. December 13, 2000. environmentaljournalists@egroups.com.

Stott, Peter A., S. F. B. Tett, G. S. Jones, M. R. Allen, J. F. B. Mitchell, and G. J. Jenkins. "External Control of 20th Century Temperature by Natural and Anthropogenic Forcings." *Science* 290 (December 15, 2000):2133–2137.

Wigley, T. M. L. "ENSO, Volcanoes, and Record Breaking Temperatures." *Geophysical Research Letters* 27 (2000):4101–4104.

Zwiers, Francis W., and Andrew J. Weaver. "The Causes of 20th Century Warming." *Science* 290 (December 15, 2000):2081–2083.

Arctic and Climate Change

Global warming is being felt most intensely in the Arctic and coastal regions of Antarctica, where a world based on ice and snow has been melting away. Arctic sea ice cover shrank more dramatically between 2000 and 2007 than at any time since detailed records have been kept. A report produced by 250 scientists under the auspices of the Arctic Council found that Arctic sea ice was half as thick in 2003 as it was 30 years earlier. If present rates of melting continue, there may be no summer ice in the Arctic by 2070, according to the study. Pal Prestrud, vice chairman of the steering committee for the report, said, "Climate change is not just about the future; it is happening now. The Arctic is warming at twice the global rate" (Harvey 2004, 1). Since that report, ice loss has accelerated, most notably during 2007 and 2008, giving rise to forecasts of an ice-free Arctic Ocean in summer between 2020 and 2040.

Melting of the Arctic Ice Cap could lead to summertime ice-free ocean conditions not seen in the area in a million years, a group of scientists wrote during 2005 in *EOS*, the transactions of the American Geophysical Union. Jonathan Overpeck, of the University of Arizona and chairman of the National Science Foundation's Arctic System Science Committee, and colleagues wrote that the Arctic during the twenty-first century is moving beyond the glacial and interglacial cycle that has characterized the last million

Ice floes in the northern Bering Sea. (National Oceanic and Atmospheric Administration Photo Library)

years. "At the present rate of change," they wrote, "a summer ice-free Arctic ocean within a century is a real possibility.... a state [that is] driven largely by feedback-enhanced global climate warming.... [into] a 'super interglacial' state" (Overpeck et al. 2005, 309).

Speaking at the annual meeting of the American Geophysical Union in San Francisco during December, 2007, Richard Alley, of Pennsylvania State University in State College, surveyed the major—and accelerating—effects that a relatively small amount of warming (compared with what is anticipated for the rest of the century) is having on the melting of ice in the Arctic and Antarctic. "If a very small warming makes such a difference," asked Alley, "It raises the question of what happens when more warming occurs." At the same meeting, Josefino Comiso of NASA's Goddard Space Flight Center in Greenbelt, Maryland, said that "[t]he tipping point for perennial sea ice has likely already been reached" (Kerr 2008b, 153).

During September 2007, the Arctic Ice Cap shrunk to its smallest extent since satellite records have been kept (in 1979), 1.59 million square miles. The previous record low was 2.04 square miles in 2005, which equated to more than a 20 percent loss of ice cover, an area the size of Texas and California combined. By 2007, the Arctic Ice Cap had lost 43 percent of its mass since 1979 (Kerr 2007d, 33).

In 2008, the Arctic Ice Cap shrunk not quite as much as during 2007, but still was 33 percent below the average since 1979. At summer's end, the ice cap was 1.74 million square miles, according to the National Snow and Ice Data Center in Boulder, Colorado. The year 2008 also was notable because two potential shipping routes, the Northwest Passage off Canada and the Northern Sea Route adjacent to Russia, opened at the same time. Walter Meier, a research scientist at the center, said that small variations from one year to the next were less significant than the long-term trajectory, which remained toward progressively more open water. "It's hard to see the summer ice coming back in any substantial way," he said (Revkin 2008i).

During 2008, NASA's ICESat satellite measured the largest annual decline in "perennial" ice (that frozen more than one year) on record. The proportion of the Arctic covered with this stable type of ice has declined from more than 50 percent during the mid-1980s to less than 30 percent in February 2008. Comiso said that Arctic Ocean temperatures appear to be rising quickly because less of the water is covered by ice, which reflects sunlight and keeps water temperatures lower (Kaufman 2008c, A-3).

Temperatures in the Arctic continued to rise into 2008, after reaching an all-time record high of more than 9°F (5°C) above average during the fall of 2007. Temperatures remained almost as high a year later, according to the annual Arctic Report Card, the product of 46 scientists from 10 countries, edited by Jackie Richter-Menge, a climate expert at the Cold Regions Research and Engineering Laboratory in Hanover, New Hampshire. While the Arctic has been warming on average, effects vary regionally. The Bering Sea, for example, is in a cooling spell, as an unusually severe winter during 2007–2008 added mass to many of Alaska's glaciers (Boyd 2008). Greenland's ice cap shrank by 88 square miles (220 square kilometers) as a result of an unusually warm spring and summer, according to Jaxon Box of the Byrd Polar Research Center in Columbus, Ohio.

Flying over the Arctic, one might perceive the sea ice cover as broad, Meier said, but that apparent breadth hides the fact that the ice is so thin. "It's a facade, like a Hollywood set," he said. "There's no building behind it" (Kaufman 2008c, A-3).

Arctic ice provides an important breeding ground for plankton, NASA senior research scientist Comiso said. Plankton make up the bottom rung of the ocean's food chain. "If the winter ice melt continues, the effect would be very profound especially for marine mammals," explained Comiso. Arctic ice sometimes melts even in winter as water warms, Comiso added. The winter ice season shortens every year ("Arctic Ice Melting" 2006).

Moving Beyond a "Point of No Return"

Scientists were shaken by the sudden retreat of the Arctic ice during the summer of 2007, which had a more profound affect than their models had anticipated. "The Arctic is often cited as the canary in the coal mine for climate warming," as Jay Zwally, a climate expert at NASA, told the Associated Press. "Now as a sign of climate warming the canary has died" (Kolbert 2007d, 44). "We could very well be in that quick slide downward in terms of passing a tipping point," said Mark Serreze, a senior scientist at the National Snow and Ice Data Center, in Boulder, Colorado. "It's tipping now. We're seeing it happen now" ("As Arctic Sea Ice Melts" 2008). Bob Corell, who headed a multinational Arctic assessment, said, "We're moving beyond a point of no return" ("As Arctic Sea Ice Melts" 2008).

By 2007, Baffin Island in the Canadian Arctic had lost half its ice in 50 years, as glaciers on its northern mountain ranges eroded. In another 50 years, what remains may be gone, according to research by Gifford Miller, professor of geological sciences at the University of Colorado at Boulder's Institute of Arctic and Alpine Research, and colleagues. "Even with no additional warming, our study indicates these ice caps will be gone in 50 years or less," he said (Anderson et al. 2008; "Baffin Island" 2008).

At the 2007 annual American Geophysical Union meeting in San Francisco, scientists reported that water temperatures near Alaska and Russia were as much as to 9°F above average. A study by scientists at the University of Washington said that the sun's heat made the greatest contribution to the record melting of the Arctic Ice Cap at the end of summer 2007. Sunlight added twice as much heat to the water as was typical before 2000. Relatively warm water entering the Arctic Ocean from the Atlantic and the Pacific oceans also was a less important factor, said Michael Steele, an oceanographer at the University of Washington. Energy from the warmer water delays the expansion of ice growth as it warms the air. Wieslaw Maslowski of the Naval Postgraduate School in Monterey, California, foresees a blue Arctic Ocean in summers by 2013 (Revkin 2007k).

In addition to the retreating extent of sea ice in the Arctic, more of the ice that remains is thin and freshly formed, unlike older, thicker ice that is more likely to survive future summer melting. The proportion of older, "durable" ice dropped drastically between 1987 and 2007, according to studies by Ignatius G. Rigor of the University of Washington, lead author of work for NASA (Revkin 2007i). The thin, freshly formed ice also is more easily melted by relatively warm air held close to the surface by cloudiness, or below by warming water.

In addition to a decrease in their extent, by 2007 large areas of Arctic sea ice were an average of only about one meter thick, half what they were in 2001, according to measurements taken by an international team of scientists aboard the research ship *Polarstern*. "The ice cover in the North Polar Sea is dwindling, the ocean and the atmosphere are becoming steadily warmer, the ocean currents are changing," said Ursula Schauer, chief scientist with the Alfred Wegener Institute for Polar and Marine Research, from aboard the *Polarstern* in the Arctic Ocean ("Arctic Ocean Ice" 2007).

Arctic Ice Cap Melts Faster Than Models Forecast

How well do observations and models agree on Arctic ice melt? Julienne Stroeve, of the National Snow and Ice Data Center, and colleagues compared more than a dozen models and found that nearly all of them underestimated the speed of ice melt. And, in many cases, they did so by large amounts. "These findings have two important implications," said a summary of this study in *Science*. "First, that the effect of rising greenhouse gases may have been more important than has been believed; and second, that future loss of Arctic sea ice may be more rapid and extensive than predicted" ("Melting Faster" 2007).

Sea ice across the Arctic Ocean has been melting much more quickly than earlier anticipated even by the most recent computer models. The latest projections indicate that the Arctic Ocean may be without ice in summer by 2020. Even projections made by the Intergovernmental Panel on Climate Change (IPCC) in its 2007 assessments (forecasting an ice-less Arctic in summer between 2050 and 2100) were outdated just weeks after they were made public, according to reports by scientists at the National Center for Atmospheric Research and the University of Colorado's National Snow and Ice Data Center. The study, which was led by Stroeve and funded by the National Science Foundation and NASA, was published during early May 2007 in the online edition of *Geophysical Research Letters.*

"While the ice is disappearing faster than the computer models indicate, both observations and the models point in the same direction—the Arctic is losing ice at an increasingly rapid pace and the impact of greenhouse gases is growing," says co-author Marika Holland of the National Center for Atmospheric Research ("Arctic Ice Retreating" 2007).

Earlier research calculated that Arctic sea ice was retreating at an average of 2.5 percent per decade between 1953 and 2006, with the greatest annual loss at 5.4 percent. The new research—which blended aircraft and ship reports with satellite measurements (that are considered more reliable than the earlier records)—showed that the September ice actually declined at a rate of about 7.8 percent per decade during the 1953–2006 period ("Arctic Ice Retreating" 2007). The new data sets indicated that actual ice loss is about 30 percent more than the IPCC's projections. More recently, from 1979 to 2006, estimated sea-ice loss in September had been 9.1 percent per decade. The accelerating loss of ice is due in large part to changes in albedo, or reflectivity, as heat-absorbing open ocean replaces the highly reflective surface of the shrinking ice cover. As the ice cap has shrunk, temperatures across the Arctic have risen from 2° to 7°F, further accelerating melting. In March, when Arctic ice is usually at its greatest extent, older models averaged a loss of about 1.8 percent per decade during 1953–2006. The new data tripled that figure.

In a separate study released in March 2006, Stroeve and her team showed that dwindling Arctic sea ice may have reached "a tipping point that could trigger a cascade of climate change reaching into Earth's temperate regions" ("Arctic Ice Retreating" 2007). "This suggests that current model projections may in fact provide a conservative estimate of future Arctic change, and that the summer Arctic sea ice may disappear considerably earlier than IPCC projections," said Stroeve ("Arctic Ice Retreating" 2007).

Thousands of Walruses Die in Stampedes

With ice receding hundreds of miles offshore of Alaska and Russia during the late summer of 2007, walruses gathered by the thousands along the shores of Alaska and Siberia. Joel Garlich-Miller, a walrus expert with the U.S. Fish and Wildlife Service, said that walruses began to gather onshore late in July, a month earlier than usual. A month later, their numbers had reached record levels from Barrow to Cape Lisburne, about 300 miles southwest, on the Chukchi Sea.

Walruses usually feed on clams, snails, and other bottom-dwelling creatures from the ice. In recent years, the ice has receded too far from shore to allow the usual feeding pattern. A walrus can dive 600 feet, but water under ice shelves in late summer is now several thousands of feet deep. The walruses have been forced to swim much farther to find food, using energy that could cause increased calf mortality. More calves are being orphaned. Russian research observers also reported many more walrus than usual onshore, tens of thousands in some areas along the Siberian coast, which would have stayed on the sea ice in earlier times (Joling 2007a).

Walrus are prone to stampedes once they are gathered in large groups, and thousands of Pacific walruses above the Arctic Circle were killed on the Russian side of the Bering Strait, where more than 40,000 had hauled out on land at Point Shmidt, as ice retreated farther northward. A polar bear, a human hunter, or noise from an airplane flying close and low can send thousands of panicked walrus rushing to the water.

"It was a pretty sobering year, tough on walruses," said Garlach-Miller. Several thousand walrus died late in the summer of 2007 from internal injuries suffered in stampedes. The youngest and the weakest animals, many of them calves born the previous spring were crushed. Biologist Anatoly Kochnev of Russia's Pacific Institute of Fisheries and Oceanography estimated 3,000 to 4,000 walruses out of a population of perhaps 200,000 died, or two or three times the usual number on shoreline haul-outs (Joling 2007b).

The reports support expectations of walrus' fate as ice retreated, said wildlife biologist Tony Fischbach of the U.S. Geological Survey. "We were surprised that this was happening so soon, and we were surprised at the magnitude of the report," he said (Joling 2007b). Walrus lacking summer sea ice that they use to dive for clams and snails, may strip coastal areas of food and then starve in large numbers.

No Recovery Cycle

In the past, low-ice years often were followed by recovery the next year, when cold winters favored accumulation, or cool summers kept ice from melting. That kind of balancing cycle stopped after 2002. "If you look at these last few years, the loss of ice we've seen, well, the decline is rather remarkable," said Serreze (Human 2004, B-2).

At the same time, Alaska's boreal forests are expanding northward at a rate of about 100 kilometers per 1°C rise in temperatures. Ice cover on lakes and rivers in the mid to high northern latitudes now lasts for about two weeks less than it did 150 years ago. During late June 2004, as temperatures in Fairbanks, Alaska, peaked in the upper 80s on some days, a flood warning was issued near Juneau, Alaska—not for rainfall, but for glacial snowmelt.

At the Abisko station, in northernmost Sweden, winter (December through February) temperatures have risen about 5.5°C in a century, eight times the Northern Hemisphere average. Within the average are increasing numbers of dramatic winter daily temperature spikes during which readings may rise 25°C or more within one day, like the "snow eaters" on the lee side of the Rockies in the United States (Schiermeier 2007a, 146). These warming spikes thaw snow, which then refreezes and denies access to food for reindeer and lemmings. Lakes in the area have become saturated with carbon and have become sources instead of sinks. The spread of trees and shrubs also darkens albedo and increases warming, as insect attacks on evergreen forests intensified by warming increase carbon output.

Every winter since 1912, when an iceberg claimed the *Titanic* in the North Atlantic, the U.S. Coast Guard has issued warnings for an average of 500 icebergs that float south of 48°N latitude, calving from Greenland's ice sheet. These icebergs threaten the shipping lanes of "Iceberg Alley," the Grand Banks southeast of Newfoundland. During the El Niño winter of 1998–1999, for the first time in 85 years, the Coast Guard did not issue a single iceberg alert for that area. "The lack of ice is remarkable," said Steve Sielbeck, commander of the International Ice Patrol (Wuethrich 1999, 37).

Rapid ice melt in the Arctic was evident to some observers by the 1980s. By 1989, the Arctic ice was reported to be thinning quickly. The Scott Polar Institute of the United Kingdom reported that ice in an area north of Greenland had thinned from an average of 6.7 meters in 1976 to 4.5 meters in 1987. At about the same time, a British Antarctic survey base reported unusual losses of ice as well (Kelly 1990, 23).

Ice-Melt Stories

Viewed from Jokulsarlon Lagoon, Iceland's ice cap smothers nearby mountains. Even in 2006, it was more than 3,000 feet thick—but it is much smaller than it was a century earlier. At that time, the lagoon did not exist; it was under 100 feet of glacial ice. The Breidamerkurjokull glacier at that time reached to within 250 yards of the North Atlantic Ocean. It now stops more than two miles from the water, flowing into an expanding lake of its own creation. This glacier and others in Iceland crack and gurgle as streams of water rend them. In places, meltwater threatens to wash through the island's major highway (Woodard 2007, 29).

Some residents of Baker Lake, Nunavut, 1,330 kilometers west of Iqaluit spotted magpies flitting around town during May 2006. These scavengers, a relative of the crow, had never been seen in Nunavut before. The magpies are not expected to become permanent residents, however, even if the climate warms, because they roost in trees. The tundra has no trees.

By the winter of 2002–2003, a warming trend was forcing hockey players in Canada's far north to seek rinks with artificial ice. Canada's *Financial Post* reported that "[o]fficials in the Arctic say global warming has cut [the] hockey season in half in the past two decades and may hinder the future of development of northern hockey stars such as Jordin Tootoo" ("Ice a Scarce" 2003). According to the *Financial Post* report, hockey rinks in northern communities were raising funds directed toward installation of cooling plants to create artificial ice because of the reduced length of time during which natural ice was available. In Rankin Inlet, Tootoo's home town on Hudson Bay in Nunavut, a community of 2,400 residents installed artificial ice during the summer of 2003 after eight years of lobbying for the funds ("Ice a Scarce" 2003).

Hockey season on natural ice, which ran from September through May in the 1970s, often now begins around Christmas and ends in March, according to Jim MacDonald, president of Rankin Inlet Minor Hockey. "It's giving us about three months of hockey. Once we finally get going, it's time to stop. At the beginning of our season, we're playing teams that have already been on the ice for two or three months," MacDonald said ("Ice a Scarce" 2003.) According to Tom Thompson, president of Hockey Nunavut, there are about two-dozen natural ice rinks in tiny communities throughout the territory, but only two with artificial ice. Both are in the capital, Iqaluit ("Ice a Scarce" 2003).

"In my lifetime I will not be surprised if we see a year where Hudson Bay doesn't freeze over completely," said Jay Anderson of Environment Canada. "It's very dramatic. Yesterday [January 6, 2003], an alert was broadcast over the Rankin Inlet radio station warning that ice on rivers around the town is unsafe. The temperature hovered around −12°C. It's usually −37 there at this time of year" ("Ice a Scarce" 2003).

Arctic Temperature Rise Documented

Air temperatures above large areas of the Arctic Ocean from January 2006 to August 2006 were about 2° to 7°F warmer than the long-term average during the previous 50 years ("Arctic Sea Ice Melt Accelerating" 2006, "Arctic Sea Ice Melt May" 2006). High winter temperatures contributed to limited ice growth, and much of the ice that did form was thinner than normal. "Unusually high temperatures through most of July then fostered rapid melt," Serreze explained. Cooler weather in August slowed the melt, he said, and storm conditions led to wind patterns that helped spread existing ice over a larger area" ("Arctic Sea Ice Melt Accelerating" 2006, "Arctic Sea Ice Melt May" 2006). A "polynya," an area of persistent open water surrounded by sea ice, formed north of Alaska. Calculations showed that in early September, the polynya was the size of the state of Indiana, a size never before seen in the Arctic. Unusual wind patterns and an influx of warmer ocean waters may have caused the polynya ("Arctic Sea Ice Melt Accelerating" 2006, "Arctic Sea Ice Melt May" 2006).

"The large area of the Arctic Ocean promises to become much warmer," said Igor Polyakov, principal investigator for Nansen and Amundsen Basins Observational Systems (NABOS) and a research professor at the International Arctic Research Center (IARC). Instruments first detected a surge of abnormally warm water, about 150 to 800 meters below the surface during February 2004 on the continental slope of the Laptev Sea, north of Siberia ("Warm Water" 2006).

"When the ice thins to a vulnerable state, the bottom will drop out and we may quickly move into a new, seasonally ice-free state of the Arctic," Serreze said. "I think there is some evidence that we may have reached that tipping point, and the impacts will not be confined to the Arctic region" ("Arctic Sea Ice Extent" 2007). The review paper by Serreze and Julienne Stroeve of University of Colorado at Boulder's National Snow and Ice Data Center (NSIDC) and Dr. Marika Holland of the National Center for Atmospheric Research titled "Perspectives on the Arctic's Shrinking Sea Ice Cover" appears in the March 16, 2007, issue of *Science*. The researchers wrote:

> Given the growing agreement between models and observations, a transition to a seasonally ice-free Arctic Ocean as the system warms seems increasingly certain.... The unresolved questions regard when this new Arctic state will be realized, how rapid the transition will be, and what will be the impacts of this new state on the Arctic and the rest of the globe. ("Arctic Sea Ice Extent" 2007)

"We're seeing more melting of multi-year ice in the summer," said Stroeve. "We may soon reach a threshold beyond which the sea ice can no longer recover." Holland added, "We have already witnessed major losses in sea ice, but our research suggests that the decrease over the next few decades could be far more dramatic than anything that has happened so far.... These changes are surprisingly rapid" ("Arctic Sea Ice Extent" 2007).

As levels of atmospheric greenhouse gases increase, Arctic sea ice will continue to decline, said Serreze. "While the Arctic is losing a great deal of ice in the summer months, it now seems that it also is regenerating less ice in the winter," Serreze explained. "With this increasing vulnerability, a kick to the system just from natural climate fluctuations could send it into a tailspin" ("Arctic Sea Ice Extent" 2007). In the late 1980s and early 1990s, shifting wind patterns from the North Atlantic Oscillation flushed much of the thick sea ice out of the Arctic Ocean and into the North Atlantic where it drifted south and eventually melted, he said.

Thinner "younger" ice that forms in its place melts more rapidly in following summers, leading to more open water and more solar radiation that is absorbed by the open ocean, raising temperatures. "This ice-flushing event could be a small-scale analog of the sort of kick that could invoke rapid collapse, or it could have been the kick itself," Serreze said. "At this point, I don't think we really know" ("Arctic Sea Ice Extent" 2007).

In addition, beginning in the mid-1990s, scientists observed pulses of relatively warm water from the North Atlantic entering the Arctic Ocean, further speeding ice melt. Serreze said, "This is another one of those potential kicks to the system that could evoke rapid ice decline and send the Arctic into a new state" ("Arctic Sea Ice Extent" 2007). "As the ice retreats, the ocean transports more heat to the Arctic and the open water absorbs more sunlight, further accelerating the rate of warming and leading to the loss of more ice," Holland explained. "This is a positive feedback loop with dramatic implications for the entire Arctic region" ("Arctic Sea Ice Extent" 2007).

Ellesmere Island's Dying Glaciers

In 1907, when Arctic explorer Robert Peary traversed the northern coast of Ellesmere Island on a dog sled, he described a "glacial fringe" along most of the area. At the time, Peary was describing a continuous ice shelf covering about 8,900 square kilometers (or about 3,500 square miles). Half a century later, a large part of that ice had disintegrated. By July 2008, only five isolated ice shelves remained: Serson, Petersen, Milne, Ward Hunt, and Markham. (The Ayles Ice Shelf had broken off in 2005.) These were the last remaining ice shelves in Canada.

On July 22, 2008, a new wave of ice-shelf disintegration began and, by late August, these ice shelves had lost another 214 square kilometers (83 square miles). The Ward Hunt Ice Shelf lost a total of 42 square kilometers (16 square miles). The Serson Ice Shelf lost 122 square kilometers (47 square miles)—60 percent of its previous area. The Markham Ice Shelf, with a total area of 50 square kilometers (19 square miles) completely broke away from the Ellesmere coast (Copland et al. 2007; "Ellesmere Island Ice" 2008).

Warming ocean waters (which absorb more of the sun's heat than ice and snow) may be playing an important role in the ice shelves' retreat, along with warming air. Air temperatures in the area have risen about 5°C in 50 years. "As sea ice retreats, the exposed water in front of the shelf warms, and as the ice in the fjord behind the shelf retreats, that water is also warmed. It's possible that warmer water is circulating under the shelf and causing basal melting" said Ted Scambos, lead scientist at the NSIDC ("Ellesmere Island Ice" 2008).

The speed of ice breakup can sometimes be astonishing. For example, because of wind and waves in warming water, the 100-foot-thick Ayles shelf of floating ice roughly 25 miles square that had extended into the Arctic Ocean from the north coast of Ellesmere Island in the Canadian Arctic for roughly 3,000 years was detached during the summer of 2005, "The breakup was observed by Laurie Weir of the Canadian Ice Service in satellite images of Ellesmere and surrounding ice on and after August 13, 2005. In less than an hour, a broad crack opened and the ice shelf was on its way out to sea" (Revkin 2006m).

In 1906, Peary first surveyed 3,900 square miles of ice shelves, of which 90 percent have now broken up, said Luke Copland, the director of the University of Ottawa's Laboratory for Cryospheric Research. "The quick pace of these changes right now is what stands out," he said (Revkin 2006m). The breakup of Ellesmere glaciers continued in August 2008, as a large piece of the 4,500-year-old Markham Ice Shelf nearly as large as Manhattan Island separated and floated into the sea. The glacier's disintegration was sudden, and unexpected by scientists, taking a matter of a day or two, during a cloudy spell that hid it from satellites.

Tundra to Forest

Given the assumptions of one model—that the Arctic tundra nearly disappears, declining from about 8 to 1.8 percent of the world's land area by the year 2100—Alaska would lose almost all of its evergreen forests and become mainly a temperate region. Land now residing beneath ice would diminish from 13.3 to 4.8 percent of the planet's area. Farther south, tropical and temperate forests could expand substantially, so that the two forest types could grow on nearly 65 percent of land surfaces instead of 44 percent in the year 2000 (Revkin 2005b).

As temperatures warm in the Arctic and ice melts in the Arctic Ocean, forests of spruce trees and shrubs are spreading over northern Canada's tundra at a rate much faster than anticipated by many climate models. A study in the *Journal of Ecology*, which used tree rings to date the year of establishment and death of spruce trees and reconstruct changes in tree-line vegetation, analyzed the density and altitude of tree-line forests in southwestern Yukon over the past three centuries ("Canadian Tundra" 2007).

"The conventional thinking on tree-line dynamics has been that advances are very slow because conditions are so harsh at these high latitudes and altitudes," explained study author Ryan Danby, a biologist with the University of Alberta. "But what our data indicates is that there was an upslope surge of trees in response to warmer temperatures. It's [as if] it waited until conditions were just right and then it decided to get up and run, not just walk" ("Canadian Tundra" 2007).

"The tundra is becoming greener with the growth of more shrubs," said Vladimir E. Romanovsky, a professor at the geophysical institute of the University of Alaska. This development is causing problems in some areas as herds of reindeer migrate. At the same time, there is some decrease in the greening of the northern forest areas, probably due to drought. The glaciers are continuing to shrink and river discharge into the Arctic Ocean is rising, Romanovsky said (Schmid 2006b).

Ribbon Seals Threatened

The ribbon seal of the Bering Sea (named after the spectacular white-circle patterns on its fur), a rare species in the best of times, may lose its habitat to global warming. Like a number of other Arctic species that have adapted to ice (such as polar bears), the ribbon seal is disappearing as Arctic ice shrinks.

Late in 2007, the Center for Biological Diversity filed a scientific petition with the U.S. National Marine Fisheries Service to protect the ribbon seal under the federal Endangered Species Act. "The Arctic is in crisis state from global warming," said Shaye Wolf, a biologist with the Center for Biological Diversity and lead author of the petition. "An entire ecosystem is rapidly melting away and the ribbon seal is poised to become the first victim of our failure to address global warming," he said ("Ribbon Seal" 2007).

During the late winter through early summer, ribbon seals rely on the edge of the sea ice in the Bering and Okhotsk Seas off Alaska and Russia as safe habitats to give birth and as nurseries for their pups. Thinning ice inhibits the seals' ability to rear their pups, many of whom die early. The ribbon seals have been threatened by increasing oil and gas development in their habitat, as well as increasing shipping in the Arctic. G. Carleton Ray of the University of Virginia Department of Environmental Sciences traveled to the Bering Sea in May 2007 and found that "[i]n Russian waters is pack ice with leads, where strong, cold winds from Siberia create thick ice with parallel leads and where, last year, we found little-known, strikingly beautiful ribbon seals," wrote Ray (2007).

FURTHER READING

"2007 Arctic Sea Ice Nearly Matches Record Low." Environment News Service, April 4, 2007. http://www.ens-newswire.com/ens/apr2007/2007-04-04-09.asp#anchor2.

Anderson, Rebecca K., Gifford H. Miller, Jason P. Briner, Nathaniel A. Lifton, and Stephen B. DeVogel. "A Millennial Perspective on Arctic Warming from 14-C in Quartz and Plants Emerging from beneath Ice Caps." *Geophysical Research Letters* 35 (2008), L01502, doi:10.1029/2007/GL032057.

"Arctic Ice Melting Rapidly, Study Says." Associated Press in *New York Times*, September 14, 2006. http://www.nytimes.com/aponline/us/AP-Warming-Sea-Ice.html.

"Arctic Ice Retreating 30 Years Ahead of Projections." Environment News Service, April 30, 2007. http://www.ens-newswire.com/ens/apr2007/2007-04-30-04.asp.

"Arctic Ocean Ice Thinner by Half in Six Years." Environment News Service, September 14, 2007. http://www.ens-newswire.com/ens/sep2007/2007-09-14-03.asp.

"Arctic Sea Ice Extent Hits Record Low." Environment News Service, August 20, 2007. http://www.ens-newswire.com/ens/aug2007/2007-08-20-01.asp.

"Arctic Sea Ice Melt Accelerating." Environment News Service, October 4, 2006. http://www.ens-newswire.com/ens/oct2006/2006-10-04-02.asp.

"Arctic Sea Ice Melt May Set Off Climate Change Cascade." Environment News Service, March 19, 2006. http://www.ens-newswire.com/ens/mar2007/2007-03-19-06.asp.

"As Arctic Sea Ice Melts, Experts Expect New Low." Associated Press in *New York Times*, August 28, 2008. http://www.nytimes.com/2008/08/28/science/earth/28seaice.html.

"Baffin Island Ice Caps Shrink by Half in 50 Years." Environment News Service, January 28, 2008. http://www.ens-newswire.com/ens/jan2008/2008-01-28-02.asp.

Boyd, Robert S. "Arctic Temperatures Hit Record High." Knight Ridder Washington Bureau, October 16, 2008. (LEXIS)

"Canadian Tundra Turning Green." Environment News Service, March 3, 2007. http://www.ens-newswire.com/ens/mar2007/2007-03-06-02.asp.

Copland, L., D. R. Mueller, and L. Weir. "Rapid Loss of the Avies Ice Shelf, Ellesmere Island, Canada." *Geophysical Research Letters* 34 (2007), L21501.

"Ellesmere Island Ice Breaks Up." NASA Earth Observatory, September 9, 2008. http://earthobservatory.nasa.gov/Study/Ellesmere.

Harvey, Fiona. "Arctic May Have No Ice in Summer by 2070, Warns Climate Change Report." *London Financial Times*, November 2, 2004, 1.

Human, Katy. "Disappearing Arctic Ice Chills Scientists; A University of Colorado Expert on Ice Worries that the Massive Melting will Trigger Dramatic Changes in the World's Weather." *Denver Post*, October 5, 2004, B-2.

"Ice a Scarce Commodity on Arctic Rinks: Global Warming Blamed for Shortened Hockey Season." *Financial Post* (Canada), January 7, 2003, A-3.

Jeffries, M. O. "Ellesmere Island Ice Shelves and Ice Islands." *Satellite Image Atlas of Glaciers of the World: Glaciers of North America—Glaciers of Canada*, J147-J164. 2002.

Joling, Dan. "Walruses Abandon Ice for Alaska Shore." Associated Press in *Washington Post*, October 4, 2007a. http://www.washingtonpost.com/wp-dyn/content/article/2007/10/04/AR2007100402299_pf.html.

Joling, Dan. "Thousands of Pacific Walruses Die; Global Warming Blamed." Associated Press, December 14, 2007b. (LEXIS)

Kaufman, Marc. "Perennial Arctic Ice Cover Diminishing, Officials Say." *Washington Post*, March 19, 2008c, A-3. http://www.washingtonpost.com/wp-dyn/content/article/2008/03/18/AR2008031802903_pf.html.

Kelly, Mick. "Halting Global Warming." In *Global Warming: The Greenpeace Report*, ed. Jeremy Leggett, 83-112. New York: Oxford University Press, 1990.

Kerr, Richard A. "Is Battered Arctic Sea Ice Down For the Count?" *Science* 318 (October 5, 2007d):33–34.

Kerr, Richard A. "Climate Tipping Points Come In from the Cold." *Science* 319 (January 11, 2008b):153.

Kolbert, Elizabeth. "Testing the Climate." (Talk of the Town) *The New Yorker*, December 24 and 31, 2007d, 43–44.

"Melting Faster." Editors' Choice. *Science* 316, no. 1 (May 18, 2007):955.

Overpeck, J. T., M. Strum, J. A. Francis, D. K. Perovich, M. C. Serreze, R. Benner, E. C. Carmack. F. S. Chapin III, S. C. Gerlach, L. C. Hamilton, L. D. Hinzman, M. Holland, H. P. Huntington, J. R. Key, A. H. Lloyd, G. M. MacDonald, J. McFadden, D. Noone, T. D. Prowse, P. Schlosser, and C. Vorosmarty. "Arctic System on Trajectory to New, Seasonally Ice-Free State." *EOS: Transactions of the American Geophysical Union* 86, no. 34 (August 23, 2005):309, 312.

Ray, G. Carlston. "Dispatches from the Bering Sea: The Icy Seascape." *University of Virginia Today*, May 11, 2007. http://www.virginia.edu/uvatoday/beringsea/bering-sea4.html.

Revkin, Andrew C. "New Climate Model Highlights Arctic's Vulnerability." *New York Times*, October 31, 2005b. http://www.nytimes.com/2005/10/31/science/earth/01warm_web.html.

Revkin, Andrew C. "Arctic Ice Shelf Broke Off Canadian Island." *New York Times*, December 30, 2006m. http://www.nytimes.com/2006/12/30/science/earth/30ice.html.

Revkin, Andrew. "Arctic Melt Unnerves the Experts." *New York Times*, October 2, 2007g. http://www.nytimes.com/2007/10/02/science/earth/02arct.html.

Revkin, Andrew C. "Grim Outlook for Polar Bears." *New York Times*, October 2, 2007i, D-4.

Revkin, Andrew C. "Arctic Update: Resilient Bears, Shrinking Ice." Dot.Earth, *New York Times* blog. December 12, 2007k. http://dotearth.blogs.nytimes.com/2007/12/12/arctic-update-resilient-bears-vanishing-ice/index.html.

Revkin, Andrew C. "Arctic Ocean Ice Retreats Less Than Last Year." *New York Times*, September 17, 2008i. http://www.nytimes.com/2008/09/17/science/earth/17ice.html.

"Ribbon Seal of the Bering Sea Losing Icy Habitat." Environment News Service, December 26, 2007. http://www.ens-newswire.com/ens/dec2007/2007-12-26-094.asp.

Schiermeier, Quirin. "On Thin Ice: The Arctic is the Bellwether of Climate Change." *Nature* 441 (May 10, 2007a):146–147.

Schmid, Randolph E. "State of the Arctic Warming, with Widespread Melting." Associated Press, November 16, 2006b. (LEXIS)

Serreze, M. C., M. M. Holland, and J. Stroeve. "Perspectives on the Arctic's Shrinking Sea-ice Cover." *Science* 315 (March 16, 2007):1533–1536.

"Warm Water Surging into Arctic Ocean." Environment News Service, September 27, 2006. http://www.ens-newswire.com/ens/sep2006/2006-09-27-01.asp.

"Winter and Summer, Arctic Sea Ice Is Shrinking." Environment News Service, November 29, 2006. http://www.ens-newswire.com/ens/nov2006/2006-11-29-09.asp.

Woodard, Colin. "Curbing Climate Change. Is the World Doing Enough?" *Congressional Quarterly Global Researcher* 1, no. 2 (February 2007):27–50. http://www.lib.iup.edu/depts/libsci/cfr/fieldnotes_files/curbing_climate_change.pdf or http://www.globalresearcher.com.

Wuethrich, Bernice. "Lack of Icebergs Another Sign of Global Warming?" *Science* 285 (July 2, 1999):37.

Zimov, Sergey A., Edward A. G. Schuur, and F. Stuart Chapin III. "Permafrost and the Global Carbon Budget." *Science* 312 (June 16, 2006):1612–1613.

Arctic Paleocene-Eocene Thermal Maximum

The first detailed analysis of the biological record from the seabed near the North Pole indicates that 55 million years ago the year-round average temperature at the Arctic Ocean was as warm as 74°F, much like the ocean off a beach in Florida in late spring. The warming, which scientists call the "Paleocene-Eocene Thermal Maximum," was massive, and sudden, by natural standards. Scientists have been studying this episode as a possible analogue to human-induced warming in our time. Very suddenly, the Arctic's ice melted during a spike in greenhouse-gas levels, as rapid changes in albedo (reflectivity) sped the change (Kintisch 2006a, 1095; Sluijs et al. 2006, 610).

"Something extra happens when you push the world into a warmer world, and we just don't understand what it is," said one lead author, Henk Brinkhuis, an expert on ancient Arctic ecology at the University of Utrecht in the Netherlands (Revkin 2006b). According to these investigations, the Arctic Ocean warmed abruptly and robustly 55 million years ago, during the Paleocene-Eocene Thermal Maximum. This warming may have been provoked by a large, sudden injection in to the atmosphere of methane and carbon dioxide. No one has yet found the "smoking gun," however. One expert, Richard B. Alley, a geoscientist at Pennsylvania State University, said, "The new research provides additional important evidence that greenhouse-gas changes controlled much of climate history, which strengthens the argument that greenhouse-gas changes are likely to control much of the climate future" (Revkin 2006b).

FURTHER READING

Kintisch, Eli. "Hot Times for the Cretaceous Oceans." *Science* 311 (February 24, 2006a):1095.

Revkin, Andrew C. "Studies Portray Tropical Arctic in Distant Past." *New York Times*, June 1, 2006b. http://www.nytimes.com/2006/06/01/science/earth/01climate.html.

Sluijs, Appy, Stefan Schouten, Mark Pagani, Martijn Woltering, Henk Brinkhuis, Jaap S. Sinninghe Damst, Gerald R. Dickens, Matthew Huber, Gert-Jan Reichart, Ruediger Stein, Jens Matthiessen, Lucas J. Lourens, Nikolai Pedentchouk, Jan Backman, Kathryn Moran, and the Expedition 302 Scientists. "Subtropical Arctic Ocean Temperatures during the Palaeocene/Eocene Thermal Maximum." *Nature* 441 (June 1, 2006):610–613.

Arctic Walker Swamped by Melting Ice

Temperatures in the Arctic rose so rapidly during the spring of 2004 that Briton Ben Saunders, 26, was forced to abandon his attempt to become the first person to trek solo across the North Pole from Russia to Canada. Saunders reached the North Pole May 11, but had to be rescued three days later about 50 kilometers on the Canadian side when open water stopped his skis in their tracks. He was airlifted to Ottawa. Saunders, from London, began his journey at Cape Artichevsky in Northern Siberia March 5. His journey ended 72 days and 965.22 kilometers later. "The weather this year was the warmest since they began keeping records," Saunders said. "And the sea ice coverage in the Arctic last year was the least ever, according to NASA. They haven't calculated it yet for this year, but we expect the sea ice coverage to be even lower" (Kristal-Schroder 2004, A-1).

Saunders said that some days of his trek the temperature rose to 15°C. as he skied without a hat or mittens. Constant thaw and freeze cycles made the ice especially hazardous. "I've never seen myself as an eco-warrior," Saunders wrote on his Web site a few days before reaching the Pole. "I've been wary of taking a stance on climate change, as I don't believe we know enough about what's going on, but it's obvious that things are changing fast. It's an issue I'll certainly be taking far more interest in" (Kristal-Schroder 2004, A-1).

FURTHER READING

Kristal-Schroder, Carrie. "British Adventurer's Polar Trek Foiled by Balmy Arctic." *Ottawa Citizen*, May 17, 2004, A-1.

Arctic Warming and Native Peoples

Global warming has become a fact of daily life for the Inuit in the Canadian Arctic. Welcome to the thawing ice-world of the third millennium. Around the Arctic, in Inuit villages now connected by the oral history of traveling hunters as well as e-mail, weather watchers are reporting striking evidence that global warming is an unmistakable reality. Weather reports from the Arctic sometimes read like the projections of the Intergovernmental Panel on Climate Change (IPCC) on fast forward. These personal stories support IPCC expectations that climate change will be felt most dramatically in the Arctic.

During the summer of 2004, several *Vespula intermedia* (yellow-jacket wasps) were sighted in Arctic Bay, a community of 700 people on the northern tip of Baffin Island, at more than 73°N latitude. Noire Ikalukjuaq, the mayor of Arctic Bay, photographed one of the wasps at the end of August. Ikalukjuaq, who said he knew no word in Inuktitut (the Inuits' language) for the insect, reported that other people in the community also had seen wasps at about the same time ("Rare Sighting" 2004).

On Earth Day 2000, the Inuit brought their accounts of dramatic warming to urban audiences in southeastern Canada. Inuit leader Rosemarie Kuptana told a press conference in Ottawa that experienced hunters have fallen through unusually thin ice. Three men had recently died this way. Never-before-seen species (including robins, barn swallows, beetles, and sand flies) have appeared on Banks Island, in the Arctic Ocean about 800 miles northeast of Fairbanks, Alaska.

Growing numbers of Inuit are suffering allergies from white-pine pollen that recently reached Kuptana's home in Sach's Harbour, on Banks Island, for the first time. "If this rate of change continues, our lifestyle may forever change, because our communities are sinking with melting permafrost and our food sources are ... more difficult to hunt," said Kuptana (Duffy 2000, A-13).

Sach's Harbour itself is slowly sinking during the summer into a muddy mass of thawing permafrost. Born in an igloo, Kuptana has been her family's weather watcher for much of her life (she was 54 years old in 2008). Her job was to scan the morning clouds and test the wind's direction to help the hunters decide whether to go out, and what everyone should wear. Now she gathers observations for international weather-monitoring organizations. During 2006, officials in Nunavut authorized the installation of air conditioners in official buildings for the first time because summertime temperatures in some southern Arctic villages have climbed into the 80°s F in recent years.

Sweating in Iqaluit

July 29, 2001, was a very warm day in Iqaluit, on southern Baffin Island, capital of semisovereign Inuit nation of Nunavut in the Canadian Arctic. The bizarre weather was the talk of the town. The urgency of global warming was on everyone's lips. Even as U.S. President George W. Bush faulted global warming for lacking "sound science," the temperature hit 25°C in a community that nudges the Arctic Circle, following a string of days that were nearly as warm. It was the warmest summer anyone in the area could remember. Travelers joked about having forgotten their shorts, sunscreen, and mosquito repellant—all now necessary equipment for a globally warmed Arctic summer.

In Iqaluit (pronounced ee-haloo-eet), a warm, desiccating northwesterly wind raised whitecaps on nearby Frobisher Bay and rustled carpets of purple Saxifrage flowers on bay-side bluffs, as people emerged from their overheated houses (which have been built to absorb every scrap of passive solar energy), swabbing ice cubes wrapped in hand towels across their foreheads. The high temperature, at 78°F, was 30°F above the July average of 48°F, comparable to a 110°F to 115°F day in New York City or Chicago. The wind raised eddies of dust on Iqaluit's gravel roads as residents swatted slow, corpulent mosquitoes.

In Iqaluit, people speak of a natural world that is being turned upside down in ways that might startle even George W. Bush. "We have never seen anything like this. It's scary, *very* scary" said Ben Kovic, Nunavut's chief wildlife manager. "It's not every summer that we run around in our T-shirts for weeks at a time" (Johansen 2001, 18). At 11:30 A.M. on a Saturday in late July, Kovic was sitting in his back yard, repairing his fishing boat, wearing a T-shirt and blue jeans in the warm wind, with many hours of Baffin's 18-hour July daylight remaining. On a nearby beach, Inuit children were building sandcastles with plastic shovels and buckets, occasionally dipping their toes in the still-frigid seawater.

Warming has extended to all seasons. In Iqaluit, for example, thunder and lightning used to be an extreme rarity. Thunderstorms are now much more common across the Arctic; the same day that the high hit 78°F in Iqaluit, the forecast for Yellowknife, in the Northwest Territories, called for a high of 85°F with scattered thunderstorms. During the summer of 2004, highs in the 80°s F became routine in Fairbanks, Alaska. The winters of 2000–2001 and 2001–2002 in Iqaluit were notable for liquid precipitation (freezing rain) in December. Tromso, in far northern Norway, had very little snow during the winter of 2001–2002, along with very mild weather, as warm as plus 15°C at one point shortly after Christmas. There was almost no snow at the time. Around the Tromso airport, which is near the ocean, residents could mow their lawns (Rudmin 2002). A few years later, thunderstorms bearing heavy rain were reported during brief temperature spikes even in February across Alaska's North Slope, over and near Iqaluit, and along the southwestern coast of Greenland.

Research supports ground-level observations of increasing Arctic storm frequency and intensity during the last half-century, attributing it to warming waters and changing atmospheric circulation patterns. The same trends play a role in provoking acceleration of arctic sea ice drift, which long has been considered by scientists as a bellwether of climate change, according to NASA researcher Sirpa Hakkinen of NASA's Goddard Space Flight Center in Greenbelt, Maryland, with colleagues Andrey Proshutinsky from the Woods Hole Oceanographic Institution in Woods Hole, Massachusetts, and Igor Ashik with the Arctic and Antarctic Research Institute in St. Petersburg, Russia. Paradoxically, even as temperatures warm, increasing turbulence in the Arctic Ocean could increase the area's capacity to absorb carbon dioxide, countering global warming to some extent (Hakkinen et al. 2008). According to Hakkinen,

> Gradually warming waters have driven storm tracks—the ocean paths in the Atlantic and Pacific along which most cyclones travel—northward. We speculate that sea ice serves as the "middleman" in a scenario where increased storm activity yields increased stirring winds that will speed up the Arctic's transition into a body of turbulently mixing warm and cool layers with greater potential for deep convection that will alter climate further. ("NASA Study" 2008)

Early in January 2004, Sheila Watt-Cloutier wrote from Iqaluit, on Baffin Island, that Frobisher Bay had just frozen over for the season at a record late date:

> We are finally into very "brrrrrr" seasonal weather and the Bay is finally freezing straight across. At Christmas time the Bay was still open and as a result of the floe edge being so close we had a family of Polar bears come to visit the town a couple of times. Also in Pangnirtung [north of Iqaluit] families from the outpost camps came into town for Christmas by boat! Imagine that the ice was not frozen at all there by Christmas. But this week our temperatures with the wind-chill reached −52°F so we are pleased. (Watt-Cloutier 2004a)

Watt-Cloutier looked out at the waters of Frobisher Bay two weeks after she had represented the Inuit Circumpolar Conference at a Conference of Parties to the United Nations Framework Convention on Climate Change in Milan, Italy, where she said, in part,

> Talk to hunters across the North and they will tell you the same story—the weather is increasingly unpredictable. The look and feel of the land is different. The sea-ice is changing. Hunters are having difficulty navigating and traveling safely. We have even lost experienced hunters through the ice in areas that, traditionally, were safe! Our Premier, Paul Okalik, lost his nephew when he was swept away by a torrent that used to be a small stream. The melting of our glaciers in summer is now such that it is dangerous for us to get to many of our traditional hunting and harvesting places.... Inuit hunters and elders have for years reported changes to the environment that are now supported by American, British and European computer models that conclude climate change is amplified in high latitudes. (Watt-Cloutier 2003a)

Snows in Iqaluit also now tend to be heavier and wetter than previously. Winter cold spells, which still occur, have generally become shorter, according to long-time residents of the area.

As a wildlife officer, Kovic sees changes that alarm him. Polar bears, for example, are often becoming shore dwellers rather than ice dwellers, sometimes with dire consequences for unwary tourists. The harbor ice at Iqaluit did not form in 2000 and 2001 until late November or December, five or six weeks later than usual. The ice also breaks up earlier in the spring—sometimes in May in places that once were icebound into early July. In Resolute Bay, bears attracted by the smell of seal meat have been known to chase children on their way to school. Some local Inuit hunters have turned profits in the thousands of dollars by leading tourists on hunts for land-bound, hungry bears that local people have come to regard as a dangerous nuisance.

Daily Life in Today's Arctic

The Arctic's rapid thaw has made hunting, never a safe or easy way of life, even more difficult and dangerous. Hunters in and around Iqaluit say that the weather has been seriously out of whack since roughly the middle 1990s. Simon Nattaq, an Inuit hunter, fell through unusually thin ice and became mired in icy water long enough to lose both legs to hypothermia, one of several injuries and deaths reported around the Arctic recently due to thinning ice.

Pitseolak Alainga, another Iqaluit-based hunter, said that climate change compels caution. One must never hunt alone, he said (Nattaq had been hunting by himself). Before venturing onto ice in fall or spring hunters should test its stability with a harpoon, he said. Such tests are especially important when a hunter travels in a snowmobile. In older times, sled dogs could sense thin ice, something a machine cannot do.

Alainga knows the value of safety on the water. His father and five other men died during late October 1994, after an unexpected late-October ice storm swamped their hunting boat. The younger Alainga and one other companion barely escaped death in the same storm. He believes that more hunters are suffering injuries not only because of climate change, but also because basic survival skills are not being passed from generation to generation as in years past, when most Inuit lived off the land.

Urbanization has caused many Inuit to lose their cultural bearings as well as their hunting skills. Less than 50 years ago, suicide was virtually unknown among the Inuit of Canada's Arctic. Now they are killing themselves at a rate six times the Canadian national average. Within two or three generations, many Inuit have become urbanized as the tendrils of industrial life infiltrate the Arctic. Where visitors once arrived by dog sled or sailing ship, they now stream into Iqaluit's small but very busy airport on Boeing 727s in which half the passenger cabin has been sequestered for freight. With no land-surface connections to the outside, freight as large as automobiles is sometimes shipped to Iqaluit by air.

The population of Iqaluit jumped from about 3,500 to 6,500 in less than three years between 1998 and 2001. Substantial suburban-style houses with mortgages worth hundreds of thousands of dollars sprang up around town, rising on stakes sunk into the permafrost and granite hillsides. In other areas, ranks of walk-up apartments marched along the high ridges above Frobisher Bay. Every ounce of building material has been imported from thousands of miles away. Inuit have been moving into the town from the backcountry. The town is awash in a sea of children.

People in Iqaluit subscribe to the same cable television services (as well as Internet) available in "the South." Bart Simpson and Tom Brokaw are well known in Iqaluit, where some homes have sprouted satellite dishes. Iqaluit now hosts a large supermarket of a size that matches stores in larger urban areas, except that the prices are three to four times higher than in Ottawa or Omaha. If one can afford the bill, however, orange-mango-grapefruit juice and ready-cooked Buffalo wings, as well as many other items of standard "southern" fare, are readily available.

Climate change has been rapid, and easily detectable within a single human lifetime. "When I was a child," said Watt-Cloutier, "We never swam in the river [Kuujjuaq] where I was born [in Nunavik, Northern Quebec]. Now kids swim there all the time in the summer." Watt-Cloutier does not remember even having worn short pants as a child (Johansen 2001, 19).

Some of the rivers to which Arctic char return for spawning have dried up, according to Kovic; the summer of 2001 brought drought as well as record warmth to Iqaluit and its hinterland. Flying above glaciers in the area, Kovic has noticed that their coloration was changing. "The glaciers are turning brown," he said, speculating that melting ice may be exposing debris, and that air pollution from southern latitudes may be a factor. Some ringed seals with little or no hair have been caught, said Kovic. Asked why seals have been losing their hair, Kovic answered, "That is a big question that someone has to answer" (Johansen 2001, 19).

Climate change has become a live issue in Inuit politics. At a press conference that was televised across Canada August 1, 2002, for example, Nunavut Premier Paul Okalik confronted Alberta Premier Ralph Klein after Klein warned that the terms of the Kyoto Protocol could reduce oil-rich Alberta's equalization contributions to poorer regions of Canada. "You can keep your money," Okalik told Klein, as he said that global warming presents a direct threat to the Inuit way of life. Alberta's government has complained that compliance with the protocol could throw thousands of Albertans out of work and take billions of dollars out his province's economy.

An Inuit Early Warning

Addressing a Senate Commerce Committee hearing on global warming August 15, 2004, Watt-Cloutier said, "The Earth is literally melting. If we can reverse the emissions of greenhouse gases in time to save the Arctic, then we can spare untold suffering." She continued, "Protect the Arctic and you will save the planet. Use us as your early-warning system. Use the Inuit story as a vehicle to reconnect us all so that we can understand the people and the planet are one" (Pegg 2004).

The Inuits' ancient connection to their hunting culture may disappear within her grandson's lifetime, Watt-Cloutier said. "My Arctic homeland is now the health barometer for the planet" (Pegg 2004). Committee chair John McCain, an Arizona Republican (and later a Republican candidate for U.S. president), said a recent trip to the Arctic showed him that "these impacts are real and consistent with earlier scientific projects that the Arctic region would experience the impacts of climate change at a faster rate than the rest of the world. We are the first generation to influence the climate and the last generation to escape the consequences" (Pegg 2004).

Alaskan natives established a Web site (www.nativeknowledge.org) to share their experiences with climate change:

> Turtles appearing for the first time on Kodiak Island, birds starving on St. Lawrence Island, thunder first heard on Little Diomede Island ... snowmobiles falling through the ice in Nenana.... Already the central Arctic is warming 10 times as fast as the rest of the planet, outpacing even our attempts to describe it. (Frey 2002, 26)

In the Eskimo village of Kaktovik, Alaska, on the Arctic Ocean roughly 250 miles north of the Arctic Circle, a robin built a nest in town during 2003—not an unusual event in more temperate latitudes, but quite a departure from the usual in a place where, in the Inupiat Eskimo language, no name exists for robins. In the Okpilak River Valley, which has heretofore been too cold and dry for willows, they are sprouting profusely. Never mind the fact that in the Inupiat language "Okpilak" means "river with no willows" (Kristof 2003). Three kinds of salmon have been caught in nearby waters in places where they were once unknown.

Correspondent Jerry Bowen, airing on the *CBS Morning News* on August 29, 2002, from Barrow, Alaska, the northernmost town in Alaska, quoted Simeon Patkotak, a native elder, as saying residents there had just witnessed their first mosquitoes. Ice cellars carved out of permafrost were melting as well, forcing local native people to borrow space in electric freezers for the first time to store whale meat. Average temperatures in Barrow have risen 4°F during the past 30 years (Bowen 2002). The average date at which the last snow melts at Barrow in the spring or summer has receded about 40 days between 1940 and the years after 2000, from early July to, some years, as early as mid or late May (Wohlforth 2004, 27).

Nick Jans, writing in *The Last Polar Bear*, sketched the ground-level effects of climate change in Barrow:

> A sharp, insistent wind sifts a rolling haze of snow along the ground, but for mid-December the temperature is positively balmy: plus 19 degrees Fahrenheit. A generation ago, 20 or even 30 below zero would have been the norm. The last few days amount to more than some freakish warm spell; this year, the sea ice has yet to become shorefast. Not so many years ago, solid freeze-up occurred by the end of October, without fail. Now it's six, seven, even ten weeks later, and the ice proves thinner and less stable. The elders shake their heads, bemused as if palm trees had sprouted on the shores of the Beaufort Sea. They're beyond debating climate change. It's staring them right in the face. (Jans 2008, 150, 153)

In the Canadian Inuit town of Inuvik, ninety miles south of the Arctic Ocean, the temperature rose to 91°F on June 18, 1999, a type of weather unknown to living memory in the area. "We were down to our T-shirts and hoping for a breeze," said Richard Binder, 50, a local whaler and hunter. Along the MacKenzie River, according to Binder, "Hillsides have moved even though you've got trees on them. The thaw is going deeper because of the higher temperatures and longer periods of exposure." In some places near Binder's village, the thawing earth has exposed ancestral graves, and remains have been reburied (Sudetic 1999, 106). Some hunters say that seals have moved farther north, killer whales are eating sea otters, and beaver are proliferating, something that would not happen if rivers and ponds were freezing to usual depths.

Coastal hunters above the Arctic Circle in Alaska say they are definitely seeing a trend: the ice regularly comes a month later than it did 20 years ago, and roughly two months later than 30 years ago. Ice also breaks up earlier than previously, so hunting seasons are becoming shorter.

Northwest Passage Nearly Open

During late August 2007, for the first time, a Northwest Passage from Baffin Bay to Northern Alaska opened during a season of record ice melt for the Arctic Ice Cap. European mariners have been seeking and failing to find such a route since 1497, when English King Henry VII sent Italian explorer John Cabot to look for a route from Europe to Asia that would avoid the southern tip of Africa. Many explorers failed at the task, including Sir Francis Drake and Captain James Cook. NASA's Advanced Microwave Scanning Radiometer aboard the *Aqua* satellite observed open water along nearly the entire route on August 22, 2007. "Although nearly open, the Northwest Passage was not necessarily easy to navigate in August 2007," the NASA report noted. "Located 800 kilometers (500 miles) north of the Arctic Circle and less than 1,930 kilometers (1,200 miles) from the North Pole, this sea route remains a significant challenge, best met with a strong icebreaker ship backed by a good insurance policy" ("Northwest Passage" 2007).

FURTHER READING

Bowen, Jerry. "Dramatic Climate Change in Alaska." CBS News Transcripts; *CBS Morning News,* 6:30 A.M. Eastern Daylight Time, August 29, 2002. (LEXIS)

Boyd, Robert S. "Earth Warming Could Open up a Northwest Passage." Knight-Ridder Newspapers in *Pittsburgh Post-Gazette,* November 11, 2002, A-1.

Duffy, Andrew. "Global Warming Why Arctic Town Is Sinking: Permafrost Is Melting under Sachs: Inuit Leader." *Montreal Gazette,* April 18, 2000, A-13.

"Eminent Scientists Warn of Disastrous, Permanent Global Warming." Environment News Service, February 19, 2007. http://www.ens-newswire.com/ens/feb2007/2007-02-19-03.asp.

Frey, Darcy, "George Divoky's Planet." *New York Times Sunday Magazine,* January 6, 2002, 26–30.

Hakkinen, Sirpa, Andrey Proshutinsky, and Igor Ashik. "Sea Ice Drift in the Arctic Since the 1950s." *Geophysical Research Letters* 35 (2008), L19704, doi:10.1029/2008GL034791.

Jans, Nick. "Living with Oil: The Real Price." In *The Last Polar Bear: Facing the Truth of a Warming World,* ed. Steven Kazlowski, 147–161. Seattle: Braided River, 2008.

Johansen, Bruce E. "Arctic Heat Wave." *The Progressive,* October 2001, 18–20.

Kristof, Nicholas D. "Baked Alaska on the Menu?" *New York Times* in *Alameda Times-Star* (California), September 14, 2003, n.p. (LEXIS)

"NASA Study Finds Rising Arctic Storm Activity Sways Sea Ice, Climate." NASA Earth Observatory. October 6, 2008. http://earthobservatory.nasa.gov/Newsroom/NasaNews/2008/2008100627645.html.

"Northwest Passage Nearly Open." NASA Earth Observatory, August 27, 2007. http://earthobservatory.nasa.gov/Newsroom/NewImages/images.php3?img_id=17752.

Pegg, J. R. "The Earth Is Melting, Arctic Native Leader Warns." Environment News Service, September 16, 2004.

"Rare Sighting of Wasp North of Arctic Circle Puzzles Residents." Canadian Broadcasting Corporation, September 9, 2004. http://www.cbc.ca/story/science/national/2004/09/09/wasp040909.html.

Rudmin, Floyd. Personal Communication from Tromso, Norway, January 17, 2002.

Sudetic, Chuck. "As the World Burns." *Rolling Stone,* September 2, 1999, 97–106, 129.

Watt-Cloutier, Sheila. Speech to Conference of Parties to the United Nations Framework Convention on Climate Change, Milan, Italy, December 10, 2003a.

Watt-Cloutier, Sheila. Personal Communication, from Iqaluit, Nunavut, January 4, 2004a.

Wohlforth, Charles. *The Whale and the Supercompter: On the Northern Front of Climate Change.* New York: North Point Press/Farrar, Strauss and Giroux, 2004.

Armadillos Spread Northward

The armadillo (Spanish for "little armored thing"), a subtropical animal with a hard shell, can be used as an indicator of climate change. The first nine-banded armadillo (the only species in the United States), which had migrated from South America over a very long period of time, was sighted in the United States during 1849. It migrated from Mexico into Texas, where armadillos were sighted in the 1850s. A century ago, the armadillo's range was restricted mainly to Mexico, southern Texas, and parts of deserts in Arizona and New Mexico. By late in the twentieth century, however, armadillos were sighted in Oklahoma, Arkansas, and Georgia. The armadillo was chosen as Texas' official state small mammal. During the Great Depression, they were known as "Hoover Hogs" by down-on-their-luck Americans who caught and ate them instead of the "chicken in every pot" that President Herbert Hoover had promised as president (Suhr 2006).

After that, armadillos' accustomed range expanded rapidly, as motorists were reported to have collided with them (they seem to have very little street sense) as far north as southern North Carolina, Kentucky, Illinois, and near Lincoln, Nebraska, where one was reported during August, 2005, licking up bugs on U.S. Route 50. "Over the last seven or eight years, we've been getting more regular reports of armadillos," said Mike Fritz, a natural heritage zoologist with the Nebraska Game and Parks Commission (Laukaitis 2005, A-1).

The nine-banded armadillo is expanding its range 10 times faster than the average rate expected for a mammal, according to Dr. Joshua Nixon, a Michigan State researcher who runs a Web site, "Armadillo Online!" (www.msu.edu/~nixonjos/armadillo). Contrary to popular assumption, the nine-banded armadillo cannot roll into a ball when threatened, but they are good swimmers, able to ford rivers by puffing air into their lungs and forging ahead, dog-paddle style. They have been discovered riding hobo-style on freight trains. Armadillos' range is restricted by temperature because they have little body fat and eat insects, and do not hibernate or store food. They can survive mild freezes in burrows or under buildings, however. Armadillos live in small colonies and give birth to identical quadruplets (Crable 2006, C-5).

Jackson County animal-control chief Lloyd Nelson said in 2006 that he had logged 13 sightings since 2003 in his southern Illinois county alone. "All the evidence, the sightings and the number of road kills would indicate that their numbers are increasing," said Clay Nielsen, a wildlife ecologist at Southern Illinois University in Carbondale. In Illinois in recent years, "there's been quite a spurt in sightings" of the nocturnal animal. A few have been sighted on the southern suburbs of Chicago. Some may be released by people who tired of having them as pets, but others seem to have migrated northward themselves (Suhr 2006).

The *Houston Chronicle* remarked that the Texas state small mammal has become a Yankee. "During periods of warm winters, they'll disperse north during the summers and won't die off in the winter, so the next summer, they'll disperse a little further," said Duane Schlitter, program leader for nongame species and rare and endangered species at the Texas Parks and Wildlife Department (Kever 2006, 1).

FURTHER READING

Crable, Ad. "For Armadillos, It's Pa. [Pennsylvania] or Bust." *Lancaster New Era* (Pennsylvania), September 5, 2006, C-5.

Kever, Jeannie. "Cuddly Critters: They're Makin' Tracks; The Armadillo, That Official Small Mammal of Texas, Is Waddling Its Way North—Maybe the Grass Really Is Greener There." *Houston Chronicle*, September 4, 2006, 1.

Laukaitis, Algis J. "Talmage Man: No Kiddo ... It's an Armadillo!" *Lincoln Journal Star*, August 18, 2005, A-1.

Suhr, Jim. "Armadillos Making Northern March." Associated Press Illinois State Wire, September 16, 2006. (LEXIS)

Arrhenius, Savante August (1859–1927)

In 1896, Savante August Arrhenius, a Swedish chemist, published a paper in the *London, Edinburgh, and Dublin Philosophical Magazine and Journal of Science* titled "On the Influence of Carbonic Acid in the Air upon the Temperature of the Ground" (Arrhenius 1896). In his paper, Arrhenius theorized that a rise in the atmospheric level of carbon dioxide could raise the temperature of the air. He was not the only person thinking along these lines at the time; Swedish geologist Arvid Hogbom had delivered a lecture on the same idea three years earlier, which Arrhenius incorporated into his article.

Arrhenius was a well-known scientist in his own time, not for his theories describing the greenhouse effect, but for his work in electrical conductivity, for which he was awarded a Nobel Prize in 1903. Later in his life, Arrhenius directed the Nobel Institute in Stockholm. His work in global-warming theory was not much discussed during his own life. Arrhenius, using the available measurements of absorption and transmission by water vapor and carbon dioxide,

Savante Arrhenius (1859–1927). (Library of Congress)

developed the first quantitative mathematical model of the Earth's greenhouse effect and obtained results of acceptable accuracy by today's standards for equilibrium climate sensitivity to carbon-dioxide changes.

Arrhenius developed his theory through the use of equations, by which he calculated that a doubling of carbon dioxide in the atmosphere would raise air temperatures about 10°F. Arrhenius thought 3,000 years would have to pass before human-generated carbon-dioxide levels would double, a miscalculation that shares something with Benjamin Franklin's belief, late in the eighteenth century, that European-American expansion across North America would take a thousand years.

Arrhenius applauded the possibility of global warming, telling audiences that a warmer world "would allow all our descendants, even if they only be those of a distant future, to live under a warmer sky and in a less harsh environment than we were granted" (Christianson 1999, 115). In 1908, in his book *Worlds in the Making*, Arrhenius wrote:

> By the influence of the increasing percentage of carbonic acid in the atmosphere, we may hope to enjoy ages with more equable and better climates, especially as regards the colder regions of the Earth, ages when the Earth will bring forth much more abundant crops than at present for the benefit of rapidly propagating mankind. (Christianson 1999, 115)

FURTHER READING

Arrhenius, Savante. "On the Influence of Carbonic Acid in the Air Upon the Temperature of the Ground." *London, Edinburgh, and Dublin Philosophical Magazine and Journal of Science*, 5th series (April 1896):237–276.

Christianson, Gale E. *Greenhouse: The 200-Year Story of Global Warming.* New York: Walker and Company, 1999.

Oppenheimer, Michael, and Robert H. Boyle: *Dead Heat: The Race against the Greenhouse Effect.* New York: Basic Books, 1990.

Asthma

Scientific evidence has been accumulating that warming temperatures "turbo-charge" the growth of plants that release allergy-causing pollens, as well as asthma, which has been on the rise in many countries. Starting during 2001, Lewis Ziska, a plant physiologist with the U.S. Department of Agriculture (Crop Systems and Global Change Laboratory at Beltsville, Maryland) planted ragweed (a major cause of hay fever during the autumn) in urban, suburban, and rural sites in and near Baltimore. He used the same seeds, and watered the plants the same way, using the same soil. Soon, the urban plants were growing much faster than the rural crop, and producing five times as much pollen (Naik 2007a, A-1).

The urban ragweed's pollen was more toxic as well, all fed by warmer temperatures and 20 percent more carbon dioxide, mainly from auto traffic. Ziska himself has allergies and asthma. He found that ragweed loves warmth. The urban plants' seeds emerged three to four days earlier at the urban site than the rural one, 40 miles away, and the urban plants were almost twice as large at the urban site as those at the rural location (Naik 2004a, A-13). The allergen that causes reactions in humans is also more potent in the city-grown plants than the rural ones. He has found that temperatures between the two sites can vary by 5° to 10°F and when the carbon-dioxide level is less than 400 parts per million (ppm) at the rural site it can be more than 470 ppm in the city.

Allergies and asthma are closely linked; more than 70 percent of people with asthma also have allergies (Naik 2007a, A-1). In its 2007 report, the Intergovernmental Panel on Climate Change (IPCC) says that spring, the most intense season for allergies, now arrives 10 to 15 days earlier than in 1975 (Naik 2007a, A-1). By 2007, about 35 million people in the United States suffered from allergies, and 20 million from asthma; the percentage of people in the United States with asthma has doubled since 1980 (Naik 2007a, A-13).

A Harvard study released in 2004 said that the asthma among preschool children rose 160 percent between 1980 and 1994, double the rate in the overall population. The study linked the rise in asthma, called "an epidemic in inner-city youth," to increases in pollen and fungal growth, both spurred by a warmer, and often wetter, climate (Naik 2007a, A-13). Dust mites, mold, and obesity also may play a role. A lack of infectious agents in childhood (which stimulate the immune system) also may be a factor. "I do think that climate change contributes," said Christopher Randolph, an allergist and clinical professor at Yale University, "but it's not the only answer" (Naik 2007a, A-13).

Ragweed, which is well known to allergy sufferers in the United States, was nearly unknown in Europe until the 1990s, where it has since spread through eastern Europe, as well as France, Italy, the Netherlands, and parts of the Nordic countries. Harvard's Paul Epstein, associate director of the Center for Health and the Global Environment, says that doubling the carbon dioxide level in a test plot of ragweed increases its pollen production 61 percent (Naik 2007a, A-13). *See also*: Hay Fever

FURTHER READING

Naik, Gautam. "Global Warming May Be Spurring Allergy, Asthma." *Wall Street Journal*, May 3, 2007a, A-1, A-13.

Australia: Heat, Drought, and Wildfires

The only major industrial country except the United States that long refused to ratify the Kyoto Protocol, the conservative government in Australia under Prime Minister John Howard was slow to recognize the economic perils of global warming. Largely because 85 percent of its electricity is generated from coal and because it hosts a large number of energy-intensive industries, such as aluminum smelting and steel manufacturing, Australia generated more greenhouse gases per capita in 2004 than any other country, 26.3 tons, except Luxembourg, at 28.1 tons (the United States comparable figure is 24.1 tons) (Wiseman 2007, 10-A).

The Howard government's inability to realize the role of climate change in the devastation of Australia's economy probably played a major role in the country's 2007 elections, when the government was replaced by Kevin Rudd of the Labor Party, who signed the Kyoto Protocol as his first official act after he was sworn into office December 3, 2007. Howard had been quite direct with his disregard of global warming; during September 2006, Al Gore had come to the country, and the prime minister refused to meet with him.

By the time Howard faced the voters, many Australian urban areas were facing severe water shortages. Inflows behind Sydney's dams in 1991–2006 were 71 percent less than their averages from 1948 to 1990 (Pincock 2007, 336). By 2006, polls in Australia indicated that a large proportion of its citizens ranked a dysfunctional climate as the number-one danger in their lives, ahead of such right-wing mainstays as international terrorism. Howard had been reelected three times as the drought intensified (1998, 2001, and 2004), but by 2007, his string of luck had run out, despite his pledge in June 2007 to implement a cap-and-trade emissions trading scheme.

While northern, subtropical Australia, which usually has high rainfall, has been getting wetter, the more arid south and west have been drier, for a combination of reasons: more frequent El Niños, which bring low rainfall and usually peak in the summer; higher temperatures, which increase evaporation of what little rain does fall; an active Indian Ocean Dipole, which reduces crucial spring rains in the agricultural areas of southeastern Australia; and changes in the Southern Annular Mode, which have been preventing rain-bearing storms from reaching southern parts of the continent in winter (Nowak 2007, 10). These last effects may be part of reorganization of atmospheric circulation—Hadley cells—that some attribute to global warming, which also may be intensifying drought in the western United States, southwestern China, and parts of Saharan Africa, notably Darfur.

A huge dust cloud rolls over the Australian town of Griffith, 400 kilometers (248 miles) southwest of Sydney, November 29, 2002, after strong winds whipped up topsoil dried from a prolonged drought. (AP/Wide World Photos)

Multiyear Reservoirs Run Dry

During the 1950s and 1960s, Australia built multiyear reservoirs that were supposed to protect against recurring droughts. These gave the country the highest storage capacity per capita in the world. Together with hundreds of miles of irrigation conduits, Australia was said, at the time, to be "drought-proof." Melbourne's water storage was 28 percent of capacity by mid-2007; Sydney's 37 percent, and Perth's, 15 percent. In May, 30-year rains hit the Hunter Valley north of Sydney, but did little to help. The land was so dry that the brief torrential rains were quickly absorbed (Nowak 2007, 10).

Six years of drought aggravated by warming temperatures from 2002 to 2008 reduced Australia's rice crop 98 percent, playing a major role in rapidly rising worldwide prices, including a doubling in three months during the first half of 2008. The price rise contributed to food riots as far away as Haiti. Rice shortages "spurred panicked hoarding in Hong Kong and the Philippines, and set off violent protests in countries including Cameroon, Egypt, Ethiopia, Haiti, Indonesia, Italy, Ivory Coast, Mauritania, the Philippines, Thailand, Uzbekistan and Yemen" (Bradsher 2008a). Police suppressed violent demonstrations in Dakar, Senegal, on March 30.

"Climate change is potentially the biggest risk to Australian agriculture," said Ben Fargher, chief executive of the National Farmers' Federation in Australia (Bradsher 2008a). Many Australian farmers have switched from rice to wine grapes, which require much less water and produce a pretax profit of about $2,000 an acre, compared with rice at about $240 an acre. In the meantime, strains of rice are being developed that bloom during cooler early hours of the day. Rice that blooms late in the day is vulnerable to warming.

With irrigation, cultivation of cotton prospered in Australia's Outback. By 2007, however, with Australia suffering record heat and its worst drought on record, production of cotton, with its thirst for water, had fallen sharply. Patrick Barta wrote in the *Wall Street Journal* that in the town of Wee Waa, Australia's self-described "cotton capital" (population 2,000) about 250 miles northwest of Sydney: "The Cotton Fields Motel that once was busy with seasonal workers now struggles to fill rooms. Elsewhere in the flat basin [of Australia's largest river system, the Murray-Darling], kangaroos hop along dry levees, and the end of giant water-transport pipes poke out over empty reservoirs" (Barta 2007b, A-1).

Australia's cotton production fell by two-thirds between 2001 and 2007 (Barta 2007b, A-12). Some fields are down to 2 percent of usual production, and reservoirs as large as 240,000 acres have dried up completely. In 2008, the drought in the Murray-Darling River Basin continued, following a record-dry June, even after parts of Australia received occasional torrential rains during a strong La Niña that briefly countered some effects of the drought.

Desalination Plants Built

Perth's first desalination plant was completed in 2006, with a wind farm meant to provide the 24 megawatts required to operate it. Perth now draws 17 percent of its water from that plant. Sydney and Melbourne are now building desalination plants. Some industries, such as BHP Billiton's copper and uranium mines in the south, also may build their own plants. These plants take a great deal of energy and leave behind a salty mush that will render whatever land or water is used for disposal useless for other purposes. Brisbane's government is considering recycling sewage after its main water supply runs dry, probably in 2009 (Nowak 2007, 11).

Desalination plants are being built not only in Australia, but also in other affluent countries that can afford the upfront costs. After construction of the plant, water costs only about $3.50 per 1,000 gallons. Such plants are common in the Middle East, and southern California has some smaller ones. The desalination plant in Perth is powered by wind from 48 turbines in the Emu Downs Wind Farm, about 100 miles to the north, which can produce as much as 80 megawatts of electricity a day, three times the plant's requirements. Similar plants are proposed for Sydney and the town of Tugun in Queensland (Mydans 2007).

During the summer of 2002–2003, wildfires pushed by raging hot, dry winds from Australia's interior seared parts of Canberra, charring hundreds of homes, killing four people, and forcing thousands to flee the area. "I have seen a lot of bush fire scenes in Australia … but this is by far the worst," said Prime Minister John Howard said ("Australia Assesses" 2003, 6-A). Flames spread through undergrowth and exploded as they hit oil-filled eucalyptus trees. The 2002–2003 drought knocked 1 percent off Australia's gross domestic product and cost $6.8 billion in exports. It reduced the size of Australia's cattle herd by 5 percent and its sheep flock by 10 percent (Macken 2004, 61). Recovery was impeded during 2004 by continuing drought.

The Australian Green Party urged the country's federal government to expand an inquiry into the devastating bushfires to consider global warming. Bob Brown, a Green Party Australian senator, said that new research indicates record daytime temperatures and unprecedented rates of water evaporation made the current drought the worst on record. The extreme dryness of vegetation arising from global warming was what made this fire season so devastating, he said ("Greens Want" 2003).

Even as drought aggravated wildfires in Australia, some parts of the country were dealing with deluges. Speaking in Sydney, Tomihiro Taniguchi, vice chairman of the Intergovernmental Panel on Climate Change (IPCC), said that recent Australian flooding and similar weather conditions around the world were further evidence that the impact of global warming was beginning to be felt. While he admitted it was difficult to directly link deluges in New South Wales to global warming, he added, "We can say with high confidence that the likelihood of flooding will increase in the future" (Maynard 2001, 17). The higher temperatures create greater seawater evaporation, which in turn produces more precipitation in coastal areas.

Vulnerable Species

Australia's Queensland state may lose half of its wet tropical highland rainforest, including many of its rarest animals because of global warming. The state's emblem, the koala, is at risk because of rising carbon-dioxide levels, which could strip the gum leaf (the koala's principal food) of its nutritional value. The same goes for many other vulnerable species. A report, released February 4, 2002, by Climate Action Network Australia, "represents one of the most comprehensive pictures yet of the local ecological effects of global warming" (Ryan 2002, 1). More than half of Australia's eucalyptus species are unlikely to survive a 3°C average temperature rise, according to the report. In addition, higher carbon-dioxide levels are expected to reduce carbohydrates and nitrogen in leaves, undermining food supplies for animals such as koalas (Ryan 2002, 1).

Ian Hume, professor emeritus of biology at Sydney University, said that increasing levels of carbon dioxide increase levels of "anti-nutrients" in the leaves, making the koalas' sole source of food toxic to them. Hume expects such problems to interfere with koala reproduction and reduce their populations substantially within 50 years. As it is, eucalyptus leaves contain very little energy; koalas cope by sleeping as much as 20 hours a day to conserve energy. Populations already have been reduced in some parts of Australia by habitat loss, including the spread of farms and suburbs ("Scientist: Warming" 2008, 7-A).

The study reviewed possible damage to Queenland's tropical areas, asserting that "90 Australian animal species, including a third of those on the endangered list, also [are] likely to suffer in the hotter, more extreme climate forecast this century" (Ryan 2002, 1). David Hilbert, principal research scientist at the Commonwealth Scientific and Industrial Research Organisation (CSIRO) Tropical Forest Research Center, said even a 1°C temperature rise would devastate half of the rainforests of north Queensland's wet tropics.

"CSIRO atmospheric scientists are now predicting 2° to 5°C of warming by the end of the century, so a 1°C change is liable to happen within 30 to 50 years," Hilbert said (Ryan 2002, 1). The Murray-Darling basin, for example, faces a 12 to 35 percent reduction in average flow by 2050 as southern regions grow hotter and drier, adding to the pressures on its overextracted and salt-affected rivers (Ryan 2002, 1). The Great Barrier Reef also runs a risk of heat-related coral bleaching, which affected 16 percent of the world's reefs in 1998, on top of existing threats such as land-based runoff (Ryan 2002, 1). In addition, Australia's alpine ecosystems are expected to shrink to six high mountaintops with continued warming, with any remaining snowfields probably disappearing within 100 years, the report asserted.

In the meantime, ring-tailed possums were falling dead out of trees in Australia's far north Queensland because the climate has become too hot for them. Green ring-tailed possums cannot survive more than five hours at an air temperature above 30°C. The ring-tailed possums are not alone. Australia's rainforest Cooperative Research Centre anticipates that half of all the unique mammals, reptiles, and birds in far north Queensland's rainforests could become extinct, given a 3.5°C rise in temperature, at the midpoint of the IPCC's anticipated range for the end of the twenty-first century. Most of these animals are found only in the Wet Tropics World Heritage Area near Cairns, at elevations above 600 meters. "It potentially takes just one day of six or seven hours of extreme heat.... and you find dozens of dead possums on the ground," said Steve Williams, a tropical biologist at Australia's James Cook University (Braasch 2007, 83).

Drought to Become Chronic?

Drought may become chronic in Australia if modeling done by CSIRO proves accurate. According to Kevin Hennessey, of the agency's Atmospheric Research Climate Impact Group,

> There is a consistency between our modeling and the reality of Australia's weather. Our modeling suggests Australia will become warmer and drier in the future as a result of global warming. By 2030 most of Australia will be between 0.5 and 2° warmer, and potentially 10 percent drier. (Macken 2004, 61)

Anecdotal evidence of warmth and drought in Australian is supported by scientific study. Neville Nicholls, of Australia's Bureau of Meteorology Research Centre (Melbourne), wrote in *Climatic Change*:

> Rainfall over nearly all of Australia during the cooler half of the year (May–October) was well below average in 2002. Mean maximum temperatures were very high during this period, as was evaporation. This would suggest that drought conditions (precipitation minus evaporation) were worse than in previous recent periods with similarly low rainfall (1982, 1994). Mean minimum temperatures were also much higher during the 2002 drought than in the 1982 and 1994 droughts. The relatively warm temperatures in 2002 were partly the result of a continued warming evident in Australia since the middle of the twentieth century. The possibility that the enhanced greenhouse effect is increasing the severity of Australian droughts, by raising temperatures and hence increasing evaporation, even if rainfall does not decrease, needs to be considered. (Nicholls 2004, 323)

Trends since issuance of the report have borne out its warnings.

FURTHER READING

"Australia Assesses Fire Damage in Capital." Associated Press in *Omaha World-Herald*, January 20, 2003, A-4.

Barta, Patrick. "Parched Outback: In Australia, a Drought Spurs a Radical Remedy." *Wall Street Journal*, July 11, 2007b, A-1, A-12.

Braasch, Gary. *Earth Under Fire: How Global Warming Is Changing the World.* Berkeley: University of California Press, 2007.
Bradsher, Keith. "A Drought in Australia, a Global Shortage of Rice." *New York Times*, April 17, 2008a. http://www.nytimes.com/2008/04/17/business/world business/17warm.html.
"Greens Want Global Warming Examined in Bushfire Inquiry." Australian Associated Press, January 21, 2003. (LEXIS)
Macken, Julie. "The Big Dry: Bushfires Re-ignite Heated Debate on Global Warming." *Australian Financial Review*, February 17, 2003, 68.
Macken, Julie. "The Double-Whammy Drought." *Australian Financial Review*, May 4, 2004, 61.
Maynard, Roger. "Climate Change Bringing More Floods to Australia." *The Straits Times* (Singapore), March 14, 2001, 17.
Mydans, Seth. "Reports from Four Fronts in the War on Warming." *New York Times*, April 3, 2007. http://www.nytimes.com/2007/04/03/science/earth/03clim.html.
Nicholls, Neville. "The Changing Nature of Australian Droughts." *Climatic Change* 63 (2004):323–336.
Nowak, Rachel. "The Continent That Ran Dry." *The New Scientist*, June 16, 2007, 8–11.
Pincock, Steve. "Showdown in a Sun-burnt Country." *Nature* 450 (November 15, 2007):336–338.
Ryan, Siobhain. "National Icons Feel the Heat." *Courier Mail* (Australia), February 4, 2002, 1.
"Scientist: Warming Threatens Koalas." Associated Press in *Omaha World-Herald*, May 8, 2008, 7-A.
Wiseman, Paul. "Australia Pushes New Climate Plan." *USA Today*, September 6, 2007, 10-A.

Automobiles and Greenhouse-Gas Emissions

Passenger vehicles in the United States by 2007 accounted for 40 percent of the country's oil consumption, and 10 percent of the world's (Kolbert 2007b, 88, 90). According to Bill McKibben, "The average American car, driven the average American distance—ten thousand miles—in an average ... year releases its own weight in carbon into the atmosphere" (McKibben 1989, 6). Global warming, in a sense, is the exhaust pipe of the "American Dream," of the bigger and better, the new and improved, the mobile life in the fast lane. Cars and trucks used in the United States burn 15 percent of the world's oil production. Transportation alone consumes one-fourth of the energy and two-thirds of oil used in the United States (Cline 1992, 200).

Times are changing, however, as people realize that combating global warming means changing the way in which people move from place to place—and, with time, even the ways our cities and towns are laid out. By 2007, 16 million of the 150 million workers in the United States telecommuted even as 76 percent of Americans drove to work alone (Vanderkam 2008, 9-A).

Climate scientist Stephen Schneider recalled sharing a stage with John Denver, the singer, in Aspen, Colorado. "John said, in introducing me to this group," Schneider recalled, "that he had been unaware of his own impact on global nature until he learned about the global warming problem. 'As soon as I learned about that,' he [Denver] confessed, 'I sold my Porsche.' Everybody applauded" (Schneider 2000).

Automobiles consume a third of the world's oil production. The number of automobiles was increasing more quickly than population, especially in countries that have just begun to experience full-scale industrial development, such as India and China.

An Exemplar of a Fossil-Fueled Culture

The automobile is an exemplar of a fossil-fuel culture—easy to use, convenient, suited to the individualism of our time, and generative of greenhouse gases in its manufacture, as well as its everyday use. Each automobile brings with it a bundle of carbon dioxide, carbon monoxide, methane, nitrous oxide, and waste heat. The tailpipes of motorized vehicles emit 47 percent of North America's nitrous oxides, a primary component of ozone in the lower atmosphere. The automobile, in addition to burning fossil fuels, is also a little greenhouse factory of its own, as the windshield captures the sun's heat and cooks the air inside its steel and glass casing.

Changing Urban Settlement Patterns

Cities in the United States and elsewhere have been remade in the image of the automobile. Any U.S. city that reached maturity after the 1930s—the last decade in which the average person did not own a car—has become an energy-gobbling sprawling mass of suburbs and freeways that virtually guarantees that most people need automobiles whether they want them or not. Outside of a few major eastern cities (New York and Boston being prominent examples), the

Heavy freeway traffic. (Courtesy of Shutterstock)

automobile has become an everyday necessity for nearly everyone. Weaning ourselves from global warming in the long run is going to require major surgery in our concepts of urban land use.

In some places, such surgery is being performed as we watch. Portland, Oregon, has impeded several freeway projects that once threatened to make the center city little more than a conduit to the suburbs. Even so, Portland has no shortage of downtown freeways, but it also has a well-used light-rail system that connects neighborhoods and its airport with the central core. Property values have risen in the center city as people renovate homes that are within bicycling or walking distance of downtown businesses. Even in Detroit, General Motors in 2007 entered a partnership to develop residences in once-abandoned buildings along the waterfront. One of the selling points for these lofts is—note that the pitch is being made by an automobile manufacturer—that people who live in them will be able to walk to work at GM's headquarters (White 2006, D-4).

Such experiments, to date, have been exceptions, however. Even as gasoline prices soared, the climate warmed, and public concern over climate change increased, most people in the United States dutifully filled their gas tanks. Gasoline demand proved notoriously inelastic (as economists say) in the face of rising prices. What choice did most people have?

Downsizing Cars

If we cannot give up our cars, could we at least downsize them? In the United States, 70 percent of car and light truck sales in 2006 had six- or eight-cylinder engines. In Europe, 89 percent had four cylinders or fewer (Spector 2007, B-1). The average travel mileage per person in the United States has increased 60 percent in 30 years. The growing numbers of sports-utility vehicles roaming the roads during the 1990s recalled the words of Henry Ford: "Mini-cars make mini-profits" (Commoner 1990, 80).

Henry Ford's wisdom proved to be rather obsolete as gasoline prices rose to record levels in 2007, and the sales of U.S.-made "maxi cars" plummeted. Soon, Toyota passed General Motors at the largest auto company on the planet, and, even then, U.S. auto executives complained that enforcing higher mileage standards might put them out of business. Actually, reliance on "maxi cars" had nearly done that already. In 2007, U.S. automakers, for the first time, sold fewer than half the cars being taken off lots in the country. By 2007, segments of the auto companies were beginning to realize this, just as the U.S. Congress sought to raise mileage standards in the United States to 35 miles a gallon by 2020, still behind most other countries and the cars they manufacture. The Toyota Prius hybrid was getting 50 miles per gallon of gasoline, and by 2006, Toyota was selling 250,000 hybrids a year in the United States.

Today's automobiles are monuments to fuel inefficiency. The average fuel economy of vehicles sold in the United States has remained nearly stagnant, at around 20 miles a gallon, for several decades. In addition to their size, only 13 percent of a car's energy reaches the wheels, and only half of that actually propels the car. The rest is lost to idling, heat, vibration, and such accessories as air conditioning. Because 6 percent converts to brake heating when the car stops, less than 1 percent of the energy the car consumes ends up propelling the driver. Amory Lovins recommends making cars much lighter, as well as developing hydrogen fuel cells. He also suggests stripping the oil industry of subsidies that make gasoline cheaper than bottled water (Lovins 2005, 74, 76, 82–83).

The X Prize Foundation is sponsoring an automotive contest, expected to carry a prize of more than $10 million, to accelerate research to lead to a car that that can travel 100 miles on a single gallon of gasoline. The same group earlier awarded $10 million to a team that built the first private spacecraft to leave the Earth's atmosphere (Bunkley 2007a). According to contest rules, the winning design must be commercially viable and production-ready, not a "concept car," such as some presented at annual auto shows. Each team must prepare a business plan for building at least 10,000 of its vehicles at a cost comparable to that of cars now available (Bunkley 2007a).

Mileage Standards: Auto Companies Don't Walk the Talk

By 2007, along with many other corporations, U.S. automakers had picked up the "green" mantra, endorsing greenhouse-gas limits in theory. When the debate moved on to practice, however, the same companies complained that enforcing meaningful mileage increases would deprive people of jobs and the companies of profits—this from companies that were swimming in red ink from years of slavishly manufacturing increasingly unsalable large, inefficient gas-guzzlers as fuel prices rose and more nimble companies in other countries (many of them Japanese) gobbled their market shares, even in the United States.

The well-paid pursuit of special interests was on display as the U.S. Congress debated energy legislation during June 2007. The U.S. Senate on June 21, 2007, dealt domestic car manufacturers a major blow by passing a bill requiring a major increase in mileage standards (from 25 to 35 miles a gallon by 2020), the first in two decades. The vote, 65 to 27, sent the bill to the House. Raising the mileage standard 10 miles per gallon will reduce U.S. oil consumption by 1.2 million barrels a day and reduce emissions of greenhouse gases by an amount equal to removing 30 million cars from the road. Getting really serious about mileage, raising the standard to 55 miles per gallon by 2020, a target for which technology is now available, would cut oil demand for transportation in the United States by half (Steinman 2007, 239).

A *New York Times* report noted that automakers' opposition to the mileage standard had been "ferocious," and added, "The clashes and impasses also provided a harbinger of potentially bigger obstacles when Democrats try to pass legislation this fall to reduce emissions of greenhouse gases tied to global warming" (Andrews 2007b).

Within days after that ferocious battle against stiffer mileage standards on Capitol Hill, Ford Motor Company and the Chrysler Group announced at a press conference in Washington, D.C., that they had joined the U.S. Climate Action Partnership (USCAP), a coalition lobbying for national legislation to impose legally binding limits on global-warming emissions. At the time, this partnership included 23 of the world's largest corporations and six of the best-known environmental groups in the United States. General Motors already had joined—even before the battle over mileage standards. "We are pleased to join USCAP at a critical stage in the conversation on climate change, energy consumption and environmental protection," said Alan Mulally, president and chief executive officer of Ford. No explanation was forthcoming as to why the same automakers seemed tooth-and-nail opposed to practical solutions to warming.

Three months before their ferocious assault on higher gas-mileage standards, in March 2007, the chief executives of the same automobile companies had pledged to support mandatory caps on carbon emissions, as long as the caps covered all sectors of the economy.

Where the short-term profit-and-loss rubber meets the public-relations road, however, the U.S. automobile industry seemed, by 2007, to be speaking with more than one voice on the mileage-standard issue. On May 31, 2007, Ron Gettelfinger, the president of the United Automobile Workers Union, and William Clay Ford, Jr., executive chairman of the Ford Motor Company, told Michigan business leaders that cutting emissions and raising fuel economy standards are critical to the future of the industry and that of Michigan. Gettelfinger said that "climate change is real" and that consumers are searching for more environmentally friendly vehicles. "If the auto industry continues to be seen as dragging its feet on environmental issues," he said, "it's going to hurt our brands and vehicles in the marketplace" (Bunkley 2007a).

The next day, a full-page advertisement ran in the Omaha *World-Herald* (and a number of other U.S. newspapers) under the sponsorship of auto manufacturers urging voters to contact their congressional representatives to oppose what it called "extreme" mileage standards, such as the 35 miles per gallon called for in the Senate legislation. The ads claimed that higher mileage standards would deny consumers choices in selection of vehicles, which the automakers promoted as an inalienable American right, along with Thomas Jefferson's life, liberty, and the pursuit of happiness.

Automobile Efficiency: U.S. States Take Action

The California Air Resources Board (ARB), defying the auto industry, voted unanimously during late September 2004 to approve the world's most stringent rules reducing automobile emissions. Under the regulations, the automobile industry must cut exhaust from cars and light trucks by 25 percent and from larger trucks and sport-utility vehicles by 18 percent. The industry will have until 2009 to begin introducing cleaner technology and will have until 2016 to meet the new exhaust standards. The auto industry went to court for relief from the new standards, and on December 12, 2007, Anthony W. Ishii, a federal judge in Sacramento, upheld the California law regulating greenhouse-gas emissions from cars and trucks.

California's plan for sharp cuts in automotive emissions of greenhouse gases eventually could lead most east- and west-coast U.S. states to require similar emissions cuts, because the state's share of the market is so large. In turn, these requirements may provoke the automakers to adopt the same standards for cleaner, more fuel-efficient vehicles across their model lines. The only way to cut global warming emissions from cars is to use less fossil fuel. Because of this limitation, proposed cuts in legally allowable emissions would, as a side effect, force automakers to increase fuel economy by roughly 35 to 45 percent.

During 2004, the governments of New Jersey, Rhode Island, and Connecticut said that they intended to follow California's automobile rules instead of the federal government's legislation. New York, Massachusetts, Vermont, and Maine already had adopted the California rules. "Let's work to reduce greenhouse gases by adopting the carbon-dioxide emission standards for motor vehicles which were recently proposed by the State of California," New York Governor George E. Pataki said in his state-of-the-state address during 2003. These seven states and California account for almost 26 percent of the U.S. auto market, according to R. L. Polk, a company that tracks automobile registrations (Hakim 2004a, C-4).

During 2007, the California ARB voted to require automobile manufacturers to affix labels listing their vehicles' smog and greenhouse-gas emissions so that the state's two million buyers a year can compare them. "This simple tool will empower consumers to choose vehicles that help the environment," said ARB Chairman Robert Sawyer. "Most Californians recognize climate change as a very serious problem. This label will help consumers make informed choices" ("California Air Board" 2007).

Hydrogen Fuel: Hype and Reality

Political correctness regarding global warming in the automobile industry has become associated with development of hydrogen fuel cells, especially after President George W. Bush used his State of the Union address in January 2003 to propose $1.2 billion in research funding to develop hydrogen-fuel technologies. With those funds, Bush said that America could lead the world in developing clean, hydrogen-powered automobiles.

In 2002, Jeremy Rifkin, a liberal social critic and author, published *The Hydrogen Economy: The Creation of the Worldwide Energy Web and the Redistribution of Power on Earth.* Rifkin believes that cheap hydrogen could make the twenty-first century more democratic and decentralized, much the way oil transformed the nineteenth and twentieth centuries by fueling the rise of powerful corporations and nation-states. With hydrogen, writes Rifkin, "Every human being on Earth could be 'empowered'" (Coy 2002, 83).

Hydrogen fuel has been hailed as a godsend. Imagine a car fueled by the most abundant element in the atmosphere that emits nothing but a little water vapor. It is a thought worthy of high praise, but our imaginations run ahead of our ability to deliver. As much as it has been touted as pollution free, hydrogen fuel is no free climatic lunch. Despite surface appearances, hydrogen in today's world is no cleaner a fuel than gasoline. The hype surrounding a "hydrogen economy" has been woefully premature, for one very important technological reason.

Hydrogen, unlike oil or coal, does not exist in nature in a combustible form. Hydrogen is usually bonded with other chemical elements, and stripping them away to produce the pure hydrogen necessary to power a fuel cell requires large amounts of energy. Unless an alternative source (such as Iceland's geothermal resource) is available, hydrogen fuel usually is produced from fossil fuels. Extraction of hydrogen from water via electrolysis and compression of the hydrogen to fit inside an automobile-size tank requires a great deal of electricity. Until electricity is routinely produced via solar, wind, and other renewable sources, the hydrogen car will require energy from conventional sources, including fossil fuels. Today, 97 percent of the hydrogen produced in the United States comes from processes that involve the burning of fossil fuels, including oil, natural gas, and coal.

In mid-2008, however, scientists at the Massachusetts Institute of Technology (MIT) announced the discovery of a cobalt-phosphorus catalyst that can split hydrogen from oxygen atoms in water to create hydrogen gas. Before that discovery, MIT chemist Daniel Nocera and colleagues reported in the *Journal of the American Chemical Society* that such a reaction was possible only using chemicals that worked under toxic conditions, or using platinum, which is much too expensive for industrial-scale processes. "If we are going to use solar energy in a direct conversion process, we need to cover large areas. That make a low-cost catalyst a must," said John Turner, an electrochemist at the National Renewable Energy Laboratory in Golden, Colorado (Service 2008c, 620).

Until such new technology comes into service on a commercial scale, however, basic problems with practicality of hydrogen power remain. Paul M. Grant, writing in *Nature*, provided an illustration:

> Let us assume that hydrogen is obtained by 'splitting' water with electricity—electrolysis. Although this isn't the cheapest industrial approach to 'make' hydrogen, it illustrates the tremendous production scale involved—about 400 gigawatts of continuously available electric power generation [would] have to be added to the grid, nearly doubling the present U.S. national average power capacity. (Grant 2003, 129–130)

That, calculated Grant, would represent the power-generating capacity of 200 Hoover Dams (Grant 2003, 129–130). At $1,000 per kilowatt, the cost of such new infrastructure would total about $400 billion.

What about producing thee 400 gigawatts with renewable energy? Grant estimated that, "with the wind blowing hardest, and the sun shining brightest," wind power generation would require a land area the size of New York State, or a layout of state-of-the-art photovoltaic solar cells half the size of Denmark (Grant 2003, 130). Grant's preferred solution to this problem is the use of energy generated by nuclear fission.

While hydrogen is not the magic wand that some of its proponents imagine, premonitions of hydrogen-based transportation systems have been emerging, small ones in out-of-the-way places. In Iceland, for example, 85 percent of the country's 290,000 people use geothermal energy to heat their homes (Brown 2003, 166); Iceland's government, working with Shell and Daimler-Chrysler, in 2003 began to convert Reykjavik's city buses from internal combustion to fuel-cell engines, using hydroelectricity to electrolyze water and produce hydrogen. The next stage is to convert the country's automobiles, then its fishing fleet. These conversions are part of a systemic plan to divorce Iceland's economy from fossil fuels (Brown 2003, 168).

In 2008, an industrial-scale hydrogen-fired power plant was being built near Venice by the Veneto regional government and Italian energy company ENEL. The new plant, which is located in the Porto Marghera industrial area on the Italian mainland across from the Venice lagoon, next to ENEL's coal-fired Fusina plant, was designed as a "zero-emission hydrogen combustion power generation system" ("Italy to Build" 2006).

Regardless of the limitations of hydrogen fuel cells, the European Union has advocated a transition to them from fossil fuels. The plan includes a $2 billion E.U. commitment, over several years, to bring industry, the research community, and government together in plans to make this transition. According to Rifkin,

> The E.U. decision to transform Europe into a hydrogen economy over the course of the next half century is likely to have as profound and far-reaching an impact on commerce and society as the changes that accompanied the harnessing of steam power and coal at the dawn of the industrial revolution and the introduction of the internal-combustion engine and the electrification of society in the 20th century. ("E.U. Plans" 2002)

Rebirth of the Electric Car

One of Henry Ford's fantasy cars was electric, but that was before petroleum took over. Electric cars may be coming back, however. In January 2007, General Motors rolled out the hybrid Chevrolet Volt, a concept car that can run 40 miles on electricity alone with a six-hour nighttime charge. Running as a hybrid, the Volt gets 150 miles per gallon of gasoline. For commuting trips under 20 miles each way, it can run on a charge from a standard 110-volt garage outlet (Griscom 2007, 60). Electricity must be generated, of course, and these days that usually involves the burning of fossil fuels.

By mid-2007, GM had committed to hiring 400 technical experts to work on fuel-saving technology. One of their goals was to make the Volt a production model within three to four years. This is a significant change from 2002, when General Motors introduced the Hummer H2, which was so fuel thirsty that the company took advantage of a loophole in federal law and refused to publish its mileage (Boudette 2007, A-8).

At the new, green GM, Lawrence Burns, GM vice president for research, development, and global planning, told the *Wall Street Journal*, "We have to have people think we are part of the solution, not part of the problem." The Volt, said Burns, is an effort to show consumers that, when it comes to climate change, "we get it" (Boudette 2007, A-8).

With the three major U.S. automakers tin-cupping Congress for bailouts late in 2008, they were being urged to build the car of the future, a plug-in hybrid. Their main problem was that neither they nor Japanese carmakers had figured out how to produce electric cars for a profit. Even Toyota, which had sold about a million of its hybrid Prius, was losing money on them. "In 10 years are they [at General Motors] going to solve the technological problems with respect to the Volt? Sure," said Maryann Keller, an automotive analyst and author of a book on GM. "But are they going to be able to stake their survival, which is really more of a now to five-year proposition, on it? I'd say they can't" (Mufson 2008c). Large inefficient cars will not lose their luster until gasoline prices rise—and late in 2008, they were diving to less than half the peak of $4.00 a gallon reached in the United States earlier that year.

"You'd think from reading the media that we have had a burial ceremony at Arlington cemetery for the last pickup truck," said James Womack, an automotive management and writer. Womack described the time required to design a new electric vehicle, produce thousands of new parts, and adjust assembly lines. "For anything that's really new it's still about four years," he said. "To get your money back, you need to make that product for eight to 10 years with only cosmetic changes" (Mufson 2008c).

New York City's Green Yellow Cabs

New York City's Yellow Taxi fleet will run entirely on gas-electric hybrids within five years, Mayor Michael Bloomberg announced May 22, 2007. "There's an awful lot of taxicabs on the streets of New York City," Bloomberg said. "These cars just sit there in traffic sometimes, belching fumes" ("New York's Mayor" 2007). Almost 400 hybrids were tested in New York City's taxi fleet over 18 months, with models including the Toyota Prius, the Toyota Highlander Hybrid, the Lexus RX 400h, and the Ford Escape. Under Bloomberg's plan, that number will increase to 1,000 by October 2008, and then will grow by about 20 percent each year until

2012, when every yellow cab—currently numbering 13,000—will be a hybrid ("New York's Mayor" 2007). The city licenses the yellow cabs, and sells licenses to individual drivers, who then purchase their own vehicles under specifications set by the Taxi and Limousine Commission. A similar hybridization of cabs was under way in San Francisco.

In 2006, the average New York taxi ran entirely on gasoline at 14 miles per gallon. By 2008, all New York taxis will be required to run at 25 miles per gallon and 30 miles per gallon by 2009. While hybrid vehicles cost more, Bloomberg said that increased fuel efficiency will reduce operating costs by about $10,000 a year per vehicle. Changing the New York City taxi fleet to hybrids is just one piece of Bloomberg's sustainability plan to reduce carbon-dioxide emissions in the city by 30 percent by 2030.

Congestion Charges

New York City Mayor Bloomberg also proposed a congestion charge for the most crowded southern half of Manhattan Island, roughly 86th Street southward. On weekdays from 6 A.M. to 6 P.M., trucks would be charged $21 a day and cars would be charged $8 in addition to premium parking fees charged by city-owned and private lots. At present, only 5 percent of the people who work in Manhattan but live outside it commute by car. Those traveling only within the zone would pay half price; taxis and livery cabs would be exempt. With uncontrolled free access to the area, studies have shown that vehicle speeds within this area average 2.5 to 3.7 miles an hour. Many times, walking is almost as fast. Anyone who wants to go anywhere with any speed in this area takes the subway. The value of time lost to congestion delays in New York City has been estimated at $5 billion a year; add wasted fuel, lost revenue, and increasing costs of doing business, and the total rises to $13 billion a year (Kolbert 2007a, 24). Bloomberg's plan was derailed by the New York State Assembly during July 2007.

Singapore was the first city to introduce such a fee. London introduced a congestion charge during 2003, followed by Milan, Italy. In London, vehicle speeds have since risen 37 percent and carbon-dioxide emissions have fallen 15 percent. London's mayor, Ken Livingstone, a major proponent of the charge, was easily reelected in 2004, and in 2006, two-thirds of London residents supported the charge. During January 2007, the congestion zone was expanded westward to include most of Kensington, Chelsea, and Westminster (Kolbert 2007a, 23–24). By early 2007, London's congestion charge had reduced private car use 38 percent and carbon emissions 20 percent in the congestion charge zone.

London improved its public transport system to offer Londoners easier alternatives; many commuters had complained that the buses were slow and expensive. By the time the London bus system was upgraded, more than six million people used it daily. The number of people commuting by bicycle in London (at no charge) soared 80 percent after the congestion charge was implemented ("Big City Mayors" 2007).

In the meantime, British Petroleum—whose corporate initials, BP, have been taken by green publicists to mean "Beyond Petroleum"—has proposed that every motorist in Great Britain sign up for a plan called "Target Neutral." Drivers can fund ventures that offset the amount of carbon dioxide that their driving adds to the atmosphere. Drivers register at a Target Neutral Web site, which calculates the estimated amount of carbon dioxide that may be produced by their driving over the coming year. Drivers then pay offsets based on the estimate. The typical family car, doing 10,000 miles a year, is likely to cost about £20 (US$39) to offset its emissions ("British Travel" 2006).

Congestion charging in downtown Stockholm was a controversial issue in the Swedish general election during the late summer of 2006. The congestion charge of up to $7 a day was narrowly approved by 52 percent in a referendum passed on September 17, 2006. The Stockholm congestion charge reduced auto traffic 20 to 25 percent, while use of trains, buses, and Stockholm's subway system increased. Emissions of carbon dioxide declined 10 to 14 percent in the inner city and 2 to 3 percent in Stockholm County. The project also increased the use of environmentally friendly cars (such as hybrids), which are exempt from congestion taxes. As in London, commuting by bicycle also increased. Following a trial period, the Stockholm congestion tax became permanent in August 2007.

The Car's Future

The need for a paradigm shift in how we fuel our cars—and how many of us drive them—can be illustrated by a comparison of car ownership in the United States, India, and China played across economic trends. In China, in 2006, there were nine personal vehicles per 1,000 people of driving age; in India, there were 11. In the United States, the same figure stood at 1,148. Between 2000 and 2006, sales of heavy trucks in China increased 800 percent. Sales of passenger cars increased 600 percent. Sales of new passenger vehicles tripled in India (from 500,000 a year to 1.5 million) between 1998 and 2008, a period during which the country built its first interstate highway system, the Golden Quadrilateral, which links Mumbai (Bombay), Delhi, Kolkata (Calcutta), and Bangalore.

If people in India and China drove at half the rate of those in the United States, world oil consumption, 86 million barrels a day in 2006, would balloon to more than 200 million. If drivers in India and China used cars as Americans do, world oil consumption would more than triple, with attendant impacts on its price, as well as greenhouse-gas emissions (Kolbert 2007b, 88).

With China's and India's economies growing at 10 percent a year, and people's incomes rising rapidly, millions of families will soon be in the market for their first private cars. Mileage for new cars in the United States has stagnated at about 20 miles per gallon—incredibly, less than that of Henry Ford's Model T when it first went on the market in 1908 (Kolbert 2007b, 90). The world's climatic future and the finite nature of oil supplies are going to demand a paradigm shift in fuel technology and economy in the coming years. In another 100 years, by necessity, the internal combustion engine may be as antique as the horse and buggy seems today.

FURTHER READING

Andrews, Edmund. "Senate Adopts an Energy Bill Raising Mileage for Cars." *New York Times*, June 22, 2007b. http://www.nytimes.com/2007/06/22/us/22energy.html.

"Big City Mayors Strategize to Beat Global Warming." Environment News Service, May 15, 2007. http://www.ens-newswire.com/ens/may2007/2007-05-15-01.asp.

Boudette, Neal E. "Shifting Gears, GM Now Sees Green." *Wall Street Journal*, May 29, 2007, A-8.

"British Travel Agents Launch Carbon Offset Scheme." Environment News Service, November 28, 2006. http://www.ens-newswire.com/ens/nov2006/2006-11-28-05.asp.

Brown, Lester R. *Plan B: Rescuing a Planet under Stress and a Civilization in Trouble.* New York: Earth Policy Institute/W.W. Norton, 2003.

Bunkley, Nick. "Seeking a Car That Gets 100 Miles a Gallon." *New York Times*, April 2, 2007a. http://www.nytimes.com/2007/04/02/business/02xprize.html.

Bunkley, Nick. "Detroit Finds Agreement on the Need to Be Green." *New York Times*, June 1, 2007b. http://www.nytimes.com/2007/06/01/business/01auto.html.

"California Air Board Adds Climate Labels to New Cars." Environment News Service, June 25, 2007. http://www.ens-newswire.com/ens/jun2007/2007-06-25-09.asp#anchor7.

Cline, William R. *The Economics of Global Warming.* Washington, DC: Institute for International Economics, 1992.

Commoner, Barry. *Making Peace with the Planet.* New York: Pantheon, 1990.

Coy, Peter. "The Hydrogen Balm? Author Jeremy Rifkin Sees a Better, Post-Petroleum World." *Business Week*, September 30, 2002, 83.

"E.U. Plans to Become First Hydrogen Economy Superpower." *Industrial Environment* 12, no. 13 (December 2002). (LEXIS)

Grant, Paul M. "Hydrogen Lifts Off—with a Heavy Load: The Dream of Clean, Usable Energy Needs to Reflect Practical Reality." *Nature* 424 (July 10, 2003):129–130.

Griscom-Little, Amanda. "Detroit Takes Charge." *Outside*, April 2007, 60.

Hakim, Danny. "Several States Likely to Follow California on Car Emissions." *New York Times*, June 11, 2004a, C-4.

"Italy to Build World's First Hydrogen-Fired Power Plant." Environment News Service, December 18, 2006. http://www.ens-newswire.com/ens/dec2006/2006-12-18-05.asp.

Kolbert, Elizabeth. "Don't Drive, He Said." *The New Yorker* (Talk of the Town), May 7, 2007a, 23–24.

Kolbert, Elizabeth. "Running on Fumes: Does the Car of the Future Have a Future?" *The New Yorker*, November 5, 2007b, 87–90.

Lovins, Amory. "More Profit with Less Carbon." *Scientific American*, September 2005, 74, 76–83.

McKibben, Bill. *The End of Nature.* New York: Random House, 1989.

Mufson, Steven. "The Car of the Future—but at What Cost? Hybrid Vehicles Are Popular, but Making Them Profitable Is a Challenge." *Washington Post*, November 25, 2008c, A-1. http://www.washingtonpost.com/wp-dyn/content/article/2008/11/24/AR2008112403211_pf.html.

"New York's Mayor Plans Hybrid Taxi Fleet." Associated Press in *New York Times*, May 22, 2007. http://www.nytimes.com/aponline/us/AP-Green-Taxis.html.

Rifkin, Jeremy. *The Hydrogen Economy: The Creation of the Worldwide Energy Web and the Redistribution of Power on Earth.* New York: Jeremy P. Tarcher/Putnam 2002.

Schneider, Stephen H. "No Therapy for the Earth: When Personal Denial Goes Global." In *Nature,*

Environment & Me: Explorations of Self in a Deteriorating World, eds. Michael Aleksiuk and Thomas Nelson. Montreal: McGill-Queens University Press, 2000.

Service, Robert F. "New Catalyst Marks Major Step in the March Toward Hydrogen Fuel." *Science* 321 (August 1, 2008c):620.

Spector, Mike. "Can U.S. Adopt Europe's Fuel-Efficient Cars?" *Wall Street Journal,* June 26, 2007, B-1.

Steinman, David. *Safe Trip to Eden: 10 Steps to Save Planet Earth from Global Warming Meltdown.* New York: Thunder's Mouth Press, 2007.

Vanderkam, Laura. "Want to Save the Planet? Stay Home." *USA Today,* May 20, 2008, 9-A.

White, Joseph R. "An Ecotopian View of Fuel Economy." *Wall Street Journal,* June 26, 2006, D-4.

B

Bangladesh, Sea-Level Rise in

Bangladesh, one of the poorest countries on Earth, is likely to suffer disproportionately from global warming, largely because 90 percent of its land lies on floodplains. Cyclones there historically have killed many people; 130,000 people died in such a storm in April 1990. Less than one-fourth of Bangladesh's rural population has electricity; the country, as a whole, emits less than 0.1 percent of the world's greenhouse gases, compared with 24 percent by the United States (Huq 2001, 1617). Bangladesh is planning to use solar energy for new energy infrastructure, but it lacks the money to build sea walls to fend off rising sea levels.

Saleemul Huq, chairman of the Bangladesh Centre for Advanced Studies in Dhaka, Bangladesh, and director of the Climate Change Programme of the International Institute for Environment and Development in London, said that the world community has an obligation to pay serious attention to the views of people who stand to lose the most from climate change (Huq 2001, 1617). A sea-level rise of half a meter (about 20 inches) could inundate about 10 percent

Flood victims carry drinking water at Bashila, on the outskirts of Dhaka, Bangladesh, August 14, 2007. (AP/Wide World Photos)

of Bangladesh's habitable land, the home, in 2004, of roughly six million people.

A one-meter water-level rise would put 20 percent of the country (and 15 million people) under water (Radford 2004, 10). In addition to sea-level rise caused by warming, large parts of the Ganges Delta are subsiding because water has been withdrawn for agriculture, compounding the problem.

FURTHER READING

Huq, Saleemul. "Climate Change and Bangladesh." *Science* 294 (November 23, 2001):1617.

Radford, Tim. "2020: The Drowned World." *London Guardian*, September 11, 2004, 10.

Bark Beetles Spread across Western North America

Nine years of intense drought and rising temperatures by 2007 were creating continued perfect conditions for bark-beetle infestations across the U.S. West. These conditions had been ravaging forests there for at least five years. By the fall of 2002, large swaths of evergreen forests in Western Montana and the Idaho Panhandle, as well as parts of California, Colorado, and Utah had fallen victim to unusually large infestations of bark beetles, including the Douglas fir bark beetle, spruce beetle, and mountain pine beetle.

The infestations were being encouraged by several factors: a warming trend, which allows the beetles to multiply more quickly and reach higher altitudes; drought, which deprives trees of sap they would usually use to keep the beetles under control; and years of fire suppression, which increased the amount of elderly wood susceptible to attack. Beetles, attacking in "epic proportions, killed many stands of trees within a few weeks" (Stark 2002, B-1).

According to one observer, "The vast tracts of Douglas fir that stood green and venerable for generations [east of Yellowstone National Park] are peppered and painted with swaths of rusty red and gray. For Douglas fir, those are the colors of death" (Stark 2002, B-1).

Trees Killed at Unprecedented Rates

Tens of millions of trees across the West were being killed at a rate never seen before. Warmer temperatures accelerate the beetles' reproduction cycle, killing trees more quickly. Some types of beetles that propagated two generations a year were reproducing three times a year. "This is all due to temperature," said Barbara Bentz, a research entomologist with the U.S. Forest Service who studies bark beetles. "Two or three degrees [of temperature] is enough to do it" (Wagner 2004). Outside Cody, Wyoming, an entire forest was killed by the drought and beetles. "It used to be a nice spruce forest," said Kurt Allen, a Forest Service entomologist. "It's gone now. You're not going to get those conditions back for 200 or 300 years. We're really not going to have what a lot of people would consider a forest" (Wagner 2004).

By 2008, so many trees were being killed by pine bark beetles that the National Center for Atmospheric Research in Boulder, Colorado, was studying how the swaths of dead trees change the weather from southern Wyoming to northern New Mexico. Trees killed by the beetles may influence rainfall, temperatures, and smog. Preliminary computer modeling indicated that the alterations could amount to temporary temperature increases between 2° and 4°F ("Pine Beetles" 2008).

Bill McEwen wrote in a letter to the *New York Times*:

> I reside in the semi-arid West, where scientists are just beginning to understand the enormous synergistic impact of global warming, atmospheric drying (drought), and the explosion in insect populations that is killing many of our forests.... On a recent vacation to the Northwest, I drive through Sun Valley, Idaho. Around Sun Valley and the nearby Salmon River Valley, entire mountainsides of forest are now being destroyed by out-of-control bark beetle infestations. (McEwen 2004, A-16)

Charles Petit described the spreading infestations, also in the *New York Times*:

> Like an army roaring out of the trenches, bark beetles overwhelm their next round of piney prey, emitting pheromones that draw more attackers to individual trees. Forest managers say defenses—quarantines, burning and other methods—are ineffective over large areas The result, in British Columbia, is the largest forest insect blight ever seen in North America, and it seems nowhere near its peak. It covers a patch running about 400 miles north-south and 150 miles across. Officials expect 80 percent of British Columbia's mature lodgepoles to be dead by 2014.
>
> Loggers are frantically cutting lifeless trees for lumber before they rot. Funguses carried by beetles stain the wood a blotchy blue. Lumber mills

promote it as stylish "denim wood." In 2002, high winds carried the beetles through the Rockies and into the Alberta plain. They appear poised to sweep east to the Atlantic through Canada's jackpine boreal forest. (Petit 2007)

In Flagstaff, Arizona, near the world's largest contiguous ponderosa pine forest, Tom Whitham wondered how much more devastation the drought and beetles would cause, and to what extent humans will contribute to it. "The thing that would make me really sad is if this was human caused," he said, glancing at the bare trees towering over his pickup truck. "If you lose a 200-year-old forest, you can't get it back" (Wagner 2004).

Deadly wildfires that burned at least 1,000 homes in Southern California during late October 2003 were aggravated not only by fierce Santa Ana winds, 100-degree temperatures, drought, and very low humidity, but also by the deaths of more than a million mature pine trees killed during the previous year by bark-beetle infestation.

According to a Canadian government projection, pine-beetle infestation presents a "worst-case scenario" that could afflict 80 percent of British Columbia pine forests by the year 2020. "Mountain pine beetle outbreaks are stopped by severe winter weather or depletion of the host. The vast spatial extent of the outbreak implies that a weather-stopping event is unlikely," said the report (Baron 2004, A-3). By 2004, the beetle infestation already had spread through a broad swath of the British Columbia interior, especially around the Prince George area, extending south to the U.S. border.

William K. Stevens of the *New York Times* described unprecedented destruction of boreal forests in Alaska by spruce bark beetles that have been whipped into a reproductive frenzy by a warming environment:

> Once these purple-gray stretches of tree skeletons were a green and vital part of the spruce-larch-aspen tapestry that makes up the taiga.... Today, in a stretch of 300 or 400 miles reaching westward from the Richardson Highway, north of Valdez, past Anchorage, and down through the Kenai Peninsula, armies of spruce bark beetles are destroying the spruce canopy or have already done so. Often the trees are red instead of gray—freshly killed but not yet desiccated. (Stevens 1999a, 178)

On the south-central coast of Alaska, cool temperatures have heretofore kept the spruce bark beetle under control. As temperatures have warmed, however, the beetles have killed much of the tree cover across three million acres, one of the largest insect-caused forest devastations in North America's history.

While warming increases many insects' reproductive energies, warming destroys many trees' reproductive capacities. According to Kevin Jardine of Greenpeace, rising temperatures can cause boreal trees' pollen and seed cones to develop too rapidly because of higher-than-usual spring temperatures, leading to reproductive failure. Boreal seeds also germinate within a specific range of soil temperatures. For example, writes Jardine, black spruce seeds germinate between 15° and 28°C. "If the soil temperature falls below 15°," writes Jardine, "the processes that cause germination come to a halt. If soil temperatures rise above 28°C, bacteria and fungi can attack and consume seeds. The probability of germination also declines rapidly for higher temperatures" (Jardine 1994).

Bark Beetles Infestations Worsen

By late 2008, British Columbia had lost 33 million acres of high-altitude lodgepole pine forest. From British Columbia's infested forests, unusually strong winds in 2006 pushed clouds of mountain pine beetles, a species of bark beetle, over the Continental Divide to northern Alberta, where they had never been observed. In years to come, the beetles could ride the winds to the Great Lakes (Robbins 2008b).

By the end of 2008, Montana, Wyoming, and Colorado each had lost more than a million acres of trees to the beetles, which were spreading in an exponential fashion, according to Clint Kyhl, director of a Forest Service incident management team in Laramie, Wyoming. Kyhl said that nearly all of Colorado's larger lodgepole pine trees, about five million acres, could die within five years "The Latin name is *Dendroctunus*, which means tree killer," said Gregg DeNitto, a Forest Service entomologist in Missoula, Montana. "They are very effective" (Robbins 2008b). The only true cure for the infestation is sustained cold winter weather, on the order of 30° to 40°F below zero, of a type that has been absent since 1980s. *See also:* Pine Beetles in Canada; Spruce Bark Beetles in Alaska

FURTHER READING

Baron, Ethan. "Beetles Could Chew up 80 percent of B.C. Pine: Report: Worst-case Scenario by 2020

Blamed on Global Warming." *Ottawa Citizen*, September 12, 2004, A-3.

Jardine, Kevin. "The Carbon Bomb: Climate Change and the Fate of the Northern Boreal Forests." Ontario, Canada: Greenpeace International, 1994. http://dieoff.org/page129.htm.

McEwen, Bill. "The West's Dying Forests." (Letter to the Editor) *New York Times*, August 2, 2004, A-16.

Petit, Charles W. "In the Rockies, Pines Die and Bears Feel It." *New York Times*, January 30, 2007. http://www.nytimes.com/2007/01/30/science/30bear.html.

"Pine Beetles Changing Rocky Mountain Air Quality, Weather," Environment News Service, October 1, 2008. http://www.ens-newswire.com/ens/oct2008/2008-10-01-091.asp.

Robbins, Jim. "Bark Beetles Kill Millions of Acres of Trees in West." *New York Times*, November 18, 2008b. http://www.nytimes.com/2008/11/18/science/18trees.html.

Stark, Mike. "Assault by Bark Beetles Transforming Forests; Vast Swaths of West are Red, Gray, and Dying; Drought, Fire Suppression, and Global, Warming are Blamed." *Billings Gazette* in *Los Angeles Times*, October 6, 2002, B-1.

Stevens, William K. *The Change in the Weather: People, Weather, and the Science of Climate.* New York: Delacourte Press, 1999a.

Wagner, Angie. "Debate Over Causes Aside, Warm Climate's Effects Striking in the West." Associated Press, April 27, 2004. (LEXIS)

Baseball Bats and Warming

The ash tree, a traditional source of major-league baseball bats, is being killed by a beetle species, the emerald ash borer, which may be encouraged by warming temperatures. Owners of bat factories in the ash country of Northwestern Pennsylvania have made emergency plans if the white ash tree (the source of the best wood) is, as the plant says, "compromised." Seeds of the same tree are being collected in Michigan in case natural species are endangered.

Asian wasps were imported and set loose in ash forests during 2007 to attack the shiny-green emerald ash borer (*Agrilus planipennis fairmaire*), itself an Asian immigrant, which has killed upward of 25 million ash trees in Michigan, Illinois, Indiana, Ohio, and Maryland after it was first found near Detroit in 2002 (Davey 2007a). By late June 2007, the ash borer was invading the choice baseball bat ash groves in Pennsylvania, near the border with New York State.

Warming temperatures may be partially to blame for the ash borer's expansion because a longer growing season is causing the white ash's wood to become softer, making it easier prey for the beetles. Changes in the wood's density also make it less suitable for baseball bats. Warmer weather may speed up the reproductive cycle of the beetle.

"We're watching all this very closely," said Brian Boltz, general manager of the Larimer & Norton company, owner of the Russell mill, which each day saws, grades, and dries scores of billets destined to become Louisville Slugger bats. "Maybe it means more maple bats. Or it may be a matter of using a different species for our bats altogether" (Davey 2007a). Barry Bonds of the San Francisco Giants used maple bats. The major leagues could turn to aluminum bats, which are used in college-level and junior-league baseball.

As with most aspects of global warming, this topic is open to debate. Dan Herms, an associate professor of entomology at Ohio State University, denies a link between the ash borer and climate change because the beetles survive in a wide range of temperatures in Asia (Davey 2007a).

FURTHER READING

Davey, Monica. "Balmy Weather May Bench a Baseball Staple." *New York Times*, July 11, 2007a. http://www.nytimes.com/2007/07/11/us/11ashbat.html.

Bicycles and Energy Efficiency

Despite its health benefits (a regular bicycle rider's health is 10 years younger than a person who regularly drives a car), bicycle transport accounts for only two-tenths of 1 percent of travel miles in the United States (one mile of every 500). Walking accounts for 0.7 percent (Hillman, Fawcett, and Rajan 2007, 49–51, 53).

Bicycle Commuting in Europe

The automobile's urban territory has been shrinking in European cities. A growing web of pedestrian malls allows tens of thousands of people to traverse downtown Stockholm on foot every day—down a gentle hill, northwest to southeast, along Drottinggatan, past the Riksdag (Parliament) and the King's Palace, merging with Vasterlanggaten, into the Old Town—for more than two miles. More and more streets across the city are gradually being placed off-limits to motor traffic (Johansen 2007, 23).

Urban life in Europe is being recast with the automobile as antithesis. Drivers are free to buy a sport-utility vehicle in Denmark, but the bill includes a registration tax up to 180 percent of the purchase price. Denmark's taxation system has become an environmental exclamation point. Imagine, for example, paying more than $80,000 in taxes (as well as $6 a gallon for gasoline) to buy and drive a Hummer H2—that, and pesky bicyclists may ridicule your elegantly pimped ride as an environmental atrocity (Johansen 2007, 23).

Bicycles have become privileged personal urban transport in many European cities. To sample bicycle gridlock, come to Copenhagen, which has deployed 2,000 bikes around the city for free use. The mayor of Copenhagen in 2007, Klaus Bondam, commuted by bicycle. Helmets are not required, despite the occasional bout of two-wheeled road rage as bicyclists clip each other on crowded streets. People ride bikes while pregnant, drinking coffee, and smoking, during rain or shine, and while using a wide array of baskets to carry groceries and briefcases.

The Copenhagen airport has parking spaces for bicycles. On weekends, more than half the admissions to the emergency room of Frederiksberg Hospital are drunken cyclists (they tend to run into poles). On a more sober note, more than a third of Copenhagen residents ride bikes to work (40 percent do so in Amsterdam), in a conscious assault on the "car culture" (Keates 2007, A-10). New bike-parking facilities have been planned at the Amsterdam's main train station that will house up to 10,000 machines. Officials from some U.S. cities, as well as bigger cities in Europe (London and Munich), have been studying Amsterdam and Copenhagen. Bicycles also account for one-eighth of urban travel in Sweden, where Stockholm is laced with many well-used bicycle paths that complement its growing web of pedestrian-only malls.

Many Danish companies offer indoor bike parking, as well as locker rooms. Employees ride company-owned bikes to off-site meetings. People tote children on extra bike seats. Dutch Prime Minister Jan Peter Balkenende rides a bicycle to work some of the time. Members of the Danish Parliament ride as well, along with chief executive officers of some major companies. Lars Rebien Sorensen, CEO of the pharmaceutical firm Novo Nordisk, conducts media interviews from his bike saddle.

On July 15, 2007, the day after Bastille Day, people in Paris woke up to discover thousands of low-cost rental bikes at hundreds of high-tech bicycle stations scattered throughout the city, part of an ambitious program to "cut traffic, reduce pollution, improve parking and enhance the city's image as a greener, quieter, more relaxed place" (Anderson 2007a, A-10). By late 2007, Paris had 20,600 bikes at 1,450 stations, one station every 250 yards across the entire city. Based on experience elsewhere, notably in Lyon, France's third-largest city, which launched a similar system during 2005, regular users of the bikes will ride them almost for free (Anderson 2007a, A-10).

"It has completely transformed the landscape of Lyon—everywhere you see people on the bikes," said Jean-Louis Touraine, the city's deputy mayor. The program was meant "not just to modify the equilibrium between the modes of transportation and reduce air pollution, but also to modify the image of the city and to have a city where humans occupy a larger space" (Anderson 2007a, A-10). The Socialist mayor of Paris, Bertrand Delano, has the same aim, said Delano's aide, Jean-Luc Dumesnil: "We think it could change Paris's image—make it quieter, less polluted, with a nicer atmosphere, a better way of life" (Anderson 2007a, A-10).

The bikes were installed after a study analyzed different trips in the city by car, bike, taxi, and foot. Bikes were the quickest form of urban transport—also the least noisy and the lightest drain on resources. The Lyon rental bikes, with their distinctive silver frame, red rear-wheel guard, handlebar basket, and bell, can also be among the cheapest ways to travel, because the first half-hour is free, and most trips are shorter than that. Anthonin Darbon, director of Cyclocity, which operates Lyon's program and won the contract to start up and run the program in Paris, said, "95 percent of the roughly 20,000 daily bike rentals in Lyon are free because of their length" (Anderson 2007a, A-10).

Cyclocity is a subsidiary of outdoor advertising company JCDecaux, which operates much smaller bike businesses in Brussels, Vienna, and the Spanish cities of Cordoba and Girona. London, Dublin, Sydney, and Melbourne are considering similar rental programs (Anderson 2007a, A-10). A number of U.S. cities, including Portland, Oregon, have experimented with community-use bicycle programs (Anderson 2007a, A-10). Cyclocity's system evolved from utopian "bike-sharing" ideas that were tried in Europe in the 1960s and 1970s. This system is patterned on

Amsterdam's "white bicycle" plan, in which volunteers repaired hundreds of broken bicycles, painted them white, and left them on the streets for free use. Many of these bikes were stolen and, with lack of continuing maintenance, again broke down.

John Ward Anderson described what happened next in the *Washington Post*:

> JCDecaux experimented with designs and developed a sturdier, less vandal-prone bike, along with a rental system to discourage theft: Each rider must leave a credit card or refundable deposit of about $195, along with personal information. In Lyon, about 10 percent of the bikes are stolen each year, but many are later recovered.... [T]o encourage people to return bikes quickly, rental rates rise the longer the bikes are out. Membership fees in Paris will be higher than in Lyon, from $1.30 for one day to about $38 for a year. (Anderson 2007a, A-10)

JCDecaux provides all of the bicycles (at a cost of about $1,300 each) and builds the pickup and drop-off stations. Each station has 15 to 40 high-tech racks connected to a centralized computer that monitors each bike's condition and location. Customers can buy a prepaid card or use a credit card at a computerized console to release a bike. The company will pay citywide start-up costs of about $115 million and employ the equivalent of about 285 full-time people to operate the system and repair the bikes for 10 years. All revenue from the program will go to the city. The company will also pay Paris a fee of about $4.3 million a year (Anderson 2007a, A-10).

By mid-2008, 20,600 Velib bicycles were being used in Paris, with more than 1,450 rental stations 300 yards apart, four times as many as Paris has subway stations. During its first year, Velib clocked 27.5 million rides, 120,000 a day, in a city of 2.1 million, many for commuting. A $46 annual fee allows a rider 30 minutes at a time, with an extra charge for more. Fifteen percent of the fleet had been stolen, despite a $240 charge on riders' credit cards. Three riders had been killed in accidents. Paris is aiming to reduce auto traffic 40 percent by 2020 (Erlanger 2008).

Noting that a quarter of trips by car each day are less than two miles, Great Britain in mid-2008 dedicated almost $200 million to improving bicycle infrastructure. The idea is to improve health, reduce greenhouse-gas emissions, and ease traffic congestion. Bristol was selected as Britain's first Cycling City, along with the following 11 Cycling Demonstration Towns: Blackpool, Cambridge, Chester, Colchester, Leighton/Linslade, Shrewsbury, Southend on Sea, Southport with Ainsdale, Stoke, Woking, and York.

By 2008, bike-sharing programs had spread to dozens of European cities, including Barcelona and Rome, in addition to cities in France, Austria, Denmark, the Netherlands, and Germany. "The critical mass of bikes on the road has pacified traffic," said Gilles Vesco, vice mayor who directs the bike-sharing program in Lyon, France. "It has become more convivial public space" (Rosenthal 2008h, A-10).

In the United States, New York City Mayor Michael Bloomberg has proposed requiring commercial buildings to maintain indoor bike parking. Some U.S. cities are bike-friendly. Boulder, Colorado has bike lanes on 97 percent of its arterials; one-fifth of people who commute do so by bike. Boulder has been spending 15 percent of its transportation budget on bike infrastructure. Davis, California, which now has a bicycle on its city logo, another university town, has 95 percent arterials with bike lanes, and had a 17 percent commuter rate in 2007.

Biking in U.S. Cities

In 2008, New York City opened protected bicycle lanes in a few parts of the city to make commutes less treacherous, after an average of 23 riders died annually in collisions with cars and trucks for seven years. Despite the risky nature of riding, bicycle use in New York City has increased 75 percent since 2000, to about 130,000 commuters a day. "We've run out of room for driving in the city. We have to make it easier for people to get around by bikes," said Janette Sadik-Khan, the city's transportation commissioner, who herself bikes to work. Manhattan-based *Vogue* magazine called bicycles "the hottest accessory" and said that "two wheels and a wicker basket become the perfect complement to the smart urban girl's spring style" (Shulman 2008, A-2).

The city installed covered bike racks that look much like bus shelters, distributed free helmets, and expanding a 400-mile network of bike lanes. Chicago created hundreds of miles of bike lanes; added a station with valet parking, showers, and indoor racks; and established penalties of as much as $500 for motorists who endanger cyclists (Shulman 2008, A-2). Both cities are emulating Paris, which by 2008 had a bicycle-sharing program with 200,000 bikes.

In May 2008, the Washington, D.C., city government joined with an advertiser, Clear Channel Outdoor, to start a bike-sharing program similar to those in Paris and other French cities. The program, called SmartBike DC, which requires a $40 annual membership fee, began with 120 bikes in the central city, with plans to expand to a thousand. Clear Channel also reached similar agreements with San Francisco, Chicago, and Portland. Also in 2008, the University of New England and Ripon College in Wisconsin gave free bikes to freshmen who promised to leave cars at home. Other colleges established free bike-sharing or rental programs. In doing so, these universities eased parking problems and reduced greenhouse-gas emissions.

FURTHER READING

Anderson, John Ward. "Paris Embraces Plan to Become City of Bikes." *Washington Post*, March 24, 2007a, A-10. http://www.washingtonpost.com/wp-dyn/content/article/2007/03/23/AR2007032301753_pf.html.

Becker, Bernie. "Bicycle-Sharing Program to Be First of Kind in U.S." *New York Times*, April 27, 2008. http://www.nytimes.com/2008/04/27/us/27bikes.html.

Erlanger, Steven. "A New Fashion Catches On in Paris: Cheap Bicycle Rentals." *New York Times*, July 13, 2008. http://www.nytimes.com/2008/07/13/world/europe/13paris.html.

Hillman, Mayer, Tina Fawcett, and Sudhir Chella Rajan. *The Suicidal Planet: How to Prevent Global Climate Catastrophe*. New York: St. Martin's Press/Thomas Dunne Books, 2007.

Johansen, Bruce. "Scandinavia Gets Serious about Global Warming." *The Progressive*, July 2007, 22–24.

Keates, Nancy. "Building a Better Bike Lane." *Wall Street Journal*, May 4, 2007, W-1, W-10.

Rosenthal, Elisabeth, "European Support for Bicycles Promotes Sharing of the Wheels." *New York Times*, November 10, 2008h, A-10.

Shulman, Robin. "N.Y. Hopes to Ensure Smooth Pedaling for Bike Commuters." *Washington Post*, May 25, 2008, A-2. http://www.washingtonpost.com/wp-dyn/content/article/2008/05/24/AR2008052401457_pf.html.

Zezima, Kate. "With Free Bikes, Challenging Car Culture on Campus." *New York Times*, October 20, 2008. http://www.nytimes.com/2008/10/20/education/20bikes.html.

Biodiversity, Decline of

Studies by many wildlife-advocacy groups support many scientists' projections of impending mass extinctions due in part to warming temperatures. Other human-induced factors often play a role as well, including exposure to toxins and destruction of habitat. These studies usually define the problem in terms of declining biodiversity. The studies do more than document present and potential extinctions of several species. They also advocate emphatic action to combat climate change. These groups call on U.S. lawmakers to help curtail greenhouse-gas emissions, for example, by enacting higher fuel-efficiency standards for ground transportation and more energy-efficient building codes (Lazaroff 2002a).

Two such reports were compiled by the National Wildlife Federation (NWF) in the United States and the International World Wildlife Fund (WWF), both of which contend that species from the tropics to the poles are at risk. Many species may be unable to move to new areas quickly enough to survive the habitat changes that rising temperatures will bring to their historic habitats. Based on a doubling of atmospheric carbon-dioxide levels, expected by many scientists within a century, the WWF report asserted that one-fifth of the world's most vulnerable natural areas may be facing a "catastrophic" loss of species (Lazaroff 2002a). "It is shocking to see that many of our most biologically valuable ecosystems are at special risk from global warming. If we don't do something to reverse this frightening trend, it would mean extinction for thousands of species," said Jay Malcolm, author of the WWF report and professor at the University of Toronto (Lazaroff 2002a).

According to the WWF report, areas most vulnerable to devastation from global warming include species in the Canadian Low Arctic Tundra; the Central Andean Dry Puna of Chile, Argentina, and Bolivia; the Ural Mountains; the Daurian Steppe of Mongolia and Russia; the Terai-Duar Savannah of northeastern India; southwestern Australia; and the Fynbos of South Africa. Among the U.S. ecosystems at risk, areas in California, the Pacific Northwest, and the Northern Prairie may be hardest hit, according to the WWF. The changes could devastate the shrub and woodland areas that stretch from Southern California to San Francisco, prairies in the northern heart of the United States, the Sierra Nevada, the Klamath-Siskiyou forest near the California-Oregon border, and the Sonoran-Baja deserts across the southwestern United States.

Banking Seeds against Extinction

In England, the Kew Royal Botanic Gardens marked International Biological Diversity Day in 2007 by banking its billionth seed in the vaults of the Millennium Seed Bank as insurance against risks of global warming. Genes of wild plants that are related to widely cultivated crops may be used to assist them in resisting pests and tolerating drought, according to a report by scientists of the Consultative Group on International Agricultural Research (CGIAR). "Kew's Millennium Seed Bank must be one of the most significant conservation projects ever," said U.K. Minister for Biodiversity, Landscape and Rural Affairs Barry Gardiner. "It is a global insurance policy against the loss of uniquely valuable plant species through land pressures or dangerous climate change" ("Saving Earth's Plant Diversity" 2007).

Climate change may threaten wild relatives of potatoes and peanuts. According to the CGIAR study's lead author, Andy Jarvis,

> Our results would indicate that the survival of many species of crop wild relatives, not just wild potato and peanuts, are likely to be seriously threatened even with the most conservative estimates regarding the magnitude of climate change.... At the moment ... existing collections are conserving only a fraction of the diversity of wild species that are out there. ("Saving Earth's Plant Diversity" 2007)

In the next 50 years, said Jarvis and his team, up to 60 percent of the 51 wild peanut species analyzed and 12 percent of the 108 wild potato species analyzed could become extinct because of warming.

"The irony here is that plant breeders will be relying on wild relatives more than ever as they work to develop domesticated crops that can adapt to changing climate conditions," said Annie Lane, the coordinator of a global project on crop wild relatives led by Bioversity International, the world's largest international research organization dedicated to the use and conservation of agricultural biodiversity. "Yet because of climate change, we could end up losing a significant amount of these critical genetic resources at precisely the time they are most needed to maintain agricultural production," said Lane ("Saving Earth's Plant Diversity" 2007).

Even during the twentieth century, warmer weather (especially milder winters) has been associated with changes in forest species. In one study, Gian-Reto Walther documented the increasing growth of broad-leaved evergreen trees in lower areas of southern Switzerland, where the growing season had increased to as long as 11 months in some areas by roughly the year 2000 (Walther 2002, 129). These areas have remarkably precise records of frost-free days. Between 1902 and 1998, for example, the onset of winter (the first day of frost) at Lugano had changed from an average of the second week in November to the first week of December. The last day of frost had changed from an average of the fourth week in March to the fourth week of February (Walthur 2002, 132).

Threats to Wildlife

Research compiled by the NWF suggests that global warming will probably pose an intensifying threat to U.S. wildlife, including more trouble with invasive species, and significant environmental changes that will jeopardize human quality of life in the near future (Lazaroff 2002a). "Global warming has come down to Earth for the wildlife right in our backyards," said Mark Van Putten, NWF president. "The effects are already happening and will likely worsen unless we get serious about reducing emissions of carbon dioxide and other heat-trapping gases to help slow global warming" (Lazaroff 2002a). The NWF's findings appeared in *Wildlife Responses to Climate Change* (2001), edited by Stephen Schneider of Stanford University and Terry Root of the University of Michigan. The book includes eight case studies by researchers that demonstrate how global warming and associated climate change is affecting North American wildlife.

The NWF report asserted that invasive species such as tamarisk shrubs in the U.S. Southwest may expand their ranges, reducing water and food available to native wildlife and humans. Imported fire ants in the Southeast also may expand their range, dominating native ant species and creating an enhanced health risk to humans (Lazaroff 2002a). Species such as the sachem skipper butterfly in the Pacific Northwest and the Bay checkerspot butterfly in California already are responding to climatic and weather changes. Changes in climate may alter habitat for grizzly bears, red squirrels, and other wildlife in the Greater Yellowstone ecosystem region of Wyoming, Montana, and Idaho by contributing to a reduction in whitebark pine trees, an important food source for the animals.

Roughly 70 percent of the plants and animals on the Fynbos of southern Africa are unique to that area, according to the WWF. In summer, the area is often so parched by drought and heat that it is ravaged by fire, with many plants dependent on regular fires to stimulate seeding. Almost half of this area will become uninhabitable for its major species within a century, according to the WWF (Browne 2002a, 15). Some species, such as the springbok of southern Africa, are attached to their habitats and are expected to die rather than move.

"Migration routes have become increasingly impassable due to human activities," wrote Walther, summarizing several scientific studies (Walther 2003, 177). At the same time, human communication facilitates invasive species worldwide. Climate change plays favorites among flora and fauna by removing climatic constraints; witness the explosion of spruce budworms throughout the pine forests of North America.

"Many habitats will change at a rate approximately 10 times faster than the most rapid changes since the last ice age, causing extinctions," said Ute Collier, head of the WWF's climate change program in the United Kingdom (Browne 2002a, 15). Washington State's Olympic Peninsula rain forest also is on WWF's list of endangered habitats. The rain forest cannot move as the climate changes. The peninsula is likely to undergo a drastic change, the report said, although scientists say it is difficult to determine how quickly that change will occur.

The Wildlife Society is an association of nearly 9,000 wildlife managers, research scientists, biologists, and educators, based in Washington, D.C. Its report, *Global Climate Change and Wildlife in North America,* anticipates that many animals may find their migratory paths blocked by cities, transportation corridors, or farmland. Predators and their prey may not move at the same time, impeding natural balances. Plants could suffer if the birds and insects that pollinate them head to cooler climes. Wetlands in the Midwest and central Canada are expected to dry up, causing some duck species to decline by as much as 69 percent over the next 75 years. Nesting habitat could be lost as wetlands become more suitable for row crops.

FURTHER READING

Browne, Anthony. "How Climate Change Is Killing off Rare Animals: Conservationists Warn that Nature's 'Crown Jewels' Are Facing Ruin." *London Observer,* February 10, 2002a, 15.

Lazaroff, Cat. "Climate Change Threatens Global Biodiversity." Environment News Service, February 7, 2002a. http://ens-news.com/ens/feb2002/2002L-02-07-06.html.

O'Malley, Brendan. "Global Warming Puts Rainforest at Risk." *Cairns Courier-Mail* (Australia), July 24, 2003, 14.

"Saving Earth's Plant Diversity from Global Warming." Environment News Service, May 22, 2007. http://www.ens-newswire.com/ens/may2007/2007-05-22-02.asp.

Walther, Gian-Reto. "Weakening of Climatic Constraints with Global Warming and Its Consequences for Evergreen Broad-Leaved Species." *Folia Geobotanica* 37 (2002):129–139.

Walther, Gian-Reto. "Plants in a Warmer World." *Perspectives in Plant Ecology, Evolution, and Systematics* 6, no. 3 (2003):169–185.

Biomass Fuel (including Ethanol)

By 2007, an ethanol boom (mainly fuel manufactured from corn) in the United States was being driven by a $0.51 per gallon federal subsidy, despite doubts about its suitability as a "green" fuel. Methane also was being extracted from organic garbage in landfills; some pig farms did the same with manure. As of 2007, this prosaic stuff was the biggest source of alternative fuel in the United States. Companies used biomass generation as part of their industrial processes. Weyerhauser, for one, generates electricity with wood waste combined with by-products from pulp mills that once were discarded as useless. Such sources have been generating power for $0.05 to $0.10 per kilowatt hour, which is competitive with fossil fuels. At St. Cloud State University, in Minnesota, vegetable oil used to fry foods in the student center, which previously was discarded, now fuels a shuttle that carries students from apartments to campus, dubbed the "Husky Fried Ride" (Jones 2008, 3-A).

Biogas from Animal Wastes and Landfills

Ingenuity is as it does; in the case of energy innovation, finding resources in what used to be waste. Witness a BMW manufacturing plant in Spartanburg, South Carolina, that by 2007 was getting more than half its energy from nearby landfills, formerly wasted methane, sparing the atmosphere several thousand tons of greenhouse gases a year.

U.S. Senator Ben Nelson, a Nebraska Democrat, during May 2007 introduced legislation creating federal tax credits, loans, and loan guarantees for "biogas"—methane fuel that can substitute for natural gas created from animal waste—notably Nebraska's more than six million cattle. This is an interesting application of Barry Commoner's idea, now several decades old, that a pollutant is a merely a resource that is out of place.

Such plants already were operating in Sweden and Denmark and one was being planned as a cooperative venture between Tyson Foods and ConocoPhillips. Such projects not only will turn what was waste into fuel, but also will reduce greenhouse-gas emissions from methane effluent. Use of cow manure as an energy source could reduce the amount of waste that contaminates water. A national biogas industry also could help pump economic activity into small towns that have been dying as the scale of farming grows.

Nelson's bill offers a federal tax credit of $4.27 for each million British thermal units (BTUs) of gas produced. In addition to loans and loan guarantees, biogas producers would receive a federal subsidy when and if natural-gas prices fall below a set level. While present technology allows for the production of biogas that is about 60 percent efficient (the remaining 40 percent is mainly carbon dioxide), Nelson's bill would fund research to a purity standard of 90 percent, which could be carried in pipelines along with natural gas (Thompson 2007, A-1).

Burgerville, a chain with 39 outlets in Oregon and Washington, by 2008 was collecting about 4,000 gallons of oil and grease from its restaurants and recycling it into 3,300 gallons of biodiesel fuel (Horovitz 2008, 2-B). By 2008, used cooking grease (as feedstock for biodiesel) was becoming so attractive as a gasoline surrogate that thieves were taking it from restaurants. Some chain burger and pizza restaurants in California and the state of Washington reported raids on their storage bins, and California police caught one man with a stash of 2,500 gallons of waste grease that could be refined as fuel (Saulny 2008). The grease is tradable as a commodity. Its price rose from $0.76 a pound in 2000 to $0.33 (about $2.50 a gallon) in mid-2008. At that price, the thief who was arrested in California with 2,500 pounds of grease had a stash worth about $6,000 (Saulny 2008).

Biogas in Sweden and Denmark

Sweden and Norway have the highest liquor taxes in the world, provoking large-scale smuggling from Denmark. Until recently, contraband seized by gold-and-blue-capped Tullverket (Swedish Customs) at Malmö (across Öresund Sound from Copenhagen) was poured down the drain. These days, however, a million illicit bottles a year are trucked to a sparkling new high-tech plant in Linköping (about 80 miles south-southwest of Stockholm) that manufactures biogas fuel. Every busted booze smuggler has been unwittingly drafted into Sweden's war against oil dependence and greenhouse gases. Signs in the Swedish Riksdag (Parliament) cafeteria advertise food scraps' second life as biogas.

The Linköping plant also accepts human and packing-plant waste. This swill produces biofuel for buses, taxis, garbage trucks, and private cars, as well as a methane-propelled biogas train that runs between Linköping and Västervik on the southeast coast. The train's boosters (not squeamish vegetarians, from the sound of it) have figured that the entrails from one dead cow, previously wasted, buys four kilometers (2.5 miles) on the train.

The Linköping plant is unusual for its omnivorous appetites. The Danish Crown slaughterhouse uses the fat of 50,000 pigs in an average week to generate biogas. The entire Danish Crown plant has been redesigned with an eye to saving energy, part of a 30-year Danish effort to eliminate waste, conserve energy, and reduce consumption of fossil fuels. Most of Denmark's energy infrastructure is owned by nonprofit cooperatives with resident shareholders. A majority of Denmark's people value high-quality health care, schools, and pensions over corporate profits, free individual choice, and low taxes.

Subsidies for Ethanol

By 2007, ethanol was becoming a lobbyist's dream in Washington, D.C. Proposals introduced in Congress during 2007 included aid for ethanol infrastructure (ethanol corrodes gasoline pipelines); the establishment of a Strategic Ethanol Reserve for years when corn harvests are reduced by droughts, similar to the national reserve for petroleum; and billions of dollars for research into cellulosic ethanol technologies, an estimated $10.8 billion by 2007 (Mufson and Morgan 2007, D-1).

In addition to the ethanol tax credit, other incentives include a $1.00 per gallon biodiesel tax credit, a subsidy for service stations that install E85 (ethanol 85%; gasoline 15%) pumps, spending by the Agriculture Department on energy programs, and various other Energy Department grants and loan guarantees (Mufson and Morgan 2007, D-1).

Even with tax subsidies, the ethanol industry found the year 2008 rather treacherous, with a severe credit contraction, as well falling prices (as oil and gasoline prices plunged) and volatile corn prices. One of the largest producers in the United States, VeraSun, which accounts for 13 percent of U.S. ethanol manufacturing, filed for bankruptcy protection after it gambled disastrously on the price of corn with commodity-price hedges, locking in corn feedstock at $7.00 a bushel as prices dove below $4.00. Other companies (such as Aventine Renewable Energy and Pacific Ethanol) found their stock prices plunging by 80 percent or more (Galbraith 2008, B-1, B-7).

How "Green" Is It?

Soon, making money became its own justification. The conductors of the ethanol gravy train stopped asking serious questions, such as: Just how "green" is it? Corn, processed into fuel, is still a carbon-based fossil fuel when it is burned to propel cars. It is not imported from dangerous locations in the Middle East (that much is true), but, depending on its source, ethanol emits only 10 to 20 percent less greenhouse gases than garden-variety gasoline. Even that savings is not what it seems. Corn must be grown on factory farms, an energy-intensive business itself. Under some circumstances (if a biomass field replaced a forest, for example), this type of fuel might actually produce a net increase in emissions of greenhouse gases, considering the entire production process.

The U.S. commitment to ethanol has become monumentally expensive. By 2008, current tax credits, grants, and loan guarantees would cost the federal treasury $140 billion in 15 years. New proposals under consideration in Congress at the time could have raised the cost to $205 billion. The biggest single prospective expense was an extension of the $0.51 per gallon ethanol tax credit that had been scheduled to expire in 2010. The extension would cost an estimated $131 billion through 2022.

In the meantime, the price of corn has been rising rapidly worldwide, feeding the inflation of food prices. Use of corn has become surprisingly widespread in our food supply, aside from the obvious products, such as corn flakes. Meat animals eat it, so it affects the price of chicken (meat and eggs), beef, and pork. The manufacture of soda pop consumes huge amounts of corn syrup, so its price also rises. A night at the movies even got pricier, as the price of popcorn rose 40 percent between 2006 and 2007.

Subsidies for ethanol have been good for framers who grow corn. The land under their farms has been rising in value. Prices for farmland in Nebraska soared 15 percent in one year, even as some farmers expressed concerns regarding ethanol's impact, especially its appetite for scarce water that may deplete aquifers.

Such prosperity comes at a price, however. The rising number of ethanol plants in the Great Plains region may drain billions of gallons of water each year from the Ogallala Aquifer, which already is depleted, according to Environmental Defense. A report authored by Martha Roberts and Theodore Toombs describes the aquifer as one of the world's largest, a vast, shallow underground pool beneath portions of Colorado, Kansas, Nebraska, New Mexico, Nebraska, Oklahoma, South Dakota, Texas, and Wyoming.

The Ogallala Aquifer already supports agriculture in the area, where the water table is declining as rates of groundwater pumping regularly exceed replacement. A gallon of ethanol requires four gallons of water in produce. "This dramatic expansion of ethanol production has substantial implications for already strained water and grassland resources in the Ogallala Aquifer region," the authors said ("Ethanol Production" 2007). Residents in Webster County, Missouri, sued to stop construction of an ethanol plant on grounds that it would use more water than the county's 33,000 residents combined. There are better ways to address the problem, however, and corn-derived fuel is a short-term answer at best.

Other concerns include truck traffic in rural areas, and air pollution with a sticky-sweet smell that resembles that of a barroom floor after a busy Saturday night (Barrett 2007, A-1). The imbalance between the amount of energy in corn ethanol and the amount that U.S. drivers consume as gasoline makes it impossible to potentially replace one with the other. Filling a sport-utility vehicle with ethanol requires 450 pounds of corn, enough to feed a person for more than a year. While sugarcane ethanol produces eight times the energy required to produce it (and gasoline produces five times), corn ethanol, energy-wise, is a wash, at 1.3 times the energy required for manufacture. Cellulosic ethanol (which distills the entire corn plant for energy, not just the kernels) has similar problems—it just does not contain enough energy to make it a suitable gasoline replacement. To replace half the 2006 gasoline consumption in the United States would require about seven times the U.S. land area planted in corn (Goodell 2007, 50, 53).

In a world of global markets, pressure on prices for one crop, such as corn, can have environmental consequences in other places. In the United States, for example, rising prices for corn caused in part by ethanol demand quickly translated into reduced planting of soybeans, which raised prices enough to provoke increasing deforestation in the Amazon basin as farmers in Brazil, the second-largest soy producer in the world, expanded their own fields. "Deforestation rates and especially fire incidence have increased sharply in the soy and beef-producing states in Amazonia," wrote William F. Laurance of the Smithsonian Tropical Research Institute in Science. "Studies suggest a strong link between Amazonian deforestation and soy demand" (Laurance 2007, 1721). Farmers purchasing large tracts of land for soy production that were being used for cattle pushed the herds farther into previously unexploited forests.

U.S. Agriculture Secretary Edward T. Schafer argues that biofuel production is responsible for only 2 to 3 percent of the increase in global food prices. He further argues that biofuels have reduced consumption of crude oil by a million barrels a day (Martin 2008a).

Once ethanol started pinching corn prices, senators' and representatives' phones lit up with protests from chicken farmers, pork producers, and the dairy industry, all of whom found their feed prices rising with corn futures. By 2008, corn had shot past $6 a bushel as farmers in the U.S. Midwest increased their plantings. The National Turkey Federation estimated that feed costs rose $600 million a year (Strassel 2007, A-16).

Some pork producers switched from feeding their pigs corn to anything else they could find in bulk—French fries, tater tots, hash browns, cheese curls, candy bars, yogurt-covered raisins, dried fruit, past-pull-date breakfast cereal mixed with chocolate powder. Sometimes farmers use burned cookies or breakfast cereal with too much sugar for human consumption, as well as noodles that fall off assembly lines while being packed and trail mix sprinkled with cardboard (Etter 2007, A-1, A-14). The manufacture of biodiesel ethanol from corn also produces distiller's grain as a by-product that can be fed to animals.

Despite doubts about ethanol as a "green" fuel, in 2007, U.S. corn acreage hit 93.6 million acres, a record, up 20 percent from 2006. One-fourth of U.S. corn went into ethanol by that time. Ethanol production capacity in the United States grew by 400 percent between 2000 and 2006, to 5.6 billion gallons a year. Ethanol is increasingly produced at large plants such as VeraSun Energy, near Charles City, Iowa, until 2008 the largest single producer in the United States. Steven Mufson of the *Washington Post* described VeraSun, which has since gone bankrupt:

> The plant is hard to miss. Its two massive concrete silos reach 150 feet into the air; each one holds half a million bushels of corn, delivered by an average of 110 brimming trucks every day. The silos are connected to a distillery with giant shiny steel vats for milling the corn, then fermenting and distilling it into 200-proof, fuel-grade ethanol. The ethanol is shipped out by train, 84 black tanker cars at a time. (Mufson 2008a, A-1)

The Ethanol Board of Nebraska, the Cornhusker State, fired back at critics of ethanol's green credentials, citing a report by several University of Nebraska researchers asserting that corn-based fuel emits 51 percent less greenhouse gases compared to comparable energy output from gasoline. The report, published in the *Journal of Industrial Ecology*, took issue with critics who had said that the energy-intensive nature of corn cultivation produced barely as much energy as it consumed in production. The ratio is at least 1.5 to 1.8 and could range as high as 2.2 with more attention to conservation methods on the farm and to energy efficiency in ethanol production.

The study suggested that 60 percent of U.S. ethanol was being produced in high-efficiency plants. Adam Liska, one of the study's co-authors, said that the proportion of ethanol from high-efficiency plants would continue to increase. The study does not factor in the environmental costs of converting tropical forest lands to production of food, including corn. By 2009, 70 percent of gasoline sold in the United States was blended with ethanol (Reed 2009, 2-A; Liska et al. 2009).

Let Them Eat Ethanol

"While many are worrying about filling their gas tanks, many others around the world are struggling to fill their stomachs, and it is getting more and more difficult every day," World Bank President Robert B. Zoellick said. "As long as you keep that ethanol industry running, grain prices will be high," says Bruce Babcock, professor of economics and the director of the Center for Agricultural and Rural Development at Iowa State University. "If you didn't have this large growth in ethanol corn, prices would be nowhere near where they are today" (Mufson 2008a, A-1).

As early as January 31, 2007, rising prices for corn were igniting demonstrations by 75,000 people in Mexico City, where the price of tortillas hit record highs as President G. W. Bush touted corn ethanol in his State of the Union message. The price of corn shot up from $1.90 to $3.75 a bushel between 2006 and 2007. Rising costs of farm goods, fed partially by demand for biofuels (including corn, sugarcane, and palm oil, among others), has been pushing up food prices globally. This rise in prices is causing distress among many poorer people in China, India, and other nations. If rising food prices are sustained, social unrest could result.

By 2008, a worldwide reaction was surging against biofuels as food prices (also beset by other supply and demand factors) had surged 83 percent in three years, according to the World Bank, provoking food riots in many poorer nations (Martin 2008a). During April 2008, food riots contributed to the dismissal of Haiti's prime minister, as many poor people there ate "cookies" of baking soda and mud. The emphases on food-to-fuel tightened supplies along with widespread droughts and affluence in developing nations with large populations, such as India and China. (Rice and wheat also have risen sharply in price, but no one is making ethanol out of them.) By 2008, one-fifth of the U.S. corn crop was going into ethanol.

In Germany, the price of beer rose as farmers abandoned barley to grow crops that could be sold as feedstock for ethanol, such as corn. The barley crop fell 5.5 percent between 2005 and 2006, and prices rose. Prices at the 2007 Oktoberfest in Munich posted a 5.5 percent price increase, raising a one-liter mug to the equivalent of $10.76. "Beer prices are a very emotional issue in Germany. People expect it to be as cheap as other basic staples like eggs, bread, and milk," said Helmut Erdmann, director of the Ayinger Brewery, in the hills of Bavaria ("Germans Blame" 2007, A-18).

Ethanol Spurs Global Warming

When the full emissions costs of producing biofuels are calculated, most of them are environmentally more expensive, causing more greenhouse gases than fossil fuels, according to studies published in *Science* early in 2008. Growth of feedstock for many biofuels, from corn to sugarcane to palm oil, destroys natural ecosystems (most notably rain forest in the tropics and South American grasslands), releasing gases as they are burned and plowed. Destruction of these older, natural ecosystems also removes carbon sinks. In addition to the greenhouse gases caused by growing biofuels, additional emissions result from refining and transporting them.

Some of the carbon footprint of biofuels may be mitigated by conservation tilling, which minimizes disturbance of the soil to retain carbon dioxide and, at the same time, leaves at least 30 percent of the surface covered with crop residues. Use of such methods can reduce carbon-dioxide emissions from the soil by as much as 50 to 60 percent. Crop residue also can provide feedstock for cellulosic ethanol (Richards 2008, 7-B).

Kenya's Tana River Delta, inhabited by 350 species of birds, lions, elephants, rare sharks, and reptiles, is about to be converted to sugarcane production over the objections of conservationists and local communities. The Kenyan government has approved a proposal by a publicly traded company based in Nairobi to covert 2,000 square kilometers of the pristine delta into irrigated sugarcane plantations.

"When you take this into account, most of the biofuel that people are using or planning to use would probably increase greenhouse gasses substantially," said Timothy Searchinger, lead author of one of the studies and a researcher in environment and economics at Princeton University. "Previously there's been an accounting error: land use change has been left out of prior analysis" (Rosenthal 2008a).

Clearance of grassland releases 93 times the amount of greenhouse gas that would be saved by the fuel made annually on that land, said Joseph Fargione, lead author of the second paper, and a scientist at the Nature Conservancy. "So for the next 93 years you're making climate change worse, just at the time when we need to be bringing down carbon emissions" (Rosenthal 2008a).

Many U.S. farmers are growing corn year-round, while they used to alternate with soybeans. More soybeans are being raised on newly cleared rain forest land in Brazil.

Joseph Fargione and colleagues wrote:

> Biofuels are a potential low-carbon energy source, but whether biofuels offer carbon savings depends on how they are produced. Converting rainforests, peatlands, savannas, or grasslands to produce food crop-based biofuels in Brazil, Southeast Asia, and the United States creates a "biofuel carbon debt" by releasing 17 to 420 times more CO_2 than the annual greenhouse gas (GHG) reductions that these biofuels would provide by displacing fossil fuels. In contrast, biofuels made from waste biomass or from biomass grown on degraded and abandoned agricultural lands planted with perennials incur little or no carbon debt and can offer immediate and sustained GHG advantages. (Fargione et al. 2008, 1235)

Timothy Searchinger and colleagues wrote:

> Most prior studies have found that substituting biofuels for gasoline will reduce greenhouse gases because biofuels sequester carbon through the growth of the feedstock. These analyses have failed to count the carbon emissions that occur as farmers worldwide respond to higher prices and convert forest and grassland to new cropland to replace the grain (or cropland) diverted to biofuels. By using a worldwide agricultural model to estimate emissions from land-use change, we found that corn-based ethanol, instead of producing a 20 percent savings, nearly doubles greenhouse emissions over 30 years and increases greenhouse gases for 167 years. Biofuels from switchgrass, if grown on U.S. corn lands, increase emissions by 50 percent. This result raises concerns about large biofuel mandates and highlights the value of using waste products. (Searchinger et al. 2008, 1238)

Critique of Corn as Fuel from a Climatic Point of View

President Bush's enthusiasm for ethanol was weak on climate security. "I am disappointed," said Senator Jeff Bingaman, a Democrat from New Mexico and chairman of the Senate Energy Committee. He said Bush was "completely silent" on energy efficiency and reduction of carbon dioxide from electric power plants, which contribute 40 percent of carbon-dioxide emissions. "If this was a real effort to solve the climate problem, it would include large stationary sources and utilities," said Eileen Claussen, president of the Pew Center on Global Climate Change, a nonpartisan policy research group.. Claussen and other environmental leaders also criticized the absence of any proposal for a mandatory cap on emissions of heat-trapping gases (Andrews and Barringer 2007).

Environmentally, ethanol is a nonstarter, according to James E. Hansen, director of NASA's Goddard Institute for Space Studies, and one of the leading climate scientists in the United States:

> A proposed national plan for 20 percent ethanol in vehicle fuels, envisaged to be derived in large part from corn, does more harm to the planet than good. It would do little to reduce CO_2 emissions, it would degrade retention of carbon in soils and forests, and it would strike hard at the world's poor through increased food prices. There are a variety of ways that renewable or other CO_2-free energies may eventually power vehicles. Governments should not dictate the nature of those solutions. Biofuels are likely to play a major part in our energy future. As a native Iowan, I like to imagine that the Midwest will come to the rescue of compatriots threatened by rising seas. Native grasses appropriately cultivated, perhaps with improved varieties, can draw down atmospheric CO_2. The prairies from Texas to North Dakota may contribute, if we get on with solving the climate problem before super-drought spreads from the west to the prairies. If we act soon, we can keep the prairies as productive land. Positive feedbacks work in both directions. (James Hansen, via e-mail, April 12, 2007)

Doing the Math: Too Much Land, Not Enough Fuel

One major problem with biofuels is the amount of land required to supply them. In Britain, author George Monbiot calculated that using the United Kingdom's most productive oil crop (rapeseed) to replace petroleum in British transport at present-day consumption levels would require four to five times the nation's entire arable land base. Commented Monbiot:

> In order to move our [British] cars and buses with biodiesel, we would require 25.9 million hectares. There are 5.7 million hectares in the United Kingdom. If this were to happen all over Europe, the consequences on food supply would be catastrophic: enough to tip the scales from being excess producers to becoming net losers. If, as some environmentalists claim, this were to be done on a world scale, most of the arable surface of the planet would have to be given over to producing food for cars, not for people. This outlook would seem, at a first glance, to be ridiculous. If the demand for food could not be covered, wouldn't the market ensure that crops be used to feed people instead of

> cars? Nothing is sure about this. The market responds to money, not to needs. (Monbiot 2006a, 158)

Researchers at the University of Minnesota have estimated that converting the entire U.S. corn crop to ethanol would replace only one-eighth of U.S. gasoline consumption (Krugman 2007, A-23). In addition, corn must be grown and transported, after which ethanol must be manufactured. Replacing a gallon of gasoline with a gallon of ethanol does not save a gallon of gasoline, because most of the energy that goes into corn comes from fossil fuels. The real savings is more like a quarter of a gallon—so make that 3 percent savings of gasoline for the entire U.S. corn crop (Krugman 2007, A-23). Then, what would we *eat*?

Ethanol from Sugarcane: Brazil's Energy Crop

Ethanol can be made through the fermentation of many natural substances, but sugarcane offers advantages over others, such as corn. For each unit of energy expended to turn cane into ethanol, 8.3 times as much energy is created, compared with a maximum of 1.3 times for corn, according to scientists at the Center for Sugarcane Technology and other Brazilian research institutes. "There's no reason why we shouldn't be able to improve that ratio to 10 to 1," said Suani Teixeira Coelho, director of the National Center for Biomass at the University of São Paulo. "It's no miracle. Our energy balance is so favorable not just because we have high yields, but also because we don't use any fossil fuels to process the cane, which is not the case with corn" (Rohter 2006). Sugarcane is generally more economical than oil with the per-barrel price under $30.

Use of ethanol in Brazil accelerated after 2003 following the introduction of "flex-fuel" engines, designed to run on ethanol, gasoline or any mixture of the two. Gasoline sold in Brazil contains about 25 percent alcohol. By 2006, more than 70 percent of the 1.1 million automobiles sold in Brazil had flex-fuel engines (Rohter 2006).

Using ethanol from sugarcane, Brazil became energy self-sufficient in 2006, even as demand for fuel grew. Brazil's full court press on ethanol was three decades old by that time. During the 1970s, Brazil's government began developing the ethanol industry by subsidizing the sugarcane industry and requiring its use in government vehicles. By the late 1990s, however, the subsidies were phased out as the cost of producing ethanol dropped to $0.80 per gallon, less than the worldwide average of $1.50 per gallon for producing gasoline (Luhnow and Samor 2006, A-1, A-8).

By 2007, sugarcane-based ethanol was supplying almost 40 percent of the energy used for ground transport in Brazil. Three-quarters of new cars sold in Brazil were flex-fuel (compared with 10 percent in the United States), and most fueling stations offered E85 fuel, compared with just 1 percent in the United States.

Brazil's government taxes ethanol at $0.09 per gallon, compared with $0.42 for gasoline, and all gasoline is legally required to contain at least 10 percent ethanol. Researchers in Brazil have decoded the genetics of sugarcane and used the knowledge to breed varieties with higher sugar content. Brazil has increased the per-acre productivity of sugarcane threefold since 1975 (Luhnow and Samor 2006, A-1, A-8).

Politicians from corn-growing states in the United States have obtained a stiff protective tariff on Brazilian sugar-derived alcohol. Brazil exported $600 million worth of ethanol in 2005, but nearly none of it went to the United States.

In the past, the residue left when cane stalks are compressed to squeeze out juice was discarded. Today, Brazilian sugar mills use that residue to generate the electricity to process cane into ethanol, and use other by-products to fertilize the fields where cane is planted. Some mills are now producing so much electricity that they sell their excess to the national grid. In addition, Brazilian scientists, with money from São Paulo State, have mapped the sugarcane genome. That opens the prospect of planting genetically modified sugar, if the government allows, that could be made into ethanol even more efficiently (Rohter 2006).

Sugarcane ethanol may be an energy winner, but local native people and fieldworkers have become victims of the boom. An Italian film *Birdwatchers* (*La Terra Degli Uomini Rossi*) describes the Brazilian Guarani-Kaiowá Indians' struggle against factory-farming biofuels that are crowding them off their land. *Birdwatchers* parses the issues (land invasion, suicides, and rebellion) against the backdrop of a love story involving the daughter of a wealthy land owner and a young Guarani who has become a shaman apprentice in the Brazilian state of Mato Grosso do Sul.

"Mato Grosso" means "thick forest" in Portuguese. Today, however, most of the trees have been felled. During the last 70 years, the Guarani and neighboring peoples have been evicted from their land by cattle ranchers, as well as sugarcane and soya planters. Many no longer work on the ranches for subsistence wages, in perpetual peonage. Many (more than 500 in 20 years, some of them only nine years of age) have killed themselves. Others have been shot to death while trying to re-occupy alienated land.

Why No Sugar-Beet Ethanol?

If sugarcane is such a good source for ethanol, why not use sugar beets? The processing equipment for sugar beets and sugarcane is somewhat similar; however, the problem with sugar beets lies in their harvest cycle in temperate regions. While sugarcane is grown nearly year-round (and thus could supply a processing plant almost all the time) sugar beets are grown on an annual cycle and harvested in the fall. Thus, according to Kenneth P. Vogel, a professor in the University of Nebraska at Lincoln's Agronomy Department, a processing plant that costs hundreds of millions of dollars would run only for a few months a year. This hurdle could perhaps be overcome with technology (not yet designed) to switch from sugar beets to corn and other types of ethanol sources. Instead, ethanol interests on the Great Plains are looking at switchgrass (which is handled like hay), corn stover (stalks), and other plant biomass. The manufacturing technology for these needs to be developed as well ("No Sugar-beet" 2007, 6-B).

Food, Fuel, or Forests?

When ethanol is processed from palm oil, the victims are usually tropical rainforests and the animals whose habitats are being turned into factory energy farms. Such plantations have an ecological and greenhouse cost; however, rainforest land is being cleared with fire, which adds carbon dioxide and methane to the atmosphere while ruining the habitat of orangutans and other increasingly scarce animals (Kennedy 2007, 515).

Palm oil is another common source of ethanol, and another one that is not likely to be a domestic product of the United States. Its high-energy efficiency per unit makes palm oil an excellent biodiesel fuel, and trucks can run entirely on it if necessary (although most run on a mixture of palm oil and oil-based products). Large tracts of land are being converted to palm-oil production in Indonesia, with plans to double production in perhaps a decade.

By 2007, one-fifth of the human-caused greenhouse gases being released into the atmosphere came from deforestation, during which carbon formerly stored in trees enters the air. Indonesia has been clearing more forests than any other country. In some areas, such as the province of Riau (on the island of Sumatra), more than half the forests have been felled in a decade, many for palm-oil plantations, often for ethanol. The burning and drying of Riau's peat lands, also to make way for palm oil plantations, also releases about 1.8 billion tons of greenhouse gases a year (Gelling 2007).

A European environment advisory panel early in April 2008 urged the European Union to suspend its goal of having 10 percent of transportation fuel made from biofuels by 2020. Europe's well-meaning rush to biofuels, the scientists concluded, according to a *New York Times* report, had created a variety of harmful ripple effects, including deforestation in Southeast Asia and higher prices for grain (Martin 2008a). The United Nations Permanent Forum on Indigenous Issues (UNPFII) estimated that increasing human rights violations, displacements, and conflicts due to expropriation of ancestral lands and forests for biofuel plantations planned by 2008 could cause 60 million indigenous people worldwide to lose their lands and livelihoods.

According to Survival International, a London-based activist group that defends indigenous peoples, palm oil is one of the most destructive crops used for biofuel, affecting millions of indigenous people in Malaysia and Indonesia. In Colombia, thousands of families, many of them indigenous, have been violently evicted from their land because of palm-oil plantations and other crops.

The World Rainforest Movement has criticized monoculture factory farming for food and fuel as destructive forests around the world. Soybean plantations in Argentina are gradually displacing the quebracho forests in the Chaco; while in Paraguay, they are replacing the Pantanal, the Mata Atlantica, and the Chaco; and in Brazil, the Pantanal, the Mata Atlantica, the Cerrado, and the Caatinga. Between 1990 and 2002, the planted area of oil palm, a source of feedstock for ethanol manufacture, increased by 43 percent worldwide, mostly in Indonesia and Malaysia. Between 1985 and 2000, oil-palm plantations have been responsible for 87 percent of deforestation in Malaysia, with plans to replace another six million hectares of forest.

In Sumatra and Borneo, roughly four million hectares of forests have been converted to oil-palm plantations. In Indonesia, thousands of indigenous people have been evicted from their lands. Deforestation often spreads via fire.

The entire region is becoming a gigantic vegetable oil field. In Uganda the destruction of tropical forests and indigenous forestlands has begun to produce palm oil and sugar.

Burning and slashing of forests to make way to plantations of oil palm releases enormous carbon reserves. Thus, the road to ethanol is hardly "carbon neutral." In marshy forests, where there is peat, once the trees are cut, the plantations dry out the soil. When the peat dries, it oxidizes and releases even more carbon dioxide than the trees.

Ethanol and the Gulf of Mexico "Dead Zone"

Yet another environmental debit of corn ethanol is its contribution to expansion of the Gulf of Mexico's seasonal "dead zone," an area of very low oxygen that kills most sea life, according to a study by Simon Donner of the University of British Columbia and Chris Kucharik of the University of Wisconsin-Madison, who modeled effects of its production. Putting it briefly: more corn ethanol requires more fertilizer, which, washed into the Gulf as nitrogen overload, aggravates the dead zone. By 2007, the dead zone by summer already covered 7,700 square miles, the size of New Jersey ("Corn-ethanol" 2008). Nitrogen and phosphorus from agricultural fertilizers provokes growth of algae that sucks up nearly all available oxygen. Animals in the water leave the affected area, or die.

The dead zone in the Gulf of Mexico is likely to be the largest on record, growing in large part because of increasing U.S. corn production. In 2008, the dead zone covered some 8,800 square miles.

Cellulosic Ethanol

For advocates of ethanol who would rather not devote most U.S. corn to fuel, "cellulosic ethanol," a fuel from plants such as switchgrass, has yet to be produced at anything close to competitive prices, failing thus-far unachieved technological advances. Switchgrass does not contain much energy, so vast amounts of land would be required to sustain a sizable fuel industry. A viable ethanol industry from cellulosic materials could require as much as 40 million additional acres devoted to growing the plant material, as well as a sprawling new infrastructure to transform the feedstock into fuel. Cellulosic ethanol, however, pending development of the right technology, has the potential of being eight times as energy efficient as corn-based ethanol because it does not need to be converted into sugar before it is manufactured into fuel.

The largest cellulosic ethanol plant in 2007 was a demonstration-scale facility in Canada built by the Ontario-based Iogen Corporation, which produces about one million gallons a year. A commercial-scale factory would require annual capacity of at least 40 million gallons, according to Robert Dineen, president of the Renewable Fuels Association, which represents ethanol producers (Andrews 2007a). In September, 2008, the Iowa Power Fund allocated $14.75 million in state constriction money to the Poet Biorefining facility in Emmettburg for a cellulosic ethanol plant that will burn corn stalks and cob as well as kernels.

"The challenge is biochemical," wrote Donald Kennedy, editor of *Science*. "Plant lignins occlude the cellulose cell walls; they must be removed, and then the enzymology of cellulose conversion needs to be worked out. The technology is complex. No commercial reactor has yet been built, although six are funded" (Kennedy 2007, 515). British Petroleum and the universities of California and Illinois have undertaken a $500 million project to cross this scientific barrier.

In 2007, U.S. Agriculture Secretary Mike Johanns attempted to assure the public that ethanol manufacture would not take food out of people's mouths. "We've already put forth a Farm Bill proposal that would increase funding for renewable energy by $1.6 billion. Without question, the President's proposals represent the most significant commitment to renewable energy that's ever been proposed in farm legislation," Johanns said. "It's focused on cellulosic ethanol, which is where we believe the next step is in terms of ethanol development." Cellulosic ethanol is not made from corn kernels but is distilled from the fermentation of sugars from the entire plant, not just the grains. Perennial grasses, corn stover, sugarcane bagasse, logging slash, and yard trimmings can all be sources of cellulosic ethanol ("Bush Orders" 2007).

During 2007, Abengoa Bioenergy, a Spanish energy company with head offices in Seville, selected the town of Hugoton in southwestern Kansas as the site of the first U.S. cellulosic ethanol plant, a $400 million project that will convert 700 tons a day of corn stover, wheat straw, milo stubble, switchgrass, and other biomass into fuel. The plant will have a capacity of 30 million gallons a year; it also will include an 85 million gallon a year corn ethanol plant. The Hugoton project will be partly funded by $76 million from the U.S. Department of Energy (DOE).

In February 2007, the DOE awarded up to $385 million to six companies, including Abengoa, for first-generation ethanol plants in Florida, Georgia, Iowa, Idaho, California, and Kansas. "These bio-refineries will play a critical role in helping to bring cellulosic ethanol to market, and teaching us how we can produce it in a more cost effective manner," said Energy Secretary Samuel W. Bodman, awarding the grants. "Ultimately, success in producing inexpensive cellulosic ethanol could be a key to eliminating our nation's addiction to oil" ("Kansas Gets First" 2007).

FURTHER READING

Andrews, Edmund L. "Bush Makes a Pitch for Amber Waves of Homegrown Fuel." *New York Times*, February 23, 2007a. http://www.nytimes.com/2007/02/23/washington/23bush.html.

Barbier, Edward B., Joanne C. Burgess, and David W. Pearce. "Technological Substitution Options for Controlling Greenhouse-gas Emissions." In *Global Warming: Economic Policy Reponses,* eds. Rutiger Dornbusch and James M. Poterba, 109–161. Cambridge, MA: MIT Press, 1991.

Barrett, Joe. "Ethanol Reeps a Backlash in Small Midwestern Towns." *Wall Street Journal*, March 23, 2007, A-1, A-8.

"Bush Orders First Federal Regulation of Greenhouse Gases." Environment News Service, May 14, 2007. http://www.ens-newswire.com/ens/may2007/2007-05-14-06.asp.

"Corn-ethanol Crops Will Widen Gulf of Mexico Dead Zone." Environment News Service, Match 11, 2008. http://www.ens-newswire.com/ens/mar2008/2008-03-11-091.asp.

"Ethanol Production Threatens Plains States With Water Scarcity." Environment News Service, September 21, 2007. http://www.ens-newswire.com/ens/sep2007/2007-09-21-091.asp.

Etter, Lauren. "With Corn Prices Rising, Pigs Switch to Fatty Snacks." *Wall Street Journal*, May 21, 2007, A-1, A-14.

"Europe to Cut Greenhouse Gases 20 Percent by 2020." Environment News Service, March 8, 2007. http://www.ens-newswire.com/ens/mar2007/2007-03-08-04.asp.

Fargione, Joseph, Jason Hill, David Tilman, Stephen Polasky, and Peter Hawthorne. "Land Clearing and the Biofuel Carbon Debt." *Science* 319 (February 29, 2008):1235–1238.

"Fueling Jets with Animal Fat." Environment News Service, July 18, 2007. http://www.ens-newswire.com/ens/jul2007/2007-07-18-09.asp#anchor7.

Galbraith, Kate. "Economy Shifts, and the Ethanol Industry Reels." *New York Times*, November 5, 2008, B-1, B-7.

Gelling, Peter. "Forest Loss in Sumatra Becomes a Global Issue." *New York Times*, December 6, 2007. http://www.nytimes.com/2007/12/06/world/asia/06indo.html.

"Germans Blame Ethanol Boom for—Oh Mein Gott!—Rising Beer Prices." Associated Press in *Omaha World-Herald*, June 3, 2007, A-18.

Goodell, Jeff. "The Ethanol Scam." *Rolling Stone*, August 9, 2007, 48–53.

Horovitz, Bruce. "Can Eateries Go Green, Earn Green?" *USA Today*, May 16, 2008, 1-B, 2-B.

Jones, Charisse. "Transit Systems Travel 'Green' Track." *USA Today*, May 8, 2008, 3-A.

"Kansas Gets First U.S. Cellulosic Ethanol Plant" Environment News Service, August 28, 2007. http://www.ens-newswire.com/ens/aug2007/2007-08-28-097.asp.

Kennedy, Donald. "The Biofuels Conundrum." *Science* 316 (April 27, 2007):515. http://www.nytimes.com/2007/12/18/washington/18ethanol.html.

Krugman, Paul. "The Sum of All Ears." *New York Times*, January 29, 2007, A-23.

Laurance, William F. "Switch to Corn Promotes Amazon Deforestation." (Letter to the Editor) *Science* 318 (December 14, 2007):1721.

Liska, Adam J., Haishun S. Yang, Virgil R. Bremer, Terry J. Klopfenstein, Daniel T. Walters, Galen E. Erickson, and Kenneth G. Cassman. "Improvements in Life-Cycle Energy Efficiency and Greenhouse Gas Emissions of Corn-Ethanol." *Journal of Industrial Ecology* (January 2009), doi: 10.1111/j.1530-9290.2008.00105.x.

Luhnow, David, and Geraldo Samor. "As Brazil Fills Up on Ethanol, It Weans off Energy Imports." *Wall Street Journal*, January 9, 2006, A-1, A-8.

Martin, Andrew. "Fuel Choices, Food Crises and Finger-Pointing." *New York Times*, April 15, 2008a. http://www.nytimes.com/2008/04/15/business/worldbusiness/15food.html.

Martin, Andrew. "Food Report Criticizes Biofuel Policies." *New York Times*, May 30, 2008b. http://www.nytimes.com/2008/05/30/business/worldbusiness/30food.html.

Monbiot, George. *Heat: How to Stop the Planet from Burning*. Toronto: Doubleday Canada, 2006a.
Mufson, Steven. "Siphoning Off Corn to Fuel Our Cars." *Washington Post*, April 30, 2008a, A-1. http://www.washingtonpost.com/wp-dyn/content/article/2008/04/29/AR2008042903092_pf.html.
Mufson, Steven, and Dan Morgan. "Switching To Biofuels Could Cost Lots of Green." *Washington Post*, June 8, 2007, D-1. http://www.washingtonpost.com/wp-dyn/content/article/2007/06/07/AR2007060702176_pf.html.
"No Sugar-beet Answer." Editorial, *Omaha World-Herald*. April 7, 2007, 6-B.
Patzek, Tad W. "The Real Biofuel Cycles" (Letter to the Editor). *Science* 312 (June 23, 2006):1747.
Reed, Leslie. "Research: Ethanol Isn't So Wasteful." *Omaha World-Herald*, January 27, 2009, 1-A, 2-A.
Richards, Bill. "A Good Combination: Biofuels, Smart Tilling." *Omaha World-Herald*, March 11, 2008, 7-B.
Rohter, Larry. "With Big Boost from Sugar Cane, Brazil Is Satisfying Its Fuel Needs." *New York Times*, April 10, 2006. http://www.nytimes.com/2006/04/10/world/americas/10brazil.html.
Rosenthal, Elisabeth. "Studies Deem Biofuels a Greenhouse Threat." *New York Times*, February 8, 2008a. http://www.nytimes.com/2008/02/08/science/earth/08wbiofuels.html.
Saulny, Susan. "As Oil Prices Soar, Restaurant Grease Thefts Rise." *New York Times*, May 30, 2008. http://www.nytimes.com/2008/05/30/us/30grease.html.
Searchinger, Timothy, Ralph Heimlich, R. A. Houghton, Fengxia Dong, Amani Elobeid, Jacinto Fabiosa, Simla Tokgoz, Dermot Hayes, and Tun-Hsiang Yu. "Use of U.S. Croplands for Biofuels Increases Greenhouse Gases Through Emissions from Land-Use Change." *Science* 319 (February 29, 2008):1238–1240.
Strassel, Kimberly. "Ethanol's Bitter Taste." *Wall Street Journal*, May 18, 2007, A-16.
Thompson, Jake. "Nelson: Don't Waste the Waste." *Omaha World-Herald*, May 9, 2007, A-1, A-2.

Birds, Butterflies, and Other Migratory Animals

A report by the United Nations Environment Program (UNEP) found that climate change is inflicting severe impacts on migratory species, from whales and dolphins to birds and turtles, and is likely to be increasingly disruptive. "Migratory species are in many ways more vulnerable than other species as they use multiple habitats and sites and many difference resources during their migrations ("Climate Change Dislocates" 2006). The National Audubon Society and the American Bird Conservancy's WatchList 2007, said that 178 bird species in the United States (a quarter of the U.S. bird species) are threatened with extinction not only by global warming, but also by invasive species and urban sprawl, up 10 percent from 2002.

The report, "Migratory Species and Climate Change: Impacts of a Changing Environment on Wild Animals," documented a wide range of present-day climate effects. European Bee-Eaters (*Merops apiaster*), birds once very rare in Germany, are now breeding regularly across the country. The Rosy-Breasted Trumpeter Finch (*Rhodopechys githaginea*) is one of many bird species once usually confined to arid North Africa and the Middle East now found in increasingly large numbers in southern Spain.

The arrival of hundreds of Bewick Swans (*Cygnus columbianus*), flying in distinctive "V" formations, used to herald the arrival of the British winter. Ornithologists report that their numbers are now down to double-digits. Warmer weather on the continent and the absence of the northeast winds that aid their migration are the likely reasons for the swans' disappearance from their traditional British wintering sites. Changing wind patterns are making it more difficult for many birds to migrate over the Caribbean Sea, where spring storms are becoming more numerous and of greater intensity ("Climate Change Dislocates" 2006).

While fears abound that rapid climate change will outstrip many species' abilities to adapt, one 47-year study of the Great Tit (*Parus major*) in the United Kingdom indicates that these wild birds not only adapted as the weather warmed, but also flourished, adjusting their eating habits and reproductive behavior. For example, the average egg-laying date of the Great Tits advanced 14 days during the study period (1961–2007) (Charmantier et al. 2008, 800). All species have their limits, of course, and one question left unanswered by this study is just what degree of warming will continue to favor the Great Tit, particularly give that climate change in the past half-century has been modest compared with what models project for the next century.

Widespread Effects of Changing Weather on Birds

The National Audubon Society report also said that "[t]his autumn [2006] several large monarch butterflies, *Danaus plexippus*, which migrate in millions every year from the USA and

Canada to Mexico, have been blown across the Atlantic to England 5,000 kilometers away" ("Climate Change Dislocates" 2006). Elsewhere, according to the report, desertification is increasing the size of the Sahara Desert, adversely affecting the ability of Afro-European migrants to cross this ecological barrier successfully ("Climate Change Dislocates" 2006).

Birds are suffering the escalating effects of climate change worldwide, according to a 2006 report by the World Wildlife Fund (WWF). The study found declines of 90 percent in some bird populations, "as well as total and unprecedented reproductive failure in others." The WWF study estimated that bird extinction rates could be as high as 38 percent in Europe, and 72 percent in northeastern Australia, if global warming exceeds 2°C above preindustrial levels. Currently warming is 0.8°C above those levels, with another degree of increase "in the pipeline" ("Climate Change Pushing" 2006). The report, "Bird Species and Climate Change: The Global Status Report," reviews more than 200 scientific articles on birds on every continent to assemble a global picture of climate change's impacts.

"Robust scientific evidence shows that climate change is now affecting birds' behavior," said Karl Mallon, scientific director at Climate Risk Pty Ltd. of Sydney, Australia, an author of the report. "We are seeing migratory birds failing to migrate, and climate change pushing increasing numbers of birds out of synchrony with key elements of their ecosystems," Mallon said ("Climate Change Pushing" 2006)

"Birds have long been used as indicators of environmental change, and with this report we see they are the quintessential 'canaries in the coal mine' when it comes to climate change," said Hans Verolme, director of WWF's Global Climate Change Program. "This report finds certain bird groups, such as seabirds and migratory birds, to be early, very sensitive, responders to current levels of climate change," explained Verolme. "Large-scale bird extinctions may occur sooner than we thought" ("Climate Change Pushing" 2006).

In Australia, the Mallee Emu-wren (*Stipiturus mallee*) is rapidly losing population, its habitat so fragmented that a single bushfire could wipe out the species. Enduring drought in the southern and western parts of the species' range has destroyed vegetation that forms the basis of this species' diet. In South Australia, this species has been reduced to about 100 birds confined to 100 square kilometers ("Warmer Climate" 2008).

The vulnerability of seabirds to climate change is illustrated by the unprecedented breeding crash of U.K. North Sea seabirds during 2004, the WWF report said. The direct cause for the breeding failure of common guillemots, Arctic Skuas, Great Skuas, Kittiwakes, Arctic Terns, and other seabirds at Shetland and Orkney colonies was a shortage of their prey, a small fish called sandeels. Warming ocean waters and major shifts in species that underpin the ocean food web are believed to be behind the major sandeel decline. Nearly 7,000 pairs of Great Skuas in the Shetlands produced only a few chicks, and WWF reports that starving adult birds ate their own young ("Climate Change Pushing" 2006).

Dutch researchers have reported that temperature increases have decoupled Pied Flycatchers migration dates with the availability of food sources. The species spends winters in West Africa and returns to woodlands in Holland each spring. Average temperatures in the spring territory have increased several degrees during the last 20 years. According to the researchers, higher spring temperatures mean that insects, the birds' food source, reach peak abundance earlier, providing birds that hatch at the same time with more food.

The researchers found that over the two decades, the Pied Flycatcher's mean laying date has advanced by a little more than a week, as selection has favored earlier laying. The dates that the birds migrate has not changed by more than a week, the researchers believe, because the birds' decision to leave their winter habitat is related to the amount of daylight, which is not affected by air temperature. The birds are arriving at the same time, but laying eggs earlier. The researchers suggest that this phenomenon may be contributing to the decline of other long-distance migrants (Fountain 2001, F-4; Both et al. 2006, 81).

Baltimore without Orioles: Anticipated Bird Extinctions

Maryland's Baltimore Orioles (*Icterus galbula*), which have been declining due to habitat loss for many years, could vanish altogether late in the twenty-first century due to changes in migration patterns strongly influenced by a warming climate. A study by the National

Wildlife Federation (NWF) and the American Bird Conservancy "suggests that the effects of global warming may be robbing Maryland and a half-dozen other states of an important piece of their heritage by hastening the departure of their state birds" (Pianin 2002a, A-3).

The report said that Earth's rising temperature "is already shifting songbird ranges, altering migration behavior and perhaps diminishing some species' ability to survive" (Pianin 2002a, A-3). Iowa and Washington State may lose the American Goldfinch, as New Hampshire's Purple Finch could become an historical relic. California could lose the California Quails, Massachusetts' Black-capped Chickadee may vanish, and Georgia could lose its Brown Thrasher (Pianin 2002a, A-3).

The life cycle of the Baltimore Oriole and other birds is tied closely to weather patterns that are changing with general warming. Seasonal changes in weather patterns tell the birds when they should begin their long flights southward in the fall and back again in the spring. Temperature and precipitation also influence the timing and availability of flowers, seeds, and other food sources for the birds when they reach their destinations (Pianin 2002a, A-3).

Peter Schultz, a global-warming expert with the nonprofit National Research Council, cautioned that long-term forecasts of disruptions in bird migration patterns are difficult. "I would be surprised if the distribution of state birds is not changed down the road," he said. "But predicting precisely where they'll be 50 years from now is very difficult, if not impossible, with the current state of knowledge" (Pianin 2002a, A-3).

Baltimore Orioles once were so numerous that the naturalist painter John J. Audubon wrote about the delight of hearing "the melody resulting from thousands of musical voices that come from some neighboring tree" (Pianin 2002a, A-3). The bird, a Maryland icon whose namesake was adopted by Baltimore's major league baseball team, was officially designated the state bird in 1947. Local legend maintains that George Calvert, the first baron of Baltimore, liked the oriole's bright-orange plumage so much that he adopted its colors for his coat of arms.

Global warming is not the only danger to the Oriole and other well-known birds. Its decline results also from diminishing breeding habitat and forests in North America (where Orioles spend summers) and in Central and South America (where they fly for the winter). "Climate change on top of fragmented habitat is the straw that breaks the camel's back," said Patricia Glick, an expert on climate change with the NWF (Pianin 2002a, A-3).

More Birds and Butterflies Affected by Warming

Roughly 50,000 Tufted Puffins that spend summers on Triangle Island, 30 miles off the coast of British Columbia, have been falling prey to periods of starvation because small changes in ocean temperature have driven away their food supply. Researchers said that the adult birds bring back far fewer sand-lance fish to their young in warm years. The sand-lance fish, the puffin's favored food, become less abundant in the waters around Triangle Island as water temperatures rise. Scientists believe the lack of food, coupled with self-preservation instincts in the adults, leads to abandonment of chicks.

"The difference between 1998 and 1999 was one of the most striking things I have ever seen in my career in ecology," said Doug Bertram, a marine bird specialist at the Canadian Wildlife Service, as he recalled how dead chicks littered the colony in 1998 (Munro 2003b, A-12). "We show that the extreme variation in reproductive performance exhibited by Tufted Puffins (*Fratercula cirrhata*) was related to changes in sea-surface temperatures both within and among seasons," the authors of a scientific study of the birds said. Such changes in ocean temperatures "could precipitate changes in a variety of oceanic processes to affect marine species worldwide" (Gjerdrum et al. 2003, 9377).

Dismayed researchers watched helplessly as chicks dropped dead of starvation during several warm summers in the 1990s. Other years, such as 1999, when the water was 1.5°C cooler, the chicks thrived (Munro 2003b, A-12). By 2003, puffin populations had returned to former levels as water temperatures cooled to near-average levels. The scientists worried, however, that the respite may be temporary if global warming causes ocean temperatures to rise again in coming years.

Jerram Brown has charted the breeding seasons of Mexican Jays in the Chiricahua Mountains of southern Arizona for 31 years. By 1998, Brown found that the jays were laying their eggs an average of 10 days earlier than they did in 1971. Camille Parmesan has analyzed records tracking the distribution patterns of 57 nonmigratory butterfly species across Europe. She found that, during the last century, two-thirds of the species have shifted their ranges northward, some of them by as much as 240 kilometers (Wuethrich 2000, 795). "We ruled out all other obvious factors, such as habitat change," said Parmesan. "The only factor that correlated was climate" (Wuethrich 2000, 795).

Parmesan and colleagues' tracking of butterfly ranges provided evidence of poleward shifts in the range of entire species. In a sample of 35 nonmigratory European butterflies, 63 percent have ranges that shifted to the north by 35 to 240 kilometers during the twentieth century, while only 3 percent of ranges have shifted to the south (Parmesan et al. 1999, 579). The study team's evaluation of its data ends with a warning about other species:

> Given the relatively slight warming in this century compared [with anticipated temperature] increases of 2.1° to 4.6°C for the next century, our data indicate that future climate warming could become a major force in shifting species distributions. But it remains to be seen how many species will be able to extend their northern range margins substantially across the highly fragmented landscapes of northern Europe. This could prove difficult for all but the most efficient colonizers. (Parmesan et al. 1999, 583)

Butterflies in Britain: Warming and Habitat Destruction

By 2000, warmer temperatures were bringing butterflies to England two weeks to a month earlier than during the 1970s. Scientists studying 35 of the estimated 60 species of British butterflies say that some, such as the Red Admiral, can now be seen a month earlier. Others, "Such as the peacock and the orange tip, are appearing between 15 and 25 days earlier than two decades ago," according to a report in the *London Times* (Nuttall 2000a). Roy and Tim Sparks, both of the Centre for Ecology and Hydrology at Monks Wood, Cambridgeshire, analyzed data from 1976 to 1998 provided by the Butterfly Monitoring Scheme, whose members check more than 100 sites each week from April to September. According to a news report in the *London Guardian*, "One [Red Admiral] was monitored after crossing the [English] Channel on New Year's Day" (Vincent and Brown 2000, 9).

Some species of British butterflies that were expected to flourish in warmer temperatures have instead declined because of severe habitat destruction. Warm summers and mild winters since the 1970s should have attracted populations of butterflies that usually steer clear of the British Isles. According to a paper by M. S. Warren and colleagues in *Nature*, however, three-quarters of the butterfly species that might have expanded northward with warmer European weather have declined. The findings come from an analysis of 1.6 million butterfly sightings by 10,000 amateur naturalists between 1995 and 1999.

Chris Thomas of York University, who coordinated the study, said, "Most species of butterflies that reach the northern edge of their geographic ranges in Britain have declined over the past 30 years, even though the climate has warmed. This is surprising because climate warming is expected to increase the range of habitats these species can inhabit" (Derbyshire 2001b, 13). "Our computer models show that climatically suitable areas are available for colonization, but most species have failed to exploit them either because they no longer contain suitable breeding sites or because breeding habitats are out of reach" (Derbyshire 2001b, 13; Warren et al. 2001, 65).

Threats of a warming climate to butterflies in Britain also were described in a study compiled by British biologists and ecologists that was published by the Royal Society. According to this study, at least 30 of Britain's butterfly species face extinction or an alarming drop in numbers because they are failing to cope with the effects of global warming. These include some rare species, such as the Large Heath and Purple Empero, the numbers of which are expected to decline by as much as three-quarters. Many face eventual extinction as populations fall below replacement levels (Mason, Bailey, and London 2002, 12).

"There's no silver lining in this data," said Richard Fox, a co-author of the study and spokesman for the Butterfly Conservation Society. "What we will see here is a retreat and, potentially, a mass extinction in the slightly longer term, of many of our familiar species" (Mason, Bailey, and London 2002, 12). According to a report describing this work in the *London Independent*, over the past few years, Red Admirals, Orange-tips and Small Tortoiseshells have been seen earlier in the spring and surviving for several weeks longer each autumn, suggesting that butterflies would generally benefit from climate change. However, the researchers, led by Jane Hill, a biologist at York University, noticed that many more butterflies had failed to spread during the 1990s, even though average temperatures were beginning to rise (Mason, Bailey, and London 2002, 12). A few species, such as the Ringlet and the Marbled White, have been prospering, moving northward and farther uphill as the summers became warmer during the 1990s, contrary to declining populations for many other species.

The butterfly study projects that, as warming accelerates in Britain, species living in northern England and Scotland, including the Western Isles, will lose two-thirds of their habitats. In southern England, butterfly habitat will decline by about a quarter. For some species, the future is particularly bleak. For example, the Black Hairstreak, a very rare species that has been confined mainly to the East Midlands, will lose at least half of its usual habitat (Mason, Bailey, and London 2002, 12). "We may get some butterflies from the south colonizing us, but what this paper shows is that our butterflies are going to become much, much rarer," Fox said (Mason, Bailey, and London 2002, 12).

FURTHER READING

Both, Christiaan, Sandra Bouhuis, C. M. Lessells, and Marcel E. Visser. "Climate Change and Population Declines in a Long-Distance Migratory Bird." *Nature* 441 (May 4, 2006):81–83.

Charmantier, Anne, Robin H. McCleery, Lionel R. Cole, Chris Perrins, Loeske E. B. Krunk, and Ben C. Sheldon. "Adaptive Phenotypic Plasticity in Response to Climate Change in a Wild Bird Population." *Science* 320 (May 9, 2008):800–803.

"Climate Change Dislocates Migratory Animals, Birds." Environment News Service, November 17, 2006. http://www.ens-newswire.com.

"Climate Change Pushing Bird Species to Oblivion." Environment News Service, November 14, 2006. http://www.ens-newswire.com/ens/nov2006/2006-11-14-01.asp.

Derbyshire, David. "Global Warming Fails to Boost Butterfly Visitors." *London Daily Telegraph*, November 1, 2001b, 13.

Fountain, Henry. "Observatory: Early Birds and Worms." *New York Times*, May 22, 2001, F-4.

Gjerdrum, Carina, Anne M. J. Vallee, Colleen Cassady St. Clair, Douglas F. Bertram, John L. Ryder, and Gwylim S. Blackburn. "Tufted Puffin Reproduction Reveals Ocean Climate Variability." *Proceedings of the National Academy of Sciences* 100, no. 16 (August 5, 2003):9377–9382.

Mason, John, Jack A. Bailey, and Ardea London. "Doomsday for Butterflies as Britain Warms Up; Dozens of Native Species at Risk of Extinction as Habitats Come under Threat." *London Independent*, September 29, 2002, 12.

"Migratory Species and Climate Change." United Nations Environmental Programme (UNEP), 2006. http://www.cms.int/news/PRESS/nwPR2006/november/cms_ccReport.htm.

Munro, Margaret. "Puffin Colony Threatened by Warming: A Few Degrees Can Be Devastating. Thousands of Triangle Island Chicks Die When Heat Drives off Their Favoured Fish." *Montreal Gazette*, July 15, 2003b, A-12.

Nuttall, Nick. "Climate Change Lures Butterflies Here Early." *London Times*, May 24, 2000a, n.p.

Parmesan, Camille, Nils Ryrholm, Constanti Stefanescu, Jane K. Hill, Chris D. Thomas, Henri Descimon, Brian Huntley, Lauri Kaila, Jaakko Kulberg, Toomas Tammaru, W. John Tennent, Jeremy A. Thomas, and Martin Warren. "Poleward Shifts in Geographical Ranges of Butterfly Species Associated with Regional Warming." *Nature* 399 (June 10, 1999):579–583.

Pianin, Eric. "A Baltimore without Orioles? Study Says Global Warming May Rob Maryland, Other States of Their Official Birds." *Washington Post*, March 4, 2002a, A-3.

Vincent, John, and Paul Brown. "Swoop to Conquer: Global Warming Brings Butterflies to Britain Earlier." *London Guardian*, May 24, 2000, 9.

"Warmer Climate Hurting Birds, New IUCN Red List Shows." Environment News Service, May 19, 2008. http://www.ens-newswire.com/ens/may2008/2008-05-20-03.asp.

Warren, M. S., J. K. Hill, J. A. Thomas, J. Asher, R. Fox, B. Huntley, D. B. Roy, M. G. Telfer, S. Jeffcoate, P. Harding, G. Jeffcoate, S. G. Willis, J.N. Greatorex-Davies, D. Moss, and C. D. Thomas. "Rapid Responses of British Butterflies to Opposing Forces of Climate and Habitat Change." *Nature* 414 (November 1, 2001):65–69.

Wuethrich, Bernice. "How Climate Change Alters Rhythms of the Wild." *Science* 287 (February 4, 2000):793–795.

Buildings and Energy Efficiency

A third of human-generated greenhouse gases are produced in shelter—homes, manufacturing plants, and offices. Considerable mitigation of global warming may be accomplished by wise use of new technology to improve the energy efficiency of dwellings, factories, and offices. Energy consumption of heating and air-conditioning systems could be reduced by as much as 90 percent in new buildings, for example, with modern insulation, triple-glazed windows with tight seals, and passive solar design (Speth 2004, 65).

Architects, including some affiliated with the Leadership in Environmental and Energy Design (LEED) standards in the United States, have been designing homes with very tight insulation, the latest in high-efficiency appliances, and alternative sources of power, such as solar panels and wind turbines, in "passive houses" that use nearly no fossil-fuel energy at all. Most of these homes have no furnaces, even in moderately cold winter climates. In Germany, according to Elisabeth Rosenthal's account in the *New York Times*, "using ultrathick insulation and complex doors and windows, the architect engineers a home encased in an airtight shell, so that barely any heat escapes and barely any cold seeps in. That means a passive house can be warmed not only by the sun, but also by the heat from appliances and even from occupants' bodies" (Rosenthal 2008). Ventilation systems avoid buildup of mold in these ultra-tight houses, and construction costs are only marginally higher (5 to 7 percent) than for other homes.

A report by consultant McKinsey & Co. maintained that some of the most basic steps to curb global-warming emissions are the most cost-effective—for example, such prosaic steps as improving energy efficiency of buildings, including lighting and air conditioning, and planting more trees. A report in the *Wall Street Journal* said that "[s]uch moves will be cheaper than high-technology methods, such as capturing the CO_2 that is emitted from power plants and then burying that CO_2 underground" (Ball 2007a, A-12). Adding green spaces helps to counter the urban heart island effect, through evaporative cooling from leaves and other vegetation. As water evaporates from the leaves of plants and trees, it cools the surrounding air much as perspiration cools human skin.

On May 16, 2007, a group of big-city mayors in several countries pledged billions of dollars worth of investments to curtail urban energy use and, thereby, emissions of greenhouse gases under the initiative of the William J. Clinton Foundation. Each participating bank committed to allocate as much as $1 billion for loans that city governments and private landlords could use to upgrade inefficient heating, cooling, and lighting systems in older buildings. The loans and interest would be repaid with savings realized through reduced energy costs. New York City Mayor Michael Bloomberg said that retrofitting older buildings was vital because 85 percent of them will still be standing (and, one might add, occupied) decades from now.

The first targets under this initiative will be municipal buildings in the participating cities, including Bangkok, Thailand; Berlin, Germany; Chicago, Illinois, Houston, Texas, and New York City, in the United States; Johannesburg, South Africa; Karachi, Pakistan; London, United Kingdom; Melbourne, Australia; Mexico City, Mexico; Mumbai, India; Rome, Italy; São Paulo, Brazil; Seoul, the Republic Korea; Tokyo, Japan; and Toronto, Canada (Revkin and Healy 2007).

Simple Conservation Measures

Some conservation methods are available now. Installing motion sensors that dim the lights by 50 percent when hallways and stairwells are not in use, for example, could save a 60,000-square-foot building 40 metric tons of carbon dioxide emissions by about 40 metric tons, enough to drive a car that gets 25 miles per gallon 110,000 miles. Hiring an electrician to install the motion sensors would cost $10,000 to $12,000, according to estimates produced by Optimal Energy Inc., a consulting company in Bristol, Vermont. The building could save that much money in lower electricity bills over two years, assuming that it uses fluorescent bulbs. Other shelter-related energy savers are just as prosaic. Assigning a specific person in an office to switch off all lights and equipment at the end of an off an office day can cut carbon emissions by reducing electricity use, while also extending equipment life and reducing maintenance costs.

At home (or the office), leaving household appliances on standby does not seem to cost much energy per unit, but adding up the bill for the 300 million people in the United States is shocking. Alan Meier, senior energy analyst at the International Agency, did just that, and estimated that U.S. households use 45 billion kilowatt-hours in standby electricity per year, at a cost of $3.5 billion, equal to the production of seventeen 500-megawatt power plants (Steinman 2007, 303–304).

A study at the University of California–Berkeley and Lawrence Berkeley National Laboratory indicated that eliminating "standby" electricity loss from home appliances could produce substantial savings on electricity bills. The study found that standby usage ranged from 6 to 26 percent of homes' annual electricity use. An average of 19 appliances per household were in standby mode (Sanders 2001).

The average desktop computer, not including the monitor, consumes between 60 and 250 watts per day. Turning a computer off after using it four hours a day could save about $70 in electricity in a year. The carbon impact would be even greater. Shutting it off would reduce the machine's carbon-dioxide emissions 83 percent, to just 63 kilograms a year.

The household equivalent of a well-tuned car is a furnace that has been checked and cleaned—and, if it is more than a decade old, replaced. If everyone in the United States upgraded furnaces with the latest energy-efficient technology, gas use for space heating would probably fall at least 25 percent, according to the U.S. Department of Energy. A new furnace may be installed as part of renovations to a basement.

Building Code Changes around the World

On January 1, 2003, Australia changed its national building code with the explicit purpose of (1) reducing energy consumption; (2) utilizing passive solar heating, where appropriate; (3) using natural ventilation and internal air movement, where appropriate to avoid or reduce the use of artificial cooling; (4) sealing houses in some climates to reduce energy loss through leakage; (5) installing insulation to reduce heat loss from water piping of central heating systems; and (6) installing insulation and sealing to reduce energy loss through the walls of ductwork associated with heating and air-conditioning systems.

British Columbia initiated a Green Building Code during 2007 containing incentives to retrofit existing homes and buildings for energy efficiency. The code requires "energy audits" of energy savings, as well as plans for "real-time, in-home smart metering," to help homeowners measure and reduce energy consumption. Separate strategies are aimed at promoting greener universities, colleges, hospitals, schools, prisons, ferries, and airports ("British Columbia" 2007).

Danish building codes enacted in 1979 (and tightened several times since) also require thick home insulation and tightly sealed windows. Between 1975 and 2001, Denmark's national heating bill fell 20 percent, even as the amount of heated space increased by 30 percent. Denmark's gross domestic product has doubled on stable energy usage during the last 30 years. The average Dane now uses 6,000 kilowatt hours of electricity a year, less than half of the U.S. average (13,300 kilowatt hours).

Surplus heat from Danish power plants is piped to nearby homes in insulated pipes, for example, via "cogeneration" or "district heating," which required tearing up streets to install the pipes. Power plants were downsized and built closer to people's homes and offices. In the mid-1970s, Denmark had 15 large power plants; it now has several hundred. By 2007, 6 in 10 Danish homes were heated this way, and it is less expensive than oil or gas (Abboud 2007, A-13).

Even without regulation, some architectural engineers are utilizing sustainable principles in their designs. The new Hearst Tower in Manhattan has natural ventilation and high-performance glass that deflects heat (Ouroussoff 2007).

The United States has no federal regulations that guarantee sustainability in new construction. Private LEED (Leadership in Energy and Environmental Design) guidelines, drawn up by the U.S. Green Building Council, a nonprofit group founded in 1993, suggest voluntary compliance. Several government agencies have adopted these guidelines, including the federal General Services Administration, which oversees the construction of federal buildings.

LEED-certified buildings emit an average 38 percent less carbon dioxide and reduce energy use 30 to 50 percent compared with other structures. The organization's intricate certification process can cost as much as $30,000, but the cost is usually recovered in a few years of energy savings (Jordan 2007a, D-2). LEED rates buildings on a four-tier scale: certified, silver, gold, and platinum. By 2007, 10,000 organizations had memberships in LEED, along with 91,000 individuals—usually contractors, developers, architects, building owners, universities, suppliers, and others (Jordan 2007a, D-1).

Architects have been finding all sorts of ways to reduce energy use; buildings are being constructed with an abundance of windows that allow passive solar energy to reduce electricity use; overhangs on exterior walls screen sun in the summer and allow it in winter; walls of limestone on sunny sides of buildings screen the sun and reduce air-conditioning use. The annual Greenbuild Conference, which includes developers, architects, contractors, suppliers, and others, drew 4,200 people to Austin in 2002. In 2007, that number rose by a factor of four, to about

18,000 at Chicago conference. By that year, "green" construction was a $12 billion annual market (Jordan 2007a, D-2).

More Stringent Energy Efficiency Standards

A study by the McKinsey Global Institute "concludes that projected electricity consumption in residential buildings in the United States in 2020 could be reduced by more than a third if compact fluorescent light bulbs and an array of other high-efficiency options including water heaters, kitchen appliances, room-insulation materials and standby power were adopted across the nation. Energy conservation over that time, if achieved, would be equivalent to the production from 110 new coal-fired 600-megawatt power plants, the researchers estimated. That would result in a significant reduction in the amounts of fossil fuel burned and carbon dioxide, the main greenhouse gas, spewed into the atmosphere" (Lohr 2007).

The study recommends market intervention to correct distortions. A person who rents an apartment, for example, may use inefficient appliances because the landlord has little incentive to upgrade. The tenant, after all, pays the power bill. In turn, the tenant, who does not own the property, has little incentive to buy new, efficient appliances. When individual parties do not have incentives to make the needed investments, the report suggests that more stringent product standards should require that all new appliances are energy-efficient models (Lohr 2007). The McKinsey report's ideas are largely borrowed from California, which, beginning during the 1970s, established requirements for appliances and building materials, among other energy-saving measures.

From 1976 to 2005, electricity consumption per capita stayed flat in California, while it grew 60 percent in the rest of the United States (Romm 2007, 23). How has California, with its increasing population and booming high-tech economy, kept its electricity consumption nearly flat? That state changed its utilities' pricing structure so that profits were not tied directly to how much power they sold. Instead, the state's electricity regulators decided to reward efficiency in a program that has since lowered energy bills of Californians $12 billion a year ($1,000 per family) and avoided emission of more than 10 million metric tons of carbon dioxide per year (Romm 2007, 164).

Hotels Go "Green"

King Pacific Lodge, a luxury resort in the Great Bear Rainforest along the British Columbia coast (where a three-night stay costs about $5,000 per person), has turned down the temperature of its showers and restricts the use of its twin-engine boats (usually used for salmon fishing) to a horsepower limit per person, filling the boats and restricting their speed with engine governors, all to restrict greenhouse-gas emissions. The lodge is acting on a plan to cut its carbon footprint by half in five years (Ball 2007c, P-1). The lodge's carbon footprint before that was enormous—1.7 tons per guest per stay (3.5 days), or about as much as an average United States citizen creates in a month. The resort's biggest carbon-producers are two 110-kilowatt diesel generators for electricity. Propane is being substituted in the kitchen.

The most carbon-intensive part of a vacation in a remote area is usually the air travel required to reach it. The *Wall Street Journal* carried an account of one couple who installed a solar hot water heater in their home and, feeling environmentally correct, then took flights around the world. Soneva Fushi, a 65-villa resort on the Maldives Islands, in the Indian Ocean, plans to add a fee to guests' bills to offset the amount of carbon-dioxide required to fly there from anywhere else. (The Maldives often are mentioned as one of the first locations that will drown in rising seas generated by melting ice worldwide). The resort announced plans to go carbon-neutral by 2010, partially by using coconut oil to reduce oil demand in its diesel generator. Devices also are being installed to cut off air conditioning if a room's door has been open for more than a minute (Ball 2007c, P-5).

"Environmental issues are one of the hottest issues within the travel industry right now," said Bill Connors, the executive director of the National Business Travel Association, which made eco-friendly elements in hotel design and operations a focus (for the first time) at its annual convention in July 2007 (White 2007). Increasing numbers of hotels have been registering for certification under the LEED program. Compact fluorescent bulbs, policies that extend the use of towels between washings, green roofs, use of nontoxic cleaning agents, and recycling bins in rooms have become more popular.

Marriott Hotels plans to reduce energy consumption in its hotels 20 percent between 2000 and 2010. Guests who drive hybrid vehicles to the Fairmont hotels in California and British Columbia receive free parking. The Habitat Suites in Austin, Texas, installed a solar hot-water heating system that cut natural gas use 60 percent. The Lenox Hotel of Boston composts 120 tons a year of restaurant waste (Bly 2007, D-2). At San Francisco's Kimpton Hotel Triton, guests may book rooms on an "Eco Floor," with low-flow toilets, use of nontoxic cleaning supplies, and shampoo and conditioner dispensed from wall-mounted canisters, instead of individual plastic bottles. At Greenhouse 26, which opened in Manhattan in 2008, the elevator will capture energy generated when it stops, much as a hybrid car recycles energy released when the brakes operate. Heating and cooling will be provided geothermally. Water from sinks and showers will be recycled for use in toilets (White 2007).

Geothermal Energy: Newly Popular at Street Level

In Iceland, 85 percent of the country's 290,000 people use geothermal energy to heat their homes (Brown 2003, 166). Iceland's government, working with Shell and Daimler-Chrysler in 2003 began to convert Reykjavik's city buses from internal combustion to fuel-cell engines, using hydroelectricity to electrolyze water and produce hydrogen. The next stage is to convert the country's automobiles, then its fishing fleet. These conversions are part of a systemic plan to divorce Iceland's economy from fossil fuels (Brown 2003, 168).

Geothermal could produce 10 percent of U.S. electricity by 2050, according to a report by a team at the Massachusetts Institute of Technology. In 2007, geothermal was feasible at $0.06 to $0.10 per kilowatt hour, without subsidies. Harrison Elementary School in Omaha, Nebraska, has been remodeled to use geothermal, just one of several examples in the same area.

Warm springs are not required for geothermal energy. Thermal contrast between the Earth and its atmosphere is enough, especially in places such as Omaha that have large contrasts in seasonal temperatures. The Earth's temperature is about 56°F year-round; by circulating water through pipes above and below ground, as much as 70 percent can be saved on heating and cooling costs. When air temperature is close to that of the Earth, the need is minimal; the greater the contrast, the greater the need, and the more energy is conserved.

Installation of a geothermal system involves major infrastructure costs. At St. Margaret Mary's Catholic School in Omaha, a soccer field was dug up to install the pipes, part of a $2.3 million project that will recoup expenses in 9 to 13 years. Such projects are feasible for schools and other institutions that will occupy the same property for decades.

Geothermal systems use underground pipes filled with fluid that pull heat from buildings in the summer and release it into underground soil. In the winter, the pipes distribute heat in the building that has been gathered underground. The principle is the same as that used by traditional residential heat pumps, but these systems are 30 to 50 percent more efficient (in that they use 30 to 50 percent less energy). The geothermal pumps use the ground, whereas the residential pumps use the air. Nationwide, installation of geothermal pumps has been growing at double-digit rates, according to John Kelly, executive director of the Geothermal Heat-pump Consortium in Washington, D.C. (Gaarder 2007b, A-2).

Electric Meters That Run Both Ways

The traditional electricity meter has entered the digital age at nearly the same time that energy infrastructure is changing to allow individual households to contribute their own power with sun, wind, and biogas, making electricity generation a two-way street. A centralized electricity-generation grid that relies on fossil fuels (meanwhile losing half of what it produces as waste heat and another 8 percent to transportation inefficiencies) will be replaced within decades by a system that will allow consumers to produce power and sell their surpluses back to the grid at market rates. Enel, an Italian electricity company, distributed 30 million "smart" meters between 2001 and 2006. By 2011, Pacific Gas and Electric, which serves northern and central California, plans to have supplied several million customers with smart meters (Butler 2007, 587).

A large number of small power sources—a "distributed grid," already being used in countries such as Denmark—increases efficiency by reducing distances between power producers and power consumers, meanwhile also better matching supply and demand. Small turbines powered by natural gas or biogas that use waste heat to provide heat and hot water to local areas convert energy with 70 to 85 percent efficiency (Butler 2007, 587). "Smart" meters provide data on how electricity is being used in any part of the grid at any moment. These meters use digital technology to make more efficient use of a system that will include many more sources of power than today's grids with but a few giant plants, forming an "energy Internet" (Butler 2007, 586).

Denmark by 2006 was distributing half of its power through distributed grids, as carbon-dioxide emissions there have fallen from 937 grams per kilowatts hour in 1990 to 517 grams per kilowatt hour in 2005 (Butler 2007, 587).

Compact Fluorescent Light Bulbs

Australia's government has required a nationwide phaseout of incandescent light bulbs in favor of compact fluorescent lights by 2010. While as much as 90 percent of energy from incandescent light bulbs is waste (mainly as heat) the compact fluorescent light bulb uses about 20 percent as much electricity to produce the same amount of light. A compact fluorescent light bulb can last between 4 and 10 times longer than the average incandescent light bulb. The new policy should reduce Australia's greenhouse-gas emissions by four million tons within two years. Household lighting costs could be reduced by up to 66 percent. In Australia, lighting currently represents around 12 percent of greenhouse gas emissions from households, and around 25 percent of emissions from the commercial sector ("Australia Screws" 2007).

In the meantime, Green Party members in Australia said that the government was taking mere baby steps toward the kind of energy conservation required to seriously challenge global warming. Screwing in new light bulbs will not do it, they said. The Greens aimed to stop coal mining. "The proposed Anvil Hill coal mine in New South Wales will generate 28 million tons of carbon dioxide emissions every year, whereas the light bulb change will reduce emissions by 800,000 tons per annum," the Greens said. Wearing "Stop Anvil Hill" banners, the activists warned that the New South Wales government has plans to approve or expand at least eight mines in the Hunter Valley, which they said would be a "climate change disaster" ("Australia Screws" 2007).

Utilities in California, which have been ordered by state regulators to sell less of their product, have paid manufacturers of compact fluorescent light bulbs to sell them for $0.25 to $0.50 each. Elsewhere, these bulbs cost several dollars each. This is part of a statewide drive to induce consumers to replace incandescent bulbs. The state's largest utility, Pacific Gas & Electric (PG&E), planned to spend $116 million of ratepayers' money on this program. At one point during the fall of 2007, PG&E enlisted the Girl Scouts and Sierra Club, unions, churches, and other groups to give away a million of these bulbs. The company subsidized 7.6 million more bulbs during the rest of 2007 (Smith 2007a, A-12).

A Proposed Chinese Eco-city

In a country that by 2007 was industrializing mainly with power generated by burning dirty coal, with 16 of the world's 20 cities with the worst air pollution, China is planning an "eco-city" on the island of Chongming (meaning "lofty brilliance"). The city, located at the mouth of the Yangtze River (one hour by ferry from Shanghai), is slated to open in 2010 with 25,000 inhabitants. In Dongtan (which translates to "eastern bank"), the first of three bedroom communities that will be linked to Shanghai by a 15.6-mile bridge and tunnel, energy will be provided by solar, wind, and biofuels, including recycled organic material, down to household vegetable peels. Grasses will be planted on rooftops for insulation, and a quarter of the land will be reserved in its natural state as an ecological buffer. The village aims to be carbon-neutral, and to produce very little waste; it will have no landfill, and sewage will be processed for irrigation and composting (MacLeod 2007, 9-A).

China's experiment was prompted by the environmental ravages of rapid industrial development that include pollution of all its major rivers, dangerously polluted urban air, acid rain over a third of its land mass, and desertification that is claiming an area the size of Connecticut in an average year,

The urban area of Dongtan will be designed for walking, public transit, and bicycles, with private cars discouraged. Plans call for the initial 25,000 population to grow to 80,000 by 2020 and 500,000 by 2030—that is, if rising seas provoked by global warming do not swamp the island, which also may be an easy target for occasional typhoons. Construction began in September 2007. Yang Ailun, a climate and energy specialist with Greenpeace, Canada, said that the eco-city could "disappear because of climate change" within decades (MacLeod 2007, 9-A).

An Operational Danish Eco-island

The 4,000 residents of largely agricultural Samso Island, in Denmark, accepted a challenge from the Danish government during the 1990s to convert to a carbon-neutral style of life and, by 2007, had largely accomplished the task, with no notable sacrifice of comfort. Farmers grow rapeseed and use the oil to power their machinery; homegrown straw is used to power centralized home-heating plants; solar panels heat water, and store it for use on cloudy days (of which the 40 square mile island has many); and wind power provides electricity from turbines in which most families on the island have a share of ownership. The island is now building more wind turbines to export power.

FURTHER READING

Abboud, Leila. "How Denmark Paved Way to Energy Independence." *Wall Street Journal*, April 16, 2007, A-1, A-13.

"Australia Screws in Compact Fluorescent Lights Nationwide." Environment News Service, February 21, 2007. http://www.ens-newswire.com/ens/feb2007/2007-02-21-01.asp.

Ball, Jeffrey. "Climate Change's Cold Economics: Industry Efforts to Fight Global Warming Will Hit Consumers' Pockets." *Wall Street Journal*, February 15, 2007a, A-12.

Ball, Jeffrey. "The Carbon Neutral Vacation." *Wall Street Journal*, July 30, 2007c, P-1, P-4-5.

Bly, Laura. "How Green Is Your Valet and the Rest?" *USA Today*, July 12, 2007, D-1, D-2.

"British Columbia to Trim Greenhouse Gases, Go Carbon Neutral." Environment News Service, February 14, 2007. http://www.ens-newswire.com/ens/feb2007/2007-02-14-02.asp.

Brown, Lester R. *Plan B: Rescuing a Planet under Stress and a Civilization in Trouble*. New York: Earth Policy Institute/W.W. Norton, 2003.

Butler, Declan. "Super Savers: Meters to Manage the Future." *Nature* 445 (February 8, 2007): 586–588.

Gaarder, Nancy. "Many Digging Deep for Cheaper Energy." *Omaha World-Herald*, May 29, 2007b, A-1, A-2.

Jordan, Steve. "Becoming Greener; Environmental Concerns, Lower Costs Push Drive for Energy-efficient Buildings." *Omaha World-Herald*, July 22, 2007a, D-1, D-2.

Lohr, Steve. "Energy Standards Needed, Report Says." *New York Times*, May 17, 2007. http://www.nytimes.com/2007/05/17/business/17energy.html.

MacLeod, Calum. "China Envisions Environmentally Friendly Eco-city." *USA Today*, February 16, 2007, 9-A.

Ouroussoff, Nicolai. "Why Are They Greener Than We Are?" *New York Times Sunday Magazine*, May 20, 2007. http://www.nytimes.com/2007/05/20/magazine/20europe-t.html.

Revkin, Andrew, and Patrick Healy. "Coalition to Make Buildings Energy-Efficient." *New York Times*, May 17, 2007. http://www.nytimes.com/2007/05/17/us/17climate.html.

Romm, Joseph J. *Hell and High Water: Global Warming—the Solution and the Politics—And What We Should Do*. New York: William Morrow, 2007.

Rosenthal, Elisabeth. "No Furnaces but Heat Aplenty in 'Passive Houses.'" *New York Times*, December 27, 2008. http://www.nytimes.com/2008/12/27/world/europe/27house.html.

Sanders, Robert. "Standby Appliances Suck Up Energy." *Cal* [California] *Neighbors*, Spring 2001. http://communityrelations.berkeley.edu/CalNeighbors/Spring2001/appliances.html.

Smith, Rebecca. "Utilities Amp Up Push to Slash Energy Use." *Wall Street Journal*, January 9, 2007a, A-1, A-12.

Speth, James Gustave. *Red Sky at Morning: America and the Crisis of the Global Environment*. New Haven, CT: Yale University Press, 2004.

Steinman, David. *Safe Trip to Eden*. New York: Thunder's Mouth Press, 2007.

U.S. Green Building Council Web site. http://www.usgbc.org/DisplayPage.aspx?CategoryID=19.

White, Martha C. "Enjoy Your Green Stay." *New York Times*, June 26, 2007. http://www.nytimes.com/2007/06/26/business/26green.html.

C

Canada and Warming-Related Stresses

More hot days will increase air-pollution problems in Canada's big cities, with children at greatest risk, according to a Canadian Institute of Child Health study of global warming's possible effects. Childhood asthma and respiratory problems will probably become worse if global-warming projections are accurate, according to the report. "Scientists predict that over the next century, Canadian children will live in a country that is 1.5° to 5°C warmer in southern Canada and as much as 5° to 7° warmer in the north," asserted Don Houston, a spokesman for the institute (Bueckert 2001, A-2). "Longer, more intense summer heat waves would result in more serious smog-and-ozone episodes," the report said. It continued: "Ground-level ozone damages the cells that line the respiratory tract, causing irritation, burning, and breathing difficulties including chest tightness and pain on inhalation" (Bueckert 2001, A-2).

Global warming could wreak havoc with Canada's prized freshwater supply over the next 100 years, sapping some of the country's hydroelectric-power potential, lowering lake levels, and playing a role in more severe droughts, according to a report described in the *Toronto Globe and Mail.* The report, "Climate Change Impacts and Adaptation: A Canadian Perspective," from Natural Resources Canada described potential problems that could result if global surface-air temperatures increase between 1.4° and 5.8°C by 2100, as projected by the Intergovernmental Panel on Climate Change (Chase 2002, A-1).

The report indicated that potential effects of climate change might include the following:

- Increased likelihood of severe drought on the Canadian prairies, parts of which were suffering through their second or third consecutive year of drought as the report was released;
- Stranded docks and harbors because of lowered lake or river levels, notably in the Great Lakes;
- A shrinking supply of potable water and more illness from contaminated water;
- Ruined fish habitat and spawning areas and possible loss of species;
- More financial pain for farmers caused by losses in production; and
- Complete evaporation of some lakes in Arctic and Sub-Arctic regions (Chase 2002, A-1).

The same report said that warming temperatures could shrink the supply of fresh water generated by melting snow during the summer, when water is in high demand. "Across southern Canada, annual mean stream flow has decreased significantly over the last 30 to 50 years, with the greatest decrease during August and September," the report said (Chase 2002, A-1). Lower water levels tend to lead to higher pollutant concentrations, whereas high-flow events and flooding increase turbidity and the flushing of contaminants into the water system. Hydroelectric power in southern Canada, where most of the population lives, also may be significantly affected by a warmer climate, the report said. It continued, "Studies suggest that the potential for hydroelectric generation will likely rise in northern regions and decrease in the south, due to projected changes in annual runoff volume" (Chase 2002, A-1).

The same study estimated that excessive heat could kill more than 800 residents in the Toronto-Niagara Falls region per year by 2080, a 40-fold increase over the present-day death toll. The number of days with temperatures above 30°C could double to 30 each summer by the 2030s. Ground-level ozone, a lung-damaging component in smog, was projected to double by 2080 across the region (Bueckert 2002b). In addition, the report anticipates that the frequency of extreme weather events such as heat waves, windstorms, and rainstorms may increase, with associated increases in injuries, illnesses, and deaths. The incidence of waterborne diseases could rise in communities that depend on wells, or in cities where sewer and stormwater drainage systems are combined, because heavy rains would increase the risk of drinking-water contamination.

Warmer weather sometimes produces more violent thunderstorms, including hail, as some Canadian apple farmers can attest. Okanagan apple grower Allan Patton said that hail used to fall about once every decade in his part of British Columbia. By 2003, however, his orchard had been pounded by hail in 7 of the past 10 years (Constantineau 2003, D-3). "We have to have beautiful perfect fruit to get anything for it in the marketplace," he said. "When you work all year and in a matter of 90 seconds, all your work is wiped out, it's very upsetting. It's tough to take" (Constantineau 2003, D-3). Patton said global warming has increased the intensity and frequency of thunderstorms at a time when farmers' crop-insurance premiums have doubled. "It's not just hail. There are bizarre wind storms where trees are blown down and a greenhouse will be lifted up but the one next to it isn't," he said (Constantineau 2003, D-3).

Freakish Weather and Wildfires Increasing

Freak weather events such the 1996 Saguenay floods and Quebec's devastating 1998 ice storm are expected to occur more frequently as the climate warms, Quebec civil-protection officials have said. Climatologist Jacinthe Lacroix said that Quebecers are likely to see long spells of dry heat in summer, followed by violent thunderstorms, showers, and hail. "This will have a huge impact on our water-drainage system, which is not designed to deal with this kind of weather," he said (Sevunts 2001, A-4).

During the summer of 2002, smoke from wildfires in Quebec, aggravated by heat, drought, and lightning, drifted as far south as Baltimore. An editorial in the *Baltimore Sun* remarked,

> Was that whiff of global warming we were breathing in on Sunday? Huge tracts of forest are burning just east of Hudson Bay in Quebec, and although Americans would normally be content to ignore a natural disaster so remote and so far away, this time it was impossible because a freakish weather pattern brought the smoke southward as far as Baltimore—and even beyond. ("Quebec's Smoky Warning" 2002, 10-A)

By the first week in July 2002, roughly 400,000 acres had burned in Quebec, twice as much as much as would usually be expected in an entire summer. Several villages had been abandoned, and 500,000 people had lost power. The Quebec spruce forests were burning near the tundra tree line, on poor soil, and would take centuries to regenerate. "Experts have predicted that global warming will mean a northward advance of the forest line," reported the *Sun*. "In Quebec, heavier precipitation will come, or maybe they are wrong. What we are witnessing today, in any case, is unprecedentedly warm and dry weather afflicting a fragile—and flammable—ecosystem" ("Quebec's Smoky Warning" 2002, 10-A). The fires themselves were pumping additional carbon dioxide into the atmosphere—a compounding effect of increasing wildfires worldwide.

A report in the *Montreal Gazette* attributed the increasing number and severity of wildfires in Eastern Canada to logging and hydroelectric development as well as global warming. "Whenever you clear a forest for any reason, you're not only changing the ecosystem on the ground, but you're also changing the local climate," said Rene Brunet, an Environment Canada meteorologist. Clear-cutting creates dry conditions. Gone are the trees and their leafy canopies that keep the undergrowth humid and the air moist (Fidelman 2002, A-4). At the same time, global warming adds to the problem, Brunet said. "When you create a dam, you're changing the landscape," he said. "Change the color and shape of the land, and as a result, retention of moisture and solar radiation all changes" (Fidelman 2002, A-4).

In Western Canada, the Vancouver, British Columbia, *Province* editorialized:

> We are running out of water, our forests are burning and where there is smoke or smog, the air is not safe to breathe. Don't think the problem will disappear with the rains as another low snow pack winter and dry summer could be in the cards. British Columbia's dangerous drought seems to be part of an intensifying global warming, which has also produced unprecedented heat waves and fires in Europe, Australia, Africa, Russia and Asia. British Columbia's Interior is a war zone with army camps in schoolyards and water bombers and helicopters in the skies. We watch the news daily to see how far the enemy fire has advanced hoping we will be safe from invasion. And we make an unimaginable sacrifice by living with a total ban on venturing into

the countryside that is punishable by fines and jail. (Editorial 2003, A-16)

Wildfires in British Columbia during August 2003 created more greenhouse gases than 1.3 million Canadians produce in one year, according to a carbon-tracking device created by scientists in Victoria, British Columbia. The wildfires, which consumed 250,000 hectares, produced 15 million tons of greenhouse gases. An average Canadian produces about 11 tons a year, and British Columbia as a whole emits about 65 million tons every year (Fong 2003, A-8).

Wildfires, drought, and pests, exacerbated by steady warming of the world's atmosphere, threaten to destroy vast tracts of forested areas in Canada and other northern regions, crippling forestry and northern tourism sectors, according to Ola Ullsten, the former prime minister of Sweden, who also has served as co-chair of the World Commission on Forests (Baxter 2002, A-5).

Ullsten has proposed adoption of a worldwide forest-capital index to measure the economic and environmental value of forests in regions and countries. He said that by quantifying the health of the forest and its economic value, politicians would be able to monitor forest conditions and take appropriate actions. Richard Westwood of the Centre for Interdisciplinary Forest Research at the University of Winnipeg said that human incursion already has had an effect on forests and the repercussions of global warming are simply "piling-on" to destroy forests (Baxter 2002, A-5). He said that drought causes fires, as well as stressing trees, leading to pest attacks, dead, dry forests, and even more fires. "We … [in Canada] are particularly vulnerable to the predicted impacts of climate change," said Terry Duguid, chair of the Manitoba Clean Environment Commission. "Forestry is Canada's largest natural-resource industry, producing a trade surplus close to the combined surpluses for agriculture, energy, fisheries, and mining," he said (Baxter 2002, A-5).

The Canadian forestry industry will suffer economically from forest fires caused by global warming in coming decades, according to University of Alberta ecologist David Schindler. "My guess is that with climate warming, forestry will take a big hit," he said. "Right now the rates of forest cutting generally don't consider burning rates at all." Schindler said that burning rates in the Western boreal forest have doubled since 1970s. The term boreal refers to forests dominated by spruce and pine, which cover vast expanses of northern Canada (Bueckert 2002a, A-3).

Wildfires have been increasing across Canada partially because average temperatures have risen about 1.5°C, Schindler said. Computer models predict that average temperatures in the region will increase another 2°C in the next 20 years. "If we go up another 2° we could see at least another doubling in the incidence of forest fire," he said (Bueckert 2002a, A-3). Schindler said fire suppression in Canada was effective from the 1940s through the 1970s, followed by many years during the 1980s and 1990s with several large, uncontrollable fires. "You could have the whole U.S. Air Force dropping water and you wouldn't be able to contain them," Schindler said (Bueckert 2002a, A-3).

Forest fires destroy valuable economic and ecological resources as they release additional carbon dioxide into the atmosphere. Schindler explained:

> We could potentially get ourselves into a jam where our rates of carbon dioxide loss from forestry could exceed what we're putting out as fossil fuels. While we can control our fossil fuels we can't control our forest fires. That is what we call a positive feedback; it would aggravate the effects of climate warming that we're already seeing from fossil fuels. I think the prudent thing to do is try to avoid that scenario as much as possible. (Bueckert 2002a, A-3)

The Canadian Arctic's Rapid Warming

Along the shore of the Beaufort Sea, pounding surf has been eating away at the melting shoreline of Tuktoyaktuk, an Inuit town of 1,000 people at the mouth of the Mackenzie Delta. In a town where explorers in 1911 found temperatures of −60°F (wind chill −110°F), residents now worry about warming that may cause its permafrost base to dissolve "like salt in the sea" (Brown A-14).

Ken Johnson, a planner for an engineering company that has studied Tuktoyaktuk, said, "There is some speculation the storms off the Beaufort Sea are created by the decreasing ice pack in the Canadian north. It is creating a larger fetch, or distance of open water, that allows the creation of large waves. Permafrost ground is much more sensitive to storm waves" (Brown 2003, A-14). Since 1934, the coast has been eroding at an average of more than six feet a year. Ten years ago, a huge storm broke off more than 42 feet of the shoreline. "The joke is," said Calvin Pokiak, assistant land administrator for Inuvialuit Regional Corporation, "in the houses over here, you have to sleep with your life jacket on" (Brown 2003, A-14).

The Mackenzie Basin Impact Study (1997) anticipated that the area is warming at three times the global rate, said Kevin Jardine, a Greenpeace climate-change campaigner (Ralston 1996). Other current and future impacts in the same area include "wildlife habitat destruction, melting permafrost which destabilizes northern homes, roads and pipelines, large-scale coastal erosion, and insect outbreaks" (Ralston 1996). The report suggested that increasing numbers of forest fires in the area may be attributed to rising temperatures.

In the Canadian Inuit town of Inuvik, 90 miles south of the Arctic Ocean near the mouth of the Mackenzie River, the temperature rose to 91°F on June 18, 1999, a type of weather unknown to living memory of anyone in the area. "We were down to our T-shirts and hoping for a breeze," said Richard Binder, 50, a local whaler and hunter (Johansen 2001, 19). Along the MacKenzie River, according to Binder, "Hillsides have moved even though you've got trees on them. The thaw is going deeper because of the higher temperatures and longer periods of exposure" (Johansen 2001, 19). In some places near Binder's village, the thawing earth has exposed ancestral graves, and remains have been reburied.

Inuit hunters in the northernmost reaches of Canada say that Ivory Gulls are disappearing, probably as a result of decreasing ice cover that affects the gulls' habitat. Sea ice in the Arctic in the year 2000 covered 15 percent less area than it did in 1978, and it has thinned to an average thickness of 1.8 meters, compared with 3.1 meters during the 1950s (Krajick 2001a, 424). A continuation of this trend could cause the Arctic Ice Cap to disappear during summers within a few decades, imperiling many forms of life that live in symbiosis with the ice, varying in size from sea algae to polar bears. Less ice means fewer ice-dwelling creatures whose carcasses end up on the ocean floor, feeding clams and other bottom-dwellers, which, in turn, are consumed by walruses (Krajick 2001a, 424).

FURTHER READING

Baxter, James. "Canada's Forests at Risk of Devastation: Global Warming Could Ravage Tourism, Lumber Industry, Commission Warns." *Ottawa Citizen*, March 5, 2002, A-5.

Brown, DeNeen L. "Hamlet in Canada's North Slowly Erodes: Arctic Community Blames Global Warming as Permafrost Starts to Melt and Shoreline." *Washington Post*, September 13, 2003, A-14.

Bueckert, Dennis. "Climate Change Linked to Ill Health in Children." Canadian Press in *Montreal Gazette*, June 2, 2001, A-2.

Bueckert, Dennis. "Forest Fires Taking Toll on Climate: CO_2 from Increased Burning Could Overtake Fossil Fuels as a Source of Global Warming, Prof Warns." Canadian Press in *Edmonton Journal*, September 19, 2002a, A-3.

Bueckert, Dennis. "Climate Change Could Bring Malaria, Dengue Fever to Southern Ontario, Says Report." Canadian Press in *Ottawa Citizen*, October 23b, 2002. http://www.canada.com/news/story.asp?id={B019135A-4FD8-4536-A2F0-908F13560CB2.

Chase, Steven. "Our Water Is at Risk, Climate Study Finds." *Toronto Globe and Mail*, August 13, 2002, A-1. http://www.globeandmail.com/servlet/ArticleNews/PEstory/TGAM/20020813/UENVIN/national/national/national_temp/6/6/23/.

Constantineau, Bruce. "Weather Wreaking Havoc on British Columbia Farms: Global Warming Has Increased the Intensity and Frequency of Weather Events, Making It Difficult for Farmers to Compete in the Marketplace." *Vancouver Sun*, March 3, 2003, D-3.

Editorial. *Vancouver Province*, September 8, 2003, A-16.

Fidelman, Charlie. "Longer, Stronger Blazes Forecast." *Montreal Gazette*, July 11, 2002, A-4.

Fong, Petti. "Greenhouse Gas Chokes Sky after Wildfires." CanWest News Service in *Calgary Herald*, September 23, 2003, A-8.

Johansen, Bruce E. "Arctic Heat Wave." *The Progressive*, October 2001, 18–20.

Krajick, Kevin. "Arctic Life, on Thin Ice." *Science* 291 (January 19, 2001a):424–425.

Ralston, Greg. "Study Admits Arctic Danger." *Yukon News*, November 15, 1996. http://yukonweb.com/community/yukon-news/1996/nov15.htmld/#study.

"Quebec's Smoky Warning." Editorial, *Baltimore Sun*, July 9, 2002, 10-A.

Sevunts, Levon. "Prepare for More Freak Weather, Experts Say." *Montreal Gazette*, May 10, 2001, A-4.

Cap and Dividend

James E. Hansen, director of NASA's Goddard Institute for Space Studies, proposed that carbon taxes be returned to people under a "Cap and Dividend" program to help individuals cope with higher energy prices. He described the concept:

> In hard economic times with high fuel costs the public will rebel against any carbon tax—unless 100 percent of the tax is returned immediately, monthly, to the public on a per capita basis. The public is fed up with politicians spending their money in cahoots with alligator-shoe-wearing toad-eating (just kidding) lobbyists. Carbon taxes will drive energy innovations and the dividend will spur the economy. Taxes can be fruitfully initiated on a national basis; any trade disadvantage should be eliminated via an import duty on products produced in other countries that do not impose a comparable carbon tax, with 100 percent of the duty added to the per capita dividends. (Hansen 2008e, quoting Barnes 2001)

FURTHER READING

Barnes, Peter. *Who Owns the Sky? Our Common Assets and the Future of Capitalism.* Washington, DC: Island Press, 2001. http://www.ppionline.org/ppi_ci.cfm?knlgAreaID=116&subsecID=149&contentID=3867.

Hansen, James E. Personal communication. September 21, 2008e.

Cap and Trade

A cap-and-trade system places restrictions on greenhouse-gas emissions and then allows buying and selling of rights to pollute. Companies that do not emit greenhouse gases to their limits can sell their rights to others, creating a market that rewards emission reductions. Over time, limits tighten, raising the price of pollution.

A cap-and-trade program is operating in Europe, and by 2008, the program had been proposed (but not enacted) for the United States. The market establishes a gradually shrinking target for Europe's carbon-dioxide emissions and divides it by country. Each country then allocates to power plants and other factories shares that reflect a reduction of their past use, thereby forcing them to "cut emissions, get credit for reducing greenhouse gases in developing countries or buy spare allowances from other firms to make up the shortfall. That creates a market, and a market price, for allowances" (Mufson 2007a, A-1).

By 2007, 10 U.S. states had started their own cap-and-trade program, called the Regional Greenhouse Gas Initiative. Starting in 2009, each state will receive a set number of carbon credits for its power plants, and each plant must have enough allowances to cover its total emissions at the end of a three-year compliance period (Fairfield 2007).

The cap-and-trade system originated in the United States as a program to reduce sulfur-dioxide emissions from acid rain. Lead was eliminated from gasoline, and ozone-depleting chemicals (principally chlorofluorocarbons [CFCs]) were phased out with similar economic incentives.

The system has proved vulnerable to abuse in Europe. Some companies with political connections have managed to win allocations for emissions that were higher than their past records. The result, described by Steven Mufson in the *Washington Post*, was that, in 2007,

> fewer companies than expected had to buy emissions this year, and the price of carbon allowances, which had topped $30 per ton of carbon about a year ago, crashed to about $1 a ton. That eased some of the pressure on electricity rates, but prices for next year, after tighter E.U. limits take effect, are still about $20 a ton. (Mufson 2007a, A-1)

The upshot has been a weird form of green capitalism. Like all markets, cap-and-trade mechanisms do not operate on altruism. Mufson wrote that

> [f]ights have erupted as countries seek to guard their interests. Eastern European nations have lobbied for more generous allocations because of their communist legacies and lower living standards. Germany, [Europe's] largest wind-energy producer, wants an E.U. mandate that each country get 20 percent of its energy from renewable resources by 2020; Poland, which uses no renewable resources, is resisting. Germany boasts that it has cut emissions to 18.4 percent below 1990 levels, the benchmark used in the Kyoto Protocol and in Europe. But nearly half the reduction was because of sagging industrial output in the former East Germany after reunification. For the 2008–2012 period, E.U. officials sliced 5 percent off Germany's emissions proposal. (Mufson 2007a, A-1)

The United States Studies Europe's Market

U.S. officials have been studying the European system. A delegation from California visited Europe for a week early in 2007; the Senate Energy and Natural Resources Committee also held a roundtable discussion with European executives, officials, and consultants. The aim was to adapt Europe's system while avoiding some of its problems (Mufson 2007a, A-1). In late September 2008, the first cap-and-trade auction for greenhouse-gas reduction in the United States raised nearly $40 million for 10 northeastern states as part of the Regional Greenhouse Gas Initiative. The states spent the money on renewable energy technologies and energy-efficiency programs.

Executives in the United States have recognized that cap-and-trade programs present an opportunity to make money in an atmosphere in which eventual emissions caps are a forgone conclusion. An increasing number of industry leaders—including executives of auto companies, FedEx, General Electric, and major utilities—have joined environmentalists in backing various versions of such a system.

An international cap-and-trade system could be used to reward nations with tropical forests to reduce deforestation, meanwhile preserving irreplaceable habitats and providing money for countries that need it. Ensuring the integrity of such a system will require rigorous monitoring, auditing, and registration (Chameides and Oppenheimer 2007, 1670).

By mid-2007, having negotiated with lobbyists for unions and industries (including some electric utilities), the U.S. Senate was preparing to consider the "Low Carbon Economy Act," which was sponsored by Senators Jeff Bingaman, a Democrat from New Mexico, and Arlen Specter, a Republican from Pennsylvania. The committee-level measure would limit the price that industries would have to pay for permits to emit greenhouse gases. The bill also contained billions of dollars to help Alaska cope with the extraordinary effects of warming there (Broder 2007). Some environmental groups asserted that the measure was half-hearted. The Bush White House opposed to any cap-and-trade plan. The proposal, which did not pass, set target emissions for 2020 at 2006 emissions levels and for 2030 at 1990 levels. It was expected to raise fuel prices passed on by oil and gas companies.

"A good sign of the legislation's limits is the fact that it's backed by some of our country's biggest coal companies and polluters," said Brent Blackwelder, president of Friends of the Earth, adding that "[t]his legislation is flawed in many ways." He said the bill would actually allow global warming emissions to increase at first and that the bill's caps would not bring emissions back to 2006 levels until 2020 (Mufson 2007d, D-1).

"This is a clear attempt to find the political middle ground on the climate debate rather than a utopian solution," said Frank O'Donnell, president of Clean Air Watch. But he added that "the bill appears to fall well short of what science says is needed." He said the proposal was "laden with goodies for coal" (Mufson 2007d, D-1).

As soon as the U.S. political system began to seriously consider taxing carbon emissions, various industries involved in the Climate Action Partnership began to jockey in favor of their interests. Although major industries supported greenhouse-gas limits in principle, specifics varied widely. The raw truth is that dealing with global warming will reward some industries and penalize others.

Depending on how a cap-and-trade market is defined, electricity bills for energy from fossil-fuel sources could increase as much as 50 percent, and gasoline prices could increase $0.50 a gallon. While everyone wants to deal with climate change, hardly anyone wants to pay the price involved in penalizing fossil-fuel use and rewarding development of renewable sources such as wind and solar. Critics argue that a cap-and-trade scheme will amount to a tax on fossil-fuel energy as a market device to reduce its use. Supporters assert that it is far better to tax undesirable activity (such as than creation of greenhouse gases) than socially beneficial activity, such as labor (as with an income tax). Sweden, for example, has replaced a large proportion of its income tax with energy taxes.

Bjorn Lomborg, a self-styled "skeptical environmentalist," said of cap-and-trade schemes:

> It would slow American economic growth by trillions of dollars over the next half-century. But in terms of temperature, the result will be negligible if China and India don't also commit to reducing their emissions, and it will be only slightly more significant if they do. By itself, Lieberman-Warner [a proposal for cap-and-trade] would postpone the temperature increase projected for 2050 by about two years. (Lomborg 2008, A-19)

The answer, explained Lomborg,

> is to dramatically increase research and development so that solar panels become cheaper than fossil fuels sooner rather than later. Imagine if solar panels became cheaper than fossil fuels by 2050: We would have solved the problem of global warming, because switching to the environmentally friendly option wouldn't be the preserve of rich Westerners. (Lomborg 2008, A-19)

A committee of the British Parliament in 2008 initiated a study of a personal version of cap and trade, a carbon debit card that would ration energy use and allow trading among individuals.

Cap and Hustle

A European cap-and-trade market that issued a limited number of permits to emit carbon dioxide provoked old-fashioned industrial lobbying for favors that enriched some of the European Union's biggest polluters and, by the end of 2008, had produced very little actual reduction in greenhouse gases. The idea: "If a company produced more gas than its permits allowed, it would be penalized by having to buy more; if it managed to reduce emissions by switching to cleaner fuels or technologies, it would be able to sell its permits to polluting companies. The marketplace would set the price" (Kanter and Mouwad, 2008).

Large utilities and basic industries convinced the European Union to give away most of the permits free, in large numbers, so the market nearly collapsed. Europe eventually issued a new system of permits that traded about $80 billion during 2008, most of which was recouped from consumers in the prices of goods and services, often as higher electric bills. However, the volume of carbon dioxide emitted by plants and factories participating in the system rose 0.4 percent in 2006 (vis-à-vis 2005) and another 0.7 percent in 2007 (Kanter and Mouwad, 2008).

Germany's power company RWE, Europe's largest industrial source of carbon dioxide, reaped $6.4 billion in three years. Companies throughout Europe took in about $374 billion at the peak of the market, as various national governments, most notably Germany, added exemptions and bonuses that fattened the take. "It was lobbying by industry, including the electricity companies, that was to blame for all these exceptional rules," said Hans Jürgen Nantke, the director of the German trading authority, part of the Federal Environment Agency (Kanter and Mouwad, 2008).

FURTHER READING

Broder, John M. "Compromise Measure Aims to Limit Global Warming." *New York Times*, July 11, 2007. http://www.nytimes.com/2007/07/11/washington/11climate.htm.

Kanter, James, and Jad Mouwad. "Money and Lobbyists Hurt European Efforts to Curb Gases." *New York Times*, December 11, 2008. http://www.nytimes.com/2008/12/11/business/worldbusiness/11carbon.html.

Lomborg, Bjorn. "A Better Way Than Cap and Trade." *Washington Post*, June 26, 2008, A19. http://www.washingtonpost.com/wp-dyn/content/article/2008/06/25/AR2008062501946_pf.html.

Mufson, Steven. "Europe's Problems Color U.S. Plans to Curb Carbon Gases." *Washington Post*, April 9, 2007a, A-1. http://www.washingtonpost.com/wp-dyn/content/article/2007/04/08/AR2007040800758_pf.html.

Mufson, Steven. "Florida's Governor to Limit Emissions." *Washington Post*, July 12, 2007d, D-1. http://www.washingtonpost.com/wp-dyn/content/article/2007/07/11/AR2007071102139_pf.html.

"Your Carbon Ration Card." Editorial, *The Wall Street Journal*, July 7, 2008, A-11.

Capitalism and the Atmospheric Commons

Can capitalism, with its appetite for pell-mell (and often environmentally destructive) growth, survive in a new world in which geophysical reality demands that we restrain our demands on the Earth? Are we ready to operate with an accounting system that brings us all to account for the toll that our activities exact on the Earth and its atmosphere? A system in which polluting the atmospheric commons is a criminal act?

Environmental writer and activist Edward Abbey frequently called capitalism the ideology of the cancer cell—an organism that grows haphazardly and eventually kills its host. The ecological implications of this metaphor have become more fully apparent as human numbers, material affluence, and global temperatures rise in tandem, along with global threats to environment.

The cold war ended with the collapse of communism as an international state-level ideology, in no small part because the nineteenth-century ideology of Karl Marx paid little attention to the effects of environmental degradation. Capitalism is now the world's dominant economic system, as the drive to profit, along with human desires for wealth and comfort are bringing hundreds of millions of people into the carbon-dioxide producing middle and upper classes worldwide. The "American Dream" (a family home, a car, appliances, and so on) is spreading around the world, as greenhouse-gas levels accelerate.

Capitalism demands growth and, like Marxism, makes little economic allowance for environmental consequences. In assuming that humankind is separate from nature (and superior to it), the cultural architecture of the fossil-fuel age has sowed some worrisome climatic seeds. The philosophic attitudes supporting this sense of superiority are much older than the industrial revolution. In the Christian Bible (*Genesis* 9:1–3), God instructs Noah and his sons to

> [b]e fruitful, and multiply, and fill the Earth. The fear of you and the dread of you shall be upon every beast of the Earth, and upon every bird of the air, and upon everything that creeps on the ground, and all the fish in the sea; into your hand, they are delivered. (McNeill 2000, 327)

Modern capitalism was born in Europe as the Protestant Reformation told the entrepreneurs of the early fossil-fueled machine age that God helps those who help themselves. Such attitudes have become entrenched in worldwide political economy. Through "globalization" during the late twentieth century, these beliefs have become a world-girding ideological currency.

Trashing the Atmospheric Commons

In addition to a belief system that alienates machine-age humankind from nature, we live in an economic and political culture that regards our environmental heritage as an ownerless commons that anyone may defile for profit without moral sanction or fiscal penalty. The atmosphere becomes the ultimate "free good," to be used and abused without cost or penalty. The Earth is a finite habitat, however, so growth (especially growth in fossil-fuel effluent) cannot continue indefinitely.

Ian H. Rowlands quoted Aristotle as the first philosopher to wrestle with the concept that private individuals will defile a freely held common good without concern for humanity's shared welfare:

> What is common to the greatest number gets the least amount of care. Men pay most attention to what is their own; they care less for what is common.... When everyone has his own sphere of interest ... the amount of interest will increase, because each man will feel that he is applying himself to what is his own. (Rowlands 1995, 4, quoted in Hardin 1997, xi)

Anita Gordon and David Suzuki's *It's a Matter of Survival* (1991) questioned several "sacred truths," assumptions of machine-age culture:

> The assumptions that we've made about how the natural world operates and what our relationship is to it are no longer tenable. These *sacred truths* that we've grown up with—"nature is infinite"; "growth is progress"; "science and technology will solve our problems"; all nature is at our disposal"; "we can manage the planet"—offer no comfort as we enter the last decade of this [the twentieth] century. In fact, we're being told that to continue to subscribe to these assumptions is no ensure the destruction of civilization as we know it. (Gordon and Suzuki 1991, 1)

Change the Character of Capitalism?

A major—perhaps *the* major question facing an Earth and its human denizens in a time of worldwide environmental crisis is whether capitalism can change its character. A sustainable environment can make good business. Witness the growth of alternative forms of energy. Can capitalism factor respect for the Earth that sustains us all into its calculus of development? If so, it may be a positive force in a new, sustainable world. If not—if it retains attributes of the cancer cell—then ultimately, in an exhausted, poisoned world, our progeny will suffer in a wrecked world.

FURTHER READING

Gordon, Anita, and David Suzuki. *It's a Matter of Survival.* Cambridge, MA: Harvard University Press, 1991.

Hardin, Garrett, and John Baden, eds. *Managing the Commons.* San Francisco: W. H. Freeman, 1977.

McNeill, J. R. *Something New Under the Sun: An Environmental History of the Twentieth-Century World.* New York: W.W. Norton, 2000.

Rowlands, Ian H. *The Politics of Global Atmospheric Change.* Manchester, UK: Manchester University Press, 1995.

Carbon Capture and Sequestration

After the U.S. federal FutureGen, clean coal plant in Illinois was called off in January 2008, similar state-level projects in Florida, West Virginia, Ohio, Minnesota, and Washington State also were canceled or stalled by regulatory problems. "It's a total mess," said Daniel M. Kammen, director of the Renewable and Appropriate Energy Laboratory at the University of California, Berkeley (Wald 2008b). Even as existing projects were canceled, in May 2008, General Electric formed a partnership with Schlumberger, an oil field services company, to develop new technology for carbon capture and sequestration (CCS).

During December 2007, a consortium of large coal companies and electric utilities had selected Mattoon, Illinois, as the site of its FutureGen plant, a 275-megawatt prototype meant to "serve as a large-scale engineering laboratory for testing new clean power, carbon capture, and coal-to-hydrogen technologies, optimistically touted as 'near zero-emission' coal-gasification technology" ("FutureGen" 2007).

"This project makes coal, one of the most abundant fossil energies in the world, available in the future in the face of growing concern over greenhouse gas emissions and climate change," said Jeff Jarrett, Department of Energy assistant secretary of fossil energy. In January 2008, however, the Energy Department canceled its 74 percent cost share in the FutureGen project, citing soaring costs and lack of results.

The plant was beset by problems even before the first spade of soil was turned at the construction site. Even the final Environmental Impact Statement (EIS) issued by the Energy Department admitted that the sequestered carbon dioxide might leak due to the "presence of undetected faults, wells penetrating the primary seal, or other subsurface pathways" ("FutureGen" 2007).

Other potential problems included the "exact quantities of materials delivered and byproducts produced, their method of transportation, and the disposition of waste" and the "exact noise profiles of power plant equipment, their proximity to nearby receptors and types and quantities of construction equipment." According to the EIS, also problematic were the "current and future water levels in potentially affected streams near the Mattoon site" ("FutureGen" 2007).

General Issues for CCS

Among the problems facing researchers who seek to design carbon-capture technology are determining which rock and soil formations best hold carbon dioxide and prevent it from reemerging. Ways are needed to design the systems so that they are not overly expensive and do not require a large part of a plant's energy output to operate. Some environmentalists fear that carbon dioxide stored in the Earth could damage groundwater.

Tenaska, Inc., an Omaha-based company, during February 2008 announced plans to build the first U.S. commercial-scale coal-fired power plant (600 megawatts, capable of serving 600,000 homes). The proposed plant will capture carbon-dioxide emissions and store them underground. Tenaska, which builds power plants nationwide, said that the stored carbon dioxide could be sold to petroleum companies that would pipe it underground to drive out oil deposits. The company plans to build the $3 billion plant, to be called the Tenaska Trailblazer Energy Center, on a 1,919-acre site north of Interstate 20 in Nolan County, near Sweetwater, Texas.

Tenaska told the *Omaha World-Herald* that the U. S. Energy Department is encouraging CCS technology, and that the National Coal Council, which advises the U.S. energy secretary, supports the technology. Plans call for capture of "as much as 90 percent" of the plant's carbon dioxide and storage in the Permian Basin, which contains oil deposits. Tenaska said that a final decision would be made on the plant during 2009, depending on government incentives, revised cost estimates, and market prices for electricity. If undertaken, the plant could provide 2,000 construction jobs and 100 permanent operating positions (Jordan 2008, D-1, D-2).

In 2008, We Energies of Wisconsin announced a pilot project to test CCS technology, and also heralded its project as a first in the United States. The initial attempt will remove just 3 percent of an existing plant's emissions, and then release them into the air (Davidson 2008, 3-A).

FURTHER READING

Davidson, Paul. "Coal Plant to Test Capturing Carbon Dioxide." *USA Today*, February 27, 2008, 3-A.

"FutureGen, World's Cleanest Coal Plant, Sited in Illinois." Environment News Service, December 18, 2007. http://www.ens-newswire.com/ens/dec2007/2007-12-18-01.asp.

Jordan, Steve. "Tenaska to Build Clean Power Plant." *Omaha World-Herald*, February 20, 2008, D-1, D-2.

Wald, Matthew L. "Mounting Costs Slow the Push for Clean Coal." *New York Times*, May 30, 2008b. http://www.nytimes.com/2008/05/30/business/30coal.html.

Carbon Cycle

Ice-core records from Antarctica dating to more than 800,000 years before the present, released by the British Antarctic Survey, indicate that present-day levels of carbon dioxide in the atmosphere are higher now than at any time in their record, and increasing at an unprecedented rate. Levels of carbon dioxide, methane, and other greenhouse gases in the atmosphere during the last years of the twentieth century were higher than at any time since humankind has walked the Earth. This circumstance that has provoked one student of the greenhouse effect to remark that "[w]e are headed for rates of temperature rise unprecedented in human history; the geological record screams a warning to us of just how unprecedented ... [the stresses on] the natural environment will be" (Leggett 1990, 22).

Adding to the complexity of the carbon cycle, the full effect of a given emission is not felt in the atmosphere for about 50 years (the exact period is open to debate). In the oceans, the full effect of emissions takes a century or two to manifest itself in actual warmth. Thus, in about 2010 air temperatures will reflect greenhouse-gas emissions from about 1960. Ocean temperatures barely reflect the first inklings of the Industrial Revolution's greenhouse-gas load.

The annual increase of carbon dioxide in the atmosphere increased from less than 1 parts per million (ppm) per year in 1958 (when the first precise measurements were initiated) to about 2 ppm per year during the 1990s and 3 ppm after 2000. The rate of increase varies by year. In business-as-usual scenarios, the annual rate of increase may reach about 4 ppm per year by the middle of the twenty-first century (Hansen 2006b, 26).

Carbon Cycle Basics Open to Debate

The basics of the carbon cycle are not fully understood, even by scientific specialists. Until recently, for example, scientists have not been able to account for all the carbon utilized in the worldwide cycle. The relationship of a certain level of greenhouse gases in the atmosphere to a given amount of warming over an assumed length of time is open to pointed debate. Other factors (the Earth's orbit, solar insolation, land-use patterns, and others) are provided more or less weight in debates regarding how quickly temperatures may rise as consumption of fossil fuels raises greenhouse-gas levels in the atmosphere. How is warming related to drought cycles? What role do clouds (including contrails, the exhaust of jet aircraft) play in the general warming of the Earth? The role of aerosols (the most important probably being black soot) are only beginning to be understood and factored into forecast models.

Kathy Maskell and Irving M. Mintzer, writing in the British medical journal *Lancet*, described the carbon cycle's natural balancing act:

> Over the past 10,000 years, the concentration of $C0_2$ and other greenhouse gases has remained fairly constant, and this represents a remarkable balancing act of nature. Every year natural processes on the land and in the oceans release to and remove from the atmosphere huge amounts of carbon, about 200 billion tons (gigatons) in each direction. Since the atmosphere contains about 700 gigatons of carbon, small changes in natural fluxes could easily produce large swings in atmospheric concentrations of CO_2 and or CH_4 [methane]. Yet for ten millennia natural fluxes have remained in remarkably close balance. (Maskell and Mintzer 1993, 1027)

Human activities are disturbing this long-standing balance in a fundamental manner, adding to concentrations of several greenhouse gases, including, according to Maskell and Mintzer, "Water vapor, the principal greenhouse gas ... [which] is expected to increase in response to higher temperatures that would further enhance the greenhouse effect" (Maskell and Mintzer 1993, 1027).

Scientists who study the future potential of human-induced warming also point to several other natural mechanisms that could cause the pace of change to accelerate, such as release of carbon dioxide and methane from permafrost and continental shelves in the oceans. The possibility of a "runaway" greenhouse effect by the year 2050 is raised in the literature, often with a palatable sense of urgency.

As scientists explore the complexities of atmospheric chemistry, some long-held assumptions have been cast aside. One such assumption has been that the Earth's carbon cycle persists in steady state over time. A. Indermuhle and several co-authors published results of ice-core tests from Alaska that indicate that "the global carbon cycle has not been in steady state during the past 11,000 years.... Changes in terrestrial biomass and sea-surface temperature were largely responsible for the observed millennial-scale changes in atmospheric CO_2 concentrations" (Indermuhle et al. 1999, 121). Carbon dioxide levels in the atmosphere have now been measured (most often from ice cores) to more than 400,000 years in the past, varying from about 180 ppm to about 285 ppm, which was roughly the level when the industrial age began. By contrast, the carbon-dioxide level by the year 2008 was about 385 ppm, about 35 percent higher than any time in the ice-core record.

The Relationship between a Given Rise in Atmospheric Greenhouse Gases and a Given Rise in Temperature

Atmospheric scientists know that the Earth's surface temperatures often have varied more or less in tandem with the atmosphere's level of carbon dioxide and other greenhouse gases. What they do not know precisely is *how much* of a temperature increase may be triggered by a given rise in the level of carbon dioxide and other greenhouse gases. Atmospheric modeling is at once very simple, because the general relationship between heat-absorbing gases and temperature is known, and very complex, because climate can be influenced by many other factors as well.

The level of carbon dioxide in the atmosphere usually rises and falls with temperature; however, paleoclimatologists have found some specific epochs in the Earth's history during which glaciation occurred when carbon-dioxide levels were several times higher than today, even at human-enhanced levels. One such example is the Late Ordovician (Hirnantian) glaciation during the early Paleozoic, a rare spike of cold weather during an otherwise balmy period when atmospheric carbon-dioxide levels sometimes reached 14 to 16 times preindustrial levels. Mark T. Gibbs and colleagues assert that positioning of the continents at that time may have contributed to this most unusual of ice ages (Gibbs et al. 2000, 386).

High Temperatures on Low CO_2?

Adding another intriguing wrinkle to the paleoclimatic puzzle, investigators have found evidence that sometimes appears to decouple carbon-dioxide levels in the atmosphere from climate change. For example, M. Pagani, M. A. Arthur, and K. H. Freeman reported in *Paleoceanography* (1999) that during the Miocene Climatic Optimum (roughly 14.5 to 17 million years ago), the Earth experienced its warmest climate in 35 million years when temperatures averaged about 6°C higher than today. This level of warmth was achieved with atmospheric carbon-dioxide levels between 180 and 290 ppm, compared with the present-day level of about 385 ppm.

The same researchers also found that rising, not falling, carbon-dioxide levels accompanied growth in the East Antarctic Ice Sheet between 12.5 and 14 million years ago. Such findings do not necessarily contradict the infrared-forcing capacities of carbon dioxide itself, but indicate that other factors (including the levels of other gases, such as methane, and effects of changes in oceanic circulation patterns) compete with carbon-dioxide levels to shape the climate of any particular place on the Earth at any given time.

By the late 1990s, scientists lacking solid theoretical models that establish a direct causal link between carbon-dioxide level (by itself) and temperature had largely stopped forecasting that a particular carbon-dioxide level would cause a specific temperature rise. Instead, increasingly sophisticated studies are attempting to describe what role greenhouse gases will play in the context of other influences, or "forcings," in the atmosphere.

As more scientists examine the dance of climate change, they find more possible influences that complicate the identification of a specific, quantifiable role for infrared forcing on its own. For example, in the April 11, 2000, edition of the *Proceedings of the National Academy of Sciences of the United States of America,* Charles D. Keeling and Timothy P. Whorf proposed that a 1,800-year oceanic tidal cycle is influencing climate:

> We propose that such abrupt millennial changes, as seen in ice and sedimentary core records, were produced at least in part by well-characterized, almost periodic variations in the strength of the global oceanic tide-raising forces caused by resonances in the periodic motions of the Earth and Moon. (Keeling and Whorf 2000, 3814)

Increased tidal turbulence influences the rate of "vertical mixing" in the oceans and, thus, their surface temperature. Greater oceanic mixing cools the surface of the oceans, and exerts an influence on global average temperatures. The last peak in this cycle is said, by Keeling and Whorf, to have coincided with the Little Ice Age which climaxed about 1600 A.D. (Keeling and Whorf 2000, 3814).

FURTHER READING

Gibbs, Mark T., Karen L. Bice, Eric J. Barron, and Lee R. Kump. "Glaciation in the Early Paleozoic 'Greenhouse:' The Roles of Paleo-geography and Atmospheric CO_2." In *Warm Climates in Earth History,* ed.

Brian T. Huber, Kenneth G. MacLeod, and Scott L. Wing, 386–422. Cambridge: Cambridge University Press, 2000.

Hansen, James E. "Declaration of James E. Hansen." *Green Mountain Chrysler-Plymouth-Dodge-Jeep, et al., Plaintiffs v. Thomas W. Torti, Secretary of the Vermont Agency of Natural Resources, et al., Defendants. Case Nos. 2:05-CV-302 and 2:05-CV-304, Consolidated. United States District Court for the District of Vermont.* August 14, 2006b. http://www.giss.nasa.gov/~dcain/recent_papers_proofs/vermont_14aug20061_textwfigs.pdf.

Indermuhle, A., T. F. Stocker, F. Joos, H. Fischer, H. J. Smith, M. Wahlen, B. Deck, D. Mastroianni, J. Tschumi, T. Blunier, R. Meyer, and B. Stauffer. "Holocane Carbon-cycle dynamics Based on CO_2 trapped in Ice at Taylor Dome, Alaska." *Nature* 398 (March 11, 1999):121–126.

Keeling, Charles D., and Timothy P. Whorf. "The 1,800-year Oceanic Tidal Cycle: A Possible Cause of Rapid Climate Change." *Proceedings of the National Academy of Sciences of the United States of America* 97, no. 8 (April 11, 2000):3814–3819.

Leggett, Jeremy, ed. *Global Warming: The Greenpeace Report.* New York: Oxford University Press, 1990, 83–112.

Maskell, Kathy, and Irving M. Mintzer. "Basic Science of Climate Change." *Lancet* 342 (1993):1027–1032.

Carbon Dioxide: An Organism to "Eat" It

J. Craig Venter, who compiled a human genetic map with private money, has decided to tackle the problem of global warming with a $100 million research endowment created from his stock holdings. Venter plans to scour deep ocean trenches for bacteria that may convert carbon dioxide to solid form with very little sunlight or other energy (Gillis 2002, E-1). If he cannot find such an organism in nature, Venter may build one himself. His idea: to devise technology that will allow humankind to continue producing energy while reducing emissions of carbon dioxide. Ideally, Venter's organism or group of organisms will be able to take in carbon dioxide, break it down, and produce both biological compounds and energy (Gillis 2002, E-1).

Venter would like to invent two synthetic microorganisms. One would consume carbon dioxide and turn it into raw materials comprising the kinds of organic chemicals that are now made from oil and natural gas. The other microorganism would generate hydrogen fuel from water and sunshine. "Other groups are

Craig Venter, president and chief scientific officer of Celera Genomics in Rockville, Maryland, stands next to a map of the human genome in his office, May 8, 2001. (AP/Wide World Photos)

considering capturing carbon dioxide and pumping it down to the bottom of the ocean, which would be insane," Venter told the *London Financial Times.* "Why risk irreversible damage to the ocean when we could be doing something useful with the carbon?" (Cookson and Firn 2002, 11).

"We've barely scratched the surface of the microbial world out there to try to help the environment," Venter said. "We're going to be searching for some dramatic new microbes" (Gillis 2002, E-1). Venter said that his ventures will be established as nonprofit corporations. "I'm not in business anymore," he said (Gillis 2002, E-1). Venter, who calls his organization the Institute for Biological Energy Alternatives, will seek grant money from the U.S. Department of Energy. His goal will be to explore whether modern science can use the power of biology to solve the world's most serious environmental crisis.

Venter has proposed installing colonies of organisms in "bioreactors" near power plants to consume emissions of carbon dioxide and turn the gas into solids such as sugars, proteins, and starches, mimicking the behavior of green plants (Gillis 2002, E-1). According to an account in the *Washington Post,*

> Venter plans to base his approach on one of the most striking developments in biology in recent years—the discovery, in deep ocean trenches and volcanic hot spots on the ocean floor, of a wide array of bacteria that can perform extensive

> chemical reactions without needing sunlight. These are thought to be descendants of the most primitive life forms that arose on the Earth, and scientists are just beginning to explore their potential. (Gillis 2002, E-1)

FURTHER READING

Cookson, Craig, and David Firn. "Breeding Bugs That May Help Save the World: Craig Venter Has Found a Large Project to Follow the Human Genome." *London Financial Times*, September 28, 2002, 11.
Gillis, Justin. "A New Outlet for Venter's Energy: Genome Maverick to Take on Global Warming." *Washington Post*, April 30, 2002, E-1.

Carbon Dioxide: Enhanced and Crop Production

Because plants require carbon dioxide to grow, folk wisdom holds that increasing the level of this trace gas in the atmosphere will accelerate plant growth. Climate-change skeptics who contend that higher carbon-dioxide levels will benefit plants often ignore the possibility that heat stress or drought-and-deluge conditions, both probable side-effects of warming, will cause problems. In addition, scientific support for the idea that elevated carbon-dioxide levels aid plant growth is far from unanimous. While a few studies indicate that in the short range higher carbon-dioxide levels and warmer temperatures may help "green" the Earth, longer-run studies cast considerable doubt on this assumption.

The case for atmospheric carbon dioxide as agricultural bromide has been advanced by some global-warming skeptics with absolute and often-simplistic statements about its assumed benefits. One such statement comes from Sylvan Wittner, director of the Michigan Agricultural Experiment Station and chairman of the agricultural board of the National Research Council:

> Flowers, trees and food crops love carbon dioxide, and the more they get of it, the more they love it. Carbon dioxide is the basic raw material that plants use in photosynthesis to convert solar energy into food, fiber and other forms of biomass. Voluminous scientific evidence shows that if CO_2 were to rise above its current ambient level of 360 parts per million, most plants would grow faster and larger because of more efficient photosynthesis and a reduction of water loss. (Klug 1997)

Like Wittner, Sherwood Idso asserts that the Earth's ecosystem is starved for carbon dioxide. Idso paints a wondrous picture of giant plants nourished by carbon-dioxide levels "in the thousands of parts per million" (Klug 1997). Idso calls such levels "a breath of fresh air for the planet's close-to-suffocating vegetation" (Klug 1997). The word "rejuvenation" is prominent in Idso's case, describing his vision of a carbon-enriched world atmosphere. Idso does not tell his audience that such carbon-dioxide levels were experienced at a time when global temperatures were at a level that would make climate over much of the Earth's surface virtually insufferable during summers for warm-blooded animals.

Idso has written:

> The whole face of the planet will likely be radically transformed—rejuvenated as it were—as the atmospheric CO_2 content reverses its long history of decline and returns, in significant measure, to conditions much closer to those characteristic of the Earth at the time when the basic properties of plant processes were originally established. (Idso 1982, 9)

A slight rise in temperatures and carbon-dioxide levels would stimulate the growth of many plants, but any gardener knows that each plant has its heat limit. Thus, studies by several researchers (see, for example, Schimel 2006, 1889–1890; Long et al. 2006, 1918–1921) reveal a considerably more complex relationship between plant growth and "enhanced" carbon dioxide. These levels vary considerably between species. Every plant has a level of temperature at which the rate of respiration surpasses the rate of photosynthesis, causing the organism to lose carbon and eventually die. Every plant also requires moisture in certain amounts. Too little (or, conversely, too much) moisture will kill many plants. Night-time temperatures also play a crucial role in relieving moisture stress for some species. If night-time temperatures are too high, some trees will die.

Levels of carbon dioxide above a certain level may actually stunt the growth of many plants. Many insects are sensitive to temperature, as well as humidity, and even a small rise can cause massive outbreaks, as witnessed by the insect population explosion in New Orleans during 1995, following five years without a killing frost.

Researchers Roseanne D'Arrigo and Gordon Jacoby of the Lamont-Doherty Earth Observatory studied tree rings near the timberline in northern and central Alaska's boreal forests. They found that growth increased with modest increases in warming during the 1930s and 1940s. After that, increased warming produced heat and drying that stressed the trees that, along with increasing activity by insects, caused tree growth to slow most years.

Growth was higher than usual during years when moisture was plentiful, but below average during relatively warm, dry years. In the meantime, insects usually flourish in warmer weather. The bark beetle, a major pest in Alaska's forests, has shortened its reproductive cycle by 100 percent due to warming, according to Jacoby and D'Arrigo (Gelbspan 1997b, 141). In the meantime, David Deming has measured soil temperatures in northern Alaska. His records indicate that the soil there has warmed by 2° to 5°C (3.6° to 9°F) during the twentieth century.

Warming Harms Crop Yields

The same fungus that caused the Irish potato famine has migrated 4,000 feet up slopes in the Andes to the town of Chacllabamba, Peru, because of warmer, wetter weather. Potato breeders are trying to outwit the fungus by developing new breeds of tubers (Halweil 2005, 18).

Hartwell Allen, a researcher with the University of Florida and the U.S. Department of Agriculture has been growing rice, soybeans, and peanuts under controlled conditions, varying temperatures, humidity, and carbon-dioxide levels. Allen and his colleagues have found that while higher temperatures (to a point) and CO_2 levels stimulate lusher and faster growth, they are deadly at flowering and pollinating stages. At temperatures above 36°C during pollination, peanut yields dropped 6 percent per degree Celsius of temperature. John Sheehy of the International Rice Research Institute in Manila found that damage to the world's major grain crops begins during flowering at about 30°C. At about 40°C, the plants' yields fall to zero. At his center, the average temperature has risen 2.5°C in 50 years, frequently reaching damaging levels. In rice, wheat, and maize, yields fall 10 percent for every degree above 30°C. Higher night-time temperatures in particular inhibit plants' ability to respirate and saps their energy (Halweil 2005, 19–20). Planting shade trees among crops may mitigate heat in the short range.

Global warming is damaging production of the world's major food crops, according to scientists at Lawrence Livermore National Laboratory and the Carnegie Institution at Stanford University. Ecologists there reported that yields of corn, wheat, and barley have declined by about 40 million tons every year since 1981 from what farms worldwide should have produced. The annual value of those lost crops is roughly $5 billion (Hoffman 2007).

While crop yields continue to improve, they would have been better during the last 20 years without warming temperatures, according to Livermore climate scientist David Lobell and Christopher Field, head of Carnegie's Department of Global Ecology at Stanford (Hoffman 2007). "At least for wheat, corn and barley, temperature trends in the last few decades have been in the direction of holding yields down," Field said. "They're still increasing, but if temperatures hadn't been warming, they would have been increasing more" (Hoffman 2007).

According to a report in *World Watch* magazine,

> For wheat and corn alone, the annual global losses are equal to the wheat and corn production of Argentina. The researchers likened the effect of global warming on crops to driving a car with the parking brake on. As a result, they said, farmers and plant scientists will have to work harder at meeting rising food demand—and harder still if farming for energy as well. (Hoffman 2007)

"I think what we're seeing is that the direct effects of climate change are negative" for some major row crops, said Lobell. "There's still this big question of what CO_2 is doing. So far, technology has kept crop yields growing so that supply keeps pace with soaring demand for food," Lobell added. "But it's a race, and I think of climate change as a sort of headwind for the supply increase," he said. "We're talking about potentially much slower increases in supply that will eventually start to lose ground to demand. The question is whether they can keep pace" (Hoffman 2007).

Reduced Plant Numbers and Diversity

After three years, according to one set of tests, elevated carbon-dioxide and nitrogen deposition together reduced plant diversity, whereas elevated precipitation alone increased it, and warmer temperatures had no significant effect. According to this study, "Results show that climate and atmospheric changes can rapidly alter biological diversity, with combined effects that, at least in some settings, are simple, additive combinations of single-factor effects" (Zavaleta et al. 2003, 7650). This research found that doubling the level of carbon dioxide reduced numbers of wildflowers by 20 percent, and cut overall plant diversity by 8 percent (Connor 2003b). Results from experiments involving potted plants have limited utility, however, and often cannot be extrapolated to the natural environment.

Scientists from Stanford University manipulated levels of carbon dioxide for 36 open-air plots of land on which wild plants grew for three years. They doubled carbon dioxide, increased available moisture by 50 percent, caused average temperatures to rise by 1.7°C, and added nitrogen to the soil. All of these conditions could become standard fare in the open air within a century, according to projections of the Intergovernmental Panel on Climate Change (IPCC). Plots that received all four treatments suffered a decline of 25 percent in volume of wild flowers; those given extra nitrogen or carbon dioxide suffered a 10 or 20 percent decline. Only increased watering produced a rise in diversity. The scientists suggested that increased carbon dioxide, temperature, and nitrogen allowed some plants to grow faster for longer periods, but impeded growth (and sometimes survival) of other plants (Connor 2003b).

Christopher Field said the team was surprised to find that increased carbon dioxide and watering caused such disparate effects, given that they were both essential for plants' growth. "One hypothesis is that elevated carbon dioxide added moisture to the soil, which tended to extend the growing season of the dominant plants, leaving less room for other species to grow," Professor Field said (Connor 2003b). Field's colleague, Erika Zavaleta, said the study demonstrated how some wild plants would suffer while other may benefit from the effects of climate change. "Certain kinds of species are much more sensitive to climate and atmospheric changes than others. It turned out that wild flowers were much more sensitive to the treatments than grasses were, no matter what combination of treatments we tried," Zavaleta said (Connor 2003b).

When enhanced temperature, nitrogen, and water were applied to a plot, the production soared by 84 percent, Shaw said. But when carbon dioxide was added to this mix, the production dropped by 40 percent. "This was unexpected," Shaw explained. "We think that by applying all four elements in combination in a realistic situation, some other nutrient becomes a limiting factor to growth" (Recar 2002; Shaw et al. 2002, 1987).

According to another study, "Well-watered and fertilized citrus trees clearly grow better when exposed to high levels of carbon dioxide, but the effect on forest trees is uncertain because most forests have limited supplies of other resources needed for growth" (Davidson and Hirsch 2001, 432). Trees fed additional nitrogen after natural supplies ran low continued to grow more quickly in a carbon dioxide-enriched atmosphere. "This gain was even larger at the poor site ... than at the moderate site," another study suggested (Oren et al. 2001, 469). "Here," concluded R. Oren and colleagues, "We present evidence that estimates of increases in carbon sequestration of forests, which is expected to partially compensate for increasing CO_2 in the atmosphere, are unduly optimistic" (Oren et al. 2001, 469). After an initial spurt of growth, trees exposed to elevated levels of carbon dioxide develop more slowly and do not absorb as much carbon dioxide from the atmosphere, according to Oren's work. This study suggested that reliance on forests to remove carbon dioxide from the atmosphere, a major part of the Kyoto Protocol, may not work. Planting trees alone is not a substitute for reducing carbon-dioxide emissions. As plants grow, they use nutrients in surrounding soils. As soils are depleted (at rates which increase with enhanced carbon dioxide) growth and ability to sequester carbon dioxide slow dramatically.

Carbon-Dioxide "Enrichment" and Nitrogen Fixation

Bruce A. Hungate led a study of carbon-dioxide "enrichment" in an oak woodland that indicated increasing nitrogen fixation during the first year. However, the effect declined in the second year and disappeared by the end of the third year. After that, through the seventh year of treatment, carbon-dioxide enrichment consistently depressed nitrogen fixation. Clearly, the relationship between carbon-dioxide level and nitrogen fixation is not simple. Hungate and colleagues found that "[r]educed availability of the micronutrient molybdenum, a key constituent of nitrogenase, best explains this reduction in N[itrogen] fixation. Our results demonstrate how multiple element interactions can influence ecosystem responses to atmospheric change" (Hungate et al. 2004, 1291).

In another study of elevated carbon-dioxide levels' affects on plants' growth, Deborah Clark and colleagues commented:

> Trees' annual diameter increments in this 16-year period (1984–2000) were negatively correlated with annual means of daily minimum temperatures.... Strong reductions in tree growth and large inferred releases of CO_2 to the atmosphere occurred during the record-hot 1997–1998 El Niño. These and other recent findings are consistent with decreased net primary production in tropical forests in the warmer years of the last two decades.... Such a sensitivity of tropical forest productivity to on-going climate change would accelerate the rate of atmospheric CO_2 accumulation. (Clark et al. 2003, 5852)

"We became interested in this when our long-term measurements of tree growth had come to cover several years, and we could see that tree growth varied greatly from year to year," Clark added. "We realized that something about yearly differences in weather patterns must be doing this, and we started to focus on what that might be" (Choi 2003). The scientists examined the annual growth of six tree species in an old-growth rainforest at the La Selva Biological Station in Costa Rica. They measured the trunk diameter of 164 adult tree species between 1984 and 2000 in an area the size of about 700 football fields. "To measure some of these trees required carrying two to three ladders cross-country through the forest in order to be able to measure the tree trunk diameter, often 3 to 10 feet above the ground to get above the large buttresses that protrude from many tropical trees," Clark said (Choi 2003).

When Clark and her team matched tree growth with local temperature readings, they found that the trees often were stunted during the hottest years—most notably during the record-warm 1997–1998 El Niño. During warmer years, atmospheric gas samples revealed that tropical regions as a whole released more carbon dioxide than they absorbed (Choi 2003). In other words, tropical forests, sometimes called the "lungs of the world" for their capacity to produce oxygen, under some conditions, actually became net emitters of carbon dioxide.

"There is now an urgent need for studies to see if what we found at La Selva is occurring generally across tropical forests," Clark said. "If the patterns we have found prove to be general, it would mean that the rate of global warming will be much greater than what has been expected based on human fossil fuel use alone." She added, "no one knows that the optimum temperature is for photosynthesis of a tropical forest. This is clearly a burning question now" (Choi 2003).

"Enhanced" CO_2 and Crops: More Quantity, Less Quality

Increased levels of carbon dioxide in the atmosphere may accelerate the growth of some crops, but at the same time reduce their nutritional quality, according to a study published in *The New Phytologist* (Jablonski et al. 2002, 9–26). Peter Curtis, a professor of evolution, ecology, and biology at Ohio State University and a co-author of the study, said:

> If you're looking for a positive spin on rising CO_2 levels, it's that agricultural production in some areas is bound to increase. Crops have higher yields when more CO_2 is available, even if growing conditions aren't perfect. But there's a tradeoff between quantity and quality. While crops may be more productive, the resulting produce will be of lower nutritional quality. ("More Carbon" 2002)

Plants produce more seeds at higher carbon-dioxide levels, but the seeds often contain less nitrogen. "The quality of the food produced by the plant decreases, so you've got to eat more of it to get the same benefits," Curtis said. "Nitrogen is a critical component for building protein in animals, and much of the grain grown in the United States is fed to livestock. Under the rising CO_2 scenario, livestock and humans would have to increase their intake of plants to compensate for the loss" ("More Carbon" 2002).

Curtis and colleagues conducted a meta-analysis, in which they combined data from a large number of studies and then summarized for common trends. The studies the researchers reviewed, published between 1983 and 2000, included information on crop and wild plant species' reproductive responses to doubled atmospheric CO_2 levels predicted by the end of the twenty-first century. The researchers' analysis documented eight ways that plants respond to higher carbon-dioxide levels: number of flowers, number of fruits, fruit weight, number of seeds, total seed weight, individual seed weight, the amount of nitrogen contained in seeds, and a plant's "reproductive allocation," a measurement of a plant's capacity to reproduce ("More Carbon" 2002).

Food crops were influenced by varying levels of carbon dioxide to a much higher degree than wild plants, according to the study. "Wild plants are constrained by what they can do with increased CO_2," Curtis explained. "They may use it for survival and defense rather than to boost reproduction. Agricultural crops, on the other hand, are protected from pests and diseases, so they have the luxury of using extra CO_2 to enhance reproduction" ("More Carbon" 2002).

Crops responded differently to increased CO_2 levels. Rice seemed to be the most responsive, as its seed production increased an average of 42 percent as carbon-dioxide levels doubled. Soybeans followed with a 20 percent increase in seed, while wheat increased 15 percent and corn 5 percent. Nitrogen levels decreased by an average of 14 percent across all plants except cultivated legumes, such as peas and soybeans ("More Carbon" 2002).

Enhanced Carbon Dioxide May Favor Faster-Growing Plants

Rising levels of carbon dioxide may favor faster-growing plants over those with slower growth gates. Referring to a set of experiments involving loblolly pines, a team of scientists, writing in *Science*, said:

> We determined the reproductive response of 19-year-old loblolly pine (*Pinus taeda*) to four years of carbon dioxide enrichment (ambient concentration plus 200 microliters per liter) in an intact forest. After three years of CO_2 fumigation, trees were twice as likely to be reproductively mature and produced three times as many cones and seeds as trees at ambient CO_2 concentration. (LaDeau and Clark 2001, 95)

Christian Korner, writing in *Science*, asserted that such "plot studies" may not fully reflect carbon fluxes in real forests because they cannot take into account all the events in the life of a forest. For example, "In a fire, carbon fixed over a period of 50 to 300 years [the lifetimes of most trees], may be emitted within a few hours" (Korner 2003, 1242).

The CO_2 levels LaDeau and Clark used to fumigate their pines will be typical of the ambient atmosphere by the year 2050, if present increases in greenhouse gases continue. The research suggested that faster-growing plants, such as pines, may have an advantage over slower-growing broadleaf trees as carbon-dioxide levels rise. Another report on the same subject, also in *Science*, speculated that "[b]ecause CO_2 is a plant nutrient as well as a greenhouse gas, some researchers argue that faster-growing trees will absorb and sequester increasing amounts of CO_2, making it unnecessary to impose new controls on the gas" (Tangley 2001, 36). Such a point of view has been adopted with enthusiasm by climate skeptics who believe that nature will correct any greenhouse-gas imbalance created by humankinds' combustion of fossil fuels at increasing levels.

Plant Growth and Increasing Rainfall

Ramakrishna Nemani and colleagues at the University of Montana School of Forestry in Missoula investigated warming-related plant growth with a grant from NASA, using climate data from 1950 to 1993. They estimated that increases in rainfall were responsible for about two-thirds of plants' additional growth. Increased rainfall allows more water for growth, while higher humidity allows plants to grow more rapidly by opening wider the pores that allow them to take in carbon dioxide, according to Steven Running of the University of Montana, one of the study's co-authors. The authors of the study found that increases in plant growth correlate with increases in rainfall across the United States (Lovett 2002, 1787). The study suggested that planting new forests to increase the effects of carbon sinks are, by themselves, "overly naïve" (Lovett 2002, 1787).

Nemani and colleagues found that

> [g]lobal changes in climate have eased several critical climatic constraints to plant growth, such that net primary production increased 6 percent (3.4 petagrams of carbon over 18 years) globally. The largest increase was in tropical ecosystems. Amazon rainforests accounted for 42 percent of the global increase in net primary production, owing mainly to decreased cloud cover and the resulting increase in solar radiation. (Nemani et al. 2003, 1560)

Although increases were most marked in tropical ecosystems, "significant growth stimulation [was recorded] in both the tropics and the northern high-latitude ecosystems" (Nemani et al. 2003, 1562).

FURTHER READING

Choi, Charles. "Rainforests Might Speed up Global Warming." United Press International, April 21, 2003. (LEXIS).

Clark, D. A., S. C. Piper, C. D. Keeling, and D. B. Clark. "Tropical Rain Forest Tree Growth and Atmospheric Carbon Dynamics Linked to Inter-annual Temperature Variation during 1984–2000."

Proceedings of the National Academy of Sciences 100, no. 10 (May 13, 2003):5852–5857.
Connor, Steve. "Global Warming May Wipe Out a Fifth of Wild Flower Species, Study Warns." *London Independent*, June 17, 2003b, n.p. (LEXIS).
Davidson, E. A. and A. I. Hirsch. "Carbon Cycle: Fertile Forest Experiments." *Nature* 411 (May 24, 2001):431–433.
Gelbspan, Ross. *The Heat Is On: The High Stakes Battle over Earth's Threatened Climate.* Reading, MA: Addison-Wesley Publishing Co., 1997b.
Halweil, Brian. "The Irony of Climate." *World Watch*, March/April, 2005, 18–23.
Hoffman, Ian. "Rising Temps—Declining Crop Yields." *Inside Bay Area* (California), March 16, 2007.
Hungate, Bruce A., Peter D. Stiling, Paul Dijkstra, Dale W. Johnson, Michael E. Ketterer, Graham J. Hymus, C. Ross Hinkle, and Bert G. Drake. "CO_2 Elicits Long-Term Decline in Nitrogen Fixation." *Science* 304 (May 28, 2004):1291.
Idso, Sherwood B. *Carbon Dioxide: Friend or Foe?* Tempe, AZ: IBR Press, 1982.
Jablonski, L. M., X. Wang, and P. S. Curtis. "Plant Reproduction under Elevated CO_2 Conditions: A Meta-analysis of Reports on 79 Crop and Wild Species. *New Phytologist* 156 (2002):9–26.
Klug, Edward C. "Global Warming: Melting Down the Facts about This Overheated Myth." CFACT Briefing Paper No. 105, November, 1997. http://www.cfact.org/IssueArchive/greenhouse.bp.n97.txt.
Korner, C. "Slow In, Rapid Out—Carbon Flux Studies and Kyoto Targets." *Science* 300 (May 23, 2003):1242–1243.
LaDeau, Shannon L., and James S. Clark. "Rising CO_2 Levels and the Fecundity of Forest Trees." *Science* 292 (April 6, 2001):95–98.
Long, Stephen P., Elizabeth A. Ainsworth, Andrew D. B. Leakey, Josef Nösberger, and Donald R. Ort. "Food for Thought: Lower-Than-Expected Crop Yield Stimulation with Rising CO_2 Concentrations." *Science* 312 (June 30, 2006):1918–1921.
Lovett, Richard A. "Global Warming: Rain Might Be Leading Carbon Sink Factor." *Science* 296 (June 7, 2002):1787.
"More Carbon Dioxide Could Reduce Crop Value." Environment News service, October 3, 2002. http://ens-news.com/ens/oct2002/2002-10-03-09.asp#anchor2.
Nemani, Ramakrishna R., Charles D. Keeling, Hirofumi Hashimoto, William M. Jolly, Stephen C. Piper, Compton J. Tucker, Ranga B. Myneni, and Steven W. Running. "Climate-Driven Increases in Global Terrestrial Net Primary Production from 1982 to 1999." *Science* 300 (June 6, 2003):1560–1563.
Oren, R., D. S. Ellsworth, K. H. Johnsen, N. Phillips, B. E. Ewers, C. Maier, K. V. Schafer, H. McCarthy, G. Hendrey, S. G. McNulty, and G. G. Katul. "Soil Fertility Limits Carbon Sequestration by Forest Ecosystems in a CO2-enriched Atmosphere." *Nature* 411 (May 24, 2001):469–472.
Recar, Paul. "Study: Elements Can Stunt Plant Growth." Associated Press online, December 5, 2002. (LEXIS)
Schimel, David. "Climate Change and Crop Yields: Beyond Cassandra." *Science* 312 (June 30, 2006): 1889–1890.
Shaw, M. Rebecca, Erika S. Zavaleta, Nona R. Chiariello, Elsa E. Cleland, Harold A. Mooney, and Christopher B. Field. "Grassland Responses to Global Environmental Changes Suppressed by Elevated CO_2." *Science* 298 (December 6, 2002):1987–1990.
Tangley, Laura. "Greenhouse Effects: High CO_2 Levels May Give Fast-Growing Trees an Edge." *Science* 292 (April 6, 2001):36–37.
Zavaleta, Erika S., M. Rebecca Shaw, Nona R. Chiariello, Harold A. Mooney, and Christopher B. Field. "Additive Effects of Simulated Climate Changes, Elevated CO_2, and Nitrogen Deposition on Grassland Diversity." *Proceedings of the National Academy of Sciences* 100, no. 13 (June 24, 2003):7650–7654.

Carbon Dioxide: Paleoclimate Levels

Data extracted from Antarctic ice cores provide details on temperature, as well as levels of atmospheric carbon dioxide and methane, during the past 800,000 years. The European Project for Ice Coring in Antarctica (EPICA) has reconstructed a record of Earth's atmospheric carbon dioxide and methane levels for the last 800,000 years with Antarctic ice cores. The upshot is that "today's concentrations of these greenhouse gases have no past analogue" to that date (Brook 2008, 291). Past levels of carbon dioxide and methane are closely tied to temperature levels; the major difference is that in the past both rose and fell slowly, forced in large part by changes in Earth's orbit. Today, the levels have gone much higher, and are rising much more quickly, because of human emissions.

Higher Levels of CO_2 in the Past

Carbon-dioxide levels have been higher in the past, however. An extreme spike in global warming at the Paleocene-Eocene Thermal Maximum about 55 million years ago probably was caused by a combination of methane outbursts from the seafloor, as well as a carbon pulse that occurred at the same time, according to research by Karla Panchuk of Pennsylvania State University and colleagues, writing in *Geology* ("Methane Didn't Act" 2008, 260).

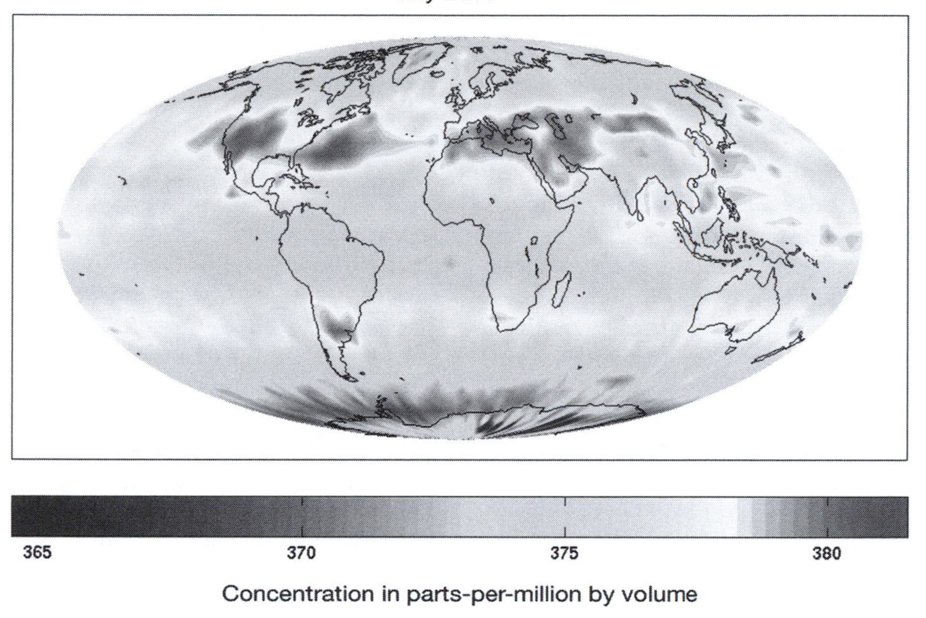

AIRS Global Map of carbon dioxide from space, July 2003. (NASA/JPL)

During much of the Paleozoic era 248 to 543 million years ago, atmospheric carbon-dioxide concentrations have been estimated by proxy methods to have been several times today's levels. These levels declined during the Carboniferous period to levels close to today's levels. Some research based on reconstruction of tropical sea-surface temperatures based on carbonate fossils indicates that the magnitude of temperature variability throughout this period was small. These results suggest that global climate sometimes may be independent of variations in atmospheric carbon-dioxide concentration (Came et al. 2007, 198). That, or the proxies are flawed.

According to research by Rosemarie Came and colleagues based on a different method, estimating sea-surface temperatures obtained from fossil brachiopod and mollusk shells using the "carbonate clumped isotope" method,

> Tropical sea surface temperatures were significantly higher than today during the Early Silurian period (443–423 million years [myr] ago), when carbon dioxide concentrations are thought to have been relatively high, and were broadly similar to today during the Late Carboniferous period (314–300 myr ago), when carbon dioxide concentrations are thought to have been similar to the present-day value. Our results are consistent with the proposal that increased atmospheric carbon dioxide concentrations drive or amplify increased global temperatures. (Came et al. 2007, 198)

FURTHER READING

Brook, Ed. "Paleoclimate: Windows on the Greenhouse." *Nature* 453 (May 15, 2008), 291–292.

Came, Rosemarie E., John M. Eiler, Ján Veizer, Karem Azmy, Uwe Brand, and Christopher R. Weidman. "Coupling of Surface Temperatures and Atmospheric CO_2 Concentrations during the Palaeozoic Era." *Nature* 449 (September 13, 2007):198–201.

"Methane Didn't Act Alone." *Nature* 453 (May 15, 2008):260.

Carbon Dioxide: Worldwide Levels

A study in the *Proceedings of the National Academy of Sciences* (PNAS) found that carbon-dioxide levels are increasing at a faster rate than at any time since they have been measured precisely starting in 1959. The National Oceanic and Atmospheric Administration (NOAA) reported at about the same time that the carbon-dioxide level in Earth's atmosphere rose 2.4 parts per million (ppm) in 2007, to an average of 385 ppm. Annual increases of about 2 ppm became common after 2000, following average rises of

1.5 ppm in the 1990s and 1 ppm in the 1980s. Growing industrialism in Asia and wetland emissions (e.g., melting permafrost) were the main sources of the acceleration. Methane also rose in 2007 by 27 million tons worldwide after having recorded flat or slightly declining readings for the previous decade.

According to United Nations figures released November 17, 2008, greenhouse-gas emissions of 40 industrial countries rose by 2.3 percent between 2000 and 2006, although they remain about 5 percent below the 1990 level. For the smaller group of industrial countries that ratified the 1997 Kyoto Protocol setting reduction targets, emissions in 2006 were about 17 percent below the Protocol's 1990 baseline, but they still grew after 2000. The pre-2000 decrease stemmed from the economic decline of transition countries in Eastern and Central Europe in the 1990s.

Human activity that produces carbon dioxide also has been increasing at an accelerating rate, as the amount of carbon required to produce a given amount of production also increases. In addition, the ability of natural "sinks" to absorb CO_2 on land and sea has been declining. Less carbon dioxide is being absorbed by our surroundings. "All of these changes characterize a carbon cycle that is generating stronger-than-expected, and sooner-than-expected climate forcing," the authors of the PNAS study wrote (Canadell et al. 2007).

A monitoring station on the summit of Hawaii's Mauna Loa has been tracking increases in the level of atmospheric carbon dioxide during the past 50 years. In 2008, the carbon-dioxide level reached 385 ppm, having risen more than 35 percent in the roughly 150 years since the advent of the industrial revolution. Measurements are taken at a National Oceanic and Atmospheric Administration (NOAA) laboratory at the Mauna Loa Observatory 11,141 feet above sea level on the island of Hawaii. The level, which before the industrial age peaked at about 280 ppm during interglacial periods, was about 315 ppm when it was first measured in the late 1950s.

Paul Pearson of the University of Bristol and Martin Palmer of Imperial College, London, reported in *Nature* that they had developed proxy methods for measuring the atmosphere's carbon-dioxide level to about 60 million years before the present. The upshot of Pearson and Palmer's studies is that carbon-dioxide levels in the year 2000 were as high as they have been in at least the last 20 million years. According to their records, however, carbon-dioxide levels reached the vicinity of 2,000 ppm during "the late Palaeocene and earliest Eocene periods (from about 60 to 52 million years ago)" (Pearson and Palmer 2000, 695).

Worldwide Carbon-Dioxide Levels Spike

Carbon-dioxide readings have continued a sharply rising trend begun in 2000, according to calculations published in the *Proceedings of the National Academy of Sciences* during 2007. The rate of increase nearly tripled over the average rate in the 1990s. Instead of rising by 1.1 percent a year, as in the previous decade, emissions grew by an average of 3.1 percent a year from 2000 to 2004. These readings indicate sharp increases in the rate at which the greenhouse gas has been accumulating in the atmosphere. The recent increases—2.08 ppm from 2001 to 2002 and 2.54 ppm from 2002 to 2003—have drawn attention of climate scientists because they deviate from historic average annual increases of around 1.5 ppm.

Carbon-dioxide emissions were 35 percent higher in 2006 than in 1990, a much faster growth rate than anticipated, researchers led by Josep G. Canadell, of Australia's Commonwealth Scientific and Industrial Research Organization, reported in the October 23, 2007, edition of the *Proceedings of the National Academy of Sciences.* Much of the increase is being traced to the reduction of the oceans' ability to remove additional carbon dioxide from the air as water temperatures increase. According to the new study, carbon released from burning fossil fuel and making cement rose from 7.0 billion metric tons per year in 2000 to 8.4 billion metric tons in 2006. A metric ton is 2,205 pounds. Methane emissions have declined, however, so greenhouse gases as a whole are not increasing as much as carbon dioxide alone.

A debate has arisen: Are these increases an aberration, or evidence of an accelerating rate of carbon-dioxide buildup? Is this accelerating rate of increase the first evidence of a "runaway greenhouse effect" stoked by a series of feedback mechanisms that will cause worldwide temperatures to rise at a much more rapid rate, along with accelerating changes of climate, melting ice caps, and quickly rising sea levels?

"Despite the scientific consensus that carbon emissions are affecting the world's climate, we are not seeing evidence of progress in managing those emissions," said Chris Field, director of the Carnegie Institution's Department of Global Ecology in Stanford, California, a co-author of the report (Raupach 2007, 10,288; Boyd 2007, A-8). The jump in the emission rate is alarming because it indicates a reversal of a long-term trend toward greater energy efficiency and away from carbon-based fuels, the report's authors said.

The CO_2 acceleration is occurring most rapidly in China and other developing areas, as populations and per capita gross domestic product rise. It is increasing more slowly in the advanced economies of the United States, Europe, and Japan, the report said. "No region is decarbonizing its energy supply," the report said (Raupach 2007, 10, 288; Boyd 2007, A-8).

NASA's Goddard Institute for Space Studies estimates that a carbon-dioxide level at about 450 ppm will cross a "tipping point" and lead to "potentially dangerous consequences for the planet" ("Research Finds" 2007). With carbon-dioxide levels at 385 ppm as of 2008, estimates vary as to when this threshold will be reached. The NASA study says that 1°C (1.8°F) additional temperature rise will place the atmosphere at peril of crossing this line. According to James Hansen, director of GISS,

> If global emissions of carbon dioxide continue to rise at the rate of the past decade, this research shows that there will be disastrous effects, including increasingly rapid sea level rise, increased frequency of droughts and floods, and increased stress on wildlife and plants due to rapidly shifting climate zones. ("Research Finds" 2007)

According to Makiko Sato of Columbia University's Earth Institute, "the temperature limit implies that CO_2 exceeding 450 ppm is almost surely dangerous, and the ceiling may be even lower" ("Research Finds" 2007).

China Number One in CO_2 Emissions

Amidst speculation that China would pass the United States as the world's leading source of carbon dioxide in 2008 or 2009, a report arrived from the Netherlands Environmental Assessment Agency in mid-June 2007 indicating that China already had taken the lead—in 2006. China, riding an economic boom fueled by low-sulfur coal and rapidly rising cement manufacturing, witnessed a rise in CO_2 emissions of 9 percent during 2006 ("China" 2007).

"There will still be some uncertainty about the exact numbers, but this is the best and most up to date estimate available," said Jos Olivier, a scientist who crunched the numbers at the Netherlands Environmental Assessment Agency. By 2008, realization that China had passed the United States in 2006 became widespread, with reports coming to this conclusion in several journals, among them, *Geophysical Research Letters* (Marland et al. 2008), *Journal of Environmental Economics and Management* (Carson et al. 2008), and *Annual Review of Environment and Resources* (Levine et al. 2008) (cited in Watson 2008, 5-A).

FURTHER READING

Boyd, Robert S. "Carbon Dioxide Levels Surge: Global Emissions Rate Is Triple That of a Decade Ago." McClatchy Newspapers in *Calgary Herald*, May 22, 2007, A-8.

Canadell, Josep G., Corinne Le Quérćc, Michael R. Raupacha, Christopher B. Field, Erik T. Buitenhuisc, Philippe Ciaisf, Thomas J. Conwayg, Nathan P. Gillettc, R. A. Houghtonh, and Gregg Marland. "Contributions to Accelerating Atmospheric CO_2 Growth from Economic Activity, Carbon Intensity, and Efficiency of Natural Sinks." *Proceedings of the National Academy of Sciences.* Published online before print October 25, 2007, doi: 10.1073/pnas.0702737104.

"Carbon Dioxide in Atmosphere Increasing." Associated Press in *New York Times*, October 22, 2007. http://www.nytimes.com/aponline/us/AP-Carbon-Increase.html.

"China Now Number One in Carbon Emissions: USA Number Two." Environment News Service, June 19, 2007. http://www.ens-newswire.com/ens/jun2007/2007-06-19-04.asp.

Pearson, Paul N., and Martin R. Palmer. "Atmospheric Carbon Dioxide Concentrations over the Past 60 Million Years." *Nature* 406 (August 17, 2000):695–699.

Raupach, Michael R., Gregg Marland, Philippe Ciais, Corinne Le Quere, Josep G. Canadell, Gernot Klepper, and Christopher B. Field. "Global and Regional Drivers of Accelerating CO_2 Emissions." *Proceedings of the National Academy of Sciences* 104, no. 24 (June 12, 2007):10,288–10,293.

"Research Finds That Earth's Climate Is Approaching 'Dangerous' Point." Press Release, NASA Goddard Institute for Space Studies, New York City, May 29, 2007.

"UN: Industrial Countries' Greenhouse Gases Rose 2000–2006." Environment News Service, November 18. 2008. http://www.ens-newswire.com/ens/nov2008/2008-11-18-01.asp.

Watson, Traci. "China Leaves U.S. in Dust as the No. 1 CO_2 Offender." *USA Today*, May 1, 2008, A-5.

Carbon-Dioxide Controls

Scientists' evaluations of global warming's impending impact are leaving less wiggle-room for future control than ever before. By 2008, a scientific article that James E. Hansen, director, NASA's Goddard Institute for Space Studies, and several co-authors prepared said that the atmosphere's carbon-dioxide level not only must stop rising from its present 385 parts per million (ppm), but also must be cut at least 10 percent (to below 350 ppm) to avoid enduring damage to the Earth's ecosystem.

Failing such a reduction, we have placed enough warming "in the pipeline" to provoke not only uncomfortably warmer temperatures, but also significantly eroding Arctic and Antarctic ice, major sea-level rises, intensifying droughts and deluges, and animal extinctions on the land and in the oceans (including much of the world's coral reefs). The oceans may acidify to the point that anything in a shell will be in peril. Of the 2°C required to provoke unstoppable feedbacks, about half is now "in the pipeline," according to Hansen's calculations. "Paleoclimate evidence and ongoing global changes imply that today's CO_2, about 385 ppm, is already too high to maintain the climate to which humanity, wildlife, and the rest of the biosphere are adapted," the authors wrote (Johansen 2008a, 5).

In a world in which climate diplomacy dawdles and China adds a new coal-fired power plant on an average of every two weeks, carbon-dioxide levels have been rising at a faster rate than at any time since detailed records have been kept. As of late 2008, reports surfaced that carbon-dioxide emissions from U.S. power plants rose almost 3 percent in 2007 over 2006, the largest increase since records have been kept.

Accepting the 2007 Novel Peace Prize in Oslo on December 12, 2007, Al Gore said, "So today we dumped another 70 million tons of global warming pollution into the thin shell of atmosphere surrounding our planet, as if it were an open sewer. And tomorrow we will dump a slightly larger amount" (Kolbert 2007d, 43).

Full-Scale Mobilization

Hansen points to World War II as an example of the full-scale mobilization required to avoid a climatic crackup, if, according to Hansen and colleagues, "humanity wishes to preserve a planet similar to that on which civilization developed" (Johansen 2008a, 5).

According to this analysis, construction of coal-fired power plants must be stopped until carbon-capture technology is available, and everyone must scour their lives for ways to conserve electricity. Efficiency (mileage) standards for vehicles must be raised dramatically, and the very structure of our cities must be changed fundamentally to reduce the need to travel long distances on a daily basis. Air travel, which has become a major source of greenhouse-gas emissions, will need to be curtailed. Nonfossil fuels (wind, solar, and others) will require development on a crisis basis. The peoples of the world will have to realize that modern machine warfare, with its huge carbon footprint, is a threat to survival of the environment. Agriculture and forestry practices will have to change to reduce production of carbon dioxide. All of these actions, and more, will require worldwide scope, and quickly.

Ice Melt to Come at the Poles

"Present-day observations of Greenland and Antarctica show increasing surface melt, loss of buttressing ice shelves, accelerating ice streams, and increasing overall mass loss," Hansen said (Johansen 2008a, 5). Hansen added that existing models of ice-sheet disintegration lack complete analysis of the physics of melting ice, including, in some cases, the fact that sea-level changes of several meters per century occur in the paleoclimate record, in response to influences that are slower and weaker than the present-day human-made forcings. "It seems likely," he wrote, "that large ice sheet response will occur within centuries, if human-made forcings continue to increase. Once ice sheet disintegration is underway, decadal changes [sea-level change within a period of 10 years] may be substantial" (Johansen 2008a, 5).

The amount of warming that we feel now is being restrained by the enormous thermal inertia of the oceans that cover two-thirds of Earth's surface. Once that warming is realized, however, it will endure for centuries, even if human consumption of fossil fuels stops completely. No additional forcing is required, according to Hansen, to raise global temperature to that of the Pliocene, two to three million years ago, "a degree of warming that would surely yield dangerous climate impacts." Equilibrium sea level rise for today's 385 ppm carbon dioxide is "at least" several meters, judging from paleoclimate history, Hansen and colleagues stated (Johansen 2008a, 5).

"Alpine glaciers are in near-global retreat," Hansen and colleagues wrote. They added:

> After a flush of fresh water, glacier loss foretells long summers of frequently dry rivers, including rivers originating in the Himalayas, Andes and Rocky Mountains that now supply water to hundreds of millions of people. Present glacier retreat, and warming in the pipeline, indicate that 385 ppm CO_2 is already a threat. (Johansen 2008a, 5)

To relieve these and other stresses, the carbon-dioxide level must be brought down to between 300 and 350 ppm and then stabilized.

Zero Carbon-Dioxide Emissions by 2050?

A modeling study published in March 2008 stiffened the prospective task of reducing carbon-dioxide levels in the atmosphere before dire consequences set in (including problems with droughts and deluges)—near-zero emissions by 2050. The studies use computer models that include deep-sea warming and other aspects of the carbon cycle. "The question is, what if we don't want the Earth to warm anymore?" asked Carnegie Institution senior scientist Ken Caldeira, co-author with H. Damon Matthews of a paper published February 27, 2008, in *Geophysical Research Letters.* "The answer implies a much more radical change to our energy system than people are thinking about" (Eilperin 2008b).

Caldeira and Matthews of the Department of Geography, Planning and Environment, Concordia University, Montreal, Quebec, Canada, wrote:

> To hold climate constant at a given global temperature requires near-zero future carbon emissions. Our results suggest that future anthropogenic emissions would need to be eliminated in order to stabilize global-mean temperatures. As a consequence, any future anthropogenic emissions will commit the climate system to warming that is essentially irreversible on centennial timescale. (Matthews and Caldeira 2008)

A goal of zero emissions was not in play even under the most optimistic political and diplomatic proposals in the United States as this study was published.

The zero-emissions by 2050 target is meant to prevent a temperature rise of more than 2°C (3.6°F), at which time many scientists agree that feedbacks will cause warming to accelerate on its own, causing widespread distortions in the hydrological cycle, eventual sea-level rises that will destroy coastal cities, as well as extinctions of flora and fauna. Oregon State University professor Andreas Schmittner said that "[t]he warming continues much longer even after emissions have declined.... Our actions right now will have consequences for many, many generations. Not just for a hundred years, but thousands of years" (Eilperin 2008b).

As scientists called for zero carbon emissions by 2050, on July 8, 2008, the United States joined other major industrial countries for the first time in a commitment to cut greenhouse gases in half by 2050. George W. Bush, who had long balked at any kind of commitment, began promoting it reluctantly during the last year of his second term. Acting as if he had made great progress, Bush was half-way to a meaningful target. "I would characterize this outcome as 'talking the talk' rather then 'walking the walk' on climate change policy," said Michael E. Schlesinger, a climatologist at the University of Illinois at Urbana-Champaign, who has cowritten many papers on climate policy (Revkin 2008g). Schlesinger was among many scientists who said that the G-8 statement nor previous climate treaties would significantly slow the rise of heat-trapping gases.

By 2007, the world debate over greenhouse-gas controls came increasingly into focus on the United States and China, with the latter poised to pass the former in total emissions. Together, by 2007, China and the United States were responsible for more than 40 percent of the world's greenhouse-gas emissions. Both countries' governments were giving little more than lip service to emission control. China's government argued that the average person in China was emitting only one-fifth of the greenhouse gases and consuming but a seventh of the energy of the average American, so limits are unfair. As of 2007, China's national plans supported ambitious theoretical targets for fuel efficiency and use of renewable energy, but without national greenhouse-gas targets.

"China's rapidly growing emissions are a serious issue," said Phil Clapp, president of the National Environmental Trust.

> But many diehard opponents of enforceable limits on global warming pollution who now can't hide

> behind the science are trying to hide behind China. Note who keeps raising the China issue: the coal industry, the oil industry, members of Congress from coal states and the auto industry. They raise threats to American competitiveness that are bogus, on the whole. (Mufson 2007b, D-1)

U.S. Supreme Court Rules that CO_2 Is a Pollutant

Anything in excess is a problem. Too much chocolate is toxic. Too many potatoes, eaten at once, can be poisonous. Too much sunshine can provoke skin cancer, although a certain amount is necessary for life as we know it. The same holds true for carbon dioxide. We need some of it to prevent the Earth from freezing, but too much, and we will fry. Witness Venus, with an atmosphere that is 95 percent CO_2, a victim of a runaway greenhouse effect. While a degree or two of additional warmth in the middle of winter may be a balm, 850°F, the surface temperature of Venus, is no one's friend.

The U.S. Supreme Court recognized this logic early in 2007 when it held, 5 to 4, in *Massachusetts v. EPA*, against the express wishes of the George W. Bush administration, that the federal government has the power to regulate carbon dioxide and other greenhouse gases from cars. Greenhouse gases are air pollutants under the landmark environmental law, Justice John Paul Stevens said in his majority opinion. The court's four conservative justices—Chief Justice John Roberts and Justices Samuel Alito, Antonin Scalia, and Clarence Thomas—dissented ("Court Rebukes" 2007).

Late in November 2006, the U.S. Supreme Court heard oral arguments regarding the authority of the U.S. Environmental Protection Agency (EPA) to regulate greenhouse gases from motor vehicles. *Massachusetts v. EPA* was brought against the EPA by 12 states, three cities, and 13 environmental groups, which argued that the EPA should regulate carbon dioxide as a pollutant. The case was filed after the Bush-led EPA during 2003 decided that it had no such legal authority under the federal Clean Air Act. Before the Court, the Bush administration argued that carbon dioxide was not a pollutant and that even if it were, the EPA has discretion over whether to regulate it ("Supreme Court" 2006).

The Clean Air Act states that the EPA shall set standards for "any air pollutant" that in its judgment causes or contributes to air pollution that "may reasonably be anticipated to endanger public health or welfare." The word "welfare," the law says, includes "climate" and "weather." The EPA made an array of specious arguments about why the act does not mean what it expressly says. But it has no right to refuse to do what Congress said it "shall" do ("Global Warming Goes to Court" 2006)

The Supreme Court, as it often does, seemed to be taking stock of a changing political consensus in the country at a time when the U.S. Congress and even some oil companies (and later on that year, even Bush himself) were paying lip service to the dangers of global warming. "In many ways, the debate has moved beyond this," said Chris Miller, director of the global warming campaign for Greenpeace, one of the environmental groups that sued the EPA. "All the front-runners in the 2008 presidential campaign, both Democrats and Republicans, even the business community, are much further along on this than the Bush administration is" ("Court Rebukes" 2007).

The court addressed three issues: First, do states have the right to sue the EPA to challenge its decision? Second, does the Clean Air Act give the EPA the authority to regulate tailpipe emissions of greenhouse gases? Third, does the EPA have the discretion not to regulate those emissions? The majority affirmed the first two questions. On the third, it ordered EPA to reevaluate its assertion that it has the discretion *not* to regulate tailpipe emissions, which it found to be out of alignment with the Clean Air Act.

Corporations Support Emissions Caps

The Democratic takeover of Congress during the 2006 elections made federal regulation of emissions more likely. Senator Barbara Boxer, a Democrat from California, who called global warming "the greatest challenge of our generation," took the place of Senator James M. Inhofe, a Republican from Oklahoma, as chairman of the Senate Environment and Public Works Committee. Inhofe called global warming a "hoax." Senator Jeff Bingaman, a Democrat from New Mexico and the incoming Energy and Natural Resources Committee chairman, said he hopes to "do something on global warming" (Mufson and Eilperin 2006, A-1).

By 2007, executives representing some of the country's largest carbon-dioxide sources came out in favor of measures to curtail carbon emissions, such as taxes and cap-and-trade systems. Paul M. Anderson, chairman of Duke Energy (the third-largest corporate coal burner in the United States), announced support of a tax on CO_2 emissions. "If we had our druthers, we'd already have carbon legislation passed," said John L. Stowell, Duke Energy's vice president for environmental policy. "Our viewpoint is that it's going to happen. There's scientific evidence of climate change. We'd like to know what legislation will be put together so that, when we figure out how to increase our load, we know exactly what to expect" (Mufson and Eilperin 2006, A-1).

During March 2007, the chief executives of America's largest automobile companies—General Motors, Ford, Chrysler, and Toyota North America—pledged to support mandatory caps on carbon emissions, as long as the caps covered all sectors of the economy. They delivered their promise to a House committee run by John Dingell—the crusty Michigan Democrat who is another convert to the cause. A group of institutional investors, including Merrill Lynch, the largest, called for mandatory emissions curbs of 60 to 90 percent by mid-century. This group called for companies to disclose investor risks associated with climate change to the Securities and Exchange Commission.

In an attempt to influence the character of these controls, many companies hired lobbyists who fashioned national legislation they could live with, thereby averting several different state plans. "We have to deal with greenhouse gases," John Hofmeister, president of Shell Oil Co., told the National Press Club late in 2006. "From Shell's point of view, the debate is over. When 98 percent of scientists agree, who is Shell to say, 'Let's debate the science?' We cannot deal with 50 different policies," said Shell's Hofmeister. "We need a national approach to greenhouse gases" (Mufson and Eilperin 2006, A-1).

States and Cities Establish their Own Emissions Limits

As the George W. Bush administration stalled on U.S. emissions limits, The Regional Greenhouse Gas Initiative, a group of seven northeastern states, established a system for a cap-and-trade regulation system. U.S. cities' collective efforts began in 2005, when Seattle Mayor Greg Nickels said that his city would adopt a policy aimed at meeting targets of the Kyoto Protocol. With the U.S. Mayors Climate Protection Agreement, Nickels challenged his fellow mayors to do the same. By 2007, 522 cities with a population of 65 million, from Boston, Massachusetts, to Portland, Oregon, had accepted his challenge. State and local efforts had an impact: U.S. fossil-fuel-related emissions fell 1.3 percent to 5.88 billion metric tons between 2005 and 2006.

Portland, Oregon, made reduction of greenhouse-gas emissions a municipal priority; by 2005, emissions there had been reduced to roughly 1990 levels as the city's economy boomed. Portland has encouraged walking and bicycle commuting, and told local companies that if they gave employees free parking they should also subsidize bus passes. One step (replacing bulbs in traffic lights with light-emitting diodes) cut the city government's electrical use by 80 percent and saved the city almost $500,000 a year in electrical costs (Kristof 2005). Water flowing through Portland's drinking system also generates hydroelectricity (Faiola and Shulman 2007, A-1). Officials in Portland said that they have been able to cut emissions in accordance with Kyoto while enjoying a healthy economy, with "less money spent on energy, more convenient transportation, a greener city, and expertise in energy efficiency that is helping local business win contracts worldwide" (Daynes and Sussman 2005, 442).

In Austin, Texas, a new policy tightened energy-efficiency standards in homes, which will be required to use 60 percent less energy by 2015. Chicago installed waterless urinals and planted several thousand trees to cool down "heat islands." Chicago also encourages the planting of lush rooftop gardens, which help to cool buildings, reducing the need for air conditioning. Keene, New Hampshire, requires parents waiting for children at schools to turn off car engines.

Seattle imposed a new parking tax, as its mayor proposed tolls on major roads to discourage driving. In Fargo, North Dakota, Mayor Dennis Walaker swapped out every traffic light bulb for a light-emitting diode (LED), which uses 80 percent less energy. Fargo also traps methane from its landfill and sells it to a soybean-processing plant as biomass fuel for its boilers (Simon 2006, A-1). In Carmel, Indiana, a suburb of Indianapolis, Mayor James Brainard is switching the entire city fleet to hybrids and vehicles that run on biofuels (Simon 2006, A-1).

By late 2006, 22 U.S. states and the District of Columbia had set standards demanding that utilities generate a specific amount of energy—in some cases, as high as 33 percent—from renewable sources by 2020. And 11 states have set goals to reduce greenhouse-gas emissions by as much as 80 percent below 1990 levels by 2050. Ten states in the U.S. Northeast established their own limits on carbon dioxide and other greenhouse gases, and set up cap-and-trade markets in which industries will trade pollution credits for carbon emissions. The markets began operating in September 2008. Typical policies cut greenhouse-gas emissions by 10 percent by 2019. California, Oregon, and Washington were designing similar limits (Eilperin 2006e).

Political leaders in California reached an agreement late in August 2006 to impose the United States' most restrictive controls on carbon dioxide emissions. Republican Governor Arnold Schwarzenegger brokered the arrangement with a legislature controlled by Democrats. The Global Warming Solutions Act requires a 25 percent reduction in carbon-dioxide emissions by 2020, a requirement that will place controls on the largest industries, including utilities, oil refineries, and cement plants, in addition to the state's existing limits on automobile emissions.

FURTHER READING

"Court Rebukes Administration in Global Warming Case." Associated Press in *New York Times*, April 2, 2007. http://www.nytimes.com/aponline/business/AP-Scotus-Greenhouse-Gase.html.

Daynes, Byron W., and Glen Sussman. "The 'Greenless' Response to Global Warming." *Current History*, December 2005, 438–443.

Eilperin, Juliet. "Cities, States Aren't Waiting for U.S. Action on Climate." *Washington Post*, August 11, 2006e, A-1. http://www.washingtonpost.com/wp-dyn/content/article/2006/08/10/AR2006081001492_pf.html.

Eilperin, Juliet. "Carbon Output Must Near Zero to Avert Danger, New Studies Say." *Washington Post*, March 10, 2008b, A-1. http://www.washingtonpost.com/wp-dyn/content/article/2008/03/09/AR2008030901867_pf.html.

Faiola, Anthony, and Robin Shulman. "Cities Take Lead on Environment as Debate Drags at Federal Level." *Washington Post*, June 9, 2007, A-1. http://www.washingtonpost.com/wp-dyn/content/article/2007/06/08/AR2007060802779_pf.html.

"Global Warming Goes to Court." Editorial, *New York Times*, November 28, 2006.

Hansen, James, Makiko Sato, Pushker Kharecha, David Beerling, Valerie Masson-Delmotte, Mark Pagani, Maureen Raymo, Dana Royer, and James C. Zachos. "Target Atmospheric CO2: Where Should Humanity Aim?" *The Open Atmospheric Science Journal* 2, no. 15 (2008):217–231. http://dx.doi.org/10.2174/1874282300802010217.

Johansen, Bruce E. "Hansen: Cut CO_2 10 Per Cent." *Nebraska Report*, May/June 2008a, 5.

Kolbert, Elizabeth. "Testing the Climate." (Talk of the Town) *The New Yorker*, December 24 and 31, 2007d, 43–44.

Kristof, Nicholas. "The Storm Next Time." *New York Times*, September 1, 2005. http://www/nytimes.com.

Matthews, H. Damon., and Ken Caldeira. "Stabilizing Climate Requires Near-zero Emissions." *Geophysical Research Letters* 35 (February 27, 2008), L04705, doi: 10.1029/2007GL032388.

Mufson, Steven. "In Battle for U.S. Carbon Caps, Eyes and Efforts Focus on China." *Washington Post*, June 6, 2007b, D-1. http://www.washingtonpost.com/wp-dyn/content/article/2007/06/05/AR2007060502546_pf.html.

Mufson, Steven, and Juliet Eilperin. "Energy Firms Come to Terms With Climate Change." *Washington Post*, November 25, 2006, A-1. http://www.washingtonpost.com/wpdyn/content/article/2006/11/24/AR2006112401361_pf.html.

Revkin, Andrew C. "After Applause Dies Down, Global Warming Talks Leave Few Concrete Goals." *New York Times*, July 10, 2008g. http://www.nytimes.com/2008/07/10/science/earth/10assess.html.

Simon, Stephanie. "Global Warming, Local Initiatives: Unhappy with Federal Resistance to World Standards, Communities Are Curbing Their Energy Use and Emissions." *Las Angeles Times*, December 10, 2005, A-1.

"Supreme Court Will Decide EPA's Authority Over Climate Gases." Environment News Service, November 27, 2006. http://www.ens-newswire.com/ens/nov2006/2006-11-27-02.asp.

Carbon Footprint

Carbon footprints—how much carbon dioxide is created by a specific activity or manufactured product—are calculated to determine energy use in order to reduce it and thus reduce fossil-fuel consumption. Many scientists assert that the best way to reduce fossil-fuel emissions is to cut unnecessary energy use. The 2008 Democratic National Convention, for example, hired environmental activist Andrea Robinson, a director of greening, as well as an official carbon adviser, to calculate the carbon footprint of every placard manufactured, every plane trip taken, every appetizer prepared, and every coffee cup thrown away, and much more.

Greenhouse-gas production of individuals in the United States varies widely, according to rankings of 100 cities released during 2008.

People living in cities tend to produce less carbon dioxide than those in rural areas (distances traveled in cars are a big factor), and west-coast cities' residents have smaller carbon footprints than those in the East and Midwest, reflecting lower heating and cooling costs on the coast. A large influence is mild climate (in California) as well as use of hydropower in lieu of coal to generate power in such states as Oregon and Washington.

The smallest carbon footprints belonged to people in the Honolulu area, however, which was ranked number one in the Brookings Institution study. The next lowest were Los Angeles and Orange Counties in California, with mild climate and progressive electricity rate-setting regulations cited as major factors. The Portland-Vancouver area gained a low rating from progressive rate-setting policies and hydropower. The New York City metropolitan area also was relatively low, a function of high-rise housing and widespread use of public transit. The worst ratings came from Midwestern locales such as Toledo and Cincinnati, Ohio; Indianapolis, Indiana; and Lexington, Kentucky, which ranked last, largely because of coal-fired power and severe climates requiring a great deal of winter heat and summer air conditioning (Barringer 2008d).

"The Washington, D.C., metro area's residential electricity footprint was 10 times larger than Seattle's footprint in 2005," the report stated. "The mix of fuels used to generate electricity in Washington includes high-carbon sources like coal while Seattle draws its energy primarily from essentially carbon-free hydropower" (Barringer 2008d).

FURTHER READING

Barringer, Felicity. "Urban Areas on West Coast Produce Least Emissions Per Capita, Researchers Find." *New York Times*, May 29, 2008d. http://www.nytimes.com/2008/05/29/us/29pollute.html.

Carbon Sequestration

Legislation was introduced (but not passed) in March 2008 in the U.S. House of Representatives that would have required new coal-fired electric generating plants to use state-of-the-art carbon capture and sequestration (CCS) technology. Representative Henry Waxman, Democrat of California, chair of the House Government Oversight Committee, and Representative Edward Markey, Democrat of Massachusetts, chair of the House Select Committee on Energy Independence and Global Warming, introduced the bill (the Moratorium on Uncontrolled Power Plants Act of 2008). The bill proposed a moratorium on permits for new coal-fired power plants without CCS from the U.S. Environmental Protection Agency (EPA) or state agencies ("National Ban" 2008).

At about the same time, several groups rallied about 100 people at the Kansas State Capitol on March 11, 2008, in support of Governor Kathleen Sebelius, who was expected to veto an energy bill that would have permitted two new 700-megawatt coal-fired power plants without CCS in the western part of that state, near Holcomb ("Kansans Rallied" 2008). The bill had narrowly passed the state House and Senate after Rod Bremby, secretary of the Kansas Department of Health and Environment, denied a permit during 2007 ("Kansans Rallied" 2008). The bill also sought to eliminate the department's role in issuing permits for power plants.

An EPA appeals board in mid-November 2008 turned aside a permit for Deseret Power Electric Cooperative's proposed coal-fired plant near Vernal, Utah, because the utility had not addressed controls for carbon-dioxide emissions. This decision could play a role in permits pending for about 100 other proposed coal-fired plants in the United States. The case was returned to an EPA regional office "to determine whether CO_2 constraints are required" (Davidson 2008, 2-B). This ruling occurs in the context of the U.S. Supreme Court's year-old ruling that carbon dioxide is a pollutant and EPA may set limits on it. The EPA did not take this action of its own accord, but only after a legal challenge by the Sierra Club, which was looking ahead to a more activist EPA once President Barack Obama took office.

The Ocean as Carbon Dump?

As some scientists warned that rising carbon-dioxide levels could raise acidity of the oceans to levels that may be lethal for some aquatic life, others advised capture of carbon dioxide emitted by coal-fired power plants—tens of billions of tons of the gas—and pumping it into the Earth or deep into the sea.

In 2006, a team of researchers proposed what they called a limitless, low-risk repository for carbon dioxide: seafloor sediments at depths and temperatures that would guarantee it would stay denser than the water above, and thus be permanently locked away (Revkin 2006f). Writing in the August 7, 2006, edition of the *Proceedings of the National Academy of Sciences*, this group identified what it believed to be the ideal conditions for carbon sequestration in the deep ocean.

The ideal location, according to this team, would be several hundred yards beneath the seabed, utilizing porous sediment in waters about 10,000 feet deep, where the temperature is usually about 35°F. Under these conditions, water pressure would cause liquid carbon dioxide to remain denser than water, the researchers said (Revkin 2006f). The lead author of the study, Kurt Zenz House, a graduate student in earth sciences at Harvard University, said that the study demonstrates "an inherently stable and permanent storage option that could bite off a huge chunk of the CO_2" (Revkin 2006f).

Thus far, this is an untested theory, partly funded by the U.S. Department of Energy. Many questions remain. For example, how much time, money, and energy would be required to design and build plants to assemble and compress the gas? How would the waste gas be conveyed so deep into the oceans, especially from inland location? Would it eventually leak back into the oceans and, from them, into the atmosphere?

FURTHER READING

Davidson, Paul. "Coal Power Plants May Have to Clean Up Their Act." *USA Today*, November 14, 2008, B-2.

"Kansans Rallied to Resist Coal-Burning Power Plants." Environment News Service, March 12, 2008. http://www.ens-newswire.com/ens/mar2008/2008-03-12-092.asp.

"National Ban on New Power Plants without CO_2 Controls Proposed." Environment News Service, March 12, 2008. http://www.ens-newswire.com/ens/mar2008/2008-03-12-091.asp.

Revkin, Andrew C. "Team Looks at Seafloor as Gas Trap." *New York Times*, August 8, 2006f. http://www.nytimes.com/2006/08/08/science/earth/08carbon.html.

Carbon Tax

Several nations have enacted taxes that discourage the production of carbon dioxide and other greenhouse gases. The U.S. Congress discussed but did not adopt such proposals during 2007 and 2008. A carbon tax reduces emissions by raising the price of fossil-fuel energy in relation to other sources. Subsidies also may be used to accelerate development of carbon-free sources of energy, such as wind, geothermal, and solar power, which enjoy several subsidies in the United States and other countries. In this and other ways, taxation is being used to channel energy development and use.

Consumer advocate and erstwhile presidential candidate Ralph Nader has proposed a global-scale carbon tax to end "trade anarchy" and a patchwork quilt of cap-and-trade schemes that have evolved around the world. China would be required to take part because, according to Fatih Birol, chief economist at the International Energy Agency, its carbon-dioxide emissions, left to grow at recent rates, within 25 years would double those of the United States, Japan, and the European Union. Nader suggested collecting a tax of at least $50 per ton of greenhouse gases or carbon-dioxide equivalent at trunk pipelines for gas, refineries for oil, railroad heads for coal, liquid natural gas (LNG) terminals, cement, steel, aluminum, and greenhouse-gas intensive chemical plants (Nader and Heaps 2008, A-17).

A carbon tax could have a significant effect on the relative costs of various types of energy that are used, for example, to generate electricity. The Electric Power Research Institute has estimated that while a $10 per ton tax on carbon-dioxide emissions would have little effect on the relative position of various power sources, at $50 a ton, wind, estimated to cost $0.095 per kilowatt hour, would become less expensive than pulverized coal (the source used most frequently in the United States today), which would rise from about $0.06 per kilowatt hour to $0.10.

Several countries have enacted carbon taxes since the 1990s, among them Denmark, Finland, Norway, and Sweden. Only in Denmark have emissions fallen significantly, down 15 percent from 1990 to 2006. Denmark was unique in that it remitted most of the tax back to industry and, at the same time, engaged in a refitting of the country's infrastructure (with such things as small-scale "district" heating to reduce transmission waste and an emphasis on insulation) to reduce energy use at the same time (Prasad 2008). Small-scale district power plants in Denmark also use power-plant waste heat to raise the temperature in nearby homes.

A carbon tax need not increase the total tax burden. As in Sweden and other countries, taxes aimed at reducing carbon-dioxide emissions may be used to replace income taxes, on a socially useful activity (labor), substituting a tax on activity that society should discouraged (producing carbon dioxide). N. Gregory Mankiw, professor of economics at Harvard University, quoted Gilbert Metcalf, a professor of economics at Tufts University,

> who has shown how revenue from a carbon tax could be used to reduce payroll taxes in a way that would leave the distribution of total tax burden approximately unchanged. He proposes a tax of $15 per metric ton of carbon dioxide, together with a rebate of the federal payroll tax on the first $3,660 of earnings for each worker. (Mankiw 2007)

By 2007, even investment bankers, energy executives, and government officials were joining environmentalists to urge that the environmental cost of producing energy should be made part of its price. When the American Council on Renewable Energy held its fourth annual conference in Washington, D.C., during December of that year, attendees included Wall Street money managers, conservatives, and even some people employed by fossil-fuel companies. For example, Katrina Landis, vice president of BP (formerly British Petroleum) Alternative Energy said she is "looking forward to a world with an industry-wide carbon price" (Gartner 2007b). Clean energy advocates should convince elected officials that promoting cleaner energy is not a risk to industry but an opportunity, said Janet Sawin, a senior researcher and director of the Energy and Climate Change Program at the Worldwatch Institute. We "need to convince political leaders that renewable energy is about creating job opportunities," she said (Gartner 2007b).

Chris Flaven of the Worldwatch Institute has proposed a worldwide carbon tax of $50 a ton. He also proposed paying 10 percent of the tax into a fund to subsidize development of new technologies that will reduce emissions of greenhouse gases. A study of greenhouse-gas emission-reduction strategies in Chicago supported a "CO_2 Fund ... which would pool resources from the state and private industry and make funds available in low-income communities in order to encourage emissions reductions" ("Energy and Equity" n.d.). In *Earth in the Balance* (1992), Al Gore supported a fossil-fuel tax, part of which would be placed in an Environment Security Trust Fund, "which would be used to subsidize the purchase by consumers of environmentally benign technologies, such as low-energy light bulbs or high-mileage automobiles" (Gore 1992, 349).

Specific Examples of Carbon Taxes

Sweden's Carbon Tax, enacted in 1991, provided an early example of the idea's use on a national scale. Finland, Norway, and Holland later enacted similar taxes. Sweden also levies an energy tax against all fossil fuels. Other taxes are used as incentives to control pollution, including a nitrogen-oxides charge, a sulfur tax, and a tax on nuclear-energy production. In many cases, energy taxes have replaced about 50 percent of the income-tax burden.

The Swedish energy tax varies widely among different products and types of fossil fuels, with by far the highest rates on gasoline. Responding to the new tax system, the use of biofuels (mainly wood-based) for "district" heating soared about 350 percent during the 1990s. Prices of wood-based fuels also fell dramatically as technology improved. Sweden's carbon tax has been credited with reducing fossil-fuel emissions about 20 percent.

In late November 2006, voters in Boulder, Colorado, home of the state's largest university, approved the first carbon tax in the United States. The tax, based on electricity usage, took effect April 1, 2007; it adds $16 a year to an average homeowner's electricity bill and $46 for businesses. The tax is collected by Boulder's main gas and electric utility, Xcel Energy, as agent for Boulder's Office of Environmental Affairs.

The tax revenue will fund efforts to "increase energy efficiency in homes and buildings, switch to renewable energy and reduce vehicle miles traveled," said Jonathan Koehn, the city's environmental affairs manager. The Boulder environmental sustainability coordinator, Sarah Van Pelt, said that residents who used alternative sources of electricity like wind power would receive a discount on the tax based on the amount of the alternative power used (Kelley 2006). A total of 5,600 residents and 210 businesses used wind power in 2006, Van Pelt said.

Oregon began to assess a 3 percent tax on electricity bills by the state's two largest investor-owned utilities during 2001. Revenue goes to the Energy Trust of Oregon, a nonprofit organization, not to state government. The trust distributes cash incentives to businesses and residents to use solar and wind power, biomass energy, and structural upgrades that improve energy efficiency (Kelley 2006).

San Francisco's Carbon Tax

San Francisco Bay Area air-quality regulators in 2008 prepared to charge greenhouse-gas-emitting businesses a fee of $0.044 per ton, the first such fee in the United States that charges industries directly. The Bay Area Air Quality Management District, which governs air quality for nine counties around San Francisco, was preparing to tax a range of businesses, raising $1.1 million a year from petroleum refineries, power plants, and cement plants to gasoline stations and large bakeries, even as representative of oil refineries said the agency's legal authority was "questionable" (Barringer 2008c).

The fee was set to take effect July 1, 2008. The fee is hardly onerous. A gas station might pay $1 a year. The biggest emitter of the gases, the Shell Oil refinery in Martinez, would pay $195,355, based on 4.4 million metric tons of emissions in 2005. Linda Weiner of the Bay Area Clean Air Task Force said, "We believe it sets a precedent as the first time that businesses and government agencies would face financial consequences for contributing to global warming" (Barringer 2008c).

Economic Effects Modeled

William D. Nordhaus modeled the economic effects of carbon taxes ranging from $5 to almost $450 per ton (Nordhaus 1991a, 56). Nordhaus provided a range of carbon-tax rates with expected reductions in fossil-fuel use and costs to gross national product in the United States: he expects a carbon tax of $13 a ton to reduce carbon emissions 6 percent (while reducing greenhouse gases 10 percent), at a negligible cost to gross national product (GNP). Nordhaus expected that a carbon tax of $98 a ton would reduce greenhouse gases 40 percent, with a half a percentage point decrease in GNP. Nordhaus' highest tax estimate, $448 per ton, would decrease greenhouse gases 90 percent, and lower GNP by more than 4 percent (Cline 1992, 168).

Nordhaus estimated that a $5 per ton tax would raise the price of coal 10 percent, the price of oil 2.8 percent, and the price of gasoline 1.2 percent. Applied worldwide, the same level of carbon taxation would, according to Nordhaus' models, reduce greenhouse-gas emissions 10 percent, provide $10 billion in tax revenue, and add $4 billion per year to the global economy. A $100 per ton tax on carbon emissions would raise the price of coal 205 percent, the price of oil 55 percent, and the price of gasoline between 23 and 24 percent, according to Nordhaus. The $100 tax would reduce greenhouse-gas emissions by an estimated 43 percent (close to the level recommended by the Intergovernmental Panel on Climate Change [IPCC] to forestall significant global warming). A $100 tax would provide $125 billion in tax revenues and would decrease global "net benefits" $114 billion, according to Nordhaus. (Nordhaus' numbers indicate that achieving the kind of greenhouse-gas reductions sought by the IPCC would impose a high cost in economic dislocation, especially if the changes are undertaken on a cash basis.)

Michael E. Mann and Richard J. Richels contend that a carbon tax of $250 a ton would be necessary to reduce carbon-dioxide emissions by 20 percent. Such a tax would add about $0.75 to the cost of a gallon of gasoline, and $30 to the cost of a barrel of oil (Cline 1992, 147).

More than 2,000 economists, including six Nobel Laureates, endorsed a statement on global warming during 1997 that called for the use of market mechanisms, including carbon taxes, to move world economies away from fossil fuels:

> As economists we believe that global climate change carries with it significant environmental, economic, social, and geopolitical risks, and that preventive steps are justified.... Economic studies have found that there are many potential policies to reduce greenhouse-gas emissions for which the total benefits outweigh the total costs. For the United States in particular, sound economic analysis shows that there are policy options that would slow climate change without harming American living standards and these measures may in fact improve U.S productivity in the longer run. ("Economists' Letter" 1997)

The U.S. tax code has been suggested as a tool to reduce fossil-fuel emissions, with such proposals as tax credits for energy-efficient appliances, changes in building codes focused on energy efficiency, energy-efficiency standards for universities and health centers that receive federal grants, and tax incentives for fuel-efficient cars.

Carbon Taxes and Technological Development

Stephen Schneider observed that any significant carbon tax probably would shape the future course of technology:

> When the price of conventional energy goes up—and will stay up because everyone knows the carbon tax or quota system will be here to stay—then a host of entrepreneurs and governmentally assisted labs will take up the challenge to develop and test an unimaginable array of more efficient and decarbonized production and end-use alternatives. The higher price of carbon-intensive fuels will spur the R[esearch] and D[evelopment] investments, which economists call induced technological changes. (Schneider 2000b)

FURTHER READING

Barringer, Felicity. "Businesses in Bay Area May Pay Fee for Emissions." *New York Times*, April 17, 2008c. http://www.nytimes.com/2008/04/17/us/17fee.html.

Bennhold, Katrin. "France Tells U.S. to Sign Climate Pacts or Face Tax." *New York Times*, February 1, 2007. http://www.nytimes.com/2007/02/01/world/europe/01climate.html.

Cline, William R. *The Economics of Global Warming*. Washington, DC: Institute for International Economics, 1992.

"Economists' Letter on Global Warming." June 23, 1997. http://uneco.org/Global_Warming.html.

"Energy and Equity: The Full Report." Illinois Environmental Protection Agency, n.d. http://www.cnt.org/ce/energy&equity.htm.

Gartner, John. "Fossil Fuels' Free Ride Is Over." Environment News Service, December 3, 2007b. http://www.ens-newswire.com/ens/nov2007/2007-11-30-02.asp.

Gore, Albert, Jr. *Earth in the Balance: Ecology and the Human Spirit*. Boston: Houghton Mifflin Co., 1992.

Kelley, Kate. "City Approves 'Carbon Tax' in Effort to Reduce Gas Emissions." *New York Times*, November 18, 2006. http://www.nytimes.com/2006/11/18/us/18carbon.html.

Mankiw, N. Gregory. "One Answer to Global Warming: A New Tax." *New York Times*, September 16, 2007. http://www.nytimes.com/2007/09/16/business/16view.html.

Nader, Ralph, and Toby Heaps. "We Need a Global Carbon Tax." *Wall Street Journal*, December, 3, 2008, A-17.

Nordhaus, William D. "Economic Approaches to Greenhouse Warming." In *Global Warming: Economic Policy Reponses*, ed. Rudiger Dornbusch and James M. Poterba, 33–66. Cambridge, MA: MIT Press, 1991.

Poterba, James. "Tax Policy to Combat Global Warming." In *Global Warming: Economic Policy Reponses*, ed. Rudiger Dornbusch and James M. Poterba, 71–98. Cambridge, MA: MIT Press, 1991.

Prasad, Monica. "On Carbon, Tax and Don't Spend." *New York Times*, March 25, 2008. http://www.nytimes.com/2008/03/25/opinion/25prasad.html.

Schneider, Stephen H. "Kyoto Protocol: The Unfinished Agenda: An Editorial Essay." Unpublished manuscript, March 18, 2000b.

Chesapeake Bay, Sea-Level Rise in

Richard A. White once planned to live the rest of his life in his waterfront home on Chesapeake Bay, which, according to Tom Pelton's account in the *Baltimore Sun*, "perches at the tip of a filament of land reaching out into Chesapeake Bay and boasts a 50-foot-long veranda with panoramic views of the sunset" (Pelton 2004, 1-A). The 60-year-old historian has been forced to drink bottled water because salt water has ruined his well. Whitecaps often froth across his lawn, which has shrunk by about 40 feet in the past 18 years. During high tides in the fall, wrote Pelton, "He pulls on tall boots, parks his car a block away on high ground and ties a rowboat beside his door, as his century-old house becomes a tiny island unto itself" (Pelton 2004, 1-A).

White is not alone. More than half of Chesapeake Bay's marshes are in danger of becoming open water, according to a University of Maryland study. Rising sea levels are flooding the marshes that filter pollutants in groundwater, guard against erosion, and provide habitat for ducks, geese, and other animals. Michael Kearney, an associate professor at the University of Maryland–College Park, said the marshes could be gone in 20 to 30 years. He said the time line could be shortened if nor'easters become more severe. Kearney and colleague J. Court Stevenson (also of the University of Maryland) said that water levels in the bay generally have been rising since 1000 C.E., but that the rate of rise accelerated sharply (to more than 10 times the previous rate) during the 1990s. Sea-level rise in the area is being made worse by subsidence of the land (McCord 2000, 1-B).

At the Blackwater National Wildlife Refuge near the shores of Chesapeake Bay, rising waters are rapidly destroying marsh habitat. "Blackwater gives us an example of what will probably occur in a lot of low-lying areas as global warming proceeds and water levels continue to rise," said Stevenson, who has been studying the swampy reserve for more than 20 years. "If we don't do anything about global greenhouse emissions, up to a third of this county where we're standing now will eventually become open water" (Shelby-Biggs 2001).

Blackwater includes about 6,900 hectares of wetlands, woodlands, and croplands. But 2,800 hectares of marshland is already flooded and the rising waters claim another 50 hectares each year. According to a Reuters report, "The rivers which run through the refuge are turning salty. Tiny island sanctuaries for bald eagles are disappearing and pines and grasses are dying. The process is accelerated by a gradual sinking of the land mass, over-pumping of underground water and over-grazing" (Shelby-Biggs 2001).

Conditions similar to those at Blackwater soon may affect much of the U.S. eastern coastline. The U.S. Environmental Protection Agency (EPA) has said that the sea level is rising more quickly along the U.S. eastern coastline than the worldwide average, and most quickly of all in Chesapeake Bay. Along the Atlantic coast and the Gulf of Mexico, the EPA forecasts a 30-centimeter rise in the sea level by 2050. That is an optimistic case; the EPA admits that waters could rise that much by 2025. Given such a sea-level rise, large parts of low-lying cities such as Boston, New York, Charleston, Miami, and New Orleans would become vulnerable to regular flooding.

FURTHER READING

McCord, Joel. "Marshes in Decay Haunt the Bay." *Baltimore Sun*, December 6, 2000, 1-B.

Pelton, Tom. "New Maps Highlight Vanishing Eastern Shore: Technology Provides a Stark Forecast of the Combined Effect of Rising Sea Level and Sinking Land along the Bay." *Baltimore Sun*, July 30, 2004, 1-A.

Shelby-Biggs, Brooke. "Hard to Swallow." Mother Jones.com. March 15, 2001. http://www.motherjones.com/commentary/columns/2001/03/newshole2.html.

China and Global Warming

China is the wildest card in the world greenhouse deck. On one hand, the world's most populous country is streamlining energy efficiency and experimenting with new fuel sources. On the other, China is undergoing an industrial revolution with a population base of about 1.5 billion, consuming rapidly increasing amounts of coal and oil even as its economy becomes more efficient.

By 2006, China and its roughly 1.5 billion people had become the world's leading national

Buildings are seen through a heavy haze in Beijing, August 8, 2007. (AP/Wide World Photos)

source of human-generated carbon-dioxide emissions, surpassing the United States, largely because of an economic boom powered by coal-fired power. The country's gross domestic product has been increasing 8 to 10 percent a year; demand for electricity, 80 percent generated from coal, grew by 14 percent during 2007 alone. Nine of the world's cities with the worst air pollution were in China, as the country struggled to reconcile its developing economy with environmental demands from inside and outside its borders.

Like Mexico City and Los Angeles, Beijing is located in a valley nearly surrounded by mountains and is prone to thermal inversions that hold pollution close to the ground. With 12 million people in 2007, a number that is steadily increasing, Beijing has been undergoing one of the world's largest construction booms, adding residential and office space that requires more electricity, heating, and cooling, nearly all of it provided by coal. Automotive pollution pours into the city as well, as it adds 400,000 cars and trucks a year. While Shanghai has held new vehicle registrations to 100,000 a year with license fees as high as $7,000 per vehicle, Beijing has allowed nearly untrammeled growth.

According to the Netherlands Environmental Agency, China's carbon-dioxide emissions increased 8 percent in 2007. This increase represented two-thirds of global growth in greenhouse gas emissions. In 2007, China's greenhouse-gas emissions exceeded those of the United States by 14 percent. Per capita, however, U.S. emissions were still four times those of China, 19.4 tons compared with 5.1 tons (Rosenthal 2008e).

By 2008, growth in China's emissions of carbon dioxide was accelerating at a rate far greater than previous estimates, according to analysis by University of California economists. Maximillian Auffhammer, a U.C. Berkeley assistant professor of agricultural and resource economics, and Richard Carson, U.C. San Diego professor of economics, calculated their estimates from pollution data in China's 30 provinces.

During 2008, the carbon emissions of China's electric-power industry jumped by 30 percent, passing those of the United States for the first time, according to the Center for Global Development. The same report expected emissions from power production in China and India to double during the ensuing 12 years. China's emissions leaped from 2.3 billion tons in 2007 to an estimated 3.1 billion in 2008, as U.S. emissions remained stable at 2.8 billion. Paul Ting, an oil analyst, said that China relies on coal for three-quarters of its energy consumption. "They cannot get away from coal," he said (Mufson 2008b, D-1). Even with China's rapid increases, electricity usage in the United States in 2007 produced roughly 9.5 tons of carbon dioxide per person, compared with 2.4 tons per person in China, and 0.6 in India.

"Making China and other developing countries an integral part of any future climate agreement is now even more important," said Auffhammer. "What we're finding instead is that the emissions growth rate is surpassing our worst expectations," he said, "and that means the goal of stabilizing atmospheric CO_2 is going to be much, much harder to achieve" ("Growth" 2008). Previous estimates by the Intergovernmental Panel on Climate Change (IPCC) said that China may see a 2.5 to 5 percent annual increase in CO_2 emissions between 2004 and 2010. The new University of California analysis put that figure at 11 percent or more. Much of this increase will come from new coal-fired electric power capacity in plants built to last for 40 to 75 years, which all but guarantees high emissions far into the future.

Reflecting a worldwide recession, in October 2008, however, China reported a 4 percent decline year over year in electricity production, the largest decline in 10 years. Until weeks before this severe decline, China's electricity production had been increasing 10 to 15 percent annually for several years (Yang 2008, A-11).

As a developing country, China is exempt from the Kyoto Protocol's requirements for reductions in emissions of global-warming gases. As China, India, and other countries develop, their emissions are expected to make up more than half the world's total within a quarter century. Demand for coal, mostly for power generation, will rise 60 percent, according to International Energy Agency (IEA) forecasts in 2006. As a result, energy-related carbon-dioxide emissions will increase 55 percent, to 44.1 billion tons in 2030 (Bradsher 2006). China has rejected greenhouse-gas limits. "You cannot tell people who are struggling to earn enough to eat that they need to reduce their emissions," said Lu Xuedu, the deputy director general of Chinese Office of Global Environmental Affairs (Bradsher 2006).

China's First Strategy on Global Warming

In mid-2007, China released its first strategy on global warming, which advocated improved energy efficiency and control of greenhouse-gas emissions, but did not include mandatory limits that could restrict growth of the country's booming economy. This program resembled a program recommended by U.S. President George W. Bush at about the same time—as each of the world's two largest greenhouse-gas sources blamed the other for taking inadequate action to address global warming. Ma Kai, head of China's powerful economic planning agency, the National Development and Reform Commission, maintained that criticism was unfair because the Chinese produce only about one-fifth the greenhouse gases per capita as the United States (Yardley 2007).

China's plans call for a 20 percent improvement in energy efficiency by 2010. China's drive for energy efficiency is compromised by its dependence on coal, much of it low in quality. China's plans call for a major expansion of nuclear power, as well as renewable energy sources. The plan calls for renewable energy to account for 10 percent of the country's power supply by 2010. China is also in the midst of a nationwide reforestation program to help absorb greenhouse gases (Yardley 2007).

The same report, according to a description in the *New York Times*,

> also painted a brief, if alarming, picture of how global warming could change China, including rising sea levels, shrinking glaciers in Tibet, rising temperatures and the likelihood of expanding deserts. By 2020, annual temperatures could rise between 1.3° to 2.1°C from 2000. In addition, projections show that Chinese agricultural output could be reduced by 10 percent by 2030. (Yardley 2007)

Winter Takes a Vacation

China's winter of 2006–2007 was unusually mild, drawing attention to the warming climate. A popular 1,400-year-old ice festival in Harbin, in northeast China about 400 miles east of the Russian border, literally melted, threatening a tourist attraction that usually draws five million people a year. Edward Cody of the *Washington Post* wrote:

> The hands had melted off a delicately entwined couple of ballet dancers crafted by an ice-sculpting team from Vladivostok. Eaves fashioned from packed snow-drooped into icicles at the Roast Meat Fire House restaurant. Authorities banned people from approaching the ice-cube tower at Ice and Snow World because big chunks kept falling off.... Heads are falling from statues and intricately sculpted ice animals are turning into shapeless blobs. (Cody 2007, A-19)

In the midst of the nonwinter of 2006–2007, the China Meteorological Administration said that temperatures probably would continue to rise by 7° to 10.8°F by the year 2100 compared with average temperatures between 1961 and 1990. In Beijing, during the Lunar New Year celebrations, the warmest since authorities began keeping records in 1951, people jogged without jackets in Ritan Park, as boys played basketball in T-shirts. On February 5, the temperature rose to a record of 61°F (Cody 2007, A-19).

In the meantime, Jiang Yu, a spokeswoman for the Foreign Ministry, said that China placed primary responsibility for global warming on richer, developed nations. "It must be pointed out that climate change has been caused by the long-term historic emissions of developed countries and their high per capita emissions," she said, adding that developed countries have responsibilities for global warming "that cannot be shirked" (Yardley 2007).

Alarms over global warming also have begun to ring in some of China's official agencies. Oceanographers at China's State Oceanic Administration argue that a sea-level rise of three feet a century could cause flooding in many of China's coastal areas, home to half of China's large cities and to 40 percent of its population.

China's Nuclear Construction Campaign

As environmental philosophers debate the role of nuclear power relative to global warming, by 2007, China had begun construction on dozens of new nuclear-power plants to address part of its growing global-warming burden from low-quality coal. China plans to spend $50 billion to build 32 nuclear plants by 2020. Some experts believe that China may build 300 more such plants by about 2050, to power what will be, by then, the largest national economy in the world. By that time, China may have half the nuclear-power capacity in the world (Eunjung Cha 2007, D-1). China also is building the world's largest repository for spent nuclear fuel in its western desert, amid the Beishan Mountains, an area that is nearly bereft of human habitation. The nuclear-construction binge represents a major change for China, where only 2.3 percent of electricity was generated from nuclear power in 2006, compared with about 20 percent in the United States and almost 80 percent in France. In 2007, China had only nine nuclear-power plants.

In part because of Chinese demand, the price of processed uranium ore increased from $10 to $120 a pound between 2003 and 2007. Expecting that China will be one of its major customers, Japan's Toshiba paid $5.4 billion during 2006 to acquire a U.S. company, Westinghouse Electric, which specializes in construction of nuclear plants. The Chinese government, emphasizing safety, has been using companies such as Westinghouse to instruct its engineers who will build and operate new nuclear plants.

China's Use of Coal

By 2005, China was using more coal than the United States, the European Union, and Japan combined. China increased coal consumption 14 percent in 2004 and 2005 in human history's largest-scale industrialization. Every week to 10 days, another coal-fired power plant opens somewhere in China that is big enough to serve all the households in Dallas or San Diego (Bradsher and Barboza 2006). Many of them use old, polluting technology, and they will be operating for an average of 75 years.

The personal effects of coal use in China were described in the *New York Times*:

> Wu Yiebing and his wife, Cao Waiping, used to have very little effect on their environment. But they have tasted the rising standard of living from coal-generated electricity and they are hooked, even as they suffer the vivid effects of the damage their new lifestyle creates. Years ago, the mountain village where they grew up had electricity for only several hours each evening, when water was let out of a nearby dam to turn a small turbine. They lived in a mud hut, farmed by hand from dawn to dusk on hillside terraces too small for tractors, and ate almost nothing but rice on an income of $25 a month Today, they live here in Hanjing, a small town in central China where Mr. Wu earns nearly $200 a month. He operates a large electric drill 600 feet underground in a coalmine, digging out the fuel that has powered his own family's advancement. He and his wife have a stereo, a refrigerator, a television, an electric fan, a phone and light bulbs, paying just $2.50 a month for all the electricity they can burn from a nearby coal-fired power plant. (Bradsher and Barboza 2006)

All new cars, minivans, and sport utility vehicles sold in China starting July 1, 2007, had to meet fuel-economy standards stricter than those in the United States. New construction codes encourage the use of double-glazed windows to reduce air-conditioning and heating costs and high-tech light bulbs that produce more light with fewer watts (Bradsher and Barboza 2006).

China continues to rely on coal for 75 percent of its energy, spewing out some 19 million tons of sulfur dioxide a year (the United States produces 11 million tons per year), and contributing mightily to acid rain. Coal consumption in China increased by an estimated 10 percent a year between 2000 and 2003. Chinese electricity generation, the main use of coal, jumped 16 percent during the first eight months of 2003 (Bradsher 2003, 1). Many Chinese homes that once used only light bulbs have acquired several appliances, including air conditioners, in recent years.

National energy conservation targets sometimes compel national leaders to say that they have closed coal mines. Nobuhiro Horii, who works with the Institute of Developing Economies in Japan, examined how China's Hunan province handled government orders to close coal mines. "He concluded," according to the *Washington Post*, "that local officials told Beijing they had shut the mines, when in fact they kept them open. Interviews with officials in other parts of China led Horii to determine this to be a nationwide problem" ("Research Casts Doubt" 2001).

Horii added that it usually takes about a decade to increase energy efficiency. China's claims that it was making inroads into carbon dioxide production in two years, or even four, are not credible, he asserted. "This is just not possible," Horii added. "Yes, China is increasing energy efficiency, but they are doing it slowly, like everyone else" ("Research Casts Doubt" 2001). A report by the U.S. Embassy in Beijing called the statistical claims of Chinese greenhouse-gas reductions "greatly exaggerated," saying they fell "outside the realm of experience of any other country in modern times." The report concluded that China's greenhouse-gas emissions "have dropped little, if at all" ("Research Casts Doubt" 2001).

In addition to massive industrial expansion, China since the year 2000 has been adding about 7.5 billion square feet of residential and commercial real estate per year, as much as all of the existing retail shopping centers and strip malls in the United States, according to the U.S. Energy Information Administration (Kahn and Yardley 2006). An increasing proportion of this space is air conditioned. In addition, most Chinese buildings, even new ones, have little or no thermal insulation, and require twice as much energy to heat or cool as similar floor space in similar European or U.S. climates, according to the World Bank. China has energy-efficiency standards, but most new buildings do not meet them (Kahn and Yardley 2006).

To light, heat, and cool all this new space (as well as industrial plants that produce so many exported goods), China in 2005 alone added 66 gigawatts of electricity, as much as Great Britain's annual demand. In 2006, it added 102 gigawatts, the total demand of France. Two-thirds of this new power is generated using coal. China has built small, inexpensive coal-fired plants that only rarely use the latest more efficient combined-cycle turbines (Kahn and Yardley 2006).

China and Automobiles

Automobile sales in China increased more than 800 percent from 2000 to 2007 (Bradsher 2008b). By 2008, more Buicks were being sold in China than in the United States, in a market where car size is closely identified with social and economic status. Some wealthy Chinese pay more than $200,000 for a Hummer.

China's highway system may soon surpass that of the United States. The 23,000 miles of highway in 2006 had doubled that of 2001. The Chinese government in 2006 announced plans to build 53,000 freeway miles by 2035. The U.S. Interstate Highway System, which is 50 years old, comprises 46,000 miles. As with the U.S. interstate system. China's goal is to consolidate the nation and to allow the easy transport of military forces between regions. Policy anticipates that western territories such as Tibet and Xinjiang (meaning "New Frontier") will be fully integrated, ethnically and economically (Conover 2006).

The number of passenger cars on the road, about six million in 2000, rose to about 20 million in 2006. Car sales are up 54 percent during the first three months of 2006, compared with the same period a year earlier. Every day, 1,000 new cars (and 500 used ones) were being sold in Beijing alone (Conover 2006).

Chinese car culture resembles that of the United States: city drivers, stuck in ever-growing jams, listen to traffic radio. They buy auto magazines with titles like *The King of Cars*, *AutoStyle*, *China Auto Pictorial*, *Friends of Cars*, and *Whaam* ("The Car—The Street—The Travel—The Racing"). Two-dozen titles now compete for space in kiosks. The McDonald's Corporation said last month that it expects half of its new outlets in China to be drive-through. Whole zones of major cities, like the Asian Games Village area in Beijing, have been given over to car lots and showrooms (Conover 2006).

As China's fleet of motor vehicles expanded, its consumption of oil also increased from roughly 2.2 million barrels a day in 1988 to 5.2 million barrels a day in 2003, or roughly 150 percent in 15 years (an average of 10 percent a year). Oil use accelerated after that to almost seven million barrels a day by 2007. The IEA issued figures from its office in Paris indicating that increases in Chinese greenhouse-gas emissions between 2000 and 2030 "will nearly equal the increase from the entire industrialized world" (Bradsher 2003, 1). During the mid-1990s, people in China owned a mere handful of private cars. Private automobile ownership grew by 26 percent between 1996 and 2000, and by 69 percent in 2003 alone (English 2004, 1).

General Motors has forecast that China will account for 18 percent of global growth in automobile sales between 2002 and 2012 (Bradsher 2003, 1). During the 1990s, motor vehicle sales in the Chinese countryside rose from about 40,000 to almost 500,000 per year (Leggett 2001, A-19). Shanghai Automotive Industrial Corporation is planning to license General Motors technology to build a basic pickup truck for China's farmers. The new vehicle, to be called "Combo," will be produced in a nonprofit government car factory. This is one of GM's effort's to tap an auto market of "one billion consumers and a fast-growing network of national highways" (Leggett 2001, A-19).

FURTHER READING

Bradsher, Keith. "China Prospering but Polluting: Dirty Fuels Power Economic Growth." *New York Times* in *International Herald-Tribune*, October 22, 2003, 1.

Bradsher, Keith. "China to Pass U.S. in 2009 in Emissions. *New York Times*, November 7, 2006. http://www.nytimes.com/2006/11/07/business/worldbusiness/07pollute.html.

Bradsher, Keith. "With First Car, a New Life in China." *New York Times*, April 24, 2008b. http://www.nytimes.com/2008/04/24/business/worldbusiness/24firstcar.html.

Bradsher, Keith, and David Barboza. "Pollution From Chinese Coal Casts a Global Shadow." *New York Times*, June 11, 2006. http://www.nytimes.com/2006/06/11/business/worldbusiness/11chinacoal.html.

Cody, Edward. "Mild Weather Takes Edge off Chinese Ice Festival: Residents of Tourist City Blame Global Warming." *Washington Post*, February 25, 2007, A-19. http://www.washingtonpost.com/wpdyn/content/article/2007/02/24/AR2007022401421_pf.html.

Conover, Ted. "Capitalist Roaders." *New York Times Sunday Magazine*, July 2, 2006. http://www.nytimes.com/2006/07/02/magazine/02china.html.

El Nasser, Haya. "'Green' Efforts Embrace Poor." *USA Today*, November 24, 2008, 3-A.

English, Andrew. "Feeding the Dragon: How Western Car-makers are Ignoring Ecological Dangers in Their Rush to Exploit a Wide-open Market." *London Daily Telegraph*, October 30, 2004, 1.

Eunjung Cha, Ariana. "China Embraces Nuclear Future: Optimism Mixes With Concern as Dozens of Plants Go Up." *Washington Post*, May 29, 2007.

"Growth in China's CO_2 Emissions Double Previous Estimates." Environment News Service, March 11, 2008. http://www.ens-newswire.com/ens/mar2008/2008-03-11-01.asp.

Kahn, Joseph, and Jim Yardley. "As China Roars, Pollution Reaches Deadly Extremes." *New York Times*, August 26, 2006. http://www.nytimes.com/2007/08/26/world/asia/26china.html.

Leggett, Karby. "In Rural China, General Motors Sees a Frugal but Huge Market: It Bets Tractor Substitute Will Look Pretty Good to Cold, Wet Farmers." *Wall Street Journal*, January 16, 2001, A-19.

Mufson, Steven. "Power-Sector Emissions of China to Top U.S." *Washington Post*, August 27, 2008b, D-1. http://www.washingtonpost.com/wp-dyn/content/article/2008/08/26/AR2008082603096_pf.html.

National Academy of Sciences. *Policy Implications of Greenhouse Warming*. Washington, DC: National Academy Press, 1991.

"Research Casts Doubt on China's Pollution Claims." *Washington Post*, August 15, 2001, A-16. http://www.washingtonpost.com/wp-dyn/articles/A10645-2001Aug14.html.

Rosenthal, Elisabeth. "China Increases Lead as Biggest Carbon Dioxide Emitter." *New York Times*, June 14, 2008e. http://www.nytimes.com/2008/06/14/world/asia/14china.html.

Yang, Jing. "China's 4 Per Cent Fall in Electricity Output May Portend Worse Economic Slump." *Wall Street Journal*, November 14, 2008, A-11.

Yardley, Jim. "China Says Rich Countries Should Take Lead on Global Warming," *New York Times*, February 7, 2007a. http://www.nytimes.com/2007/02/07/world/asia/07china.html.

Yardley, Jim. "China Releases Climate Change Plan." *New York Times*, June 4, 2007c. http://www.nytimes.com/2007/06/04/world/asia/04cnd-china.html.

Chlorofluorocarbons, Relationship to Global Warming

When chlorofluorocarbons (CFCs) were banned in the late 1980s, most experts expected damage to stratospheric ozone over the Antarctic to heal. A single atom of chlorine from CFC can destroy more than 100,000 ozone molecules; the expectation was that declining levels of CFCs would cause the decline in ozone levels to reverse.

By 2000, continuing to 2008, however, ozone depletion still was a major problem, and an ozone "hole" began to open over the Arctic as well. The nature of science has evolved to explain how the capture of heat near the Earth's surface by greenhouse gases accelerates cooling in the stratosphere, causing even a decreasing level of CFCs to destroy more ozone, and thus playing an important role in continuing ozone depletion at that level. As greenhouse gases retain heat in the lower atmosphere, the stratosphere cools, enabling what CFCs remain there to become more active, consuming greater volumes of ozone. Thus, because CFCs are "cold-activated," the healing of the stratospheric ozone layer depends, to some degree, on reduction of greenhouse-gas levels in the lower atmosphere. The stratospheric ozone shield is important because it protects plants and animals at the surface from ultraviolet (UV) radiation, which can cause skin cancer, cataracts, and damage to the immune systems of human beings and other animals. Thinning of the ozone layer also may alter the DNA of plants and animals.

The energetic nature of UV-B radiation can break the bonds of DNA molecules. While plants and animals are generally able to repair damaged DNA, on occasion, damaged DNA molecules can continue to replicate, leading to dangerous forms of skin cancer in humans. The probability that DNA can be damaged by UV radiation varies with wavelength, shorter wavelengths being the most dangerous. "Fortunately," wrote one observer,

> at the wavelengths that easily damage DNA, ozone strongly absorbs UV and, at the longer wavelengths where ozone absorbs weakly, DNA damage is unlikely. But given a 10-percent decrease in ozone

in the atmosphere, the amount of DNA-damaging UV would be expected to increase by about 22 percent." (Newman 1998)

History of CFCs

CFCs initially raised no environmental questions when they were first marketed by DuPont Chemical during the 1930s under the trade name Freon. Freon was introduced at a time when environmental questions usually were not asked. At about the same time, asbestos was being proposed as a high-fashion material for clothing, and radioactive radium was being built into timepieces so that they would glow in the dark.

By 1976, manufacturers in the United States were producing 750 million pounds of CFCs a year, and finding all sorts of uses for them, from propellants in aerosol sprays, to solvents used to clean silicon chips, to automobile air conditioning, and as blowing agents for polystyrene cups, egg cartons, and containers for fast food. They were very useful, wrote Anita Gordon and Peter Suzuki. Inexpensive to manufacture, not flammable, and chemically stable. By the time scientists discovered during the 1980s that CFCs were thinning the ozone layer over the Antarctic, they found themselves taking on a $28-billion-a-year industry.

By the time their manufacture was banned internationally during the late 1980s, CFCs had been used in roughly 90 million car and truck air conditioners, 100 million refrigerators, 30 million freezers, and 45 million air conditioners in homes and other buildings. Because CFCs remain in the stratosphere for up to 100 years, they will deplete ozone long after industrial production of the chemicals ceased.

Warm at the Surface, Cold Aloft

The Antarctic ozone "hole" formed earlier and endured longer during the September and October of 2000 than ever before—and by a significant amount. Figures from NASA satellite measurements showed that the area of depleted ozone covered an area of approximately 29 million square kilometers in early September, exceeding the previous record from 1998. The record size persisted for several days. Ozone levels, measured in Dobson units, fell below 100 DU for the first time. The area cold enough to produce ozone depletion also grew by 10 to 20 percent more surface area than any other year. The ozone-depletion zone was coming closer to New Zealand, where usual springtime ozone levels average about 350 DU. During the spring of 2000, ozone levels reached as low as 260 DU, as atmospheric circulation patterns nudged the Antarctic zone northward. Scientists usually regard an area of the stratosphere as ozone-depleted when its DU level falls below 220.

During early September 2003, the area of depleted ozone over Antarctica approached near-record size again. By the end of the month, the area of severely depleted ozone was the second-largest on record, at about the size of North America. The coupling of global warming near the surface with declining stratospheric temperatures (and increasing activity of CFCs) continued in 2006, as the loss of ozone over Antarctica reached a new record, according to scientists with the European Space Agency (ESA). "Such significant ozone loss requires very low temperatures in the stratosphere combined with sunlight," said ESA atmospheric engineer Claus Zehner. "This year's extreme loss of ozone can be explained by the temperatures above Antarctica reaching the lowest recorded in the area since 1979," the beginning of recordkeeping.

Measurements by the ESA's Envisat satellite indicated that ozone loss over Antarctica totaled 40 million metric tons in 2006, surpassing the record loss of 39 million tons recorded in 2000. The loss is calculated by measuring the area and the depth of the stratospheric ozone that has been depleted. Late in September 2006, the World Meteorological Organization reported that the size of that year's ozone hole had expanded to 10.6 million square miles (28 million square kilometers), larger than the previous record ozone hole size during 2000. This represented an area of depleted ozone larger than the surface area of North America.

FURTHER READING

Aldhous, Peter. "Global Warming Could Be Bad News for Arctic Ozone Layer." *Nature* 404 (April 6, 2000):531.

Crutzen, Paul J. "The Antarctic Ozone Hole, a Human-caused Chemical Instability in the Stratosphere: What Should We Learn from It? In *Geosphere-Biosphere Interactions and Climate,* ed. Lennart O. Bengtsson and Claus U. Hammer, 1–11. Cambridge: Cambridge University Press, 2001.

Flannery, Tim. *The Weather Makers: How Man Is Changing the Climate and What It Means for Life on Earth.* New York: Atlantic Monthly Press, 2005.

Hartmann, Dennis L., John M. Wallace, Varavut Limpasuvan, David W. J. Thompson, and James R. Holton. "Can Ozone Depletion and Global Warming Interact to Produce Rapid Climate Change?" *Proceedings of the National Academy of Sciences of the United States of America* 97, no. 4 (February 15, 2000):1412–1417.

Rowland, Sherwood, and Mario Molina. "Stratospheric Sink for Chlorofluoromethanes: Chlorine Atom-Catalyzed Destruction of Ozone." *Nature* 249 (June 28, 1974):810–812.

Shindell, Drew T., David Rind, and Patrick Lonergan. "Increased Polar Stratospheric Ozone Losses and Delayed Eventual Recovery Owing to Increasing Greenhouse-Gas Concentrations." *Nature* 392 (April 9, 1998):589–592.

Christianity and Global Warming

The Bible's content is diverse enough to be quoted in almost any context. The same Good Book that commands us to multiply and subdue the Earth also may be quoted to commend stewardship of the natural world. The U.S. Conference of Catholic Bishops has done as much in its new "plea for dialogue, prudence, and the common good," its consensus statement on "global climate change" (Catholic Bishops 2001). The statement continued: "How are we to fulfill God's call to be stewards of creation in an age when we may have the capacity to alter that creation significantly, and perhaps irrevocably? We believe our response to global climate change should be a sign of our respect for God's creation" (Catholic Bishops 2001).

The bishops' statement continued:

> Global climate is by its very nature a part of the planetary commons. The Earth's atmosphere encompasses all people, creatures, and habitats.... Stewardship [is] defined in this case as the ability to exercise moral responsibility to care for the environment.... Our Catholic tradition speaks of a "social mortgage" on property and, in this context, calls us to be good stewards of the Earth.... Stewardship requires a careful protection of the environment and calls us to use our intelligence "to discover the earth's productive potential and the many different ways in which human needs can be satisfied." (Catholic Bishops 2001, quoting John Paul II)

According to this statement, Catholic response to the challenge of climate change "must be rooted in the virtue of prudence." While some uncertainty remains, it said, most experts agree that "something significant is happening to the atmosphere. Human behavior and activity are, according to the most recent findings of the international scientific bodies charged with assessing climate change, contributing to a warming of the earth's climate" (Catholic Bishops 2001). This statement included an ecological twist on beliefs in American exceptionalism: "Because of the blessings God has bestowed on our nation and the power it possesses, the United States bears a special responsibility in its stewardship of God's creation to shape responses that serve the entire human family" (Catholic Bishops 2001).

This statement concluded that responsibility weighs more heavily upon those with the power to act, because the threats are often greatest for those who lack similar power, namely vulnerable poor peoples as well as future generations. According to reports of the Intergovernmental Panel on Climate Change (IPCC), significant delays in addressing climate change may compound the problem and make future remedies more difficult, painful, and costly. On the other hand, said the bishops, the impact of prudent actions today can potentially improve the situation over time, avoiding more painful but necessary actions in the future (Catholic Bishops 2001).

The bishops believe that passing along the problem of global climate change to future generations as a result of our delay, indecision, or self-interest would be easy. However, the statement said, "We simply cannot leave this problem for the children of tomorrow. As stewards of their heritage, we have an obligation to respect their dignity and to pass on their natural inheritance, so that their lives are protected and, if possible, made better than our own" (Catholic Bishops 2001).

"Grateful for the gift of creation," said this statement,

> [w]e invite Catholics and men and women of good will in every walk of life to consider with us the moral issues raised by the environmental crisis.... These are matters of powerful urgency and major consequence. They constitute an exceptional call to conversion. As individuals, as institutions, as a people, we need a change of heart to preserve and protect the planet for our children and for generations yet unborn. (Catholic Bishops 2001, quoting "Renewing the Earth" n.d.)

Evangelical Christian Leaders Debate Global Warming

Early in 2006, despite opposition from some of their colleagues, 86 evangelical Christian leaders decided to support an initiative to combat global warming, as they wrote: "millions of people could die in this century because of climate change, most of them our poorest global neighbors" (Goodstein 2006). Signers included presidents of 39 evangelical colleges, leaders of aid groups and churches, including the Salvation Army, and pastors of some megachurches, including Rick Warren, author of the best seller *The Purpose-Driven Life*. "Many of us have required considerable convincing before becoming persuaded that climate change is a real problem and that it ought to matter to us as Christians. But now we have seen and heard enough" (Goodstein 2006).

The statement advocates federal legislation to legally require reductions in carbon-dioxide emissions through cost-effective, market-based mechanisms. This initiative was regarded as the first stage of an "Evangelical Climate Initiative" to include television and radio public-service advertising, as well as other educational publicity.

"We have not paid as much attention to climate change as we should, and that's why I'm willing to step up," said Duane Litfin, president of Wheaton College, an influential evangelical institution in Illinois. "The evangelical community is quite capable of having some blind spots, and my take is this has fallen into that category."

Meanwhile, 22 well-known evangelical leaders, signed a letter in January 2006 declaring, "Global warming is not a consensus issue" (Goodstein 2006). Among the signers were Charles W. Colson, the founder of Prison Fellowship Ministries; James C. Dobson, founder of Focus on the Family; and Richard Land, president of the Ethics and Religious Liberty Commission of the Southern Baptist Convention. E. Calvin Beisner, associate professor of historical theology at Knox Theological Seminary in Fort Lauderdale, Florida, helped organize the opposition into a group called the Interfaith Stewardship Alliance. He said that "the science is not settled" on whether global warming was actually a problem or even that human beings were causing it. He also said that the solutions advocated by global warming opponents would only cause the cost of energy to rise, with the burden falling most heavily on the poor (Goodstein 2006).

FURTHER READING

Catholic Bishops, U.S. Conference. "Global Climate Change: A Plea for Dialogue, Prudence, and the Common Good: A Statement of the U.S. Catholic Bishops." Ed. William P. Fay. June 15, 2001. http://www.ncrlc.com/climideas.html.

Goodstein, Laurie. "Eighty-six Evangelical Leaders Join to Fight Global Warming." *New York Times*, February 8, 2006. http://www.nytimes.com/2006/02/08/national/08warm.htm.

John Paul II. *On the Hundredth Anniversary of Rerum Novarum (Centesimus Annus)*. No. 32. Washington, DC: United States Catholic Conference, 1991.

Renewing the Earth: An Invitation to Reflection and Action on Environment in Light of Catholic Social Teaching. Washington, DC: United States Catholic Conference, n.d.

Cities Organize against Global Warming

During May 2007, at the "C40 Large Cities Climate Summit" in New York City, a group of the world's largest cities committed to addressing climate change. Mayors from across the United States and around the world attended, including Bangkok, Berlin, Bogotá, Chicago, Copenhagen, Delhi, Houston, Istanbul, Johannesburg, Mexico City, Rio, Rome, São Paulo, Seoul, Sydney, Tokyo, Toronto, and Vancouver. The mayors, their senior staff members, and business leaders shared projects showing how they were reducing greenhouse-gas emissions and conserving energy.

In Los Angeles, Mayor Antonio Villaraigosa, in partnership with the Los Angeles City Council and environmental leaders, unveiled "GREEN LA—An Action Plan to Lead the Nation in Fighting Global Warming." Villaraigosa pledged to reduce his city's carbon footprint by 35 percent below 1990 levels by 2035, the most

Seattle Mayor Greg Nickels speaks during the opening session of the U.S. Conference of Mayors "Climate Protection Summit," Thursday, November 1, 2007, in Seattle. (AP/Wide World Photos)

ambitious goal set by a major American city. Los Angeles also planned to increase its use of renewable energy to 35 percent by 2020, much of it through changes at its municipal electrical utility, the largest in the country ("Big City Mayors" 2007).

By mid-2008, 850 cities in the United States, with a population of more than 85 million, from Boston, Massachusetts, to Portland, Oregon, had pledged to meet Kyoto Protocol standards, as local officials sped ahead of federal policy on global warming. A third of Americans regarded global warming as the world's single most pressing environmental problem by that time, double the number a year earlier, according to a poll conducted by the *Washington Post*, ABC News, and Stanford University. Greg Nickels, Seattle's mayor, started the U.S. Mayors Climate Protection Agreement in 2005, after a season without a winter ruined the Pacific Northwest's skiing season. In the meantime, U.S. fossil-fuel-related emissions fell 1.3 percent to 5.88 billion metric tons between 2005 and 2006.

At the Seattle Climate Protection Summit during November 2007, more than 100 U.S. mayors called for a federal partnership against energy dependence and global warming. "We are showing what is possible in light of climate change at the local level, but to reach our goal of 80 percent reductions in greenhouse gases by 2050, we need strong support from the federal government," said Nickels ("U.S. Mayors Seek" 2007).

New York Mayor Michael Bloomberg said, "Climate change presents a national security imperative for us, because our dependence on foreign oil has entangled our interests with tyrants and increased our exposure to terrorism. It's also an economic imperative, because clean energy is going to be the oil gusher of the 21st century" ("U.S. Mayors Seek" 2007). Bloomberg called for a pollution fee (a carbon tax) to discourage activities that generate greenhouse gases. "As long as greenhouse gas pollution is free, it will be abundant," Bloomberg said. "If we want to reduce it, there has to be a cost for producing it. The voluntary targets suggested by President Bush would be like voluntary speed limits—doomed to fail" ("U.S. Mayors Seek" 2007).

City Initiatives

In Austin, Texas, energy efficiency standards were raised for homes, requiring a 60 percent reduction in energy use by 2015. Chicago was trying waterless urinals and has planted several thousand trees. Philadelphia has been replacing black tarpaper roofs atop old row houses with snow-white, high-reflection composites. Keene, New Hampshire, requires parents waiting for children at schools to turn off car engines. In Portland, Oregon, carbon emissions had been reduced to 1990 levels by 2007. Water flowing through Portland's drinking-water system also generates hydroelectricity (Faiola and Shulman 2007, A-1). Mayors of at least 134 U.S. cities by 2007 were using more energy-efficient lighting in public buildings, streetlights, parks, traffic signals, and other places. Many city governments' auto fleets had converted to alternative fuels or hybrid-electric technology.

The Chicago Climate Action Plan, announced during September 2008, aimed to cut greenhouse-gas emissions by 25 percent of 1990 levels by 2020. The plan requires retrofitting of commercial and industrial buildings, increased energy efficiency in residences, and increased use of electricity from renewable sources of electricity. The Task Force aims to reduce Chicago's emissions to 80 percent below its 1990 levels by 2050. Buildings, which emit 70 percent of Chicago's carbon dioxide, are the major target of the Climate Action Plan. Chicago City Hall already has a green roof, designed as a model for as many as 6,000 buildings citywide. By 2007, Chicago's "Smart Bulb Program" had distributed 500,000 free compact fluorescent light bulbs to residents.

Dallas, Texas' municipal government decided early in 2008 to purchase 40 percent of its power from renewable energy sources, primarily wind power, which has been expanding rapidly in Texas. The city government also reduced its energy use 5 percent for each of the preceding five years by using lighting upgrades, solar panels, high-efficiency heating and air-conditioning systems, and automated building controls. Nearly half of the streetlights in Dallas by 2008 used renewable energy. Three to 4 million cubic feet of methane also was being captured daily at the McCommas Bluff Landfill. The methane was purified and then sold to Atmos Energy, as biogas to heat homes and businesses, replacing natural gas. Plans were afoot to increase this source by 300 percent in five years. Municipal buildings of more than 10,000 square feet are required to meet LEED (Leadership in Energy and Environmental Design) standards. The police department was one of the first to be upgraded. Upgrades at City Hall are saving $1.5 million in energy costs per year ("Dallas Toots" 2008).

Using the AlbuquerqueGreen Program, that city reduced natural-gas consumption 42 percent, and cut greenhouse-gas emissions 67 percent between 2000 and 2006. AlbuquerqueGreen promotes growth of green-tech companies, bicycle use, and pedestrian-friendly streets. Albuquerque requires that all new buildings be designed to be carbon neutral, with architecture suitable for 100 percent renewable energy use by 2030 ("U.S. Mayors" 2007).

Several cities have targeted poor neighborhoods with subsidies and grants for insulation of older homes that often leak heat in winter. Such programs also allow some people to acquire insulation, buy more efficient compact fluorescent light bulbs, and replace older, inefficient basic electrical appliances, such as refrigerators, washers, and dryers. NeighborWorks America, a nonprofit group, was coordinating efforts along this line in 230 local organizations by late 2008. Workers with Greenprint Denver, for example, have gone door-to-door in low-income neighborhoods offering energy audits and help with goods and services. Rays of Hope in Austin, Texas, offers basic services as well as solar panels.

Tree-Planting Programs Taking Root

A number of major U.S. cities have launched sizable tree-planting programs, including Washington, Baltimore, Minneapolis, Chicago, Denver, and Los Angeles. Still, the decline in tree cover has been accelerating since the 1970s, especially on private property and new development, according to American Forests, an environmental group that uses satellite imagery to document tree cover across the United States (Harden 2006, A-1). "This is like a creeping cancer," said Deborah Gangloff, the group's executive director. "In the two dozen cities we have studied, we have noticed about a 25 percent decline in tree canopy cover over the past 30 years. This is a dramatic trend that is costing cities billions of dollars" (Harden 2006, A-1). Washington, D.C., is among the cities with the largest reduction in dense tree cover, with a 64 percent decline from 1973 to 1997, according to American Forests.

Three shade trees strategically planted around a house can reduce home air-conditioning bills by about 30 percent in hot, dry cities such as Sacramento, and a nationwide shade program similar to the one there could reduce air-conditioning use by at least 10 percent, according to Energy Department research (Harden 2006, A-1). A study of greenhouse-gas reduction potential in Chicago endorsed urban tree-planting projects to reduce air pollution in cities,

About 375,000 shade trees have been given away to Sacramento, California, city residents, where the city hoped to plant a total of four million. To receive up to 10 free trees, residents simply call the Sacramento Municipal Utility District (SMUD), a publicly owned power company. "A week later, they are here to tell you where the trees should be planted and how to take care of them," said Arlene Willard, a retired welfare caseworker who with her husband John has planted four SMUD trees in the back yard of their east Sacramento house (Harden 2006, A-1).

By planting 10 million trees as well as installing lighter-colored roofs and pavement, Los Angeles could begin to reverse an urban "heat island" effect caused by concrete, asphalt, and heat-retaining buildings that has been increasing for a hundred years, according to a simulation study by the Department of Energy's Lawrence Berkeley National Laboratory. It found that Los Angeles could lower its peak summertime temperature by five degrees, cut air-conditioning costs by 18 percent, and reduce smog by 12 percent (Harden 2006, A-1).

In 2006, Los Angeles started a campaign to plant a million trees, part of a free-tree program following the Sacramento model. For every dollar it spends on trees, the city expects to realize a $2.80 return from energy savings, pollution reduction, stormwater management, and increased property values, said Paula A. Daniels, a commissioner on the Board of Public Works (Harden 2006, A-1).

FURTHER READING

"Big City Mayors Strategize to Beat Global Warming." Environment News Service, May 15, 2007. http://www.ens-newswire.com/ens/may2007/2007-05-15-01.asp.

"Chicago Sets Goals for a Cooler City." Environment News Service, September 19, 2008. http://www.ens-newswire.com/ens/sep2008/2008-09-19-092.asp.

"Dallas Toots Its Green Horn." Environment News Service, January 21, 2008. http://www.ens-newswire.com/ens/jan2008/2008-01-21-093.asp.

Faiola, Anthony, and Robin Shulman. "Cities Take Lead on Environment as Debate Drags at Federal Level." *Washington Post*, June 9, 2007, A-1. http://

www.washingtonpost.com/wp-dyn/content/article/2007/06/08/AR2007060802779_pf.html.
Harden, Blaine. "Tree-Planting Drive Seeks to Bring a New Urban Cool: Lower Energy Costs Touted as Benefit." *Washington Post*, September 4, 2006, A-1. http://www.washingtonpost.com/wp-dyn/content/article/2006/09/03/AR2006090300926_pf.html.
"U.S. Mayors Seek Federal Help to Protect Climate." Environment News Service, November 5, 2007. http://www.ens-newswire.com/ens/nov2007/2007-11-05-01.asp.

Climate Change, Speed of

Climatologists have been sharing the disquieting idea that small shifts in global conditions may lead to sudden and abrupt climate changes. The National Academy of Sciences has warned that global warming could trigger "large, abrupt and unwelcome" climatic changes that could severely affect ecosystems and human society (McFarling 2001d, A-30). "We need to deal with this because we are likely to be surprised," said Richard Alley, a Pennsylvania State University climate expert. "It's as if climate change were a light switch instead of a dimmer dial," Alley said (McFarling 2001d, A-30). The report, which was commissioned by the U.S. Global Change Research Program, includes a plea for more research on the links between the land, oceans, and ice that may trigger abrupt change. Alley also suggested that many of today's models of climate change are too simple because they do not include such changes (McFarling 2001d, A-30).

In *Abrupt Climate Change* (2002), Alley, who has become an expert on abrupt climate change, wrote that climate may (and, in the past, has) changed rapidly "[w]hen gradual causes push the Earth system across a threshold" (Alley 2002a, v). The temperature record for Greenland, according to Alley's research, more resembles a jagged row of very sharp teeth than a gradual passage from one climatic epoch to another. According to Alley, "Model projections of global warming find increased global precipitation, increased variability in precipitation, and summertime drying in many continental interiors, including grain belt regions. Such changes might produce more floods and more droughts." (2002a, 114).

"Although abrupt climate change can occur for many reasons, it is conceivable that human forcing of climate change is increasing the probability of large, abrupt events," wrote a team led by Alley (Alley et al. 2003, 2005). At times, they wrote, regional temperature changes one-third to one-half as large as those associated with 100,000-year ice-age cycles have taken place on a time scale of a decade (Alley 2000b, 1331; Alley et al. 2003, 2005). An intense drought that played a major role in destroying classic Mayan civilization may be an example of such a change (Alley et al. 2003, 2005). Regional temperature changes of 8° to 16°C have been known, from the paleoclimatic record, to have occurred within a few years. Such changes have, in the past, been most likely during the beginning or end of ice ages (Alley et al. 2003, 2005).

Alley has described "threshold transitions" comparable to leaning over the side of a canoe: "Leaning slightly over the side of a canoe will cause only a small tilt, but leaning slightly more may roll you and the craft into the lake" (Alley et al. 2003, 2005). At just the right point, a small "forcing" may set into motion a very large climatic change. Thermohaline Circulation and El Niño Southern Oscillation (ENSO) changes may be notable stress points in the global system, they asserted.

James E. Hansen, director of NASA's Goddard Institute for Space Studies, agreed with Alley and his colleagues:

> The occurrence of abrupt climate changes this century is practically certain, if we continue with business-as-usual greenhouse gas emissions. This assertion is based on the magnitude and speed of the human-induced changes of atmospheric composition (which dwarf natural changes) and the rate of global warming that will result from such atmospheric changes (which exceeds any documented natural global warming event). The magnitude of the expected total climate change under business-as-usual is so large that it almost surely passes a number of thresholds. Although, it is impossible to say when these thresholds will be passed, we give examples here of abrupt changes that seem highly likely under business-as-usual and we discuss additional more speculative possibilities. (Hansen 2006b, 30)

During May 1997, 21 nationally prominent ecologists warned President Clinton that rapid climate change due to global warming could ruin ecosystems on which human societies depend. In the United States, the scientists said,

> Rapid climate change could mean the widespread death of trees, followed by wildfires and ... replacement of forests by grasslands. National parks and

> forests could become inhospitable to the rare plants and animals that are preserved there—and where the parks are close to developed or agricultural land, the species themselves may disappear for lack of another safe haven. Worldwide, fast-rising sea levels could inundate the marshes and mangrove forests that protect coastlines from erosion and serve as filters for pollutants and nurseries for ocean fisheries. "The more rapid the rate [of change] the more vulnerable to damage ecosystems will be," the scientists told the president. "We are performing a global experiment [with] little information to guide us." (Basu 1997)

The abrupt nature of climate change can be enhanced by several feedback mechanisms that compound the effects of any single "forcing." Arctic snow and sea ice, for example, cover a large portion of the Earth's surface with a light, reflective surface that reflects sunlight and heat back into space. If large parts of the Arctic Ice Cap melt in the summer, the darker liquid sea surface will accelerate warming.

The possibility of such abrupt changes complicates the task of policymakers in two ways. It could mean that the amount of time available to adjust to climate change will be much shorter than many government officials have believed. It also increases the uncertainty of predictions, indicating that future climate cannot simply be projected forward in a straight line from the present. "We're a little spoiled by the last 30 years," said John M. Wallace, an atmospheric scientist at the University of Washington. "Many years, we're just barely breaking the previous record" (McFarling 2001d, A-30).

A Fundamental Change in the Earth System

"We know from ice-core records and deep-sea sediment records that the earth's climate is capable of changing much more quickly than we had previously thought," said Jeff Severinghaus of the University of California's Scripps Institute of Oceanography (Webb 1998d). "In some cases," said Severinghaus, "[t]he climate warmed abruptly in less than 10 years ... up to possibly 10°C" (Webb 1998d). Severinghaus's findings were presented at the 1998 climate-change conference in Buenos Aires.

Severinghaus continued:

> It is possible that by increasing greenhouse gases, we will induce such a change and that, instead of the smooth warming that's being anticipated over the next 50 years, that we'll instead go along for a while with very little warming and then all of a sudden in a matter of three or five or ten years we'll have a very large catastrophic warming. (Webb 1998d)

According to Dean Edwin Abrahamson, a rapid rise in temperatures could change Earth's ecosystem fundamentally:

> One must go back in time 5 to 15 million years to the late Tertiary to find a time that was 3° or 4°C warmer than now. During periods when there was no permanent pack ice in the Arctic, climatic and vegetational region and boundaries were displaced as much as 1,000 to 2,000 kilometers north of their present position (a displacement which we may replicate during the next 100 years). (1989, 15)

During the period Abrahamson describes, intense aridity was the norm from present-day North and South Dakota to Missouri and Alabama, as well as throughout Central and Southern Africa. These changes may be similar to those that will be experienced by generations to come.

Warming More Rapid Than Expected

Early in the year 2000, scientists working for National Atmospheric and Oceanic Administration (NOAA) released compilations of global temperatures for the last half of the twentieth century that reveal a speed of warming that most climatologists had not expected until late in the twenty-first century. The rate of warming (1°F over the entire century) increased to a rate of 4°F during the century's last quarter, according to calculations of Tom Karl and associates, published in the March 1, 2000, edition of *Geophysical Research Letters.* This is roughly the rate of increase which several climate models at that time had forecast for the second half of the twenty-first century. "The next few years could be very interesting," Karl told the *Los Angeles Times.* "It could be the beginning of a new increase in temperatures" ("Analysis" 2000, 12). Tom Wigley, a senior scientist at the National Center for Atmospheric Research in Boulder, Colorado, countered, saying that warming was strengthened by frequent El Niño events, which he said are not human-induced. "Those months were unusual," he said, "but they weren't unusual due to human influences" ("Analysis" 2000, 12).

The Speed of Change in the Arctic

Evidence from ice cores taken in Greenland has altered the scientific view regarding the speed with which climatic change takes place at higher latitudes. The older assumption that climate changes occur slowly in the polar regions is built into the English language when we say that something moves with "glacial speed." On the contrary, the ice-core record indicates that several rapid warmings and coolings have convulsed the arctic during the last 100,000 years.

The last ice age may have ended much more quickly than previously thought. Severinghaus of the Scripps Institute described a new method of analyzing gases trapped in Greenland's ice sheet that indicates temperatures in the area rose about 16°F within a decade or two at the end of the last great ice age. "The old idea was that the temperature would change over a thousand years, but we found it was much faster," said Severinghaus ("Report" 1999, 15). "We know that over the next one hundred years the Earth will probably warm because of the greenhouse effect," Severinghaus continued. "There is a remote possibility that we might trigger one of these abrupt climate changes. This certainly gives us pause" ("Report" 1999, 15).

Past rapid changes in regional climates in the past took place mainly without human provocation. Factors contributing to these rapid changes may have included variations in the sun's radiance, changes in angle of the Earth's axis, natural variations in atmospheric levels of carbon dioxide and methane (among other "trace" gases), changes in atmospheric water-vapor levels, and dust ejected from volcanoes and other sources. Adams and colleagues commented that

> climate has a tendency to remain quite stable for most of the time and then suddenly "flip," at least sometimes over just a few decades, due to the influence of ... various triggering and feedback mechanisms.... Such observations suggest that even without anthropogenic climate modification there is always an axe hanging over our head, in the form of random very large-scale changes in the natural climate system; a possibility that policy makers should perhaps bear in mind with contingency plans and international treaties designed to cope with sudden famines on a greater scale than any experienced in written history. By starting to disturb the system, humans may simply be increasing the likelihood of sudden events which could always occur.... To paraphrase W. S. Broecker; "Climate is an ill-tempered beast, and we are poking it with sticks." (Adams, Maslin, and Thomas 1999, 1)

Repeated Records Set

Karl and colleagues began a statistical analysis of recent global temperature trends with the observation that between May 1997 and September 1998, 16 consecutive months, global temperatures set observational (e.g., century-scale) monthly records. Their analysis of more than a century of records (roughly 1880 to 2000) indicated that the rate of warming tends to surge upward, then relent a little, and then surge again. "The increase in global mean temperatures is by no means constant" (Karl, Knight, and Baker 2000, 719). Karl and colleagues concluded that "[t]he warming rate over the past few decades [since the mid-1970s] is already comparable to that projected during the twenty-first century based on IPCC business-as-usual scenarios of anthropogenic climate change" (Karl, Knight, and Baker 2000, 719).

> We interpret the results to indicate that the mean rate of warming since 1976 is clearly greater than the mean rate of warming averaged over the late nineteenth and twentieth centuries. It is less certain whether the rate of temperature change has been constant since 1976 or whether the recent string of record-breaking temperatures represents yet another increase in the rate of temperature change.... Moreover, these results imply that if the climate continues to warm at present rates of change, more events like the 1997 and 1998 record warmth can be expected. (Karl, Knight, and Baker 2000, 720, 721)

FURTHER READING

Abrahamson, Dean Edwin. "Global Warming: The Issue, Impacts, Responses." *The Challenge of Global Warming,* 3–34. Washington, DC: Island Press, 1989.

Adams, Jonathan, Mark Maslin, and Ellen Thomas. "Sudden Climate Transitions during the Quaternary." *Physical Geography* 23 (1999):1–36.

Alley, Richard B. "Ice-core Evidence of Abrupt Climate Changes." *Proceedings of the National Academy of Sciences of the United States of America* 97, no. 4 (February 15, 2000b):1331–1334.

Alley, Richard B., ed. *Abrupt Climate Change: Inevitable Surprises.* Committee on Abrupt Climate Change, Ocean Studies Board, Polar Research Board, Board on Atmospheric Sciences and Climate, Division of Earth and Life Sciences, National Research Council. Washington, DC: National Academy Press, 2002a.

Alley, R. B., J. Marotzke, W. D. Nordhaus, J. T. Overpeck, D. M. Peteet, R. A. Pielke Jr., R. T. Pierrehumbert, P. B. Rhines, T. F. Stocker, L. D. Talley, and

J. M. Wallace. "Abrupt Climate Change." *Science* 299 (March 28, 2003):2005–2010.
Alley, Richard B., Peter U. Clark, Philippe Huybrechts, and Ian Joughin. "Ice-Sheet and Sea-Level Changes." *Science* 310 (October 21, 2005):456–460.
"Analysis: Climate Warming at Steep Rate." *Los Angeles Times* in *Omaha World-Herald*, February 23, 2000, 12.
Basu, Janet. Ecologists' Statement on the Consequences of Rapid Climatic Change, May 20, 1997. http://www.dieoff.com/page104.htm.
Hansen, James E. "Declaration of James E. Hansen." *Green Mountain Chrysler-Plymouth-Dodge-Jeep, et al., Plaintiffs v. Thomas W. Torti, Secretary of the Vermont Agency of Natural Resources, et al., Defendants.* Case Nos. 2:05-CV-302 and 2:05-CV-304, Consolidated. United States District Court for the District of Vermont. August 14, 2006b. http://www.giss.nasa.gov/~dcain/recent_papers_proofs/vermont_14aug20061_textwfigs.pdf.
Karl, Thomas R., Richard W. Knight, and Bruce Baker. "The Record-breaking Global Temperatures of 1997 and 1998: Evidence for an Increase in the Rate of Global Warming." *Geophysical Research Letters* 27 (March 1, 2000):719–722.
McFarling, Usha Lee. "Scientists Now Fear 'Abrupt' Global Warming Changes: Severe and 'Unwelcome' Shifts Could Come in Decades, not Centuries, National Academy Says in an Alert." *Los Angeles Times*, December 12, 2001d, A-30.
Warrick, Jody. "Earth at Its Warmest in Past 12 Centuries: Scientist Says Data Suggest Human Causes." *Washington Post*, December 8, 1998. http://www.asoc.org/currentpress/1208post.htm.
Webb, Jason. "World Temperatures Could Jump Suddenly." Reuters, November 4, 1998d. http://bonanza.lter.uaf.edu/~davev/nrm304/glbxnews.htm.

Climatic Equilibrium (E-folding Time)

The climate system is never actually in thermodynamic equilibrium. Rather, it is forever playing catch-up with the daily and seasonal variations of incident sunlight, as the ground tries to come into thermal equilibrium with changes in solar radiation. This is where the heat capacity of the ground, the atmosphere, and the ocean come into play, as well as the thermal opacity of the atmosphere, which regulates how readily heat energy from the ground can be radiated to space.

The time it takes the system to get to a new equilibrium is expressed in terms of a time constant, or "e-folding time"—that is, the length of time that it takes the system to reach approximately 63 percent of its final equilibrium temperature. Mathematically, it takes forever to reach true equilibrium, but practically, after a few e-folding times, the system can be said to be in effective equilibrium. The e-folding time of the atmosphere is a few months, the ocean mixed layer a few years, and the total (deep) ocean a few hundred years. Present climate has accumulated about 0.5 watts per meter squared of unrealized warming.

In 1989, Veerabhaadran Ramanathan stated that

> The rate of decadal increase of the total radiative heating of the planet is now about five times greater than the mean rate from the early part of this century. Non-CO_2 trace gases in the atmosphere are now adding to the greenhouse effect by an amount comparable to the effect of CO_2 increase.... The cumulative increase in the greenhouse forcing until 1985 has committed the planet to an equilibrium warming of about 1° to 2.5°C. (Ramanathan 1989, 241)

Ramanathan continued:

> The climate system cannot restore the equilibrium instantaneously, and hence the surface warming and other changes will lag behind the trace-gas increase. Current models indicate that this lag will range [from] several decades to a century. However, analyses of temperature records of the last 100 years as well as proxy records ... [of] paleoclimate changes indicate that climate changes can also occur abruptly instead of a gradual return to equilibrium as estimated by models. The timing of the warming is one of the most uncertain aspects of the theory. (Ramanathan 1989, 245)

At about the same time, Peter Ciborowski projected that "within 50 years, we will be committed to a mean global temperature rise of 1.5°C to 5°C. And if no attempt is made to slow the rate of increase, we could be committed to another 1.5°C to 5°C in another 40 years" (Ciborowski 1989, 227–228). *See also:* Feedback Loops

FURTHER READING

Ciborowski, Peter. "Sources, Sinks, Trends, and Opportunities." In *The Challenge of Global Warming,* ed. Edwin Abrahamson, 213–230. Washington, DC: Island Press, 1989.
Ramanathan, V. "Observed Increases in Greenhouse Gases and Predicted Climatic Changes." In *The Challenge of Global Warming,* ed. Edwin Abrahamson, 239–247. Washington, DC: Island Press, 1989.

Clouds and Evaporation

Increasing warmth causes more evaporation, which may affect patterns of cloud formation.

Will additional clouds trap more heat, or reduce warming by deflecting sunlight? Will areas with fewer clouds heat further because they receive more sunlight? Climate-change skeptic Richard Lindzen, who advocates the "Iris effect," has argued that increasing evaporation will allow heat to escape into space, effectively nullifying warming caused by rising carbon-dioxide levels in the lower atmosphere. "The more scientists look at the indirect effects of human emissions on clouds, the more convinced they are that the effects are large," said James E. Hansen, the director of NASA's Goddard Institute for Space Studies (Revkin 2001c, F-4).

According to satellite studies and computer models, increasing evaporation associated with warming may cause the tropics' canopy of clouds to shrink. Scientists at NASA Langley Research Center report that clearer tropic skies will allow more sunlight to reach the Earth's surface. "The result is that the 'Iris effect' slightly warms the Earth instead of strongly cooling it," said Bruce Wielicki of NASA (Wielicki et al. 2002, 841).

Robert J. Charlson and colleagues wrote in *Science*:

> Man-made aerosols have a strong influence on cloud albedo, with a global mean forcing estimated to be of the same order (but opposite in sign) as that of greenhouse gases, but the uncertainties associated with the aerosol forcing are large. Recent studies indicate that both the forcing and its magnitude may be even larger than anticipated.... making the largest uncertainty in estimating climate forcing even larger. (Charlson et al. 2001, 2025–2026)

Human activity, while increasing emissions of greenhouse gases, also alters ways by which some clouds form, perhaps causing them to screen the sun and partially cool the Earth by changing the size and density of water droplets. Researchers at the California Institute of Technology and the University of Washington have described their work along this line. "Almost no work is actually being done to model in detail how clouds respond to the polluted climate," said O. Brian Toon, an atmospheric scientist at the University of Colorado who was not involved in this study (Revkin 2001c, F-4). Clouds exert their strongest cooling influence when they are dense and composed of small droplets. A cloud with a large number of droplets blocks more light. Many scientists have assumed that particles in the air serve as "seeds" on which water vapor can condense. This view, taken to its logical conclusion, would mean that clouds would not form in an atmosphere bereft of "condensation nuclei" provided by sea salt, dust, and pollution.

An analysis led by John H. Seinfeld, a professor of chemical engineering at Caltech, and Robert J. Charlson, of the University of Washington, asserted that certain acids and organic compounds can limit the growth of individual cloud droplets. "If each droplet is constrained from growing ... the same amount of condensing water ends up spread out among a higher number of smaller droplets. The result is a more reflective cloud," the study said (Revkin 2001c, F-4). Nitric acid, a common by-product of emissions from vehicles, is one of the strongest shapers of this process; others include several organic molecules produced when fuels and wood are burned. "The work has been restricted to computer models, so the team did not try to calculate just how large the cooling effect of these other kinds of pollution might be relative to the warming effect of heat-trapping gases" (Revkin 2001c, F-4).

Reduced Cloud Cover in the Tropics

Global warming reduced cloud cover over the tropics during the 1990s, NASA researchers have found. After examining 22 years of satellite measurements, the researchers concluded that more sunlight entered the tropics and more heat escaped to space in the 1990s than during the 1980s because less cloud cover blocked incoming radiation and trapped outgoing heat. "Since clouds were thought to be the weakest link in predicting future climate change from greenhouse gases, these new results are unsettling," said Bruce Wielicki of NASA's Langley Research Center. "It suggests that current climate models may, in fact, be more uncertain than we had thought," Wielicki added. "Climate change might be either larger or smaller than the current range of predictions" ("Warming Tropics" 2002).

The observations capture changes in the radiation budget—the balance between Earth's incoming and outgoing energy—that controls the planet's temperature and climate. A research group at NASA Goddard Institute for Space Studies developed a new method of comparing the satellite observed changes to other meteorological data. "What it shows is remarkable," said Wielicki. "The rising and descending motions of air that cover the entire tropics, known as the Hadley and Walker circulation cells, appear to increase in strength from the 1980s to the 1990s. This suggests that the tropical heat engine increased its speed." The faster circulation reduced the amount of water vapor that is required for cloud formation in the upper troposphere over the most northern and southern tropical areas. Less cloudiness formed allowing more sunlight to enter and more heat to leave the tropics ("Warming Tropics" 2002).

Interactions between aerosols and clouds are important drivers of climate change, but have been poorly understood, provoking much scientific study. The effects of biomass burning on clouds in the Amazon valley have come under particular scrutiny. One study examined the suppression of boundary-level clouds (popularly called fair-weather cumulous) in smoky areas during the dry season. Other studies (Andreae et al. 2004; Graf 2004) examined clouds that seem to emit smoke during the transition between dry and wet seasons, as a by-product of biomass burning. The transport of such pollution upward in the atmosphere may make thunderstorms more violent.

FURTHER READING

Andreae, M. O., D. Rosenfeld, P. Artaxo, A. A. Casta, G. P. Frank, K. M. Longo, and M. A. F. Silva-Dias. "Smoking Rain Clouds over the Amazon." *Science* 303 (February 27, 2004):1337–1342.

Charlson, Robert J., John H. Seinfeld, Athanasios Nenes, Markku Kulmala, Ari Laaksonen, and M. Cristina Facchini. "Reshaping the Theory of Cloud Formation." *Science* 292 (June 15, 2001):2025–2026.

Graf, Hans-F. "The Complex Interaction of Aerosols and Clouds." *Science* 303 (February 27, 2004):1309–1311.

Revkin, Andrew C. "Both Sides Now: New Way That Clouds May Cool." *New York Times*, June 19, 2001c, F-4.

"Warming Tropics Show Reduced Cloud Cover." Environment News Service, February 1, 2002. http://ens-news.com/ens/feb2002/2002L-02-01-09.html.

Wielicki, Bruce A., Takmeng Wong, Richard P. Allan, Anthony Slingo, Jeffrey T. Kiehl, Brian J. Soden, C. T. Gordon, Alvin J. Miller, Shi-Keng Yang, David A. Randall, Franklin Robertson, Joel Susskind, and Herbert Jacobowitz. "Evidence for Large Decadal Variability in the Tropical Mean Radiative Energy Budget." *Science* 295 (February 1, 2002):841–844.

Coal and Climatic Consequences

Ninety percent of Earth's remaining fossil-fuel reserves are in the form of coal, the most dangerous fossil fuel from a greenhouse point of view, because most coals produce roughly 70 percent more carbon dioxide per unit of energy generated than natural gas, and about 30 percent more than oil. Coal is also the most plentiful fossil fuel, especially in places with large populations, such as China, which controls 43 percent of remaining reserves. During the 1980s, China passed the Soviet Union as the world's largest coal producer. China also built 114,000 megawatts of coal-fired power in 2006 and 95,000 more in 2007. Coal poses environmental problems other than carbon emissions. The mining of coal also produces methane; its combustion produces sulfur dioxide and nitrous oxides, as well as carbon dioxide. Transport of coal also usually requires more energy than any other fossil fuel.

Many scientists agree that greenhouse-gas emissions worldwide will have to be reduced by at least 70 percent by 2050 to stabilize temperature rises at less-than-dangerous levels. Increasingly, a consensus has been forming that the single most compelling policy change to forestall dangerous warming will be to halt construction of coal-fired power plants pending development of adequate technology to clean and sequester (store) their emissions.

Coal is the most widely used fuel for electricity generation worldwide. Almost 40 percent of carbon-dioxide emissions in the United States result from coal-fired electricity generation. Between 2002 and 2008, worldwide consumption of coal rose by 30 percent, two-thirds of which went into power generation, the rest went into industries such as steel and cement manufacturing. China was adding one to two coal-fired electricity plants per week. Other countries also were adding coal-fired power. For example, as of late 2008, Eskom, South Africa's national electric utility, was building six coal-fired power plants of 800 megawatts each.

A Moratorium on Coal-fired Electricity Generation

James E. Hansen, director of NASA's Goddard Institute for Space Studies, has proposed a moratorium on construction of new coal-fired power plants until technology for carbon-dioxide capture and sequestration is available. About a quarter of power plants' carbon-dioxide emissions will remain in the air "forever," that is, more than 500 years, long after new technology is refined and deployed. As a result, Hansen expects that all power plants without adequate sequestration will be obsolete and slated for closure (or at least retrofitting) before mid-century (Hansen 2007b).

Hansen believes that

> [c]oal will determine whether we continue to increase climate change or slow the human impact. Increased fossil fuel CO_2 in the air today, compared to the pre-industrial atmosphere, is due 50 percent to coal, 35 percent to oil and 15 percent to [natural] gas. As oil resources peak, coal will determine future CO_2 levels. Recently, after giving a high school commencement talk in my hometown, Denison, Iowa, I drove from Denison to Dunlap, where my parents are buried. For most of 20 miles there were trains parked, engine to caboose, half of the cars being filled with coal. If we cannot stop the building of more coal-fired power plants, those coal trains will be death trains—no less gruesome than if they were boxcars headed to crematoria, loaded with uncountable irreplaceable species. (Hansen 2007c)

By the beginning of 2008, the European Commission was weighing whether to require new power stations to include facilities that will retrofit to store greenhouse-gas emissions via carbon capture and sequestration (CCS) technology when it is available, the first legal move of this type in the world, a large step toward making CCS a commercial reality. The requirement as written does not contain a date on which actual CCS would be required. Installation of CCS technology, now still in its infancy, could reduce global carbon-dioxide emissions by one-third by 2050, if widely deployed (Schiermeier 2008a, 232). At present, Norway, Britain, China, and the United States are planning CCS pilot plants.

Direct political action has been heating up over coal-fired power. In the United Kingdom, the "Kingsnorth Six" were acquitted by a Crown Court jury during September 2008. They were among 23 Greenpeace volunteers who attempted to shut down the Kingsnorth coal-fired power plant, which plans say will become Britain's first new coal-fired plant in more than a decade. The six painted the smokestack with "Gordon Bin It" before they were interrupted by police. Their defense (upheld by the jury) was "lawful excuse," that they were protecting property of greater value (the Earth!) from the impact of climate change. With briefs submitted by James Hansen and others, the defendants argued that coal has a dominant role in a warming climate that poses a clear and present danger.

Environmental groups tightened their focus on proposed coal-fired power plants during 2007. Environmental Defense and the Natural Resources Defense Council assembled "strike forces" to mobilize opposition to new plants state by state. These strike forces played a role in obtaining cancellation of plants in Florida and Texas. In New Mexico, for example, the groups intervened into a dispute over whether to constrict a new power plant on the Navajo Nation, where the state government, which is opposed to the project, has no direct power to prevent it. The plant's carbon footprint would equal 1.5 million average automobiles. Coal-fired electricity contributes more than half of the 57 million tons of annual carbon-dioxide emissions in New Mexico. Together, two existing plants in that state emit 29 million tons (Barringer 2007b).

On June 30, 2008, Thelma Wyatt Cummings Moore, a Fulton County, Georgia, Superior Court judge blocked construction of the first coal-burning power plant proposed in Georgia in more than 20 years, ruling that the plant must limit its emissions of the heat-trapping gas carbon dioxide. This was the first time that a court had applied an April 2007 ruling of the U.S. Supreme Court recognizing that carbon dioxide is a pollutant under the federal Clean Air Act to an industrial source. The judge overturned the Georgia Environmental Protection Division's approval of an air-pollution permit for Dynegy's proposed Longleaf power plant south of Columbus, Georgia.

"In a case that is being watched across the country, Judge Moore has sent a message that it is not acceptable for the state to put profits over public health," said Justine Thompson, executive director of GreenLaw, the Atlanta public interest law firm that represented the environmental groups ("Georgia Judge" 2008).

Opposition to New Coal Plants Accelerates

In 2007, opposition to new coal plants accelerated. In early August, Missoula, Montana's mayor, John Engen (Democrat), won city council support to buy electricity from a new coal-fired plant starting in 2011, to save the city money. He then was inundated by hundreds of e-mails and phone calls from protesting constituents.

Late in July, Senate Majority Leader Harry M. Reid told chief executives of four power companies he would "use every means at [his] disposal" to stop plans to build three coal-fired plants in Nevada. "There's not a coal-fired plant in America that's clean. They're all dirty," Reid said, urging a turn to wind, solar, and geothermal power in an to slow climate change (Mufson 2007f, D-1).

In June, a unanimous vote of Florida's Public Service Commission rejected a Florida Power & Light proposal to build a coal-fired plant near Lake Okeechobee. Florida Governor Charlie Crist said approvingly that the Public Service Commission "sent a very powerful message" and that the state "should look to solar and wind and nuclear as alternatives to the way we've generated power in the Sunshine State" (Mufson 2007f, D-1).

In July, Citigroup downgraded the stocks of all coal companies. "Prophesies of a new wave of coal-fired generation have vaporized, while clean coal technologies ... remain a decade away, or more," their report said. The Citigroup analysts said that by 2008 "election politics are likely to turn progressively more bestial for coal. Candidates are already stepping up to 'ban coal'." The Citigroup report said that coal producers' earnings would probably be hurt by "new regulatory mandates applied to a group perceived as landscape-disfiguring global warming bad guys" (Mufson 2007f, D-1).

By mid-2007, several new coal-powered generating plants were being canceled or postponed across the United States. By that time, 645 coal-fired plants were producing about half of the country's electricity; as recently as May 2007 more than 150 new ones had been planned to meet electricity demand that was rising at a 2.7 percent annualized rate. A private equity deal worth $32 billion involving TXU Corporation canceled 8 of 11 planned coal plants, as similar plants were scuttled in Florida, North Carolina, Oregon, and other states. In the meantime, late in August 2008, Xcel Energy of Minneapolis became the first builder of coal-fired power plants that agreed with New York State Attorney General Andrew Cuomo to provide investors with an analysis of global warming risks posed by its business.

Climate-change concerns were often cited as coal plants were canceled, especially in Florida, where rising sea levels from melting ice in the Arctic, Antarctic, and mountain glaciers is already eroding coastlines. Florida's Public Service Commission is now legally required to give preference to alternative energy projects over new fossil-fuel generation of electricity. The states of Washington and California have been moving toward similar requirements. Xcel Energy and Public Service of Colorado were allowed to go ahead with a 750-megawatt coal-fired power plant only after it agreed to obtain 775 megawatts of wind power.

In March 2008, the federal government suspended its loan program for new coal plants in rural communities through 2009 due in part to uncertainty over a suit filed by Earthjustice during July 2007. The suit cited the government's failure to consider global warming as it financed new coal plants.

Even as climate experts called for reduction of coal use, Enel, Italy's largest electricity producer, raised its reliance on coal to 50 percent. Italy as a whole, switching from expensive oil to cheaper and dirtier coal, increased its use from 14 percent to 33 percent (Rosenthal 2008). Europe as a whole was turning to coal with plans for 50 new coal-fired plants over the next five years. Scattered protests have arisen in Civitavecchia, Germany, the Czech Republic, and the Kingsnorth power station in Kent, England.

"In order to get over oil, which is getting more and more expensive, our plan is to convert all oil plants to coal using clean-coal technologies," said Gianfilippo Mancini, Enel's chief of generation and energy management. "This will be the cleanest coal plant in Europe. We are hoping to prove that it will be possible to make sustainable and environmentally friendly use of coal" (Rosenthal 2008). "Clean coal" reduces visual pollution but not carbon-dioxide emissions. The kind of CCS that Hansen advocates is not now commercially available.

"Building new coal-fired power plants is ill conceived," said Hansen. "Given our knowledge about what needs to be done to stabilize climate, this plan is like barging into a war without having a plan for how it should be conducted, even though information is available." He added, "We need a moratorium on coal now with phase-out of existing plants over the next two decades" (Rosenthal 2008).

Emission-Reducing Technologies

Many technologies have been developed to remove carbon dioxide from the emissions of coal-fired plants; one popular approach is the Integrated Gasification Combined Cycle (IGCC), which creates hydrogen and CO_2 to be sequestered. Some technologies remove the carbon dioxide from the flue stream after combustion. All of these technologies (and others) require energy and cost money and, as a result, would probably raise electric bills 40 to 50 percent, barring an unforeseen technological breakthrough.

Additionally, underground sequestration will succeed only in areas where geology is suitable. Thus, space is limited, and not always located in the same areas as the power plants. Errant carbon dioxide can be dangerous. In 1986, 300,000 tons of naturally occurring carbon dioxide that had been trapped in Cameroon's Lake Nyos rose to the surface, suffocating 1,700 people (Goodell 2006).

During January 2008, the U.S. Energy Department canceled the country's biggest carbon-capture demonstration project, in Illinois, because of expense. The costs of the project, undertaken in 2003 with a budget of $950 million, had spiraled to $1.5 billion that year, and it was far from complete (Rosenthal 2008).

One experimental approach at the Warrior Run power plant in Cumberland, Maryland, captures carbon dioxide from its boiler and sells it to beverage gas distributors for use mainly in soft drinks. "If you've had a Coke today, you've probably ingested some of our product," said plant manager Larry Cantrell. The problem is that four megawatts of the plant's output (about 200 megawatts) goes to remove 5 percent of its carbon-dioxide emissions (Kintisch 2007b, 185). *See also:* Fossil Fuels

FURTHER READING

Barringer, Felicity. "Navajos and Environmentalists Split on Power Plant." *New York Times*, July 27, 2007b. http://www.nytimes.com/2007/07/27/us/27navajo.html.

"Georgia Judge Yanks Coal Power Permit on Climate Concerns." Environmental News Service, June 30, 2008. http://www.ens-newswire.com/ens/jun2008/2008-06-30-091.asp.

Goodell, Jeff. "Our Black Future." *New York Times*, June 23, 2006. http://www.nytimes.com/2006/06/23/opinion/23goodell.html.

Hansen, James E. "Political Interference with Government Climate Change Science." Testimony of James E. Hansen, 4273 Durham Road, Kintnersville, PA, to Committee on Oversight and Government Reform United States House of Representatives, March 19, 2007b.

Hansen, James E. "Coal Trains of Death." James Hansen's e-mail List, July 23, 2007c.

Hansen, James E. "Dear Chancellor [Merkel, of Germany], Perspective of a Younger Generation." James Hansen's e-mail list, January 23, 2008a.

Hansen, James E. "Written Testimony, Kingsnorth Case." Accessed September 11, 2008d. http://www.columbia.edu/~jeh1/mailings/20080910_Kingsnorth.pdf.

Kintisch, Eli. "Making Dirty Coal Plants Cleaner." *Science* 317 (July 13, 2007b):184–186.

Mufson, Steven. "Coal Rush Reverses, Power Firms Follow Plans for New Plants Stalled by Growing Opposition." *Washington Post*, September 4, 2007f, D-1. http://www.washingtonpost.com/wp-dyn/content/article/2007/09/03/AR2007090301119_pf.html.

Schiermeier, Quirin. "Europe to Capture Carbon." *Nature* 451 (January 17, 2008a):232.

Coelacanth

The coelacanth, a "living fossil" fish that has been swimming the seas for 400 million years, has been threatened by changes in ocean temperatures that were leading to destruction of life-nurturing coral-reef systems. "The coelacanths are vulnerable and global warming could affect them adversely," said Horst Kleinschmidt, deputy director general of South Africa's Department of Environmental Affairs and Tourism (Stoddard 2001, B-4). Previously assumed to be extinct, a colony of living coelacanth have been found in waters near South Africa, living in the reefs of Sodwana, where bleaching prompted by warming ocean temperatures is threatening the ecosystem.

FURTHER READING

Stoddard, Ed. "Global Warming Threatens 'Living Fossil' Fish: Coelacanths Have Existed 400 Million Years." Reuters in *Ottawa Citizen*, July 14, 2001, B-4.

Consensus, Scientific

While a lively debate in political circles and the media still questions whether human activity is significantly warming the Earth, scientific evidence has been accumulating steadily in support of the idea. Much of this evidence, unobscured by special economic interests that sometimes cloud popular debate, is not at all ambiguous. With the exception of a small minority, the human role in a rapid warming of the Earth has become nearly incontrovertible.

To a consensus of scientists who make up the Intergovernmental Panel on Climate Change (IPCC), a United Nations committee, global warming is no longer a matter of whether, but how much, how soon, and with how much damage to the Earth's flora and fauna, including humankind. Many of the scientists who are closest to the evidence are urging policymakers to decide what is to be done, and, failing remedies on a global scale, what price will be paid by future generations.

In 1997, a group of prominent scientists issued this warning:

> Our familiarity with the scale, severity, and costs to human welfare of the disruptions that the climatic changes threaten leads us to introduce this note of urgency and to call for early domestic action to reduce U.S. emissions via the most cost-effective means. We encourage other nations to join in similar actions with the purpose of producing a substantial and progressive global reduction in greenhouse-gas emissions, beginning immediately. We call attention to the fact that there are financial as well as environmental advantages to reducing emissions. More than 2,000 economists recently observed that there are many potential policies to reduce greenhouse-gas emissions for which total benefits outweigh the total costs. ("Scientists' Statement" 1997)

Global warming has been no secret to atmospheric scientists. Articles on the subject began to appear occasionally in the scientific literature of meteorology and other fields during the middle 1970s. A scientific consensus was forming as early as October 1985, at a conference in Villach, Austria, where a consensus statement began: "As a result of the increasing concentrations of greenhouse gases, it is now believed that in the first half of the next century a rise in global mean temperature could occur which is greater than any in man's history" ("Scientific Consensus" 1989, 63)

Other conferences on global warming were convened in Villach, Austria (October 1987), and Bellagio, Italy (November 1987). An executive summary of these meetings said that "[i]t is now generally agreed that if present trends of greenhouse-gas emissions continue during the next hundred years, a rise global mean temperature could occur that is larger than any experienced in human history" (Jager 1989, 97). This summary also projected that the most extreme increases would come in the northern latitudes during the winter, a forecast that was being borne out by observations decades later.

FURTHER READING

Jager, Jill. "Developing Policies for Responding to Climate Change." In *The Challenge of Global Warming*, ed. Dean Edwin Abrahamson, 96–109. Washington, DC: Island Press, 1989.

"Scientific Consensus: Villach (Austria) Conference." In *The Challenge of Global Warming*, Edwin Abrahamson, 63–67. Washington, DC: Island Press, 1989.

"Scientists' Statement on Global Climatic Disruption." Statements on Climate Change by Foreign Leaders at Earth Summit, June 23, 1997. http://uneco.org/Global_Warming.html.

Contrarians (Skeptics)

Three main lines of disagreement exist between those who believe that global warming is a major problem (and will become more serious in the future) and those who do not. No one is arguing seriously over whether it has been getting warmer in the short term. The temperature record worldwide during the last few decades is unambiguous on that score.

The pace of increase has slowed during the last decade, however, which is probably due to a strong El Niño pattern being replaced by a cooling La Niña, a short-term effect that raises and lowers global temperatures year by year. The rise in temperatures is never linear; greenhouse gases are important, but other factors influence climate as well. Some contrarians even herald a new ice age.

One point of departure in the debate involves whether the increase in temperatures will be long term or short term. Another involves the degree to which human beings are complicit in the warming trend. The most important disagreement, however, involves the relative importance of various "forcings"—climate science's term for influences—in the workings of Earth's atmosphere.

Most scientists assign greenhouse gases (of which carbon dioxide is the most plentiful) a major role in holding atmospheric heat. Methane has a role, as do several other "trace" gases.

Carbon dioxide, methane, and other greenhouse gases have no party affiliation. They have no hopes, nor morals. They are not arguing with us over how soon their abundance in the atmosphere will push the Earth past some ominous "tipping point," toward their runaway dominance of our climate. Carbon dioxide and methane do not worry whether their overabundance will kill polar bears. All they do is retain heat. The more of them we emit, the more heat they absorb.

The contrarians, as a rule, deny that greenhouse gases contributed by human activity have an important role in climate change. They believe that solar activity, oscillations in Earth's orbit, and other "forcings" have more influence. These influences, after all, have done a good job of putting Earth through its cyclical paces, from massive ice sheets to nearly ice-free states during the last few million years without a whiff of human intervention.

The Skeptics' Parallel Climatic Universe

Contrarians (they still prefer "skeptics") gathered at the International Conference on Climate Change in New York City, sponsored by the Heartland Institute, March 2–4, 2008. Summoned by large advertisements in the *New York Times*, the event drew about 400 people (including 98 speakers) to the Marriott New York Marquis Times Square Hotel, drawn by the theme that no consensus exists on the causes or likely consequences a of global warming.

Participants at this conference enter a parallel universe, in which humanity bears no blame for rising temperatures, and consequences, if any, will be benign. Temperatures may be rising, but it is a natural cycle, so goes the mantra, and a corrective decline—an ice age, perhaps!—will soon prove the alarmists stoked by Al Gore massively wrong. This is the party line. Extreme weather patterns (including hurricanes) have nothing to do with a hydrological cycle addled by heat, and, whatever you do, do not blame any problems on the rising atmospheric levels of carbon dioxide and methane.

The sun is this group's designated driver of climate change, and they believe that it soon will change its spots, and end what the contrarians believe is a short-term rise in temperatures. With greenhouse gases relegated to a minor role in climate change, the contrarians feel no reason to worry about thermal inertia, feedback loops, or looming mass extinctions of flora and fauna.

"Laughing in the Face of Impending Apocalypse"

The conference's sessions featured deconstruction of climate models galore, bright prospects for polar bears, and many analyses of the Kyoto Protocol's uselessness in doing much about climate change. The sessions boost solar cycles' influence on Earth's climate and other variations in climate provoked by just about anything other than human combustion of fossil fuels, and provide coaching on how to impede climate-change legislation.

It is all attended by a certain self-assured cockiness—even with its own comedian, Tim Slagle, "Laughing in the Face of Impending Apocalypse," and movies with such titles as *The Great Global Warming Swindle* and *Global Warming, or Global Governance*, which invoke the John Birch Society's nightmare framing the United Nations in a plot to enslave the United States, updated to include the Intergovernmental Panel on Climate Change (IPCC), a U.N. body. If carbon dioxide had a sense of humor, it would get a laugh or two out of all of this.

During the three-day event, Robert Balling, professor of climatology at Arizona State University, spoke on "The Increase in Global Temperatures: What It Does and Does Not Tell Us." Iain Murray of the Competitive Enterprise Institute discoursed on "Assuming the Worst: A No-Regrets Approach to Global Warming." Gerd-Rainer Weber, Ph.D., consulting meteorologist, Essen, Germany, held forth on inconsistencies in IPCC Projections.

There was much more. Myron Ebell, director of energy and global warming policy, Competitive Enterprise Institute, talked about "The Alarmist Consensus: Global Warming Is Not a Crisis." Stan Goldenberg, meteorologist, Hurricane Research Division, NOAA, told listeners about "The Mythical Link between Hurricanes and Global Warming." Physicist Russell Seitz exonerated coal-fired electric power generation from complicity with carbon overload in "Coal Power Not at Odds with Reducing Carbon Emissions."

The Heartland Institute raises "Unanswered Questions," especially about the usefulness of models, as well as paleoclimatic proxies. Undermining scientific tools tends to erode the basis of argument for future warming. The Institute's climate contrarians have formed a countergroup to the IPCC that calls itself the Nongovernmental International Panel on Climate Change (NIPCC). This group issued a report on March 3, 2008, pitching natural cycles as the primary cause of recent warming. The NIPCC document listed 23 authors from 15 nations. "We reached the opposite conclusion," said Fred Singer, who edited the report (and has written widely on the same theme). The effect of human-generated greenhouse gas emissions, he said, is "not a cause for concern, at least not yet. If the evidence changes, I'll change my opinion, okay?" (Eilperin 2008a, A-16).

Heartland Institute President Joseph L. Bast said he believes skeptics are now on the verge of overturning the idea that humans are driving climate change. "Yeah, I think we're at a tipping point that's going in exactly the opposite direction," Bast said. "That point of view has had its moment. That moment has passed" (Eilperin 2008a, A-16).

Global Warming as Mass Hysteria?

Bret Stephens, "Global View" columnist on the *Wall Street Journal's* editorial page (the world capital of global-warming denial), has got global warming figured out. It is, he wrote July 1, 2008, "a 20-year-old mass hysteria phenomenon" (Stephens 2008, A-15). He was expressing a popular point of view among climate-change contrarians, who believe that increasing greenhouse-gas levels ion the atmosphere are not an environmental threat.

"Science" comes up once in Stephens' piece, before the words "has been discredited." He said James E. Hansen, NASA scientist who has warned often of global warming's dangers, is doing nothing more than "banging the gong."

"Mother nature," wrote Stephens, "has her own opinions." "Global warming," opined Stephens, "isn't science." "The real place where discussions of global warming belong is in the realm of belief" (Stephens 2008, A-15). First, he argues, it is a socialist, anticonsumerist rebuke to capitalism, a "leftist nostrum," along with "population, higher taxes, a vast new regulatory regime, global economic redistribution, an enhanced role for the United Nations." His second explanation is theological: western religions have a preference for eternal damnation, penance for our fossil-fueling sins. "Global warming," he concluded," "is sick-souled religion" (Stephens 2008, A-15).

Questioning the Motives of "Global Warmists"

Some of the skeptics do not restrict themselves to debating points of scientific veracity. Occasionally, the debate descends to questions of personal or political motivation. The IPCC, as a deputy body of the United Nations, sometimes becomes the target of a general fear, prominent in conservative U.S. political circles since the days of the John Birch Society, that climate diplomacy, and especially the Kyoto Protocol, is a cover for an international plot to strip the United States of its national sovereignty. Other critics argue that the members of the IPCC and other "global warmists" are ringing imaginary alarm bells to gain fame.

In the foreword to Robert C. Balling's book, *The Heated Debate: Greenhouse Predictions Versus Climate Reality* (1992), S. Fred Singer wrote that "[t]hese [forecast] disasters are not grounded in fact, but spring from the feverish imagination of activists and their ideological desire to impose controls on energy use and to stop—or at least micromanage—economic growth" (Balling 1992, vii)

Singer's point of view has been borrowed, with his usual rhetorical flourish, by Rush Limbaugh, who insists that the whole global warming case is fiction invented by "environmental whackos." Since this point of view conceives of global warming as a problem of (at most) a few tenths of a degree's worth of temperature and a couple of inches of sea level, Singer and Limbaugh believe that warming will be good for the economy. It will, they say, increase agricultural production.

The *real* problem, wrote Singer, is that the Earth is overdue for a new ice age. Burning more carbon-based fuels may help human civilization escape this new ice age, Singer asserted.

Contrarians' Distrust of Climate Models

Nearly all the prominent skeptics distrust climate modeling implicitly. A favorite article of faith among skeptics is: if they can't forecast the weather five days in advance, what about a century? Additionally, many skeptics argue that the surface-temperature record is woefully incomplete, so incomplete that indications of warming may be bogus. During the late 1990s, the most sophisticated climate models tallied surface temperature readings of 20 to 40 percent of the Earth's land surface. No readings are taken in many large regions of the oceans.

Comparisons of real-world observations to models reveal that, if anything, they tend to be conservative about projections of global warming. Stefan Rahmstorf of the Potsdam Institute for Climate Impact Research, having undertaken such an exercise, commented:

> We present recent observed climate trends for carbon dioxide concentration, global mean air

> temperature, and global sea level, and we compare these trends to previous model projections as summarized in the 2001 assessment report of the Intergovernmental Panel on Climate Change (IPCC). The IPCC scenarios and projections start in the year 1990, which is also the base year of the Kyoto Protocol, in which almost all industrialized nations accepted a binding commitment to reduce their greenhouse gas emissions. The data available for the period since 1990 raise concerns that the climate system, in particular sea level, may be responding more quickly to climate change than our current generation of models indicates. (Rahmstorf et al. 2007, 709)

Much of Michael L. Parsons' *Global Warming: The Truth Behind the Myth* (1995) is a detailed attack on climate modeling. Parsons argues that the models are not sophisticated enough to tell us much of anything about future climate trends. Parsons subscribes to the viewpoint that "Models are like sausages. You don't want to know what goes into them." Parsons accuses "warmists" of using climate models to confirm predetermined conclusions.

Thomas R. Karl, a senior scientist at NOAA's National Climatic Data Center in Asheville, North Carolina, has analyzed weather data for the United States (excluding Alaska and Hawaii) over this century to determine whether observed changes are consistent with models for greenhouse warming. Karl created a greenhouse climate response index (GCRI) to measure how well actual observations fit the models. This index is the average of four indicators: the percentage of the United States with much-above-average minimum temperatures, the percentage with much-above-average precipitation during the cold season, the percentage of the area in extreme drought during the warm season, and the percentage of area with a much-greater-than-average proportion of precipitation derived from extreme one-day events.

Karl found that since 1976, GCRI values have been higher than the average for previous years in the century, indicating that "[t]he late-century changes in the U.S. climate are consistent with the general trends anticipated from a greenhouse-enhanced atmosphere.... There is only a 5 to 10 percent chance that the increase in GCRI results stem from natural variability" (Hileman 1995). Karl continued:

> Night-time temperatures have generally increased more than daytime temperatures. Climatic variability, or the frequency of extreme events, has increased in some regions, although it is not known whether these have risen on a worldwide scale. For example, heavy rainfall events in the U.S. have increased in intensity, and behavior of El Niño—the warming of the eastern equatorial Pacific that sometimes brings severe droughts to some regions and heavy rains to other regions around the world—has been unusual since 1976. These occurrences fit well with complex mathematical models of global climate change. (Hileman 1995)

Increasingly sophisticated climate models strictly conserve energy, mass, momentum, and water substance as they calculate atmospheric motions, temperatures, clouds, precipitation, and evaporation at thousands of points around the globe in response to changing solar insolation, greenhouse gases, and aerosols (Hansen et al. 1983). These models are "coupled" (integrated) with similar ones for the oceans to simulate real-world influences of one upon the other. By comparison, computing the spacecraft trajectory to the outer reaches of the solar system is a rather simple task.

FURTHER READING

Balling, Robert C. *The Heated Debate: Greenhouse Predictions versus Climate Reality.* San Francisco: Pacific Research Institute, 1992.

Eilperin, Juliet. "Global Warming Skeptics Insist Humans Not at Fault." *Washington Post,* March 4, 2008a, A-16. http://www.washingtonpost.com/wp-dyn/content/article/2008/03/03/AR2008030302781_pf.html.

Hansen, James E., G. Russell, D. Rind, P. Stone, A. Lacis, S. Lebedeff, R. Ruedy, and L. Travis. "Efficient Three-dimensional Global Models for Climate Studies: Models I and II." *Monthly Weather Review* 111 (1983):609–662.

Hileman, Bette. "Climate Observations Substantiate Global Warming Models." *Chemical and Engineering News,* November 27, 1995. http://pubs.acs.org/hotartcl/cenear/951127/pg1.html.

Karl, Thomas R., Neville Nicholls, and Jonathan Gregory. "The Coming Climate: Meteorological Records and Computer Models Permit Insights into Some of the Broad Weather Patterns of a Warmer World." *Scientific American* 276 (1997):79–83. http://www.scientificamerican.com/0597issue/0597karl.htm.

Rahmstorf, Stefan, Anny Cazenave, John A. Church, James E. Hansen, Ralph F. Keeling, David E. Parker, and Richard C. J. Somerville. "Recent Climate Observations Compared to Projections." *Science* 216 (May 4, 2007):709.

Stephens, Bret. "Global Warming as Mass Neurosis." *Wall Street Journal,* July 1, 2008, A-15.

Coral Reefs on the Edge of Disaster

When an aquatic environment nurtures coral reefs, they are among the most productive, diverse ecosystems on Earth. Coral reefs are sometimes called the rainforests of the sea. Coral reefs cover much less than 1 percent of the world's ocean floors, while at the same time hosting more than a third of the marine species presently described by science, with many species remaining undocumented. Some of these organisms may provide new sources of anticancer compounds and other medicines. Coral reefs protect shorelines from erosion by acting as breakwaters that, if healthy, can repair themselves.

A worldwide survey of 845 zooxanthellate reef-building coral species was assessed using International Union for Conservation of Nature Red List Criteria. The findings, published in 2008, indicated that "[t]he viability of the world's major coral reefs is endangered both by direct human disturbance and by disease and bleaching events brought on by climate change." Of the 704 species that could be assigned conservation status, 32.8 percent were listed in categories with "elevated risk of extinction" (Carpenter et al. 2008, 560). According to this study,

> The proportion of corals threatened with extinction has increased dramatically in recent decades and exceeds that of most terrestrial groups. The Caribbean has the largest proportion of corals in high extinction risk categories, whereas the Coral Triangle (western Pacific) has the highest proportion of species in all categories of elevated extinction risk. Our results emphasize the widespread plight of coral reefs and the urgent need to enact conservation measures. (Carpenter et al. 2008, 560)

Mortal Damage by 2050 to 2100

Atmospheric carbon dioxide concentration at 500 parts per million (ppm) and global air temperatures 2°C above levels of the year 2000 are expected between 2050 and 2100 to mortally damage many of the world's coral reefs. These carbon-dioxide levels and temperatures "significantly exceed those of at least the past 420,000 years during which most extant marine organisms evolved." Such conditions will acidify the oceans and "compromise carbonate accretion, with corals becoming increasingly rare on reef systems. The result will be less diverse reef communities and carbonate reef structures that fail to be maintained," O. Hoegh-Guldberg and colleagues concluded in a literature survey on threats to corals in *Science.*

Climate change is one of several human-provoked stresses on corals, including declining water quality and overexploitation of key species that are "driving reefs increasingly toward the tipping point for functional collapse." This array of threats may be "the final insult to these ecosystems." A level of carbon dioxide in the atmosphere at 380 ppm, more than 80 ppm above any level measured in ice cores—that is, perhaps the highest level in 20 million years—is going to make the demise of many corals (along with "increasingly serious consequences for reef-associated fisheries, tourism, coastal protection, and people") very difficult to avoid without what Hoegh-Guldberg and colleagues call "decisive action on global emissions" (2007, 1737).

Coral reefs are extremely intolerant of small rises in water temperature, which destroy the ecosystem of the reef, leaving only a lifeless, bleached exoskeleton. By the early 1990s, an increasing number of the Earth's coral reefs were turning white, a sign that they have been killed because of excessive heat. Most corals die at levels above 30° to 32°C. Bleaching results when symbiotic *zooxanthellae* algae are expelled from coral reefs. The algae provide reefs with most of their color, carbon, and ability to deposit limestone. Once bleached, the coral does not receive adequate nutrients or oxygen. If temperatures fall, the reefs may recover, but if thermal stress continues, a large part of any given reef will die.

Pollution, disease, ultraviolet radiation, too much shade, and changes in the ocean's salinity also may contribute to coral bleaching. Warming seas are one among several human-induced changes that threaten coral reefs. Other problems include overfishing, coastal development, nutrient runoff from agriculture and sewage, and sedimentation from logging on streams that feed into coastal waters. By one estimate, 58 percent of the world's coral reefs were being threatened by human activity by the late 1990s (Bryant 1998).

The scope of corals' devastation from climate change and other human impacts rivals the losses endured by the flora and fauna of the world's great rainforests. According to a number of estimates, half of the world's coral reefs may be lost by 2025 unless urgent action is taken to save them from the ravages of pollution, dynamite fishing, and warming waters. Many of the coral reefs that are falling prey to human-induced destruction are among the largest living structures on Earth. Many are more than 100 million years old. Coral that has lost its ability to sustain plant and animal life turns white, as if it had been doused in bleach. Afterward, the dead coral often is subsumed by a choking shroud of gray algae. Since coral polyps and their calcium carbonate skeletons "are the foundation of the entire ecosystem, fish, mollusks, and countless other species, unable to survive in this colorless graveyard, rapidly disappear, too" (Lynas 2004a, 107).

Aside from the obvious ravages of fishermen who blast the reefs and pour cyanide on them, the reefs are also threatened by ocean temperature spikes that many marine biologists attribute to global warming and short-term climate phenomena such as El Niño. Most corals live close to the upper limit of their temperature tolerances. Temperature rises of only a few degrees over a sustained period cause the death of symbiotic microorganisms living within the coral tissue that provide energy for the colony (Hirsch 2002, 14).

The Scope of Corals' Devastation

Descriptions of the world oceans' coral holocaust (occurring and impending) have been widespread in the scientific literature. The journal *Science*, for example, devoted a cover story to the subject August 15, 2003, wherein T. P. Hughes and colleagues concluded that "[t]he diversity, frequency, and scale of human impacts on coral reefs are increasing to the extent that reefs are threatened globally" (Hughes et al. 2003, 929). Prominent among these impacts is anthropogenic warming of the atmosphere and oceans that may exceed limits under which corals have flourished for a half-million years. Some types of corals are more vulnerable to warming than others, however, so "reefs will change rather than disappear entirely" (Hughes et al. 2003, 929). Rising ocean temperatures, however, will certainly reduce biological diversity among corals.

Clive Wilkinson, representing the Australian Institute of Marine Science, said during 2000 that 27 percent of reefs worldwide, crucial nurseries for fish and plants, had died or were in serious trouble following record sea temperatures two years earlier. Another 25 percent will die during the next 25 years if existing trends continued, he said. Wilkinson, the coordinator of the Global Coral Reef Monitoring Network, cited slight cause for optimism, however, thanks to the new priority given to coral conservation by some governments. "Until recently, governments were in denial. That's no longer the case but we have a long way to go to reverse the trend," Wilkins said. He added, however, that "[t]he climate-change models suggest things are going to get a lot worse for corals in most regions" (Nuttall 2000c).

Individual governments may be able to curtail destructive fishing practices and pollution, but the gradual, pervasive warming of the oceans that kills corals is another matter. Worldwide, coral bleaching accelerated in 2002 at its worst pace since 1998, paralleling a rise in global temperatures. The year 2002 became the second-worst year for bleaching after the major El Niño year of 1998. More than 430 cases of coral bleaching were documented that year, notably from the Great Barrier Reef in Australia as well as from reefs in such countries as the Philippines, Indonesia, Malaysia, Japan, Palau, Maldives, Tanzania, Seychelles, Belize, Ecuador, and off the Florida coast of the United States ("New Wave" 2002).

Reports from the World Fish Center and the International Coral Reef Action Network (ICRAN) all documented the increasing toll on the world's corals. Coral bleaching has become endemic around the world. A survey of reefs in the Seychelles in January 1999, for example, showed that 80 percent of the coral was dead, with more than 95 percent mortality in some areas (Leggett 2001, 324). ReefBase (which can be viewed at www.reefbase.org) features more than 3,800 records going back to 1963 that include information on the severity of bleaching.

Callum M. Roberts and colleagues, writing in *Science*, sketched the scope of possible extinctions faced by the world's coral reefs:

> Analyses of the geographic ranges of 3,235 species of reef fish, corals, snails, and lobsters revealed that between 7.2 percent and 53.6 percent of each taxon have highly restricted ranges, rendering them vulnerable to extinction.... The 10 richest centers of endemism cover 15.8 percent of the world's coral reefs (0.012 percent of the oceans) but include between 44.8 and 54.2 percent of the restricted-range species. Many occur in regions where reefs are being severely affected by people, potentially leading to numerous extinctions. (Roberts et al. 2002, 1280)

A study involving more than 5,000 scientists and divers who monitored coral reefs adjacent to 55 countries for five years found only one reef out of more than 1,100 that was judged to be in near-pristine condition. "Coral reefs have suffered more damage over the last 20 years than they have in the last one thousand," said Gregor Hodgson, author of the report and a University of California at Los Angeles visiting professor who heads Reef Check, a monitoring program based at UCLA's Institute of the Environment. "It is the rate of decline and the global extent of the damage that is so alarming, with species reasonably abundant 30 years ago now on the verge of extinction" (Hirsch 2002, 14). Hodgson said the Reef Check project is designed to be an early-warning system and that he hopes other scientists will examine the issues raised in his report. "Although it is a big study, we have looked at only a tiny fraction of the world's reefs," he said (Hirsch 2002, 14).

World Atlas of Coral Reefs

The threats to coral reefs are similar around the world. More than 80 percent of Indonesia's coral reefs, for example, have been threatened mainly due to blast-fishing practices and bleaching, according to the United Nations Environment Program's (UNEP) *World Atlas of Coral Reefs*. Indonesia, along with the Philippines, Malaysia, and Papua New Guinea, with between 500 and 600 species of coral in each of these countries, is home to the world's most diverse range of corals ("Over 80" 2001).

"Our new atlas clearly shows that coral reefs are under assault," said Klaus Toepfer, UNEP's executive director. He continued:

> They are rapidly being degraded by human activities. They are over-fished, bombed and poisoned. They are smothered by sediment, and choked by algae growing on nutrient rich sewage and fertilizer run-off. They are damaged by irresponsible tourism and are being severely stressed by the warming of the world's oceans. Each of these pressures is bad enough in itself, but together, the cocktail is proving lethal. ("Over 80" 2001)

The newly published *World Atlas of Coral Reefs* provides detailed descriptions of the coral reefs in every country in the world. "And," wrote Mark Spalding, lead author of the *World Atlas*, "in every country the same problems are present" (Spalding 2001a, 8). Traveling in the Indian Ocean, Spalding wrote in the *London Guardian*,

> Over the next six weeks we watched the corals of the Seychelles die. Corals are to reefs what trees are to forests. They build the structure around which other communities exist. As the corals died they remained *in situ* and the reefs became, to us, graveyards. Fine algae grows over a dead coral within days, and so the reefs took on a brownish hue, cobwebbed. In fact, the fish still teemed and in many ways it still appeared to be business as usual, but as we traveled—over 1,500 kilometers across the Seychelles—the scale of this disaster began to sink in. Everywhere we went was the same, and virtually all the coral was dying or already dead.... What I witnessed in the Seychelles was repeated in the Maldives and the Chagos Archipelago. In these Indian Ocean islands alone, 80 to 90 percent of all the coral died (Spalding 2001a, 8)

Warmer sea-surface temperatures killed 90 percent of the coral reefs near the surface of the Indian Ocean in only one year, while the remaining 10 percent could die in the next 20 years, according to work by Charles Sheppard of England's University of Warwick. During 1998, according to Sheppard, rapid warming devastated shallow-water corals to 130 feet below the surface. Some of these corals began to recover in subsequent years, but the risk continues, according to Sheppard. "It's like a forest," he said. "If you kill off 90 percent, there might be just enough left to sustain some life around it, such as squirrels and so on. But if you have the same impact again and again there's no clear line as to when it's alive or not as a forest" (Arthur 2003b).

In research published by *Nature*, Sheppard used a computer model for global warming developed by Britain's Hadley Centre for Forecasting to anticipate how sea temperatures will affect coral survival (Sheppard 2003, 294–297). "Most corals don't mature until [they are] five years old and, five years since the 1998 event, most sites have recovered only marginally," Sheppard said (Arthur 2003b). According to Sheppard's analysis, a proportion of given coral dies with each annual peak in the sea temperature. "The warming trend is only a fraction of a degree each year, which is swamped by the annual change," Sheppard said. "But south of the Equator the probability of the temperature rising enough to kill all the corals means they could be wiped out as soon as 2020" (Arthur 2003b).

A History of Water Temperature Rises

A scientific team led by Thomas J. Goreau plotted water temperatures at a number of locations in the tropical western Atlantic and Caribbean, finding slow rises beginning during the 1980s in such places as the Bahamas and Jamaica. In some cases, these temperature increases have been significant enough to dramatically accelerate the bleaching of coral reefs. The rate of bleaching correlates to general air temperature rises detected by global surveys of temperatures on land. "The fact that the high water temperatures of the late 1980s and mass coral bleaching have not been seen before in the Greater Caribbean during nearly 40 years of continuous study of coral reef ecosystems implies that temperatures have only recently exceeded tolerance limits of the corals" (Goreau et al. 1993, 251). Goreau and the co-authors of this paper cite similar results in Hawaiian waters. This paper includes an extensive list of references that detail exactly what is happening to coral reefs around the world (Goreau et al. 1993, 253–255).

Some areas, including the Cook Islands and several locations in the Philippines, have been afflicted with massive coral bleaching since 1983. In 1995, warming in the Caribbean produced coral bleaching for the first time in Belize, as sea-surface temperatures surpassed 29°C (84°F). In 1997, Caribbean sea-surface temperatures reached 34°C (93°F) off southern Belize, and coral bleaching was accompanied by the deaths of many starfish and other sea life.

The central barrier reef near Belize suffered a mass coral die-off following record-high water temperatures in 1998. According to Richard Aronson, a marine biologist at the Dauphin Island Sea Laboratory in Alabama, this die-off represents the first complete collapse of a coral reef in the Caribbean Sea from bleaching (Connor 2000b, 12). Aronson, working with colleagues at the Smithsonian Institution in Washington, D.C., said in the journal *Nature* that lettuce coral was the most abundant species living in the area until 1998.

Later surveys showed that "virtually all living colonies had been bleached white at almost every depth investigated" (Conner 2000b, 12). Core samples from the corals indicate that such a mass bleaching is unprecedented for at least the last 3,000 years. The research team found that "[c]omplete bleaching was also evident in almost all ... species down to the lagoon floor at 21 meters" (Aronson et al. 2000, 36).

By the late 1990s, coral bleaching had become epidemic in many tropical seas and oceans around the world. A World Wildlife Fund report found that massive coral bleaching occurred during the 1990s, in response to unusually high water temperatures, particularly during the El Niño years of 1997 and 1998. Sites of large-scale coral mortality included the Pacific Ocean, Indian Ocean, Red Sea, Persian Gulf, as well as the Mediterranean and Caribbean seas (Mathews-Amos and Berntson 1999).

With more warming anticipated, the extent and severity of damage to coral reefs worldwide is expected to intensify. An Australian marine scientist at the Australian Institute of Marine Science, Dr. Terry Done, told Murray Hogarth of the *Sydney Morning Herald* that he regards coral reefs as the "canaries in the coalmine" (Hogarth 1998). During 1998, large coral reefs up to 1,000 years old died, and in some parts of the world up to 90 percent of reefs have been devastated, Done said.

The World Wildlife Fund sketched the worldwide nature of coral bleaching caused by rising ocean temperatures:

> According to NOAA, coral bleaching was reported throughout the Indian Ocean and Caribbean in 1998, and throughout the Pacific, including Mexico, Panama, Galapagos, Papua New Guinea, American Samoa, and Australia's Great Barrier Reef starting in 1997.... The severity and extent of coral bleaching in 1997–1998 was widely acknowledged among coral reef scientists as unprecedented in recorded history. (Mathews-Amos and Berntson 1999)

The International Society for Reef Studies concluded that 1997 and 1998 had witnessed the most geographically widespread coral bleaching in the recorded histories of at least 32 countries and island nations. Reports of bleaching came from sites in all the major tropical oceans of the world, some for the first time in recorded history. The 1997–1998 bleaching episode was exceptionally severe, as a large number of corals died. According to one study, parts of Australia's Great Barrier Reef have been so severely affected that many of the usually robust corals, including one dated more than 700 years old, were badly damaged or had died during 1997 and 1998 (Mathews-Amos and Berntson 1999).

Massive Bleaching in the Indian Ocean

Some of the most severe coral bleaching took place in the tropical Indian Ocean, where water temperatures during the 1998 El Niño reportedly rose to between 3° and 5°C above long-term averages in some areas. According to Clive Wilkinson and colleagues writing in the Swedish environmental journal *Ambio,*

> Massive mortality occurred on the reefs of Sri Lanka, [the] Maldives, India, Kenya, Tanzania, and [the] Seychelles, with moralities of up to 90 percent in many shallow areas.... Coral death during 1998 was unprecedented in severity.... Coral reefs of the Indian Ocean may prove to be an important signal of the potential effects of global climate change, and we should heed that warning. (Wilkinson et al. 1999, 188)

According to another observer, "In some parts of the Indian Ocean ... reefs in the Maldives, Sri Lanka, Kenya and Tanzania were devastated with shallow reefs looking like graveyards" (Perry 1998).

In the Asia-Pacific region, the worst-hit coral reefs during the same El Niño episode were near Japan, Taiwan (China), the Philippines, Vietnam, Thailand, Singapore, Indonesia, and the islands of Palau. A scientists' statement, released at the same time in the United States and Australia, relied heavily on new satellite data from America's National Oceanic and Atmospheric Administration, which indicated a degree of warming that they had not anticipated, especially in the tropics. During May 1998, marine scientists said Australia's Great Barrier Reef (the world's largest living reef stretching 1,300 miles in length) experienced its worst case of coral bleaching in recorded history. Australian marine scientists said bleaching had hit more than 60 percent of its 3,000 coral reefs and aerial surveys showed that 88 percent of inshore reefs were bleached, with 25 percent of those reefs severely bleached (Perry 1998).

By 2000, some coral reefs in the Indian and Pacific oceans had recovered somewhat from the rapid bleachings of 1998. Scientific studies indicated that some young corals in these areas had survived bleaching and were helping to rebuild some of the reefs. Done said that reefs may be "more resilient than we had thought." According to a report in *Science,* Done explained that many corals may "not be able to mature and recover from the repeated bleaching forecast to accompany global warming" (Normile 2000, 941). Done is best acquainted with some of the Indian Ocean corals, which were characterized as looking like graveyards after the 1998 bleaching (Normile 2000, 942). He noted that most coral reefs require at least a decade to recover from bleaching. "This recovery won't do the reefs much good," Done said. "They'll no sooner get one or two years old before they'll be wiped out again" (Normile 2000, 942).

Warming and Diseases

Corals are not the only living things imperiled by rising ocean temperatures. C. D. Harvell and colleagues reported in *Science* that "[i]n the past few decades, there has been a world-wide increase in the reports of diseases, affecting marine organisms" (Harvell et al. 1999, 1505). Harvell and colleagues related this rapid increase in disease-related mortality among plants and animals of the seas and oceans to global warming, which acts to expand the range of many diseases, as well as to human pollution and other forms of habitat degradation. Many of the diseases surveyed by these researchers are new to science.

"The current trend toward a warming climate could result in modifications of many marine organisms' basic biological properties, thereby making them more susceptible to disease," the research team asserted. (Harvell et al. 1999, 1507). One of the team's primary examples of warming's synergy with disease is the mass bleaching of tropical corals worldwide during the 1997–1998 ENSO (El Niño Southern Oscillation). The authors of this article found that warmer water often promoted diseases, along with the heat, and explained that the "[d]emise of some corals is likely to have been accelerated by opportunistic infections" (Harvell et al. 1999, 1507). Harvell and colleagues called corals an "indicator species of a heightened disease load" throughout the oceans (Harvell et al. 1999, 1509). The same authors cited increased mortality of oysters in Chesapeake Bay, which they ascribed to warmer winters that have "decreased parasite mortality, resulting in oysters retaining heavy infections" (Harvell et al. 1999, 1507).

Caribbean Coral Catastrophe

Four-fifths of the corals on Caribbean reefs have disappeared within 25 years. Human provocations, including but not limited to warming seas, are responsible for most of the destruction. The scope of the loss has astonished even those scientists who have been studying the global decline of the corals. The scope of destruction has been unmatched for several thousand years, according to a study in the journal *Science* (McCarthy 2003a, 3; Gardner et al. 2003). The team leader, Isabelle Cote, a French-Canadian specialist in tropical marine ecology, said that causes of the corals' decline include industrial, agricultural, and other human pollution, overfishing, diseases, stronger storms, and higher sea temperatures (McCarthy 2003a, 3).

The work was conducted by researchers at the University of East Anglia, United Kingdom, and its associated Tyndall Centre for Climate Change Research, using data from 263 Caribbean sites, including Mexico, Barbados, Cuba, Panama, the Florida Keys, and Venezuela. "We report a massive region-wide decline of corals across the entire Caribbean basin," the scientists wrote in *Science* (Gardner et al. 2003). The reefs of the Caribbean, seriously weakened by human depredation, may now be unable to withstand future warming. "The ability of Caribbean coral reefs to cope with future local and global environmental change may be irretrievably compromised," the team reported (McCarthy 2003a, 3).

The study involved hard corals, the tiny animals that slowly build coral reefs from the calcium carbonate that they excrete. The scientists found that in 1977, the start of the survey period, a typical Caribbean reef was 50 percent covered in live corals, which is considered healthy. By 2002, however, a typical reef was 10 percent covered, which is regarded as potentially fatal (McCarthy 2003a, 3). "The end result surprised us, as well as all the people who gave us data," said Cote. "The rate of decline we found exceeds by far the well-publicized rates of loss for tropical forests" (McCarthy 2003a, 3; Gardner et al. 2003).

To provide one of many examples, *London Independent* environmental writer Geoffrey Lean described the fate of corals off one stretch of Jamaica's coastline:

> Discovery Bay in Jamaica was once one of the finest in the Caribbean but now its beauty and rich wildlife have been smothered by algae as a result of a complex mixture of man-made and natural catastrophes. These began [with] over-fishing ... soon after the island's colonization ... overfishing [undermined the health of the reef] ... laying it open to a series of catastrophes in the 1980s. The first affected a species of sea urchin which had grazed on the algae and kept it in check after the fish were depleted. It was hit by a mystery disease in 1983–1984 (brought, according to one theory, in ballast water in a ship passing through the Pacific Canal) and began dying out. Then the coral itself was hit by diseases and a hurricane did even greater damage. Now, scientists say, it can hardly be called a reef at all. (Lean 2001a, 7)

Corals may recover if conditions change. This study, while pessimistic about the prospects of the corals as a whole, found that some areas of degraded coral in the Caribbean appear to be recovering. "The bad news, however, is that the new coral communities seem to be different from the old ones," said Cote. "At this point, we do not know how well these new assemblages will be able to face new challenges, such as rising sea levels and temperatures as a result of global warming" (McCarthy 2003a, 3). "Given current predictions of increased human activity in the Caribbean," the study concluded, "The growing threat of climate change on coral mortality and reef framework-building, and the potential synergy between these threats, the situation for Caribbean coral reefs does not look likely to improve in either the short or the long term" (McCarthy 2003a, 3).

The Condition of Coral Reefs Near Fiji

David Bellamy, a British botanist, has warned that many of Fiji's spectacular coral reefs are being ruined as a result of bleaching caused at least partially by global warming. Bellamy visited Fiji during 2001 to inaugurate a project by a British charity to survey reefs in the Mamanuca islands off the west coast of the main island of Viti Levu. Coral Cay Conservation has been invited by members of the tourism industry in Fiji to undertake a three-month pilot project. "Hoteliers contacted us saying they wanted to find out what had gone wrong, because without reefs there's nothing to attract high-spending divers," said Bellamy (Miles 2001, 4). In addition to bleaching, reefs near 13 Mamanuca resorts have been threatened by overfishing, which results in an imbalance in the ecosystem, Bellamy said. Andrea Dehm, manager of Ovalau Watersports on Viti Levu, said the bleaching had caused diving in the area to become increasingly disappointing. "Our divers used to comment on the beautiful colors, but now it's all white," she said. "People don't want to see that and some operators are closing" (Miles 2001, 4).

Is Coral Reefs' Bleaching a Defense Mechanism?

Andrew C. Baker wrote in *Nature* that bleaching may be a sign of a coral reef's defenses in coping with environmental change, including a warming habitat. "Coral bleaching can promote rapid response to environmental change by facilitating compensatory change in algal symbiont communities," he wrote (Baker 2001, 765). Bleaching helps the corals adjust to new conditions: "Reef corals are flexible associations that can switch or shuffle symbiont communities in response to environmental change" (Baker 2001, 765). There are costs—some symbionts that are unsuited to new conditions do die. As described by Baker, "Bleaching is an ecological gamble in that it sacrifices short-term benefits for long-term advantage" (Baker 2001, 766). Coral symbionts are, according to Baker, individually fragile, but the entire reef's flexibility gives it remarkable endurance: "[B]leaching may ultimately help reef corals to survive the recurrent and increasingly severe warming events projected by current climate models" (Baker 2001, 766).

A lively debate has developed in scientific journals regarding how quickly some corals may adapt to thermal stress (Baker et al. 2004, 741; Rowan 2004, 742). Ove Hoegh-Guldberg and colleagues take issue with Andrew C. Baker's assertions that coral "bleaching favours new host-symbiont combinations that guard populations of corals against rising sea temperature" (Hoegh-Guldberg et al. 2002, 602). They disagreed with Baker, who replied that

> far fewer corals in the far-eastern Pacific Ocean died after the 1997–1998 El Niño event (0 to 26 percent) than after the 1982–1983 El Niño event (52 to 97 percent) even though the magnitude and duration of sea-surface temperature anomaly in the region in 1997–1998 exceeded those of 1982–1983. (Baker, reply to Hoegh-Guldberg et al. 2002, 602)

Coral Reefs' Possible Rehabilitation

Once protected from various human insults, some coral reefs may recover quickly. Coral reefs have been present for 100 million years, and they can be surprisingly resilient. "Ecologists have been surprised at how quickly reef fish populations have rebounded at successful marine parks," Hodgson said. "Two years is enough to replenish many species. Some types of corals also grow very quickly—like corn—but others take decades to reach a large size" (Hirsch 2002, 14).

Proposals have been made to establish networks of marine reserves where fishing is prohibited. By minimizing the stresses of overfishing, the corals may be able to cope with other stresses, such as global warming, researcher Callum Roberts said (Radford 2002a, 12). Roberts' study asserted that protection of 10 coral reef "hot spots" around the world could save a large number of marine species. The 10 reefs account for 0.017 percent of the oceans' surface area, but are home to 34 percent of all species with limited ranges (Radford 2002a, 12; Roberts 2002, 1280–1284).

Roberts and his colleagues looked at 18 areas with the greatest concentrations of species found nowhere else, and selected the 10 most vulnerable. These areas are located in the Philippines, the Gulf of Guinea, the Sunda islands in Indonesia, the southern Mascarene islands in the Indian ocean, eastern South Africa, the northern Indian Ocean, southern Japan, Taiwan and southern China, the Cape Verde islands, the western Caribbean, and the Red Sea and Gulf of Aden.

Eight of the 10 reefs that Roberts and colleagues would like to protect already are being dramatically altered by human activity through fishing, logging, and farming. The felling of forests, for example, means that soils easily erode, often depositing mud that can choke reefs. Farming releases nutrients that encourage seaweeds to grow where corals once would have flourished (Radford 2002a, 12).

"One of the arguments is that there is nothing we can do, it is all going to go to hell, and that coral reefs are doomed. The other argument is that we should work very hard to try and do something about protecting them," Roberts said. "The question then is how? Where are we going to focus our efforts, given that we don't have the resources to do all that we would like? We cannot save all coral reefs everywhere" (Radford 2002a, 12).

FURTHER READING

Aronson, Richard B., William F. Precht, Ian G. MacIntyre, and Thaddeus J. T. Murdoch. "Coral Bleach-out in Belize." *Nature* 405 (May 4, 2000):36.

Arthur, Charles. "Temperature Rise Kills 90 Per Cent of Ocean's Surface Coral." *London Independent*, September 18, 2003b, n.p. (LEXIS)

Baker, Andrew C. "Reef Corals Bleach to Survive Change." *Nature* 411 (June 14, 2001):765–766.

Baker, Andrew C., Craig J. Starger, Tim R. McClanahan, and Peter W. Glynn. "Corals' Adaptive Response to Climate Change." *Nature* 430 (August 12, 2004):741.

Bryant, D., L. Burke, J. McManus, and M. Spaulding. *Reefs at Risk: A Map-based Indicator of Threats to the World's Coral Reefs.* Washington, DC: World Resources Institute, 1998.

Carpenter, Kent E., Muhammad Abrar, Greta Aeby, Richard B. Aronson, Stuart Banks, Andrew Bruckner, Angel Chiriboga, Jorge Cortès, J. Charles Delbeek, Lyndon DeVantier, Graham J. Edgar, Alasdair J. Edwards, Douglas Fenner, HÈctor M. Guzm·n, Bert W. Hoeksema, Gregor Hodgson, Ofri Johan, Wilfredo Y. Licuanan, Suzanne R. Livingstone, Edward R. Lovell, Jennifer A. Moore, David O. Obura, Domingo Ochavillo, Beth A. Polidoro, William F. Precht, Miledel C. Quibilan, Clarissa Reboton, Zoe T. Richards, Alex D. Rogers, Jonnell Sanciangco, Anne Sheppard, Charles Sheppard, Jennifer Smith, Simon Stuart, Emre Turak, John E. N. Veron, Carden Wallace, Ernesto Weil, and Elizabeth Wood. "One-Third of Reef-Building Corals Face Elevated Extinction Risk from Climate Change and Local Impacts." *Science* 321 (July 25, 2008):560–563.

Connor, Steve. "Global Warming Is Blamed for First Collapse of a Caribbean Coral Reef." *The London Independent*, May 4, 2000b, 12.

Gardner, Toby A., Isabelle M. Cote, Jennifer A. Gill, Alastair Grant, and Andrew R. Watkinson. "Long-Term Region-Wide Declines in Caribbean Corals." *Science* 301 (August 15, 2003):958–960. http://www.scienceexpress.org.

Goreau, Thomas J., Raymond L. Hayes, Jenifer W. Clark, Daniel J. Basta, and Craig N. Robertson. "Elevated Sea-surface Temperatures Correlate with Caribbean Coral Reef Beaching." In *A Global Warming Forum: Scientific, Economic, and Legal Overview*, ed. Richard A. Geyer, 225–262. Boca Raton, FL: CRC Press, 1993.

Harvell, C. D., K. Kim, J. M. Buckholder, R. R. Colwell, P. R. Epstein, D. J. Grimes, E. E. Hofmann, E. K. Lipp, A. D. M. E. Osterhaus, R. M. Overstreet, J. W. Porter, G. W. Smith, and G. R. Vasta. "Emerging Marine Diseases—Climate Links and Anthropogenic Factors." *Science* 285, no. 3 (September 3, 1999):1505–1510.

Hirsch, Jerry. "Damage to Coral Reefs Mounts, Study Says: Broad Survey Cites Human Causes such as Overfishing and Pollution." *Los Angeles Times*, August 26, 2002, 14.

Hoegh-Guldberg, Ove, Ross J. Jones, Selina Ward, and William K. Loh. "Is Coral Bleaching Really Adaptive?" *Nature* 415 (February 7, 2001):601–602.

Hoegh-Guldberg, O., P. J. Mumby, A. J. Hooten, R. S. Steneck, P. Greenfield, E. Gomez, C. D. Harvell, P. F. Sale, A. J. Edwards, K. Caldeira, N. Knowlton, C. M. Eakin, R. Iglesias-Prieto, N. Muthiga, R. H. Bradbury, A. Dubi, and M. E. Hatziolos. "Coral Reefs under Rapid Climate Change and Ocean Acidification." *Science* 318 (December 14, 2007):1737–1742.

Hogarth, Murray. "Sea-warming Threatens Coral Reefs." *Sydney Morning Herald* (Australia), November 26, 1998. http://www.smh.com.au/news/9811/26/text/national13.html.

Hughes, T. P., A. H. Baird, D. R. Bellwood, M. Card, S. R. Connolly, C. Folke, R. Grosberg, O. Hoegh-Guldberg, J. B. C. Jackson, J. Kleypas, J. M. Lough, P. Marshall, M. Nystr̂m, S. R. Palumbi, J. M. Pandolfi, B. Rosen, and J. Roughgarden. "Climate Change, Human Impacts, and the Resilience of Coral Reefs." *Science* 31 (August 15, 2003):929–933.

Lean, Geoffrey. "Quarter of World's Corals Destroyed." *London Independent*, January 7, 2001a, 7.

Leggett, Jeremy. *The Carbon War: Global Warming and the End of the Oil Era.* New York: Routledge, 2001.

Lynas, Mark. *High Tide: The Truth about Our Climate Crisis.* New York: Picador/St. Martins, 2004a.

Mathews-Amos, Amy, and Ewann A. Berntson. "Turning Up the Heat: How Global Warming Threatens Life in the Sea." World Wildlife Fund and Marine Conservation Biology Institute, 1999. http://www.worldwildlife.org/news/pubs/wwf_ocean.htm.

McCarthy, Michael. "'Rainforests of the Sea' Ravaged: Overfishing and Pollution Kill 80 Percent of Coral on Caribbean Reefs." *London Independent*, July 18, 2003a, 3.

Miles, Paul. "Fiji's Coral Reefs are Being Ruined by Bleaching." *London Daily Telegraph*, June 2, 2001, 4.

"New Wave of Bleaching Hits Coral Reefs Worldwide." Environment News Service, October 29, 2002. http://ens-news.com/ens/oct2002/2002-10-29-19.asp#anchor1.

Normile, Dennis. "Some Coral Bouncing Back From El Niño." *Science* 288 (May 12, 2000):941–942.

Nuttall, Nick. "Coral Reefs 'On the Edge of Disaster.'" *London Times*, October 25, 2000c, n.p. (LEXIS)

"Over 80 per cent of Indonesia's Coral Reefs under Threat." *Jakarta Post*, September 13, 2001, n.p. (LEXIS)

Perry, Michael. "Global Warming Devastates World's Coral Reefs." Reuters, November 26, 1998. http://www.gsreport.com/articles/art000023.html.

Radford, Tim. "Ten Key Coral Reefs Shelter Much of Sea Life: American Association Scientists Identify Vulnerable Marine 'Hot Spots' with the Richest Biodiversity on Earth." *London Guardian*, February 15, 2002a, 12.

Roberts, Callum M., Colin J. McClean, John E. N. Veron, Julie P. Hawkins, Gerald R. Allen, Don E. McAllister, Cristina G. Mittermeier, Frederick W. Schueler, Mark Spalding, Fred Wells, Carly Vynne, and Timothy B. Werner. "Marine Biodiversity Hotspots and Conservation Priorities for Tropical Reefs." *Science* 295 (February 15, 2002):1280–1284.

Rowan, Rob. "Thermal Adaptation in Reef Coral Symbionts." *Nature* 430 (August 12, 2004):742.

Sheppard, Charles R. C. "Predicted Recurrences of Mass Coral Mortality in the Indian Ocean." *Nature* 425 (September 18, 2003):294–297.

Spalding, Mark. "Coral Grief: Rising Temperatures, Pollution, Tourism and Fishing have All Helped to Kill Vast Stretches of Reef in the Indian Ocean. Yet, with Simple Management, says Mark Spalding, the Marine Life Can Recover." *London Guardian*, September 12, 2001a, 8.

Spalding, Mark. *World Atlas of Coral Reefs*. Berkeley: University of California Press, 2001b.

Wilkinson, Clive, Olof Linden, Herman Cesar, Gregor Hodgson, Jason Rubens, and Alan E. Strong. "Ecological and Socioeconomic Impacts of 1998 Coral Mortality in the Indian Ocean: An ENSO Impact and a Warning of Future Change?" *Ambio* 28, no. 2 (March 1999):188–196.

Corporate and Academic Sustainability Initiatives

Some companies are appointing "chief sustainability officers." Eileen Claussen, president of the Pew Center on Global Climate Change, said, "Environmental vice presidents usually spend company money, but this new breed is helping companies make money." The upshot, said Geoffrey Heal, a business professor at the Columbia Business School, is that "what started out as a compliance job has evolved into one that guards the value of the brand" (Deutsch 2007a).

These officers, who may go under the title "vice president for environmental affairs," join with vendors and customers to create and market green products. Dow Chemical's first chief sustainability officer, David E. Kepler, has been meeting with Dow's technology, manufacturing, and finance leaders about alternative fuels and green products. "We usually agree," Kepler said. "But if a critical environmental issue is in dispute, I'll prevail" (Deutsch 2007a).

By 2008, about 25 percent of Fortune 500 companies had a board committee overseeing environmental issues (including global warming), compared with only 10 percent five years previously, according to Mindy Lubber, president of Ceres, a national coalition of activists, investors, and others who study environmental issues. The number of shareholder proposals related to environmental issues also doubled between 2004 and 2008. Some companies (for example, American Electric Power [AEP], a leading supplier of coal-fired electricity) have gone so far as to dock executives if their firms are cited for violations of environmental regulations more often than guidelines allow. Michael G. Morris, chief executive officer of AEP, was docked $80,045 in 2006 from a projected bonus of about $2 million after the company received nine violation citations, when the "allowance" was five (Lublin 2006, B-6).

By late 2008, a corporate alliance, USCAP (the United States Climate Action Partnership), had formed and included the following companies and environmental groups: Alcoa, AIG, Boston Scientific, BP America, Caterpillar, ConocoPhillips, Chrysler, John Deere, Dow, Duke Energy, DuPont, Environmental Defense Fund, Exelon, Ford, FPL Group, General Electric, General Motors, Johnson & Johnson, Marsh, National Wildlife Federation, Natural Resources Defense Council, NRG Energy, The Nature Conservancy, PepsiCo, Pew Center on Global Climate Change, PG&E, PNM Resources, Rio Tinto, Shell, Siemens, World Resources Institute, and Xerox ("Big U.S." 2008).

Linda J. Fisher, the chief sustainability officer at DuPont, weighed in against purchase of a company that was not in a "sustainable" business. "We're building sustainability into the acquisition criteria," she said. When two business chiefs at General Electric (GE) opposed the cost of developing environmentally friendly products, Jeffrey R. Immelt, GE's chairman, gave Lorraine Bolsinger, vice president of GE's Ecomagination business, the research money. "I have an open door to get projects funded," she said (Deutsch 2007a).

Meanwhile, Owens-Corning named Frank O'Brien-Bernini, its research chief since 2001, to the post of chief research and development and sustainability officer. He now uses what he calls the "lens of sustainability" to prioritize research. He helped husband development of a machine that makes it easier to insulate attics. Owens-Corning developed it after research showed that drafty attics are prime culprits in greenhouse-gas emissions, but that the "hassle factor" kept homeowners from addressing the situation. "I drive innovation around products and processes," O'Brien-Bernini said. "And I make sure that our claims are backed by deep, deep science" (Deutsch 2007a). Elizabeth A. Lowery, vice president for environment, energy, and safety policy at General Motors, helped to brief reporters, shareholders, and environmentalists on the Chevy Volt, an electric car.

The "Al Gore of Citicorp"

Stephen Lane, who jokes that he is the "Al Gore of Citigroup," is the company's executive vice president whose full-time job is coaxing energy savings out of the 340,000-employee, worldwide financial services giant. His tasks range from an inventory of energy use in all of the company's facilities "What you can't measure, you can't manage," he says (Carlton 2007, B-1). Lane sets policies that govern everything from installing solar energy and timed lighting in bank branches to convincing employees to cut off unused lights and climb stairs instead of using escalators, which are now stopped during nonbusiness hours. Other banks are taking similar measures. HSBC, for example, has opened a "green" prototype branch in Greece, New York, which its 400 branches in the United States soon will emulate. Lane oversees a $10 billion Citigroup plan to reduce its carbon footprint to 10 percent below 2005 levels by the year 2011 (Carlton 2007, B-8).

By the end of 2007, more than 2,400 companies in the United States were reporting their carbon emissions and energy costs through the Carbon Disclosure Project, a nonprofit group of 315 institutional investors that control $41 trillion worth of assets. Some are household names, such as Coca-Cola and Wal-Mart (which has begun to require its suppliers to report and reduce their carbon footprints). Dell, the computer maker, announced in 2008 that it would begin to neutralize the carbon impact of its operations around the world ("Investors Sizing Up" 2007, D-1).

Also by the end of 2007, some of the world's largest multinational companies, among them Procter & Gamble, Unilever, Tesco (the British grocery chain), and Nestle SA were requiring their suppliers to disclose carbon-dioxide emissions and global-warming mitigation strategies. The companies are among several that have formed the Supply Chain Leadership Coalition, which cooperates with the London-based Carbon Disclosure Project. Eventually, products may be labeled with carbon-emission information. In 2007, Cadbury Schweppes was making plans to print such information on its chocolate bars. At about the same time, Wal-Mart began a similar project by asking Oakhurst Dairy, of Portland, Maine, to measure the carbon footprint of a case of milk (Spencer 2007, A-7).

Early in 2007, Sir Terry Leahy, chief executive of the Tesco supermarket chain, which sells a quarter of Britain's groceries, told Forum for the Future that the company would cut its energy consumption in half by 2010, sharply cut the number of products it ships by air, and place a carbon label on each of the 70,000 products it sells, taking into account, he said, "its complete life cycle, from production to distribution to consumption" (Specter 2008, 44).

A year later, Tesco had produced one carbon label, for a house brand of Walkers crisps (potato chips). Calculating the carbon footprint of potato chips is one of the simpler food products, but still a daunting task, involving the amount of energy required to raise the potatoes and sunflower oil, plus the fertilizers and pesticides involved in growing them, the tractors that harvest them, the manufacture of the chips, as well as their packaging, transport to stores, and disposal of packaging. The company and Britain's Carbon Trust calculated that 75 grams of greenhouse gases were produced for each bag.

Waste to Renewable Energy

Many companies have been finding ways to turn what used to be waste into renewable energy. Kraft Foods is powering a plant in New York with methane produced by adding bacteria to whey, a by-product of cream cheese. Central Vermont Public Service, which supplies 158,000 customers with electric power, has been obtaining some of it from the manure of dairy cows. About 4,000 customers pay a premium on their bills to foster the biomass technology at four Vermont dairy farms.

A Frito-Lay plant in Casa Grande, Arizona, has been implementing plans to take itself nearly completely off the power grid by the year 2010, using several strategies. One is new filters that recycle most of the water used to wash and rinse potatoes and corn destined for snack chips. The remaining sludge will then be used as a source of methane to power the factory's boiler. The factory also plans to install 50 acres of solar concentrators, to be used for power, along with a biomass generator burning agricultural waste. Plans are to reduce electricity and water consumption by 90 percent, natural gas usage by 80 percent, and greenhouse-gas emissions by 50 to 75 percent. These strategies, once adapted at the giant Casa Grande plant (the size of two football fields) will be considered at Frito-Lay's 37 other processing plants in the United States and Canada (Martin 2007, A-22).

Academic Initiatives

Corporations sometimes show their "green" credentials by funding university sustainability centers. Four multinational companies (ExxonMobil,

General Electric, Schlumberger, and Toyota) provided the seed capital for the Stanford University Global Climate and Energy Project, with an emphasis on innovative energy technologies. The Shell Oil Foundation has financed Rice University's Shell Center for Sustainability since 2002. Wal-Mart in 2007 pledged funding for the Applied Sustainability Center at the University of Arkansas (Deutsch 2007b). Dow Chemical gave $10 million, spread over five years, to the University of California–Berkeley, to set up a sustainability center, one of a growing trend of interdisciplinary study centers at academic institutions that cross many field boundaries.

"We give professors a chance to step beyond their usual areas of expertise, and we give students exposure to the worlds of science and business," said Daniel C. Esty, the new director of the Yale Center for Business and the Environment, combining resources from the school of Management with Forestry and Environmental Studies. The University of Tennessee consolidated environmental-research programs within an Institute for a Secure and Sustainable Environment. Arizona State University in 2004 initiated a degree-granting Global Institute of Sustainability, funded in part with private donations. Some "sustainability" centers (such as the Kenan-Flagler Center for Sustainable Enterprise at the University of North Carolina) study global cultural issues as well as business ethics and corporate social responsibility in the context of environmental issues (Deutsch 2007b).

Some of this may be connotative wordplay as much as a paradigm shift in academic priorities. "We are seeing more centers framed as sustainability, but they may not be qualitatively different from the ethics, innovation or globalization centers of 15 years ago," said Jonathan Fink, Arizona State University's sustainability officer. "Universities realize that you can discuss sustainability with a CEO and not get laughed out of the room." Others have specific environmental focuses. The Yale center hosts an "eco-services clinic" to aid corporations with environmental issues, while Duke's Corporate Sustainability Initiative (with resources from earth sciences, business, and environmental policy) has developed a residential-scale wind turbine, and advises local businesses to reduce their carbon footprints (Deutsch 2007b).

At Oberlin University's sustainability house (SEED, Student Experiment in Ecological Design), students reduce carbon footprints in their own lives by doing such things as timing showers. Oberlin spent $40,000 to renovate this house in 2007. Students maintain patrol to turn off lights and reduce thermostats to 60°F in winter. They even turned off the refrigerator that stores their main source of food (Rimer 2008).

Middlebury College is building an $11 million wood-chip-powered plant as it pledges to become carbon-neutral by 2016, part of campus goals that include serving local food in dining halls and emphasizing, biking, bus riding, and hybrid cars. "This is a generation that is watching the world come undone," said David Orr, a professor of environmental studies at Oberlin. Projects like the Oberlin house, he said, are "helping them [the students] understand how to stitch the world together again" (Rimer 2008).

FURTHER READING

"Big U.S. Corporations Urge Quick Carbon Cap-and-Trade Legislation." Environment News Service, November 18, 2008. http://www.ens-newswire.com/ens/nov2008/2008-11-19-091.asp.

Carlton, Jim. "Citicorp Tries Banking on the Natural Kind of Green." *Wall Street Journal*, September 5, 2007, B-1, B-8.

Deutsch, Claudia. "Companies Giving Green an Office." *New York Times*, July 3, 2007a. http://www.nytimes.com/2007/07/03/business/03sustain.html.

Deutsch, Claudia. "A Threat So Big, Academics Try Collaboration." *New York Times*, December 25, 2007b. http://www.nytimes.com/2007/12/25/business/25sustain.html.

"Investors Sizing Up 'Carbon Footprints.'" Cox News Service in *Omaha World-Herald*, September 30, 2007, D-1.

Lublin, Joann S. "Environmentalism Sprouts Up on Corporate Boards." *Wall Street Journal*, August 11, 2006, B-6.

Martin, Andrew. "In Eco-friendly Factory, Low-Guilt Potato Chips." *New York Times*, November 15, 2007, A-1. A-22.

Rimer, Sara. "How Green Is the College? Time the Showers." *New York Times*, May 26, 2008. http://www.nytimes.com/2008/05/26/education/26green.html.

Spencer, Jane. "Big Firms to Press Suppliers on Climate." *Wall Street Journal*, October 9, 2007, A-7.

Corporations and Global Warming

While most corporations dismissed the impact of global warming during the 1990s, within 10 years, by 2007, rising temperatures, accumulating scientific evidence, and Democratic control of the U.S. Congress was changing the business landscape.

By early in 2007, 10 major companies, including utilities, manufacturing, petroleum, chemicals, and financial services—household names such as General Electric, DuPont, Duke Energy, Caterpillar, British Petroleum (BP), Lehman Brothers, and Alcoa—had joined several national environmental groups in support of nationwide limits on carbon-dioxide emissions in a cap-and-trade market system. The corporations reacted to the threat that states acting on their own might enact a complex quilt of state and local controls. The companies also wanted to shape legislation before it was imposed on them. Many feared a stiff carbon tax.

An account in the *New York Times* sketched the corporate agenda:

> In addition, the chief executives agreed after some discussion, to "strongly discourage further construction of stationary sources that cannot easily capture" carbon dioxide. This comes close to a rejection of almost all new coal-fired power plants on the drawing boards, including the 11 plants recently proposed by TXU, a Texas utility. The technology that would isolate carbon dioxide emissions and bury them is still in the earliest phases of development, so this near-repudiation of existing coal technology would have a disproportionate impact on utilities that depend largely on coal, like TXU and the Southern Company. (Barringer 2007a)

Many companies also undertook detailed audits of their own energy use. The earliest included some surprising names. By 2000, for example, the DuPont Company, once the world's largest producer of ozone-consuming chlorofluorocarbons (CFCs), announced that an ambitious corporate program had eliminated half of the company's greenhouse-gas emissions since 1990, nearly all of it without losing sales or profits.

Atlantic Richfield Oil's chief executive, Mike Bowlin, gave a speech to an oil-industry audience in January 1999, during which he said that his industry was entering "the last days of the age of oil." Oil and gas companies face a crucial choice, he said. "Embrace the future and recognize the growing demand for a wide array of fuels; or ignore reality and slowly but surely be left behind" (Leggett 2001, 329).

Green Talk at the Ford Motor Company

During the first week of October 2000, the chief executive officer of Ford Motor Company, speaking before a Greenpeace business conference in London, predicted the demise of the internal-combustion engine. William Clay Ford, 43, great-grandson of Henry Ford, described an all-out race by automakers to design the first mass-produced hydrogen fuel-cell vehicles, anticipating that major Japanese and American automakers would be using fuel-cell vehicles to supersede gasoline engines by the year 2003 (nothing of the sort happened).

At that time, according to one newspaper account, Ford called global warming "the most challenging issue facing the world." Furthermore, said Ford, "I believe fuel cells will finally end the 100-year reign of the internal-combustion engine" (McCarthy 2000a, 10). Ford continued, "The climate appears to be changing, the changes appear to be outside natural variation, and the likely consequences will be serious. From a business-planning point of view, that issue is settled. Anyone who disagrees is, in my view, still in denial" (McCarthy 2000a, 10).

Ford at one point assembled a well-financed corporate division that promoted alternatives to the internal-combustion engine, including hydrogen fuel cells, electric commuter cars, and even bicycles. "The global temperature is rising and the evidence suggest that the shift is being affected by human activity, including emissions related to fossil fuels used for transportation," Ford told Paul McKay of the *Ottawa Citizen*. "We believe it is time to take appropriate action" (McKay 2001, A-1). Other automakers also have paid at least lip service to the issue. At about the same time, Daimler-Chrsyler's chairman of the board, Juergen Schrempp, also said that he supported the goals of the Kyoto Protocol.

Ford Motor earned some notice among environmentalists for its annual "corporate citizenship reports," which included frank discussion of subjects such as the impact of sports-utility vehicles (SUVs) on global warming. Ford's first corporate-responsibility report, issued in 2000, took issue with the company's rising production of gas-thirsty, carbon-belching SUVs.

In its second "corporate citizenship report," released during 2001, Ford said that it had formed a committee of executives to examine ways to reduce the company's contribution to global warming. "On the issue of climate change, there's no doubt that sufficient evidence exists to move from argument to action," said Jacques Nasser, Ford's chief executive, in a preamble to the report (Bradsher 2001, C-3). The 2002 report, Ford's third, was released at a time of financial distress for the company. In that report, Ford said "difficult business conditions make it harder to achieve the goals we set for ourselves in many areas, including corporate citizenship.... But that doesn't mean we will abandon our goals or change our direction" (Hakim 2002, C-4).

In the talk-the-talk department, Ford committed itself to a 25 percent improvement in gas mileage for its SUVs in five years, a goal that, by 2008, had not even been seriously breathed on in that part of reality where the rubber meets the road. Lip service is about all that environmentalists got from Ford; gas mileage on for its 2003 model SUVs was worse than for the 2002s. Still, Ford executives were very good at dreaming about fuel efficiency. On August 4, 2004, for example, the company's highest executives, including William Clay Ford, gathered at corporate headquarters. They came away, according to a *New York Times* report two months later, glowing to each other about a projected 80 percent improvement in fuel efficiency by the year 2030. The *Times* report added that "[t]he company had not planned to publicize the strategy because it is a long-term objective subject to change and because the company has recently been under attack from environmental groups for falling short of previously stated environmental objectives" (Hakim 2004b, 11).

The *Montreal Gazette* complained editorially that the automotive industry seemed intent on selling the car-buying public ever-larger SUVs no matter what the Intergovernmental Panel on Climate Change (IPCC) and the industry's own spin-doctors said about their effects on climate.

> It's easy to mistake the latest automotive news for a joke. Consumers next year will be able to buy a passenger vehicle that is a meter longer than today's largest-sport utility vehicle and twice—repeat, twice—as heavy. It sounds like a joke, but it's true—and all too symptomatic of North America's breezy insouciance toward gas guzzling and climate change. Appropriately, the vehicle's maker will be Freightliner, the Daimler-Chrysler subsidiary best known for turning out 18-wheelers.... The vehicle's ad slogan resonates: "You don't need roads, when you can make your own." It perfectly captures the ethos that nature should not get in the way. ("Global Warning" 2001, B-2)

Shareholders Raise the Issue

A minority of shareholders at some companies have become restive with regard to climate change. Shareholders, for example, have filed global-warming resolutions at General Motors and Ford annual meetings that attempt to increase pressure on the automakers to reduce their greenhouse-gas emissions. A coalition of shareholders—often members of various Catholic orders—asked that the automakers report carbon-dioxide emissions from their plants and vehicles. "We believe that both General Motors and Ford face material and reputational risk in their current failure to address and reduce carbon dioxide emissions," said Sister Patricia Daly, executive director of the Tri-State Coalition for Responsible Investment (Eggert 2002).

A resolution calling for a report on how ExxonMobil will respond to the growing pressure to develop renewable energy won support from 21 percent of shareholders at the ExxonMobil annual general meeting May 28, 2003. The 21 percent shareholder support for the renewables resolution represented $42.34 billion worth of ExxonMobil stock ("One in Five" 2003). The same day, almost a quarter of Southern Company shareholders (one of the largest U.S. electric utilities) voted to require the company to evaluate potential financial risks associated with its emissions of fossil fuels. Earlier, 27 percent of shareholders at American Electric Power (AEP) Company had voted to support a similar resolution. Southern, AEP, Xcel Energy Co., Cinergy Co., and TXU Corporation are the top five emission sources of carbon dioxide in the United States. A similar resolution at Chevron-Texaco won 32 percent of shareholders' votes in 2003, up from 9.6 percent in 2001 (Seelye 2003, C-1).

Some Businesses Accept and Profit from Carbon Curbs

In some cases, private businesses in the United States have taken actions ahead of the federal government. "We accept that the science on global warming is overwhelming," said John W. Rowe, chairman and chief executive officer of Exelon Corp. "There should be mandatory carbon constraints" (Carey and Shapiro 2004). Exelon, the United States' largest operator of commercial nuclear-power plants, probably would benefit from such constraints.

Shell Company's chairman Sir Philip Watts has called for global warming skeptics to accept action to limit greenhouse-gas emissions "before it is too late" (Macalister 2003, 19). Watts said that "[w]e can't wait to answer all questions [on global warming] beyond reasonable doubt," adding that "there is compelling evidence that climate change is a threat" (Macalister 2003, 19). Watts made his remarks at the opening of a new Shell Center for Sustainability at Rice University in Houston. Executives at Shell have "seen and heard enough" to believe that the burning of fossil fuels poses a problem, Watts said. "We stand with those who are prepared to take action to solve that problem ... now ... before it is too late ... and we believe that businesses, like Shell, can help to bridge differences that divide the U.S. and Europe on this issue" (Macalister 2003, 19).

Shell Oil has developed a roster of alternative energy projects through a $500 million investment in a subsidiary, Shell Renewables. Shell constructed a four-megawatt wind farm off the coast of northeastern England that was supplying 3,000 homes with electricity by 2001. British Petroleum Solar by 2000 had become the largest solar-power company in the world. In 2000, in an early initiative, it sold 40 megawatts of new capacity, a large amount at that time. ExxonMobil, on the other hand, spent $500 million on renewable energy and then "decided that the practical [obstacles to] turning renewables into profitable energy sources were too great to warrant further funding" (Schrope 2001, 519).

AEP, which burns more coal than any other U.S. electric utility, once resisted the idea of combating climate change. During the late 1990s, however, then–chief executive officer E. Linn Draper Jr. pushed for a strategy shift, preparing for limits instead of denying that global warming existed. "We felt it was inevitable that we were going to live in a carbon-constrained world," said Dale E. Heydlauff, AEP's senior vice president for environmental affairs (Carey and Shapiro 2004). AEP has invested in renewable energy projects in Chile, retrofitted school buildings in Bulgaria for greater efficiency, and explored ways to burn coal more cleanly.

Florida Power & Light Co. by 2004 had 42 wind-power facilities and had promoted energy efficiency, reduced emissions, and eliminated the need to build 10 midsize power plants, according to Randall R. LaBauve, vice-president for environmental services (Carey and Shapiro 2004). Private U.S. companies in 2004 formed a Climate Group to share information about climate-related business planning, including everything from new technologies to plugging leaks. In the meantime, British Petroleum (whose advertising styles "BP as "Beyond Petroleum") reduced its greenhouse-gas emissions 10 percent and saved $650 million during three years by performing a detailed energy inventory of its operations.

FURTHER READING

Barringer, Felicity. "A Coalition for Firm Limit on Emissions." *New York Times*, January 19, 2007a. http://www.nytimes.com/2007/01/19/business/19carbon.html.

Bradsher, Keith. "Ford Tries to Burnish Image By Looking to Cut Emissions." *New York Times*, May 4, 2001, C-3.

Carey, John, and Sarah R. Shapiro. "Consensus Is Growing among Scientists, Governments, and Business That They Must Act Fast to Combat Climate Change. This Has Already Sparked Efforts to Limit CO_2 Emissions. Many Companies Are Now Preparing for a Carbon-constrained World." *Business Week*, August 16, 2004, n.p. (LEXIS)

Eggert, David. "Shareholders File Global Warming Resolutions at Ford, GM." Associated Press, December 11, 2002. (LEXIS)

"Global Warning on Climate." *Montreal Gazette*, February 27, 2001, B-2.

Hakim, Danny. "Ford Stresses Business, but Disappoints Environmentalists." *New York Times*, August 20, 2002, C-4.

Hakim, Danny. "Ford Executives Adopt Ambitious Plan to Rein in Global Warming." *New York Times* in *International Herald-Tribune*, October 5, 2004b, 11.

Leggett, Jeremy. *The Carbon War: Global Warming and the End of the Oil Era*. New York: Routledge, 2001.

Macalister, Terry. "Shell Chief Delivers Global Warming Warning to Bush in His Own Back Yard." *London Guardian*, March 12, 2003, 19.

McCarthy, Michael. "Ford Predicts End of Car Pollution: Boss Predicts the End of Petrol." *London Independent*, October 6, 2000a, 10.

McKay, Paul. "Ford Leads Big Three in Green Makeover." *Ottawa Citizen*, May 28, 2001, A-1.

"One in Five ExxonMobil Shareholders Want Climate Action." Environment News Service. May 28, 2003. http://ens-news.com/ens/may2003/2003-05-28-09.asp#anchor3.

Schrope, Mark. "Global Warming: A Change of Climate for Big Oil." *Nature* 411 (May 31, 2001):516–518.

Seelye, Katharine. "Environmental Groups Gain as Companies Vote on Issues." *New York Times*, May 29, 2003, C-1.

Creation Care

Religious people have been turning their attention more frequently and avidly toward environmental affairs, under the rubric of "creation care," or as the Bible phrases it, stewardship of the Earth. Large, drafty churches with high ceilings have been adding insulation through a ministry named Interfaith Power and Light. The vice president of the National Association of Evangelical bought a Toyota Prius as more than 50 other evangelical Christian leaders pledged to neutralize their "carbon footprints" through energy conservation (Cooperman 2007b, A-1).

Such stewardship of the Earth is being applied to various types of greenhouse-gas-creating activities. Witness Leslie Lowe, energy and environmental director of the Interfaith Center on Corporate Responsibility, who characterizes the building of coal-fired power plants as "under any faith-based view of the world, a blasphemy and a sin. It is to profit from the destruction of the natural world ... and from the poisoning of other human life on this planet" (Jordan 2007b, D-1).

By 2008, Interfaith Power and Light had enlisted 5,000 congregations across 28 states in the United States to actively cut greenhouse-gas emissions. Richard Cizik of the National Association of Evangelicals said that stewardship of the Earth is not solely a left-wing concept. "Aren't conservatives supposed to be conservers?" he asked (Brinton 2008, 11-A). The Sierra Club by 2008 was actively courting religious groups.

Mike Huckabee, an ordained Southern Baptist minister and former governor of Arkansas, ran for president in 2008 as a Republican populist. Before the caucuses, he neatly defined the growing trend toward "creation care," saying that he believed in mandatory curbs for greenhouse-gas emissions because he lives "in God's house." Huckabee said, "I believe that God is the creator of the Earth. He owns it. I'm his guest. I have a responsibility to take care of it" (Tysver 2007, A-1). "My faith is my life," Huckabee told voters in Iowa as they geared up for caucuses in early January 2008. "It defines me. My faith doesn't influence my decisions. It drives them; when it comes to the environment, I believe in being a good steward of the Earth" (Tysver 2007, A-1). Otherwise, however, Huckabee is no liberal. He does not believe in "coddling Islamic extremists" and told voters that the war in Iraq was the frontline against religious evil. He opposed abortion and universal health care as well, is a fervent advocate of free-market capitalism, and has doubts about evolution.

During 2007, Evangelical Christian leaders openly split over the issue of global warming as a spiritual calling. Repudiating Christian radio commentator James C. Dobson, of the National Association of Evangelicals, the board of directors reaffirmed its position that environmental protection, which it calls "creation care," is an important moral issue. Dobson, founder of Focus on the Family, and several other conservative Christian leaders, including Gary L. Bauer, Tony Perkins, and Paul M. Weyrich, had sent the board a letter denouncing the association's vice president, Reverend Cizik, for urging attention to global warming (Cooperman 2007a, A-5).

The letter asserted that evangelicals were divided on whether climate change is a real problem. The more right-wing leaders believed that pursuing global warming as a priority would draw resources away from issues that they regarded as more important, such as abortion and same-sex marriage. During March 2007, the association's board had approved a 12-page statement on terrorism and torture. Anderson said that Cizik gave a report to the board on his work in Washington, D.C., as vice president for governmental affairs and that there was no effort to reprimand him for his stance on global warming. "I think there was a lot of support from me, from the executive committee and from the board for Rich Cizik," Anderson said (Cooperman 2007a, A-5).

One of the most pointed exercises of "creation care" involves investment managers of theologically based assets, many of them quite small, who purchase small blocks of energy and utility companies so they can raise global warming as a moral issue at annual meetings. Patricia Daly, for example, who manages finances for a Catholic order, Sisters of Saint Dominic of Caldwell, New Jersey, purchased 300 shares of ExxonMobil (of 5.5 billion shares outstanding), the world's largest and most profitable energy company, for the express purpose of introducing shareholder resolutions on the subject. Daly and other Catholic activists began filing global-warming resolutions in 1997, usually asking the company to account for its greenhouse-gas emissions. The first such resolution was approved by only 4.6 percent of proxy votes that year. By 2007, Daly's resolution won approval by 31 percent, or about 1.4 billion shares (Slater 2007, 22–27). Gary Stuard, executive director of the Interfaith Environmental Alliance, said

> Given the now overwhelming evidence for the reality of global warming and the growing clarity of what is at stake for humanity and for all life on this planet if we still refuse to address the threat facing us, the time for action to be taken to prevent the worse consequences from climate change from happening is now. ("Extreme Rainstorms" 2007)

"People of faith from the various religious traditions ... and around the world are calling upon ALL clergy and lay women and men, politicians, and business leaders to take on the moral leadership to courageously and selflessly address this pressing crisis," said Stuard ("Extreme Rainstorms" 2007).

The Vatican agreed during 2007 to fund the planting of trees on 37 acres of weedy land near the Tisza River in Hungary in an effort to offset its carbon emissions "As the Holy Father, Pope Benedict XVI, recently stated, the international community needs to respect and encourage a 'green culture'," said Cardinal Paul Poupard, leader of the Pontifical Council for Culture. "The Book of *Genesis* tells us of a beginning in which God placed man as guardian over the earth to make it fruitful," he added (Rosenthal 2007c). This a new twist on the Biblical passage describing Adam and Eve's expulsion from the Garden of Eden to go forth, multiply, and subdue the Earth.

The Vatican, which also has installed solar panels atop ancient buildings, has not yet embraced reduction of population as a strategy to implement its new green consciousness, however. After the Vatican agreement was announced, Monsignor Melchor Sánchez de Toca Alameda, an official at the Council for Culture at the Vatican, told the Catholic News Service that buying credits was like doing penance. "One can emit less CO_2 by not using heating and not driving a car, or one can do penance by intervening to offset emissions, in this case by planting trees," he said (Rosenthal 2007c).

Two Republican candidates for president in 2008, appealing to religious conservatives, former Governor Mike Huckabee of Arkansas and Senator Sam Brownback of Kansas, called for strong actions to address global warming, calling for change in spiritual terms (Santora 2007).

Scientists and Creation Care

James E. Hansen, NASA climate scientist, has referenced creation in several of his statements, including the following:

> [M]y feeling about that topic developed during a meeting with evangelical leaders on a Georgia plantation. We found no reason for conflict between science and religion, but many reasons for working together. We all felt strongly about the need for stewardship, for passing on to our children and grandchildren the planet that we received, with its remarkable forms of life. (Hansen 2007e)

Hansen added:

> Burning all fossil fuels, if the CO_2 is released into the air, *would destroy creation*, the planet with its animal and plant life as it has existed for the past several thousand years, the time of civilization, the Holocene, the period of relative climate stability, warm enough to keep ice sheets off North America and Eurasia, but cool enough to maintain Antarctic and Greenland ice, and thus a stable sea level. We cannot pretend that we do not know the consequences of burning all fossil fuels. (Hansen 2007e)

Hansen relates climate activism to a "strategy" for saving creation:

> Actions needed to stabilize climate. Two fossil fuel facts define the basic actions that are required to preserve our planet's climate: (1) it is impractical to capture CO_2 as it is emitted by vehicles (the mass of emitted CO_2 is about three times larger than the mass of fuel in the tank), and (2) there is much more CO_2 contained in coal and unconventional fossil fuels than in oil and gas. As a consequence, the *strategy for saving creation* must have two basic elements. First, and this is 80 percent of the solution, coal use must be phased out except where the CO_2 is captured and sequestered. Thus there should be a moratorium on construction of new coal- fired power plants until the technology for CO_2 capture and sequestration is ready. (Hansen 2007e)

FURTHER READING

Brinton, Henry G. "Green, Meet God." *USA Today*, November 10, 2008, 11-A.

Cooperman, Alan. "Evangelical Body Stays Course on Warming: Conservatives Oppose Stance." *Washington Post*, March 11, 2007a, A-5. http://www.washingtonpost.com/wp-dyn/content/article/2007/03/10/AR2007031001175_pf.html.

Cooperman, Alan. "Eco-Kosher Movement Aims to Heed Tradition, Conscience." *Washington Post*, July 7, 2007b, A-1. http://www.washingtonpost.com/wp-dyn/content/article/2007/07/06/AR2007070602092_pf.html.

"Extreme Rainstorms a New Texas Trend." Environment News Service, December 3, 2007. http://www.ens-newswire.com/ens/dec2007/2007-12-04-095.asp.

Hansen, James E. Personal communication, November 27, 2007e.

Jordan, Steve. "Coal-emission Cleanup a Challenge for Utilities." *Omaha World-Herald*, October 8, 2007b, D-1, D-2.

Rosenthal, Elisabeth. "Vatican Penance: Forgive Us Our Carbon Output." *New York Times*, September 17, 2007c. http://www.nytimes.com/2007/09/17/world/europe/17carbon.html.

Santora, Marc. "Global Warming Starts to Divide G.O.P. Contenders." *New York Times*, October 17, 2007. http://www.nytimes.com/2007/10/17/us/politics/17climate.html.

Slater, Dashka. "Public Corporations Shall Take Us Seriously." *New York Times Sunday Magazine*, August 12, 2007, 22–27.

Tysver, Robynn. "Populist Huckabee Gains Ground with Religious Approach." *Omaha World-Herald*, December 2, 2007, A-1, A-2.

Cretaceous and Sea-Surface Temperatures

Data presented at the 1996 American Association for the Advancement of Science annual meeting in St, Louis by climate modeler Karen Bice of Woods Hole Oceanographic Institution from seafloor cores (analysis of shells) indicated that sea-surface temperatures during the Cretaceous (65 million to 145 million years ago) may have reached 42°C, 14°C higher than at present and 5°C higher than former estimates for that time. Atmospheric carbon-dioxide levels during 20 million years of that period were two to six times the present level of 380 million parts per million (ppm). This research, published in *Paleoceanography*, brought "into question whether there are any limits to the temperature in the tropics in the future," said paleoclimate specialist Mark Chandler of Columbia University (Kintisch 2006a, 1095).

Bice said that her results did not fit the climate models she was using unless she assumed that methane was 30 times as abundant as now. Failing that far-fetched conclusion, present models are underestimating the sensitivity of temperature to rising levels of carbon dioxide. If her models hold up, Bice said that global warming "is going to be more dramatic than what is shown in these models" (Kintisch 2006a, 1095).

FURTHER READING

Kintisch, Eli. "Hot Times in the Cretaceous Oceans." *Science* 311 (February 24, 2006a):1095.

Crichton, Michael, Author of *State of Fear*

During 2004, Michael Crichton, seller of 100 million books to date (including *Jurassic Park*), threw *State of Fear*—a 603-page novel published by HarperCollins with a 14-page bibliography and a five-page "factual" antiwarming screed—into the weather wars. *State of Fear* creates a world in which environmentalists fake global warming and murder contrarians. Two million copies of this fantasy hit bookstores just before Christmas.

Crichton's fake take puts a new angle on the Green Party. Crichton's greenies are unnaturally powerful people who are capable of provoking earthquakes, underwater landslides, and even a tsunami to cow the gullible into believing that global warming is a threat. And what do earthquakes have to do with global warming? Ask Crichton. Sound science, this most certainly isn't.

As described by Michiko Kakutani in the *New York Times*, Crichton's villains are

> tree-hugging environmentalists, believers in global warming, proponents of the Kyoto Protocol. Their surveillance operatives drive politically correct, hybrid Priuses; their hit men use an exotic, poisonous Australian octopus as their weapon of choice. Their unwitting (and sometimes, witting) allies are—natch!—the liberal media, trial lawyers, Hollywood celebrities, mainstream environmental groups (like the Sierra Club and the Audubon Society) and other blue-state apparatchiks. (Kakutani 2004, E-1)

This "ham-handed" novel, Kakutani continued, "[r]eads like a shrill, preposterous right-wing answer ... [to the] shrill, preposterous but campily entertaining global warming disaster movie *The Day After Tomorrow*" (Kakutani 2004, E-1).

FURTHER READING

Kakutani, Michiko. "Beware! Tree-Huggers Plot Evil to Save World." *New York Times*, December 13, 2004, E-1.

Cuckoo Numbers Decline in Great Britain

The cuckoo, England's herald of spring, a bird well-known for the male's distinctive call (as well as for laying its eggs in nests assembled by other species) is disappearing from the British countryside. Cuckoo numbers have declined by 30 percent during the past 30 years in urban areas, while in woodland areas, cuckoos have declined by as much as 60 percent. The cuckoo migrates to all parts of the country from winter feeding grounds in Sub-Saharan Africa. David Marley of the Woodland Trust said that the cuckoo's preferred woodland habitat was particularly vulnerable to global warming, which could be affecting its breeding season and food supplies (Smith 2002).

"The cuckoo is one of the most amazing birds you can come across in Britain, but it is declining by a staggering amount and we are getting reports from across the country that people just aren't hearing it any more," Marley said. Other theories for the 30-year slump in cuckoo populations include habitat loss and the spread of intensive farming practices. At its wintering places in Africa, Cuckoos also may be suffering from indiscriminate use of agricultural chemicals as well as widespread drought (Smith 2002).

The British Trust for Ornithology (BTO) believes that cuckoos in Britain number between 12,000 and 24,000 pairs, down from 17,500 to 35,000 in 1970. The Woodland Trust has launched a study that will record cuckoo sightings and build up data to document the changing ecology that may explain why cuckoos are in sharp decline. Reports will be added to data collected since 1726 to monitor the impact of climate change. Volunteers will watch for cuckoos, as well as other wildlife, including ladybirds, bumble bees, and swallows (Smith 2002).

While some birds were declining in England as weather warmed, others were flourishing. For example, the number of wild parrots was rising rapidly by 2004, with 100,000 expected by the end of the decade. The parrots, which are large and aggressive birds, were competing with domestic species, such as starlings, jackdaws, and small owls, for food and territory (Prigg 2004, A-9).

FURTHER READING

Prigg, Mark. "Despite All the Heavy Rain, That Was the Hottest June for 28 Years." London *Evening Standard*, July 1, 2004, A-9.

Smith, Lewis. Falling Numbers Silence Cuckoo's Call of Spring." *London Times*, March 6, 2002, n.p. (LEXIS)

Cyanobacterial Algal Blooms

Rising temperatures favor the growth of cyanobacterial algal blooms, which have flourished in over-enriched waters as a by-product of urban, agricultural, and industrial development. As is so often the case, one human activity aggravates others in degrading the environment. In this case, these blooms "increase the turbidity of aquatic ecosystems, smothering aquatic plants and thereby suppressing important invertebrate and fish habitats" (Paerl and Huisman 2008, 57). Lake Victoria (in Africa), Lake Eire in North America, Lake Taihu in China, and Europe's Baltic Sea have been afflicted by excessive growth of these algae, which produce toxins that contribute to human liver, digestive, and neurological skin diseases that sometimes have been fatal. Expanding blooms of this type may threaten supplies of drinking water and edible fish.

The growth of cyanobacterial accelerates as temperatures rise top above 25°C. Thus, they enjoy an advantage over other types of algae in a warmer environment. Warming surface waters also impede vertical mixing of lake waters, allowing even more growth for the toxic blooms on and near the surface. Increasing cyanobacterial absorb more heat, altering their environment still further to increase their growth. Increased growth in turn increases their buoyancy and ability to crowd out phytoplankton that are less buoyant and thrive less well in warmer water.

Extremes in the hydrological cycle assist cyanobacterial dominance by pumping nutrients into bodies of water in bursts, feeding the algae. The toxic algae then consolidate their dominance during periods of drought. These algae are more tolerant of salt than most freshwater species, and so may be aided by other human activities, such as the use of deicing agents on roads. As temperatures warm, these organisms are being observed farther north in Europe and North America.

FURTHER READING

Paerl, Hans W., and Jef Huisman. "Blooms Like It Hot." *Science* 320 (April 4, 2008):57–58.

D

Darfur

Stephan Faris, writing in the April 2007 edition of *The Atlantic*, traced the conflict between ethnic groups in Darfur to decreasing rainfall and increasing aridity that has increased competition for land, a likely product of climate change. "By the time of the Darfur conflict four years ago, scientists had identified another cause [in addition to ethnic conflict]," wrote Faris.

> Climate scientists fed historical sea-surface temperatures into a variety of computer models of atmospheric change. Given the particular pattern of ocean-temperature changes worldwide, the models strongly predicted a disruption in African monsoons. "This was not caused by people cutting trees or overgrazing," says Columbia University's Alessandra Giannini, who led one of the analyses. The roots of the drying of Darfur, she and her colleagues have found, lay in the changes to the global climate.... With countries across the region and around the world facing similar pressures, some see Darfur as a canary in the coal mine, a foretaste of climate-driven political chaos. (Faris 2007, 69)

"There is a very strong link between land degradation, desertification and conflict in Darfur," said a United Nations Environmental Program report, which noted that rainfall in northern Darfur has decreased by a third over the last 80 years. "Exponential population growth and related environmental stress have created the conditions for conflicts to be triggered and sustained by political, tribal or ethnic differences," the report said, adding that Darfur "can be considered a tragic example of the social breakdown that can result from ecological collapse" (Polgreen 2007).

FURTHER READING

Faris, Stephan. "The Real Roots of Darfur." *The Atlantic*, April 2007, 67–69.

Polgreen, Lydia. "New Depths: A Godsend for Darfur, or a Curse?" *New York Times*, July 22, 2007. http://www.nytimes.com/2007/07/22/weekinreview/22polgreen.html.

Deforestation

The Earth had less than five million square miles of old-growth forest remaining by about 2005. An estimated deforestation rate of 62,000 square-miles a year, the amount being logged in the late 1990s, could reduce that figure to zero within two human lifetimes. More than 50 percent of the world's forests have been destroyed within the past 100 years. An area equivalent to 57 soccer fields falls each minute (Webb 1998c).

Trees are roughly 50 percent carbon. Alive, they remove carbon dioxide from the atmosphere and replace it with oxygen. Dying or dead trees (or those being burned as fuel) become sources, not absorbers ("sinks"), of carbon dioxide and other greenhouse gases. The destruction of forests around the world is no trivial matter for the atmosphere's carbon budget. While the atmosphere contains about 750 billion tons of carbon dioxide, forests contain about 2,000 billion tons. Roughly 500 billion tons is stored in trees and shrubs and 1,500 billion tons in peat bogs, soil and forest litter (Jardine 1994).

By 2007, about one-fifth of the human-caused greenhouse gases being released into the atmosphere came from deforestation, during which carbon stored in trees enters the air. Indonesia has been clearing more forests than any other country. In some areas, such as the province of Riau (on the island of Sumatra), more than half the forests have been felled in a decade, many for palm oil plantations, often for ethanol, a supposedly Earth-friendly fuel that replaces oil-based products, such as gasoline. The burning and drying of Riau's peatlands, also to make way for

Monteverde Golden Toad. (Charles H. Smith/U.S. Fish and Wildlife Service)

palm oil plantations, releases about 1.8 billion tons of greenhouse gases a year (Gelling 2007).

Stephen Schwartzman, a senior scientist with the International Program of the Environmental Defense Fund, sketches the scope of tropical deforestation:

> An area of forest bigger than Belgium, Holland and Austria put together, or about 40 percent of California, was cut down and burned every year between 1980 and 1995, some 62,000 square miles per year. NASA's Landsat satellite photographs show that more than 200,000 square miles, an area about the size of France, has been cleared and burned in Brazil alone. All of this has happened since the 1970s. (Schwartzman 1999, 60)

Deforestation increased rapidly during the 1980s. Three million acres of forest were being lost per year in Indonesia, providing 70 percent of the world's plywood and 40 percent of its tropical hardwoods. By the early 1980s, in the tropical forests of Indonesia, roughly 1.2 percent of remaining coverage was being felled each year. In 1988, 48,000 square miles of Amazon rain forest were burned to clear land for farming. Nearly overnight, most of the burned land was turned from a net producer of oxygen to a source of carbon dioxide and methane (Gordon 1991,118–119).

As Indonesia's forests were being felled, government policy encouraged a homesteading program on Sumatra and Borneo for farmers moving from overcrowded Java. Indonesia was using some of the same government incentives Brazil was using to "develop" the Amazon Valley. During the 1980s, hundreds of thousands of Brazilians moved to the Amazon Valley in a fashion resembling the homesteading of the North American West a century earlier. Some of this migration was financed by international development agencies, such as the World Bank.

Deforestation has become a major problem (and political issue) in Mexico as well as in Central America. During the spring of 1999, fires raged out of control throughout the mountains of southern Mexico, sweeping through parts of the Chiapas highlands, as well as the Sierra Tarahumara of Chihuahua. The fires cloaked Mexican skies in an acrid haze from its borders with Guatemala and Texas. Among the causes of these fires are private timber operations on *ejido* (native-held) lands that compound the deforestation caused by *campesinos* (farm workers) clearing lands for their *milpas* (fields). In October 1997, deforestation took a deadly toll when Hurricane Paulina hit the Sierra Madre del Sur in the Mexican states of Guerrero and Oaxaca. Mudslides cascading down denuded mountainsides left scores dead and thousands homeless, especially in Zapotec country.

The main cause of tropical deforestation—perhaps two-thirds of it—is slash-and-burn agriculture. Much of the deforestation is caused, in turn, by small farmers and ranchers who are threatened by the spread of commercial farming and ranching enterprises. Driven out of more populous areas, subsistence farmers are forced to destroy large amounts of tropical forest to create new farmland. According to a study by the *Global Futures Bulletin*, the pattern is the same in much of Africa, Asia, and Latin America. Increasing human population and pressure on the land is the ultimate cause of the deforestation of slash-and-burn agriculture, according to the report.

Warming and Tropical Mountain Forests

Tropical mountain forests depend on predictable, frequent, and prolonged immersion in moisture-bearing clouds. Clearing lowland forests alters surface energy budgets in ways that influence dry-season cloud fields (Lawton et al. 2001, 584). Cloud forests form where mountains force trade winds to rise above the condensation level, the point of orographic cloud formation. "We all thought we were doing a great job of protecting mountain forests [in Costa Rica]," said Robert O. Lawton, a tropical forest ecologist at the University of Alabama in Huntsville, Alabama. "Now we're seeing that deforestation outside our mountain range, out of our control, can have a big impact" (Yoon 2001b, F-5).

The clearing of forests, sometimes many miles from the mountains, alters this pattern, often raising the elevation at which clouds form. Lawton and colleagues used Landsat and Geostationary Operational satellite images to measure such changes in the Monteverde cloud forests of Costa Rica. They found that "Our simulations suggest that conversion of forest to pasture has a significant impact on cloud formation" (Lawton et al. 2001, 586). Patterns found in Costa Rica resemble those in other tropical areas, including parts of the Amazon Valley. "These results suggest that current trends in tropical land use will force cloud forests upward, and they will thus decrease in area and become increasingly fragmented—and in many low mountains may disappear altogether," they explained (Lawton et al. 2001, 587).

Deforestation's effects probably extend further than most observers have heretofore believed. "Mountain forests ... may be affected by what's happening some distance away," said Lawton. Each year about 81,000 square miles of tropical forests are cleared, said Gary S. Hartshorn, president of the Organization for Tropical Studies, a consortium of rain forest researchers at Duke University (Polakovic 2001b, A-1). Thus, the weather in the lush cloud forests of Costa Rica is changing because of land-use changes, including deforestation, many miles away. As trees on Costa Rica's coastal plains are removed and replaced by farms, roads, and settlements, less moisture evaporates from soil and plants, in turn reducing clouds around forested peaks 65 miles away.

At risk is the Monteverde Cloud Forest Reserve, an ecosystem atop a Central American mountain spine, "a realm of moss and mist, the woodland in the clouds—a type of rain forest—[that] is home to more than 800 species of orchids and birds, as well as jaguar, ocelot and the Resplendent Quetzal, a plumed bird sacred to the Mayans" (Polakovic 2001b, A-1). The same area also is a watershed that supplies farms, towns, and hydropower plants in the lowlands.

Similar localized weather changes have been observed in deforested parts of the Amazon basin. Scientists say that cloud forests in Madagascar, the Andes, and New Guinea also are at risk. According to this study, the scientists concluded that "[t]hese results suggest that current trends in tropical land use will force cloud forests upward and they will thus decrease in area and become increasingly fragmented and in many low mountains may disappear altogether" (Polakovic 2001b, A-1). "It's incredibly ominous that over such a distance deforestation can alter clouds in mountains. This is a very serious concern," said Hartshorn. "This is confirmation of what we have predicted for a long time," said Stanford University ecologist Gretchen Daily. "The implications are very serious for the tropics and other parts of the world" (Polakovic 2001b, A-1).

Using data collected from satellites and computer models, scientists examined how forest clearing along the Caribbean coastline, where more than 80 percent of lowland forests have been cleared for farms and towns, influences weather downwind in the Cordillera de Tilaran mountain range. Evaporation from lowland vegetation is a principal source of moisture for the 4,000- to 5,000-foot mountains during the dry season of January to mid-May.

As Gary Polakovic explained in the *Los Angeles Times*:

> The researchers found that the moisture content of the clouds over the mountains has declined by about half since intensive land clearing began in the 1950s. Also, the cleared land is warmer, pushing the base of clouds nearly a quarter of a mile higher on some days, meaning they pass over the mountain range dropping little moisture. In contrast, clouds were more abundant over forested lowlands just across the border in Nicaragua, where forest still blankets much of the coastal plain. (Polakovic 2001b, A-1)

Tropical rainforests are typically the last areas colonized by people, a final refuge of biodiversity in places where lowlands have been cleared and developed. "Many cloud forest organisms have literally nowhere to go," said Dr. Nalini M. Nadkarni, an ecologist at Evergreen State College in Olympia, Washington. "They're stuck on an island of cloud forest. If you remove the cloud, it's curtains for them" (Yoon 2001b, F-5).

Many watersheds fed by wet highland forests also are threatened. "We always knew that if you have a town and a mountain behind it, you protect the mountain forests to protect the water," said James O. Juvik, a tropical ecologist at the University of Hawaii at Hilo. "This says, even if you leave the cloud forest intact like good conservationists, if you clear the lowland forests, you can diminish the cloud forest and affect your water flow" (Yoon 2001b, F-5).

The Monteverde Rainforest and the Golden Toad

Begun as a colony of expatriate North American Quakers, Monteverde has been a major attraction for American biologists, many of whom have devoted their careers to the flora and fauna of the region. Carol Kaesuk Yoon described the area in the *New York Times:* "A quick look at the forest's inhabitants makes clear why—from the record-breaking diversity of orchids to creatures like the resplendent quetzal, a glittering green, kite-tailed bird truly worthy of its name, and the so-called singing mice that whistle and chirp like birds" (Yoon 2001b, F-5). In addition to its value as a shelter for plants and animals, Monteverde also has become a center for ecotourism. Nearly 50,000 tourists visit each year to walk its many trails, "stopping for a bite to eat or perhaps picking up a golden toad mug or Monteverde T-shirt along the way." Scientists say that if the cloud forest falters, much business will be lost along with the unique species (Yoon 2001b, F-5).

Deforestation and a warming environment may help to explain the disappearance of *Bufo periglenes*, the golden toad of the Monteverde tropical rainforest, which was found nowhere else on Earth. No one knows how long the golden toad had lived in the cloud forest that runs along the Pacific coast of Costa Rica. By 1987, however, the level of mountain clouds had risen, reducing the frequency of mists during the dry season, and probably playing a role in a massive population crash that affected most of the 50 species of frogs and toads in the forest. No fewer than 20 species became locally extinct (Moss 2001, 18).

According to Yoon,

> A flashy orange creature last seen in the late 1980's, the golden toad has become an international symbol of the world's disappearing amphibians. Seeing how even pristine and protected forests such as those at Monteverde can lose their crucial mists and clouds, researchers say it becomes less mysterious how a water-loving creature like the golden toad could vanish even from a forest where every tree still stands. (Yoon 2001b, F-5)

No Longer the "Lungs of the World"

In Indonesia and Brazil, new landowners have been migrating into the tropical forests year by year, altering the landscape (and the Earth's carbon balance) by felling trees, then burning what they have cut. The rainforest canopy, which had heretofore produced a surplus of oxygen, is replaced by cattle, human beings, and motor vehicles. A survey conducted by the United Nations in 1982 estimated that tropical forests were being felled at a rate of 70,000 square miles (180,000 square kilometers) per year. The report estimated that 300 million subsistence farmers around the world, most of them in the tropics, were turning an area of forest the size of Denmark into a net producer of carbon every three months. In 1987, 25,000 square miles of Brazilian rainforest were felled; a year later, the area that was deforested doubled. In the 1980s, lowland tropical forests in Indonesia, the Philippines, Malaysia, and parts of West Africa shrank quickly as well.

According to the Brazilian government's own figures, the amount of Amazonian forest lost to fire doubled between 1997 and 1998. The Woods Hole Research Center in Massachusetts reported that 7,800 square miles of Amazonian rain forest caught fire in 1997. About 80 percent of the fires were set on purpose, according to the United Nations; 42 million acres of rain forest (an area the size of Florida) were being lost per year by the 1990s.

Amazon No Longer a Carbon Sink

Given deforestation and other land-use changes in the Amazon Valley, the area no longer functions generally as "the lungs of the world." The area acts as a carbon sink (absorber) only when temperature and precipitation patterns are healthy for the forest. In years of unusual drought (such as during the El Niño episodes of the middle and late 1990s), unusually hot, dry weather turns the Amazon into a net producer of carbon dioxide. In a usual year, the Amazon absorbs about 700 million tons of carbon dioxide, but in the recent El Niño year of 1995, the Amazon forest *added* 200 million tons of carbon dioxide to the atmosphere. According to A. Lindroth, A. Grelle, and A. S. Moren, writing in *Nature*, a decrease in soil moisture can turn a carbon-dioxide consuming forest into one that adds the gas to the atmosphere; this variation occurs in temperate-zone forests as well as those of the tropics, the authors asserted (Lindroth, Grelle, and Moren 1998).

By the 1990s, according to R. A. Houghton and colleagues, "This large area of tropical forest is nearly balanced with respect to carbon" (Houghton et al. 2000, 301).

> The combined effects of deforestation, abandonment, logging, and fire may thus yield sources of carbon.... These fluxes are similar in magnitude (but opposite in sign) to the sink calculated recently for natural ecosystems in the region. Taken together, the sources (from land-use change and fire) and the sinks (in natural forests) suggest that the net flux of carbon between Brazilian Amazonia and the atmosphere may be nearly zero, on average. (Houghton et al. 2000, 303–304)

Threats to the last remnants of a nation's forests sometimes provoke action. According to Schwartzman,

> China, not a world leader in green consciousness, last year [1998] banned all logging in its few remaining natural forests after disastrous flooding wreaked havoc along heavily populated rivers. In so doing, China hoped to save remnants of forest cover on the upper headwaters. But so much forest is already gone that it may not make much difference. (Schwartzman 1999, 61)

Deforestation has taken the trees from the world's tropical forests to some notable venues. Vintage ships of the British Royal Navy, for example, were refurbished during the 1990s with large amounts of mahogany harvested illegally from an indigenous reserve in the Amazon River Basin. The harvest raised a ruckus among English environmentalists, including Friends of the Earth, as the defense ministry's mahogany purchase was questioned on national television. More than 7,000 cubic feet of mahogany was cut, moving by 10 to 12 trucks a day from the Kayapo indigenous reserve in Para state to the town of Redencao, home of the Juary logging company, harvester of the lumber. Each log was harvested illegally under Brazilian law. Despite the illegality of the harvest (aided by slack enforcement), the mahogany was purchased for the British ministry of defense by an English supplier, Parker Kislingbury.

Deforestation and Flooding

About half of the rain that falls on a rainforest is produced by the forest's own humidity. Deforestation over large areas influences regional climate because there are fewer plants to produce the moisture that returns to the Earth as rain. More water runs off, carrying more topsoil, leaving less forest cover. Deforestation thus breaks a forest's hydrology cycle. As a tropical forest declines, regional weather becomes hotter, drier, and more prone to fire within surviving stands of trees. Increasing wildfires (such as those during the late 1990s in Mexico, Brazil, and Indonesia) add even more carbon to the atmosphere. By the late 1990s, the burning of tropical forests was contributing about 20 percent of the human-induced carbon-dioxide buildup in the atmosphere. The burning of the Amazon rainforest alone contributes about a quarter of the worldwide total.

In February 1999, deforestation was implicated as a major culprit when dozens of people died and hundreds of millions of dollars in property was destroyed in massive floods that shut down the industrial capital of South America, São Paulo. That same year, flooding (also aggravated by deforestation) provoked unprecedented destruction in Caracas, Venezuela. Deforestation also intensified the death and destruction wrought in Central America by Hurricane Mitch.

According to Stephen Schwartzman,

> The Woods Hole Research Center has found that for every acre cleared and burned in the Amazon, at least another acre burns in ground fires under the forest canopy or is degraded by selective logging (not picked up by the satellites). The frequency and extent of these ground fires skyrocket in El Niño events, which can then cause drought in some tropical forests ... such fires are likely to increase in frequency and intensity with global warming. (Schwartzman 1999, 63)

Deforestation is especially dangerous because once an area is denuded, local soil conditions (and sometimes weather) changes in such a way as to make the regrowth of trees more difficult. In many areas where deforestation has been extensive, previously forested areas have been replaced by sparse grasses, stunted shrubs, and bare, eroded earth.

Fire in the Amazon Valley

"The scale of the problem is mind-boggling," said Daniel C. Nepstad of the Woods Hole Research Center, who has been surveying deforestation in the Amazon Valley. Nepstad estimated that 400,000 square kilometers of rainforest, an area 20 times that of Massachusetts, became vulnerable to fire during 1998, compared with an average of 15,000 square kilometers a year that had been cleared and burned during previous few years. "Once burned," wrote Nepstad, "Amazonian forests become more susceptible to future burning. Wildlife is killed or dispersed (except for some species that thrive on disturbance). Timber, forest medicines and forest fruits are destroyed, and a large part of the carbon contained in the trees is gradually released into the atmosphere" (Nepstad 1998).

A major problem, according to Nepstad, "is that the rainforest is available in abundance and is therefore very cheap. When forest is cheap and labor and capital are scarce, it is the forest itself that becomes the fertilizer, the pesticide, the herbicide, the plow" (Nepstad 1998). Nepstad suggested "[a] new model of rural development ... in Amazonia, which restricts access to most (about 70 percent) of the region's forests, and which creates a situation of land scarcity, higher land prices, and higher investments in agricultural production systems" (Nepstad, 1998). Nepstad believes that people will value the forest only when it becomes an economic good in human financial terms. People will quit burning the forests when they can earn more money by sustaining it. About half of the area burned each year in Amazonia is accidental, according to Nepstad, who recommended prohibition of burning at times when fires may easily spread out of control.

Destruction of the Amazon rainforest accelerated toward the end of the twentieth century. According to studies by Nepstad, the forest was being destroyed two to three times more quickly than previous estimates. Nepstad conducted his research with scientists at the Institute of Environmental Research in Belem, Brazil. They measured forest losses at 1,104 sample points from a light plane and on the ground. They also interviewed more than 1,500 mill operators and landholders in the Amazon region. Nepstad and his associates assert that satellite imagery, on which most previous estimates have been based, "often fails to distinguish between pristine forest and burned or cut forest land newly covered with fast-growing brush" ("Amazon Destruction" 1999).

Research published during October 1998 by the Hadley Centre for Climate Prediction and Research, part of the United Kingdom Meteorological Office, anticipated that large areas of "closed" tropical forests will die as the world warms during the next century. Instead of soaking up pollution, these forests will dump more carbon dioxide into the atmosphere than all of the world's power stations and cars have produced during the past 30 years. "Closed" tropical forests, according to the Hadley Centre report, contain trees covering a large proportion of the ground where grass does not form a continuous layer on the forest floor. In Africa, forests of this type declined from 289,700 thousand hectares in 1980 to 241,500 in 1990. In Latin America, tropical forests declined from 825,000 to 753,000 hectares during the same period, and in Asia, tropical forest coverage fell from 334,500 hectares in 1980 to 287,500 in 1990 (National Academy of Sciences 1991, 9).

Deforestation sometimes places additional debits on the world account of greenhouse gases. When tropical forests are cut and slashed, for example, the resulting piles of dead wood attract termites, whose populations explode as they feast on a food supply that has suddenly increased. Termites very efficiently digest carbon, turning it into methane, a greenhouse gas (McKibben 1989, 17).

Amazon Valley Desiccated by 2050?

The 1998 Hadley Centre report indicated that large parts of the Amazon Valley could become desiccated by 2050, an event which could threaten the world with an unstoppable greenhouse effect. The Hadley report asserted that land temperatures will rise by an average 10°F over the next century. In addition to intensifying aridity in parts of tropical South America, large parts of tropical Africa also are expected to become desiccated by 2050, according to the Hadley Centre researchers. These climate changes will force roughly 30 million people to face intense hunger, the report asserted.

By about the year 2050, according to another study, "a decrease in annual rainfall of up to 500 millimeters in some key areas, combined with temperature rises of up to 7°C, will begin to kill tropical forests, creating deserts" (Nuttall 1998). At this point, the study said that these forests will become carbon sources rather than absorbers, as their remains burn or rot. This study found that

> Up to 2050, the land surface takes up carbon through an increased growing season, which is being measured at the moment. But when we move to 2050, the tropical dieback releases so much carbon to the atmosphere it causes the concentrations to increase. This may enhance the build-up of carbon dioxide in the atmosphere. (Nuttall 1998)

Hurricane Mitch and Deforestation

Five years before flooding from Hurricane Mitch devastated Honduras, J. Almendares, a Honduran medical doctor, warned readers of the British medical journal *Lancet* that deforestation was making the country more vulnerable than ever to deadly flooding. Almendares presented evidence that "desiccation and soil erosion caused by cattle grazing and sugarcane and cotton cultivation have altered the regional hydrological cycle" (Freier et al. 1993, 1464). These changes have led to fewer rainy days, but more intense downpours.

Almendares presented temperature statistics from one deforested area of Honduras indicating that the average ambient air temperature had risen 7.5°C between 1972 and 1990 (Freier et al. 1993, 1466). In neighboring Nicaragua, during the final months of the Sandinista revolution (1979), similar temperature rises were reported in Managua after dictator Anastasio Somoza ordered the wholesale destruction of trees in the city to deny Sandinistas places to hide during gun battles.

Central America had 200,000 square miles of forest in 1900. By the 1980s, only 36,000 square miles survived, and the rate of deforestation was increasing, especially in the Miskito region of Nicaragua and Honduras. A 1998 report by United Nations agencies and nongovernmental organizations documented a regional deforestation rate of 958,360 acres (1,500 square miles) a year (Weinburg 1999, 51).

Honduras lost a third of its forests between 1964 and 1990. Honduran forests continued to be felled at an annual rate of 80,000 hectares a year during the 1990s, a rate that, if sustained, would strip a quarter of remaining forested land per decade. In the meantime, people who can no longer wrest a living from denuded (or corporate-controlled) land have been moving to Honduran cities, where malaria has become endemic.

During October 1998, Hurricane Mitch made landfall on Central America's Miskito Coast, with winds as strong as 178 miles an hour, dropping as much as three feet of rain. Crossing the mountains, Mitch turned northwest, moving slowly through El Salvador, Guatemala, and southern Mexico. By the time the storm reached the sea again, it had killed 10,000 people and left nearly three million others homeless. In Honduras, thousands of people who survived the storm lost their jobs in devastated banana plantations.

The devastation wrought by Mitch raised questions in Central America not only about potential increases in hurricane severity caused by global warming, but also about changes in land use across the region that make many areas more prone to severe flooding during heavy rains. The same questions were raised in Caracas, Venezuela, after devastating floods there late in 1999.

Bill Weinburg described the conditions that have made hurricane flooding so devastating in Central America:

Fire and flood fuel each other in a vicious cycle. Landless peasants colonize the agricultural frontier, or clear-forested slopes for their *milpas* (fields). The more forest is destroyed, the more the hydrologic cycle is disrupted; with no canopy for transpiration, local rainfall, and cloud cover decline; aridity makes the surviving forest vulnerable to wildfires. Then, when the rains do come, sweeping in from the Caribbean on the trade winds, there are no roots to hold the soil and absorb the water. Millennia [of] accumulated wealth of organic matter is swept from the mountainsides in deluges of mud. Tlaloc, the revered Nahua rain god who the Maya called Chac Mool, brings destruction instead of abundance. (Weinburg 1999, 51–52)

In Nicaragua, 2,000 people died in the municipality of Posoltega, as 10 communities were buried in mudslides when the Casita volcano crater collapsed. Three-quarters of a million people were left homeless in the area. Posoltega lies in an area that has been almost completely deforested. Hurricane Mitch's trail of death and damage in eastern Nicaragua was intensified by deforestation of the upper Rio Coco watershed. More than 20 inches of rain caused the river to rise more than 60 feet within a few days.

Preservation of forests became an environmental issue in eastern Nicaragua following Hurricane Mitch's devastation. Local native peoples and ecologists united the same year to evict Korean-owned timber giant Solcarsa, which had won a government contract in a large area of the Miskito rainforest. According to Weinburg, at least 370,500 acres were being deforested annually in Nicaragua at the time of Hurricane Mitch. Nicaragua has lost 60 percent of its forest cover within the last two generations.

Hurricane Mitch left Nicaragua with an enduring legacy, as described by Weinburg:

[I]n September 1999, almost a year after the disaster … President Aleman declared a national emergency over a plague of rats in … Nicaragua. Their numbers exploded from an overabundance of dead meat—both human and animal—following the hurricane. The rats overran fields and homes, decimating crops. Poison had to be distributed in mass quantities to beat the infestation. (Weinburg 1999, 54)

Former Honduran President Rafael Callejas attributed the extreme death and damage wrought by Hurricane Mitch to "mudslides that were the result of uncontrolled deforestation and therefore could have been prevented" (Weinburg 1999, 54). Ecological activism can be as dangerous in Honduras as the winds, rains, and mudslides of a major hurricane. A few weeks before Mitch made landfall, Carlos Luna, a local opponent of logging in the central mountains near Tegucigalpa, the country's capital and largest city, was gunned down by unknown assailants.

FURTHER READING

Almendares, J., and M. Sierra. "Critical Conditions: A Profile of Honduras." *Lancet* 342 (December 4, 1993):1400–1403.

"Amazon Destruction More Rapid Than Expected." April 10, 1999. http://www.gsreport.com/articles/art000099.html.

Freier, J. E., D. J. Rogers, M. J. Packer, N. Nicholls, and J. Almendares. "Vector-borne Diseases." *Lancet* 342 (December 4, 1993):1464–1469.

Gelling, Peter. "Forest Loss in Sumatra Becomes a Global Issue." *New York Times*, December 6, 2007. http://www.nytimes.com/2007/12/06/world/asia/06indo.html.

Gordon, Anita, and David Suzuki. *It's a Matter of Survival.* Cambridge, MA: Harvard University Press, 1991.

Houghton, R. A., D. L. Skole, Carlos A. Nobre, J. L. Hackler, K. T. Lawrence, and W. H. Chomentowski. "Annual Fluxes of Carbon from Deforestation and Regrowth in the Brazilian Amazon." *Nature* 403 (January 20, 2000):301–304.

Jardine, Kevin. "The Carbon Bomb: Climate Change and the Fate of the Northern Boreal Forests." Ontario, Canada: Greenpeace International, 1994. http://dieoff.org/page129.htm.

Lawton, R. O., U. S. Nair, R. A. Pielke Sr., and R. M. Welch. "Climatic Impact of Tropical Lowland Deforestation on Nearby Montane Cloud Forests." *Science* 294 (October 19, 2001):584–587.

Lindroth, A., A. Grelle, and A. S. Moren. "Long-term Measurements of Boreal Forest Carbon Balance Reveal Large Temperature Sensitivity." *Global Change Biology* 4 (April 1998): 443–450.

McKibben, Bill. *The End of Nature.* New York: Random House, 1989.

Moss, Stephen. "Casualties." *London Guardian*, April 26, 2001, 18.

National Academy of Sciences. *Policy Implications of Greenhouse Warming.* Washington, DC: National Academy Press, 1991.

Nepstad, Daniel C. "Report from the Amazon: May 1998." Woods Hole Research Center. http://terra.whrc.org/science/tropfor/fire/report2.htm.

Nuttall, Nick. "Global Warming 'Will Turn Rainforests into Deserts.'" *London Times*, November 3, 1998. http://bonanza.lter.uaf.edu/~davev/nrm304/glbxnews.htm.

Polakovic, Gary. "Deforestation Far Away Hurts Rain Forests, Study Says: Downing Trees on Costa Rica's

Coastal Plains Inhibits Cloud Formation in Distant Peaks." *Los Angeles Times*, October 19, 2001b, A-1.

Schwartzman, Stephen. "Reigniting the Rainforest: Fires, Development and Deforestation." *Native Americas* 16, no. 3/4 (Fall/Winter 1999):60–63.

Webb, Jason. "World Forests Said Vulnerable to Global Warming." Reuters. November 4, 1998c. http://bonanza.lter.uaf.edu/~davev/nrm304/glbxnews.htm.

Weinburg, Bill. "Hurricane Mitch, Indigenous Peoples and Mesoamerica's Climate Disaster." *Native Americas* 16, no. 3/4 (Fall/Winter 1999):50–59. http://nativeamericas.aip.cornell.edu/fall99/fall99weinberg.html.

Yoon, Carol Kaesuk. "Something Missing in Fragile Cloud Forest: The Clouds." *New York Times*, November 20, 2001b, F-5.

Dengue Fever

In the Caribbean, cases of dengue hemorrhagic fever have climbed from a few hundred a year in the 1980s to as many as 8,000 a year since the early 1990s. There are concerns that rising cases of dengue could affect the economically important tourism industry. In one case study by the Assessments of Impacts and Adaptations to Climate Change (AIACC), researchers estimated that a 2°C temperature rise in the Caribbean by the 2080s could triple the cases of dengue fever. This AIACC study found, for example, that pupae of the dengue-carrying mosquito favor breeding in 40 gallon drums commonly used for outside water storage by informal settlements and poor households. Education on the disease and its transmission, targeted to these households, is suggested as one adaptation strategy, alongside measures to eliminate the mosquito breeding grounds ("Vulnerable Communities Worldwide" 2007).

During the summer of 2007, more than 100 people in the Italian village of Castiglione di Cervia fell ill with a tropical disease, chikungunya, similar to dengue fever, with high fever, exhaustion and excruciating bone pain, as well as fever as high as 104°F. The disease usually is spread by tiger mosquitoes in countries on the Indian Ocean. Aided by shipping and warming temperatures that allowed the mosquitoes to thrive, Castiglione di Cervia became the first site in modern Europe to experience such an outbreak. By 2007, the tiger mosquito had reached France and Switzerland. The joint pain was so bad that many people could not raise spoons to their mouths nor get out of bed. Joint pain and arthritis persisted for months after initial attacks (Rosenthal 2007e).

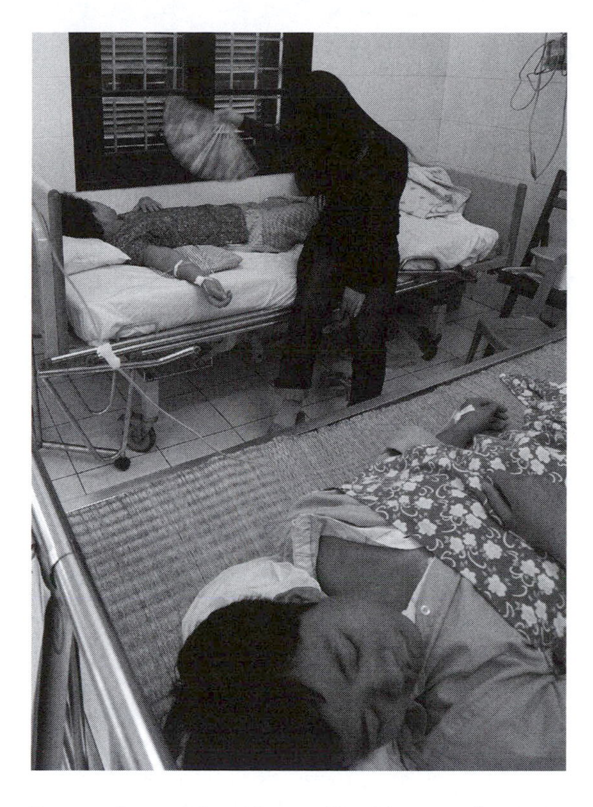

Dengue fever patient Nguyen Van Toan, 22, lays on a bed (foreground), as Nguyen Hong Nhung, fans her mother Nguyen Thi Mai Hien, 67, also a dengue fever patient (background), at Dong Da Hospital in Hanoi, Vietnam, October 25, 2007. (AP/Wide World Photos)

FURTHER READING

Rosenthal, Elisabeth. "As Earth Warms Up, Tropical Virus Moves to Italy." *New York Times*, December 23, 2007e. http://www.nytimes.com/2007/12/23/world/europe/23virus.html.

"Vulnerable Communities Worldwide Adapt to Climate Change." Environment News Service, December 5, 2007. http://www.ens-newswire.com/ens/dec2007/2007-12-05-01.asp.

Desertification

Climate change is playing a role in turning large amounts of fertile land to desert—enough within the next generation to create an "environmental crisis of global proportions," as well as large-scale migrations, especially in parts of Africa and Central Asia, according to the United Nations. "The costs of desertification are large," said Zafar Adeel of the United Nations University (Rosenthal 2007b).

"Already at the moment there are tens of millions of people on the move," Adeel said. "There's internal displacement. There's international migration. There are a number of causes. But by and large, in Sub-Saharan Africa and Central Asia this movement is triggered by degradation of land." Overuse of limited water resources has been made worse by climate-change driven drought (Rosenthal 2007b). In addition, populations are rising and many rivers are being diverted for irrigation and short-term gain.

"Today, those migrants who are escaping dry lands are mostly moving around far from the developed world," said Janos Bogardi of the United Nations University in Bonn, Germany, a technical adviser on the report. "Those who end up on boats to Europe are the tip of an iceberg" (Rosenthal 2007b). "The numbers we now find alarming may explode in an uncontrollable way," Bogardi said. "Because if you look at land use now and dry land, there is the potential that we are nearing a tipping point" (Rosenthal 2007b).

Africa has been plagued with expanding deserts, as the Sahara pushes the populations of Morocco, Tunisia, and Algeria northward toward the Mediterranean. At the southern edge of the Sahara, in countries from Senegal and Mauritania in the west to Sudan, Ethiopia, and Somalia in the east, demands of growing human populations and livestock numbers are converting land into desert. Mohammad Jarian, who directs Iran's Anti-Desertification Organization, reported in 2002 that sandstorms had buried 124 villages in the southeastern province of Sistan-Baluchistan (Brown 2006).

Spreading Deserts in Africa's Sahel

During the first 20 years of the twenty-first century, about 60 million people are expected to leave the Sahelian region of Africa, a region of northern Africa that borders the fringe of the Sahara Desert, if desertification is not halted, United Nations Secretary-General Kofi Annan said on June 17, 2002, the day set aside each year by the United Nations as the World Day to Combat Desertification and Drought. In northeast Asia, "dust and sandstorms have buried human settlements and forced schools and airports to shut down," Annan said, "while in the Americas, dry spells and sandstorms have alarmed farmers and raised the specter of another Dust Bowl, reminiscent of the 1930s." In southern Europe, "lands once green and rich in vegetation are barren and brown," he said ("Global Climate Shift" 2002).

Australian government researcher Leon Rotstayn has compiled evidence indicating that air pollution probably has contributed to catastrophic drought in the Sahel. Sulfate aerosols, tiny atmospheric particles, have contributed to a global climate shift, he said. "The Sahelian drought may be due to a combination of natural variability and atmospheric aerosol," said Rotstayn. "Cleaner air in future will mean greater rainfall in this region," he continued ("Global Climate Shift" 2002).

"Global climate change is not solely being caused by rising levels of greenhouse gases. Atmospheric pollution is also having an effect," said Rotstayn, who is affiliated with Commonwealth Scientific and Industrial Research Organisation (CSIRO), the Australian government's climate-change research agency. Using global climate simulations, Rotstayn found that sulfate aerosols, which are concentrated mainly in the Northern Hemisphere, make cloud droplets smaller. This makes clouds brighter and longer lasting, so they reflect more sunlight into space, cooling the Earth's surface below ("Global Climate Shift" 2002). As a result, the tropical rain belt, which migrates northward and southward with the seasonal movement of the sun, is weakened in the Northern Hemisphere and does not move as far north. This change has had a major impact on the Sahel, which has experienced devastating drought since the 1960s. Rainfall was 20 to 49 percent lower than in the first half of the 20th century, causing widespread famine and death ("Global Climate Shift" 2002). *See also:* Deforestation

FURTHER READING

Brown, Lester R. "The Earth Is Shrinking." Environment News Service, November 20, 2006. http://www.ens-newswire.com.

"Global Climate Shift Feeds Spreading Deserts." Environment News Service, June 17, 2002. http://ens-news.com/ens/jun2002/2002-06-17-03.asp.

Rosenthal, Elisabeth. "Likely Spread of Deserts to Fertile Land Requires Quick Response, U.N. Report Says." *New York Times*, June 28, 2007b. http://www.nytimes.com/2007/06/28/world/28deserts.html.

Disaster Relief, Global Warming's Impact on

International disaster aid will not be able to keep up with the impact of global warming, the Red Cross has said, as it reported a sharp increase in weather-related disasters during the late 1990s. In its annual *World Disasters Report,* the International Federation of Red Cross and Red Crescent Societies said that floods, storms, landslides, and droughts, numbering about 200 a year before 1996, rose to 392 in 2000. Recurrent disasters "are sweeping away development gains and calling into question the possibility of recovery," the report said (Capella 2001, 15). Roger Bracke, Red Cross director of disaster relief operations, blamed the trend on global warming: "It is probable that these kind of disasters will increase even more spectacularly [in the future]" (Capella 2001, 15).

Floods have accounted for more than two-thirds of the average 211 million people a year affected by natural disasters during the 1990s, according to the Red Cross. Famine caused by drought was responsible for about 42 percent of the deaths caused by natural disasters, according to this report, which said that the poor are most vulnerable to disasters; 88 percent of those affected and two-thirds of those killed in the past 10 years lived in the most impoverished countries (Capella 2001, 15).

The same report was critical of the international charity infrastructure, asserting that emergency international aid to the poorest countries declined during the late 1990s, as the amount sent elsewhere rose. The report complained that many donors focus on high-profile projects to rebuild infrastructure, not people's daily livelihoods. A large amount of aid also often is spent on studies rather than actual street-level aid. For example, the report said that nearly two-thirds of the funds spent on a flood-action plan in Bangladesh from 1990 through 1995 left the country to pay foreign-aid consultants, "thereby undermining the local economy" (Capella 2001, 15).

Anticipated climatic changes related to global warming probably will deepen the gap between the richest and poorest nations, especially in parts of Africa, South Asia, and South America, according to assessments of warming's anticipated effects on food production. Nations in tropical climates, including India, Brazil, and much of Sub-Saharan Africa, probably will experience the largest proportional losses in food production (McFarling 2001b, A-3). Most of the world's poorest peoples live in the tropics, where the effects of warming on agriculture may be the most catastrophic, with widespread starvation and malnutrition, according to the report. By contrast, people who live in cooler climates may experience gains in crop yields as higher temperatures lengthen growing seasons (McFarling 2001b, A-3). For the poorest nations, "there is no margin for loss," said Mahendra Shah, one of the report's authors and a United Nations advisor and expert on land use from Austria's International Institute for Applied Systems Analysis (McFarling 2001b, A-3).

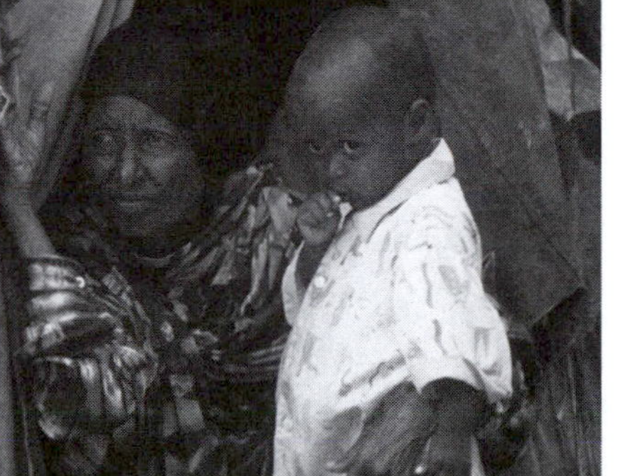

Refugee woman and child in an emergency shelter during the floods in Somalia. (World Food Programme/ Tom Haskell)

Although poor nations would bear the heaviest burden as a group, some of the largest "developing" nations (China, Indonesia, Mexico, Chile, Zaire, and Kenya are examples) would probably see increased production. Some industrial countries, including Britain, the Netherlands, and Australia, could see crop yields decline as warmer, wetter weather increases diseases and pests (McFarling 2001b, A-3).

FURTHER READING

Capella, Peter. "Disasters Will Outstrip Aid Effort as World Heats Up: Rich States Could Be Sued as Voluntary Assistance Falters, Red Cross Says." *London Guardian*, June 29, 2001, 15.

McFarling, Usha Lee. "Warmer World Will Starve Many, Report Says." *Los Angeles Times*, July 11, 2001b, A-3.

Diseases and Climate Change

A warming climate tends to expand the ranges of insects that convey diseases to human beings worldwide. The most pervasive example on the planet is malaria, which has been expanding northward and southward into temperate zones as well as to higher altitudes in tropical mountains that once escaped its ravages. Other diseases also may spread, including influenza, which is already year-round in the tropics.

Animal diseases are likely to reflect temperature increases, according to Bernard Vallat, director general of the World Organization for Animal Health, based in Paris. "As a result of globalization and climate change we are currently facing an unprecedented worldwide impact of emerging and reemerging animal diseases and zoonoses—animal diseases transmissible to humans," Vallat said ("Global Warming Alarms" 2007).

Availability of clean water is crucial to combating many diseases. Hundreds of millions of Africans and tens of millions of Latin Americans will be short of water in less than 20 years. By 2050, more than a billion people in Asia could face water shortages. By 2080, water shortages could threaten 1.1 billion to 3.2 billion people, depending on the level of greenhouse gases that cars and industry spew into the air ("Top Scientists Warn" 2007). Death rates for the world's poor from conditions worsened by the changes global warming brings, like malnutrition and diarrhea, will rise by 2030. By 2080, 200 million to 600 million people could be hungry because of global warming's effects ("Top Scientists Warn" 2007).

A World Health Organization report in 2000 found that warming had caused malaria to spread from three districts in western Kenya to 13 and led to epidemics of the disease in Rwanda and Tanzania. In Sweden, cases of tick-borne encephalitis have risen in direct correlation to warmer winters. Asian tiger mosquitoes, the type that carry dengue fever, have been reported recently as far north as the Netherlands (Struck 2006c, A-16).

Cholera, which is waterborne, afflicted parts of South America in 1991 for the first time in the twentieth century. Aided by poverty and poor public health, it swept from Peru across the continent and into Mexico, killing more than 10,000 people (Struck 2006c, A-16).

West Nile Virus

Susan Harrison, then 45, prepared a Labor Day barbecue in 2005 with her husband and two daughters on the deck of their small house in Toronto. She was bitten by a mosquito, but shrugged it off. In a few days, she felt a shooting pain in her legs. Within two weeks, she could not get out of bed. Harrison's husband, Phil, rushed her to the hospital, where she was put on a respirator and spent three months in intensive care. Months later, she maneuvered around her narrow house in a wheelchair, her legs and right arm paralyzed by West Nile virus. Tears welled in her eyes as she spoke of her daughters, Allison, 10, and Tara, 13. "I used to do things with them, take them places," she said. Her husband, a waiter, struggles to fill the role of two parents (Struck 2006c, A-16).

Valere Rommelaere, 82, a Saskatchewan farmer who survived the D-Day invasion in Normandy, died of the West Nile virus, one of its most northerly victims, within weeks, one casualty in spreading ranges of mosquitoes, ticks, mice, and other disease vectors that are surviving warmer winters and expanding their ranges. By early spring 2006, West Nile virus, first detected in North America in 1999, had infected more than 21,000 people in the United States and Canada and killed more than 800 (Struck 2006c, A-16).

"We didn't even know West Nile virus existed here," said Maria Bujak, 63, of Toronto. Her husband, Andrew, contracted the disease in their garden in 2002. He never fully recovered, she said, and died two years later (Struck 2006c, A-16). A common mosquito, *Culex pipiens,* carries the West Nile virus and thrives in heat and drought.

Paul Epstein, a physician who worked in Africa and is now on the faculty of Harvard

Medical School, said that scientists weren't worried enough about the problem. "Things we projected to occur in 2080 are happening in 2006. What we didn't get is how fast and how big it is, and the degree to which the biological systems would respond," Epstein said in an interview in Boston. "Our mistake was in underestimation." Diseases are also expanding in a surprisingly complex dance with their environment, taking advantage of the swings from deluge to drought made more frequent by global warming, Epstein said (Struck 2006c, A-16).

Warming Waters, Torrential Rains, Sewage, and Disease

During 1991, a cholera outbreak killed thousands in Peru during a period when coastal waters were unusually warm, provoking inquiries regarding whether warming will bring increases in waterborne diseases. Much of this increase is related to exploding growth in parasites and algal blooms that flourish in warming waters, as well as growth in mosquito populations that may carry the West Nile virus, malaria, and dengue fever.

Sewage overflows prompted by heavier rainfall also contribute to disease outbreaks. According to an account in the *Washington Post*,

> [t]he consequences will be particularly severe in the 950 U.S. cities and towns—including New York, the District of Columbia, Milwaukee and Philadelphia—that have "combined sewer systems," archaic designs that carry stormwater and sewage in the same pipes. During heavy rains, the systems often cannot handle the volume, and raw sewage spills into lakes or waterways, including drinking-water supplies. (Lydersen 2008, A-8)

During a deluge in Chicago on September 13, 2008, the city reduced urban flooding by releasing runoff laced with raw sewage into Lake Michigan. A month later, a study in the *American Journal of Preventive Medicine* predicted a 50 to 120 percent increase of such releases into the lake by the end of the twenty-first century, given warmer temperature and increased storm intensity (Lydersen 2008, A-8). "One of the strongest indicators from climate models is more intense rains," said co-author Stephen Vavrus. "They don't agree on everything, but they do agree on that. A warmer atmosphere holds more moisture, so as we get more moisture in the air, when we do have a storm situation, you get more total rainfall" (Lydersen 2008, A-8).

Torrential rains in Milwaukee in 1993 triggered a sewage release that exposed 403,000 people to cryptosporidium, a protozoan parasite transmitted in fecal matter. Fifty-four people died. "Raw sewage got sucked back into the clean water supplies," said Paul Epstein, associate director of the Center for Health and the Global Environment at Harvard Medical School. "Cryptosporidium is a parasite that chlorine doesn't kill, so it escaped water treatment" (Lydersen 2008, A-8). During summer 2004, on Ohio's South Bass Island in Lake Erie at least 1,450 residents and tourists contracted gastrointestinal illnesses that were associated with above-average rainfall over several weeks that contaminated local drinking water. *See also:* Dengue Fever

FURTHER READING

"Global Warming Alarms Infectious Disease Experts." Environment News Service, May 23, 2007. http://www.ens-newswire.com/ens/may2007/2007-05-23-03.asp.

Lydersen, Kari. "Risk of Disease Rises with Water Temperatures." *Washington Post,* October 20, 2008, A-8. http://www.washingtonpost.com/wp-dyn/content/article/2008/10/19/AR2008101901533_pf.html.

"Top Scientists Warn of Water Shortages and Disease Linked to Global Warming." Associated Press in *New York Times,* March 12, 2007. http://www.nytimes.com/2007/03/12/science/earth/12climate.html.

Struck, Doug. "Climate Change Drives Disease to New Territory: Viruses Moving North to Areas Unprepared for Them, Experts Say." *Washington Post,* May 5, 2006c, A-16.

Diseases in Marine Wildlife and Global Warming

Global warming may be associated with a significant increase of unusual diseases among marine life. Many of these marine illnesses—such as *aspergillosis,* a coral-killing fungal infection—are not yet well understood, and they may pose large economic problems in fisheries and threaten human health.

Speaking at a U.S. House of Representatives Oceans Caucus Forum during October 2003, scientists associated these diseases with discharge of sewage into waters and increased groundwater runoff. Scientists described a rise in influenza A and B in seals and other marine mammals, as well as *morbillivirus*, which is similar to feline distemper. Such diseases are common in land mammals, but until now had not been noted in sea creatures. The scientists also noted an increase of the herpes virus among sea turtles and skin disorders among dolphins (Burnham 2003).

"The worst part is that we don't really have any effective tools for fighting this," said C. Drew Harvell, a professor of ecology and evolutionary biology at Cornell University (Burnham 2003). Harvell noted a growing body of evidence suggesting that global climate change is a major reason for increased marine pathogen development, disease transmission, and host susceptibility. "The level of impacts we're seeing from these diseases are new and unusual," Harvell said. "We need to figure out the human sources of these [pathogens] and stop the inputs," he said (Burnham 2003).

As a result of warmer temperatures, harmful algal blooms have been increasing worldwide, affecting the entire food chain, said panelist Rita Colwell, president of the National Science Foundation. Harmful algal blooms, often called "red tides," result from rapid reproduction of phytoplankton in warm water. With this, a small number of species produce potent neurotoxins that can be transferred up the food chain and harm higher life forms, such as shellfish, fish, birds, and even humans who are exposed to the toxins (Burnham 2003).

FURTHER READING

Burnham, Michael. "Scientists Link Global warming with Increasing Marine Diseases." Greenwire, October 7, 2003. (LEXIS)

Drought

As temperatures increase, the hydrological cycle becomes more prone to extremes. A warmer atmosphere holds more moisture, provoking more precipitation in areas that experience it. At the same time, warmth also increases evaporation. Areas prone to drought can become even drier as the amount of water vapor in the atmosphere increases on average.

During three decades ending in 2005 (a time of rising temperatures over most of the Earth), the number of "very dry areas" on Earth, as defined by the widely used Palmer Drought Severity Index, has more than doubled, to about 30 percent of global land. The National Center for Atmospheric Research (NCAR) has concluded that "[t]hese results provide observational evidence for the increasing risk of droughts as anthropogenic global warming progresses and produces both increased temperatures and increased drying" (Romm 2007, 55). In March 2006, Phoenix, a dry area, set a record with more than 140 consecutive rainless days. In June, 45 percent of the contiguous United States was in a moderate-to-extreme state of drought. By July, that figure was 51 percent (Romm 2007, 56).

Climate models and satellite observations both indicate that the total amount of water in the atmosphere will increase at a rate of 7 percent per degree Kelvin of surface warming. However, the climate models predict that global precipitation will increase at a much slower rate of 1 to 3 percent per degree Kelvin. A recent analysis of satellite observations does not support this prediction of a muted response of precipitation to global warming. Rather, the observations suggest that precipitation and total atmospheric water have increased at about the same rate over the past two decades (Wentz et al. 2007, 233).

Some climate models anticipate that areas afflicted by drought could double during the twenty-first century, threatening the livelihoods of millions of people worldwide. British climate scientists have warned that this estimate may be too conservative because it does not include the potential for carbon feedbacks in the climate system that are likely to accelerate warming. The study was funded by the British government and carried out by climate scientists at the Met Office Hadley Centre for Climate Prediction and Research ("Global Warming Could" 2006).

"This report is jaw-dropping," said Andrew Simms, spokesman for the Climate Clinic, a coalition of United Kingdom environmentalists, businesses, and the Energy Saving Trust. "The new projections on drought from the Hadley Centre are like being told that this is the day the earth catches fire." The researchers examined climate records for the second half of the twentieth century and found that global drought increased 25 percent in the 1990s ("Global Warming Could" 2006).

A study published in 2005 by U.S. climate scientists at the NCAR found that the amount of land suffering from severe drought has more than doubled in the past 30 years. Some 30 percent of the Earth's surface experienced drought in 2002, according to the NCAR study, compared with 10 to 15 percent in the early 1970s ("Global Warming Could" 2006).

The Met Office study said that "severe drought may affect 40 percent of the Earth's land up from 8 percent, and moderate drought, which currently affects about 25 percent of the world's surface, will rise to 50 percent" ("Global Warming Could" 2006). The same research forecasts a particularly distressing future for much of the developing world, where drought is already affecting many of the world's poorest communities. "Lack of access to clean water and sanitation already kills 1.6 million children under five years old every year," Simms said. "Small scale farming in Africa provides most of the continents food and 70 percent of employment, virtually all is dependent on regular rainfall" ("Global Warming Could" 2006).

FURTHER READING

"Global Warming Could Spread Extreme Drought." Environment News Service, October 5, 2006. http://www.ens-newswire.com/ens/oct2006/2006-10-05-01.asp.

Romm, Joseph. *Hell and High Water: Global Warming—the Solution and the Politics—and What We Should Do.* New York: William Morrow, 2007.

Wentz, Frank J., Lucrezia Ricciardulli, Kyle Hilburn, and Carl Mears. "How Much More Rain Will Global Warming Bring?" *Science* 317 (July 2007): 233–235.

Drought and Deluge: Anecdotal Observations

Extreme weather is not a novelty in much of the world. In recent years, however, many daily weather reports have indicated an accelerating trend of wild weather. During August 2002, for example, Italy's grape harvest was pummeled by hail (Townsend 2002, 15). During July 2001, a 15-year-old girl was killed by a rare lightning strike about 100 miles from Oslo, Norway, in an area where such storms are rare. In mid-August 2003, the Norwegian village of Atnadalen was flooded by a thunderstorm that dumped 116 millimeters of rain in 24 hours, twice the village's monthly average in August. During July 2003, temperatures in Norway were 3.1°C above average, the highest for July in a record dating to 1866.

During the summer of 2004, as if on cue to support climate models, the usual monsoon in southern Asia brought deluges that killed more than 2,000 people and left millions homeless in Bangladesh, Vietnam, China, India, and Nepal. After five years of intense drought, southern California was battered in late December 2004 and January and February 2005 by incredible amounts of rain and snow. Los Angeles received 17 inches of rain between December 27, 2004, and January 10, 2005, a record amount for two weeks. Some areas in the Sierra Nevada reported more than 10 feet of snow during the same period. At the same time, the usually wet Pacific Northwest experienced its worst drought in almost 30 years. On June 9, 2008, as large parts of California suffered its worst drought on record (since 1895), flooding rains repeatedly doused large parts of the Midwest, including Indiana, which was one-third under water.

"If you warm up the air, [it] can hold more moisture," explained David Easterling of the U.S. National Climatic Data Center. "And the amount it increases is not linear; it goes up exponentially" (Barringer 2007d).

Violent weather has been hitting unusual areas with increasing frequency. Major tornadoes, for example, have been sighted in Michigan and Japan. An F-3 tornado killed several people in Michigan on October 18, 2007—very unusual so far north for the third week of October. The storm killed two people in Ingham County's Locke Township, near Lansing, when their home was ripped from its foundation and dumped into a pond 100 yards away. At least nine people were killed, at least 20 injured, and others trapped under rubble when a tornado hit northern Japan on November 6, 2006, a location where such stores had been all but unknown. Officials said the tornado demolished two prefabricated buildings at a construction site, about eight private houses, and a factory, according to an official at the fire department in Engaru, on the northeastern part of Japan's northern island of Hokkaido. Fire engines and ambulances were at the site, where crushed cars could be seen along with lumber and other scattered debris. Several houses were reduced to rubble.

Intensity of Rainfall Increases

The frequency of torrential rainstorms in the U.S. Midwest has jumped 20 percent since the late 1960s, according to Amanda Staudt, a climate scientist with the National Wildlife Federation. "Global warming is making tragedies like these more frequent and more intense," Staudt

A half-submerged car down a heavily flooded street in Oxford, England, July 25, 2007, during Great Britain's worst floods in 60 years. (AP/Wide World Photos)

said. "As climate continues to warm and we have even more moisture in the air, the trend toward increasingly intense weather events will continue. These are not random events," she said. "We are getting a systematic pattern of floods larger and more frequent than estimated by those calculations" (Pegg 2008). Compared with historical averages, the Midwest has experienced two 500-year floods in 15 years, 1993 and 2008.

Around the world, growing seasons have been steadily lengthening, and many areas that suffered occasional droughts and heat waves now experience them more often. The broad deserts of Australia are spreading toward the thickly populated southeastern coast, as heat waves and wildfires singe the urban area of Sydney, riding hot, dry winds from the interior that are much like southern California's Santa Anas. Meanwhile, the Japanese are contending with rapidly rising temperatures in and near Tokyo, caused by general warming as well as an intensifying urban heat-island effect.

"It's a situation of the poor getting poorer and the rich getting richer when it comes to rainfall," said Yochanan Kushnir of the Lamont-Doherty Earth Observatory of Columbia University. "From a climate perspective, these changes are quite dramatic" (Kaufman 2007c). Warm air from the equatorial tropics rises 8 to 12 miles to the stratosphere, where it spreads north and southward, remaining aloft until it passes 10 to 30 degrees north and south latitude, where it cools and descends. The computer models show this zone of descending air, which indicates lack of rainfall, expanding north and southward.

Extremes, Not the Averages, Cause Damage

"It's the extremes, not the averages, that cause the most damage to society and to many ecosystems," said lead author Claudia Tebaldi, a scientist with the National Center for Atmospheric Research (NCAR). "We now have the first model-based consensus on how the risk of dangerous heat waves, intense rains, and other kinds of extreme weather will change in the next century" ("Scientists Predict" 2006).

James E. Hansen, director of NASA's Goddard Institute for Space Studies, said that global warming is expected to cause an increase in the extremes of the hydrological cycle, and thus in the intensity of droughts and forest fires, on the one hand, and on the intensity of heavy rainfall events and floods, on the other. Hansen explained:

> Specifically a tendency for increased drought in subtropical regions including the American Southwest is expected because of a slowdown in the tropospheric overturning circulation. An increase in the strength of storms driven by latent heat of vaporization, which includes tropical storms, is expected. Melt-back of mountain glaciers is expected and, overall, a decrease of the snow-pack in most mountain ranges. (Hansen 2006b 24)

The world—especially the U.S. West, the Mediterranean region, and Brazil—is likely to suffer more extended droughts, heavy rainfalls, and longer heat waves over the next century. These forecasts were part of a major international report on climate change from NCAR that describes nine computer models. It was published in the December 2006 issue of *Climatic Change.* "It's going to be a wild ride, especially for specific regions," said the study's lead author, Claudia Tebaldi, a scientist at the federally funded academic research center. Tebaldi described the U.S. West, Mediterranean nations, and Brazil as "hot spots" that will get the most extreme weather (Borenstein 2006b).

Seasonal snowmelt in the Sierra Nevada started in mid-March 2004, one of its earliest dates in almost 90 years of recordkeeping. Stream gauge data of California's Merced River from 1916 to the present show a shift in the onset of the melt from mid-spring to late winter or early spring over the last two decades, said Dan Cayan, director of the Climate Research Division at Scripps and a researcher with the United States Geological Survey. In 10 of the last 20 years, the melt began on or before April 1. Philip Mote, a research scientist at the University of Washington, said that snow packs had declined an average of 11 percent across the West since the 1970s (Hymon 2004, B-1).

Joel Achenbach, writing in the *Washington Post,* emphasized that global warming is often one factor among many in drought-and-deluge events. He used floods in Iowa and other Midwestern states as an example:

> Humans surely contributed to the calamity: Farmland in the Midwest has been plumbed with drainage pipes; streams have been straightened; most of the state's wetlands have been engineered out of existence; land set aside for conservation is being put back into corn production to meet the demands of the ethanol boom. This is a landscape that's practically begging to have 500-year floods every decade.... Was climate change a factor in the floods? Maybe. A recent report from the National Oceanic and Atmospheric Administration said that heavier downpours are more likely in a warming world. This will be a wet year (as was last year), but Iowa may not set a rainfall record. The wettest year on record was 1993. The second wettest: 1881. The third wettest: 1902. (Achenbach 2001, B-1)

Achenbach ignores the reality of Midwest weather: one extreme, flood-provoking wet period and one devastating drought can average, over a year, to an unremarkable precipitation total.

The intensity of downpours has increased substantially across the United States as temperatures have risen, according to data developed by Environment America. Heavy rainfall episodes in 2006 were more than 50 percent more frequent in Louisiana, New York, Massachusetts, Rhode Island, and Vermont than 60 years previously, according to the report. In the Dallas-Fort Worth area, such storms increased 42 percent during the same period. On average, extreme rainstorms increased 22 to 26 percent across the country (Barringer 2007d, A-21; "Extreme Rainstorms" 2007).

The report, titled *When it Rains, It Pours: Global Warming and the Rising Frequency of Extreme Precipitation in the United States,* used data from 3,000 weather stations and a methodology originally developed by scientists at the National Climatic Data Center and the Illinois State Water Survey. While the report concentrates on Texas, it also extrapolates its findings to the United States as a whole. It says that extreme precipitation events have increased 24 percent across the continental United States since 1948 ("Extreme Rainstorms" 2007).

Deluges Alternate with Droughts

At times, the swift passage from drought to deluge can mimic Robert Frost's legendary duality of fire and ice. During November 2003, for example, the Los Angeles area was scorched by its worst wildfires on record, driven by hot, desiccating Santa Ana winds that pushed temperatures to near 100°F. Less than two weeks later, parts of the Los Angeles basin were pounded by a foot of pearl-size hail. The island of Hispaniola (the Dominican Republic and Haiti) was seared by drought during 2003 and then drowned in floods that killed at least 2,000 people during May 2004.

Examples abound of increasing extremes in precipitation. November 2002, December 2002, and January 2003 were Minneapolis-St. Paul's driest in recorded history. These months followed the wettest June through October there in more than 100 years. In December 2002, Omaha recorded its first month on record with no measurable precipitation. In March 2003, having endured its driest year in recorded history during 2002, Denver, Colorado, recorded 30 inches of snow in *one* storm. Snowfall on the drought-parched Front Range totaled as much as eight feet in the same storm.

Fifteen months later, Denver's weather let loose again; on June 9, 2004, suburbs north and west of the city received as much as three *feet* of hail. Residents used shovels to free their cars. The summer of 2003 was unusually dry in the Pacific Northwest; during the third week in October, however, Seattle recorded its wettest day on record, with 5.02 inches of rain. The Pacific Northwest is accustomed to rain, but it has not heretofore been like this: a storm December 1–3, 2007 dumped a foot of water on parts of western Washington and coastal Oregon, along with winds gusting to 129 miles an hour.

On the night of July 27, 2004, Dallas, Texas, recorded a foot of rain and widespread flooding, even as the U.S. West continued to endure its worst multiyear drought in at least 500 years. At the same time, in December 2006, the western suburbs of Denver recorded four feet of snow, even though the city's average temperature for the month was 1.4°F above average.

The Pacific Northwest may experience a paradox of longer drought spells punctuated by heavier, sometimes flooding, rainfall. As the world warms, there is likely to be more rain in the tropical Pacific Ocean, and that will change the airflow for certain areas, in much the same way that El Niño weather oscillations now do, said the study's co-author, Gerald Meehl, a computer modeler. The Pacific Northwest occasionally receives flooding rains from the "pineapple express" that arrives from the tropical Pacific.

Roughly half the United States was under serious drought conditions during the summer of 2002. The drought was occasionally punctuated by torrential rains, however. On September 13, 2002, for example, drought-stricken Denver was inundated by floods from a fast-moving thunderstorm that caused widespread flooding. Similar events took place south of Salt Lake City. Ten days later, a flooding cloudburst inundated similarly drought-stricken Atlanta.

At the same time, deluge alternated with drought. A year that ended in Omaha with the most intense drought since the Dust Bowl years also included (on August 7), a one-day storm that dumped 10.5 inches of rain on Omaha, the most intense deluge in the city's history. Beginning in early September, however, Omaha and its hinterland endured a rainless spell of between 44 days (in Omaha) and 49 days (in Lincoln), the fourth-longest on record (Range 2000, 18). Other weather extremes played out in the same area. A storm that destroyed large parts of Aurora, Nebraska, produced hail that was among the largest ever reported in the United States, as well as a very large tornado that stood virtually in one place for half an hour, devastating the town of Deshler.

As many as 80 million piñon trees (the state tree of New Mexico) died in that state and Arizona between 2001 and 2005 because of intense drought. That represents about 90 percent of the area's piñon trees (Carlton 2006, A-1). Four million have died in Santa Fe alone. Filmmaker Jerry Wellman wrote and self-published a 110-page book, *What to Do with a Dead Piñon.* The trees also have fallen victim to the piñon ips, a rice-grain-size bark beetle that feasts on dying diseased trees. In some cases piñon nuts (from the tree's cones) have been imported from China. Some people have bought chippers to turn the fragrant dead trees into some of the world's most expensive wood-chip mulch, worth $6 to $12 per tree.

Even as much of the Northeast experienced extreme rainfall and flooding at the end of June 2006, at the same time, 45 percent of the contiguous United States was in moderate to extreme drought, an increase of 6 percent from May. Dry conditions aided more than 50,000 wildfires, burning more than 3 million acres in the continental United States, according to the National Interagency Fire Center ("First Half" 2006). A month later, on July 27, 2006, in the southern suburbs of Chicago, 5.5 inches of rain fell in 90 minutes. On August 19, 2002, Las Vegas, Nevada, in the midst of a desert, was hit by a thunderstorm that dumped three inches of rain in 90 minutes. The storm, characterized by local recordkeepers as a 100-year event, produced widespread flooding.

After one of its driest summers on record, Seattle recorded its wettest month on record (15.63 inches at Seattle-Tacoma airport) during November 2006. After an El Niño set in at Christmas of the same year, the weather in Seattle again became unusually dry. In El Paso, Texas, after receiving one inch of rainfall since January 2006, three days of heavy rains produced flooding (July 30–August 1, 2006), followed by a monsoon of unprecedented proportions. Similarly, in Greece during August 2007, 100-degree temperatures and hot, dry winds played a role (along with arson) in provoking wildfires that ravaged the country. Two weeks later, torrential rains pelted the same areas, eroding hillsides stripped of their foliage by the fires.

During the week of May 3–10, 2003, 562 tornadoes were reported in the United States, the largest weekly total since records of this kind began during the 1950s. Two years later, during the month of August 2004, another record was set for tornado observations in the United States. Even months not usually noted for tornadic activity seemed to be getting more—September 2004, for example, also set a record for tornado sightings in the United States.

Drought and Deluge: Europe and Asia

Similar reports of an intensifying hydrological cycle have been plentiful outside the United States. The summer of 2002 featured a number of climatic extremes, especially regarding precipitation. Excessive rain deluged Europe and Asia, swamping cities and villages and killing at least 2,000 people, while drought and heat scorched the U.S. West and Eastern cities. Climate skeptics argued that weather is always variable, but other observers noted that extremes seemed to be more frequent than before.

In Europe, on September 10, 2002, six months' worth of rain fell in a few hours in the Gard, Herault, and Vaucluse departments in the south of France, drowning at least 20 people. In the village of Sommieres, near Nimes, an usually tiny stream exploded to a width of 300 meters, cutting off road traffic. By the summer and early fall of 2004, the U.S. east coast, which had experienced intense drought three years earlier, was drowning in record rainfall, part of which arrived courtesy the remains of four hurricanes that had devastated Florida.

In India, with its annual monsoon dry season that usually alternates with heavy rains, the country has adapted to a drought-deluge cycle. About 90 percent of India's precipitation falls between June and September during an average year, so heavy rain in Mumbai (Bombay) during late July is hardly unusual. On July 26 and 27, 2005, however, 37.1 inches of rain fell in Mumbai during 24 hours, the heaviest on record for an Indian city in one diurnal cycle. The deluge contributed to more than 1,000 deaths in and near Mumbai and surrounding Maharashtra State ("Record Rainfall" 2005, A-12).

Metropolitan Mumbai, a city of 17 million, was shut down by the rain, as several people drowned in their cars. Mass transit and telephone services stopped. Other people were electrocuted by wires falling onto flooded streets. Tens of thousands of animals also died. Two years later, some of the worst monsoon rains in India in memory killed at least 2,800 people in India, Bangladesh, Nepal, and Pakistan during 2007. Several million people lost their homes.

During August 2002, Prague, in the Czech Republic, was inundated by flooding rains that forced 200,000 people to evacuate—the worst flooding in 200 to 500 years, depending on who was keeping the tally. The rains followed springtime drought in the same area. Debate ranged over whether this was part of global warming, provoking more intense rainstorms, or a chance natural event. During the same month, torrential rains killed 900 people in China, and another 700 in Southeast Asia, India, and Nepal. Deforestation and the spread of pavement were cited as reasons for increasing flooding in urban areas.

Also during the summer of 2002, near the Black Sea, a large tornado and heavy rains left at least 37 people dead and hundreds of vacationers stranded. During the same week, in China's southern province of Hunan, 70 people died after rains caused landslides and floods. The Republic of Korea mobilized thousands of troops after a week that saw two-fifths of the average annual total rainfall (Townsend 2002, 15).

More Extreme Precipitation Events in 2007

Summer 2007 provided many more examples of extreme precipitation. As the U.S. Southwest and Southeast baked in drought, with intense wildfires, several locations in the middle of the country had record rains. Parts of Iowa, about 150 miles east of Omaha, got 10 inches in one night, and Omaha had only sprinkles. A location near Marble Falls, Texas, in Burnet County, 40 miles northwest of Austin, received 19 inches of rain in 24 hours on June 26–27. Omaha received 10 inches of rain during May 2007 and only a quarter of an inch in June.

On July 21, 2007, the town of D'Hanis, Texas, near San Antonio, was severely flooded after 17 inches of rain fell in 12 hours. The same day, locations in southern and central England witnessed their worst flooding on record after a month's worth of rain, as much as five inches, fell in one hour, following the wettest June in England's history. One such location was Shakespeare's Stratford-on-Avon, which led some residents to remark that it had become Stratford-under-Avon. The floods displaced thousands of people and fouled water for half a million. Once-in-a-century rains also flooded large parts of China, destroying 3.6 million homes and killing at least 500 people. At the same time, the U.S. West was scorched by a record number of wildfires provoked by heat and drought. At about the same time, on June 12, 2007, heavy thunderstorms dumped 11 to 15 inches of rain in one night on parts of parched southwestern Nebraska, surpassing the entire previous year's worth of precipitation. The next week, similar amounts of rain drenched northern Texas, killing four people in flash floods.

During the summer of 2007, Europe experienced the worst of all weather worlds—massive flooding in England as well as searing heat, drought, and wildfires south of the storm track. In the Canary Islands, fires fanned by 40-mile-an-hour winds and temperatures of 110°F forced at least 13,000 tourists to run for their lives. Two-thirds of the Palmitos wildlife park on Gran Canaria Island was destroyed by a wildfire as many toucans, orangutans, and other animals perished. That fire was set by Juan Antonio Navarro Armas, an employee of the park who had lost his job.

Wildfires provoked by record heat also scorched the Balkans and areas eastward. Temperatures in drought-plagued Romania and Greece reached 110°F (Anderson 2007b, A-11). An unusually sharp boundary between very hot and very cold air caused temperatures to fall suddenly from 50°F to 5°F in the highland Alps within a few hours, killing six mountain climbers who were caught at high elevations in light clothing. In Hungary, 500 elderly people died of heat stress.

During April 2007, Australia considered banning irrigation in the fertile Murray-Darling River Basin, which grows 40 percent of the country's agricultural produce. Australia was in the midst of its worst drought on record. If substantial rain did not arrive in eight weeks, the irrigation ban could ruin farmers heavily in debt following six consecutive dry years. At the same time, on April 15, 2007, a late-season nor'easter dumped 7.46 inches of rain on Central Park, New York City, the second-wettest day since recordkeeping began there in 1869. (The wettest day was 8.28 inches on September 23, 1882.)

Sometimes, even deep snow can be tied to an intensifying hydrological cycle addled by warmth. Oswego County, New York, received almost no lake-effect snow during December 2006 and January 2007, but was buried under more than 110 inches in seven days during February, as a relentless cold wind crossed an unfrozen Lake Ontario. Within eight days, some areas near Oswego were blitzed by 10 feet of snow. In nearby Redfield, the National Weather Service reported that 141 inches had fallen in 10 days, a state record for a single storm, spurred by very cold air traversing relatively warm lake water before hitting cold earth.

FURTHER READING

Achenbach, Joel. "Global Warming Did It! Well, Maybe Not." *Washington Post*, August 3, 2001, B-1. http://www.washingtonpost.com/wp-dyn/content/article/2008/08/01/AR2008080103014_pf.html.

Anderson, John Ward. "Europe's Summer of Wild, Wild Weather: Fires, Droughts, and Floods Leave Wake of Destruction." *Washington Post*, August 2, 2007b, A-11. http://www.washingtonpost.com/wp-dyn/content/article/2007/08/01/AR2007080102347_pf.html.

Ayres, Ed. *God's Last Offer: Negotiating a Sustainable Future*. New York: Four Walls Eight Windows, 1999.

Barringer, Felicity. "Precipitation across U.S. Intensifies over 50 Years." *New York Times*, December 5, 2007d. http://www.nytimes.com/2007/12/05/us/05storms.html.

Borenstein, Seth. "Future Forecast, Extreme Weather: Study Outlines a Climate Shift Caused by Global Warming." Associated Press, October 21, 2006b, A-2. http://www.washingtonpost.com/wp-dyn/content/article/2006/10/20/AR2006102001454_pf.html.

Carlton, Jim. "Some in Santa Fe Pine for Lost Symbol, but Others Move On." *Wall Street Journal*, July 31, 2006, A-1, A-8.

"Extreme Rainstorms a New Texas Trend." Environment News Service, December 3, 2007. http://www.ens-newswire.com/ens/dec2007/2007-12-04-095.asp.

"First Half of Year Was Warmest on Record for U.S." Associated Press, July 14, 2006. (LEXIS)

Hansen, James E. "Declaration of James E. Hansen." *Green Mountain Chrysler-Plymouth-Dodge-Jeep, et al., Plaintiffs v. Thomas W. Torti, Secretary of the Vermont Agency of Natural Resources, et al., Defendants.* Case Nos. 2:05-CV-302 and 2:05-CV-304, Consolidated. United States District Court for the District of Vermont. August 14, 2006b. http://www.giss.nasa.gov/~dcain/recent_papers_proofs/vermont_14aug20061_textwfigs.pdf.

Hymon, Steve. "Early Snowmelt Ignites Global Warming Worries: Scientists Have Known Rising Temperatures Could Deplete Water Sources, but Data Show It May Already Be Happening." *Los Angeles Times*, June 28, 2004, B-1.

Kaufman, Marc. "Southwest May Get Even Hotter, Drier: Report on Warming Warns of Droughts." *Washington Post*, April 6, 2007c, A-3. http://www.washingtonpost.com/wp-dyn/content/article/2007/04/05/AR2007040501180_pf.html.

Pegg, J. R. "Warming Climate Adds to U.S. Flood Fears." Environment News Service, July 2, 2008. http://www.ens-newswire.com/ens/jul2008/2008-07-02-10.asp.

Range, Stacey. "Climatologists Say Midlands in Dust Bowl-like Drought." *Omaha World-Herald*, January 12, 2000, 18.

"Record Rainfall Floods India." *New York Times*, July 28, 2005, A-12.

"Scientists Predict Future of Weather Extremes." Environment News Service, October 20, 2006.

Townsend, Mark. "Monsoon Britain: As Storms Bombard Europe, Experts Say That What We Still Call 'Freak' Weather Could Soon Be the Norm." *London Observer*, August 11, 2002, 15.

Drought and Deluge: Scientific Issues

As frequent and intense as such incidents may be, reports describing wild weather are episodic, not systematic. Atmospheric scientist Kevin E. Trenberth and colleagues have advocated "creation of a database of frequency and intensity using hourly precipitation amounts" (Trenberth et al. 2003, 1213). "Atmospheric moisture amounts are generally observed to be increasing ... after about 1973, prior to which reliable moisture soundings are mostly not available" (Trenberth et al. 2003, 1211). Annual mean precipitation amounts over the United States have been increasing at 2 to 5 percent per decade, with "most of the increase related to temperature and hence in atmospheric water-holding capacity.... There is clear evidence that rainfall rates have changed in the United States.... The prospect may be for fewer but more intense rainfall—or snowfall—events" (Trenberth et al. 2003, 1211, 1212). Individual storms may be further enhanced by latent heat release that supplies more moisture.

Thomas Karl, director of the National Climatic Data Center in the U.S. government's National Oceanic and Atmospheric Administration, said, "It is likely that the frequency of heavy and extreme precipitation events has increased as global temperatures have risen." This, he said "[i]s particularly evident in areas where precipitation has increased, primarily in the mid- and high latitudes of the Northern Hemisphere" (Hume 2003, A-13). Studies at the Goddard Institute for Space Studies and Columbia University indicated that the frequency of heavy downpours has indeed increased and suggest that trend will intensify. In the U.S. corn belt, for example, the average number of extreme precipitation events is predicted to jump by 30 percent over the next 30 years and by 65 percent over the next century.

The first statistical support linking tropical deluges to a warming climate was published August 7, 2008, by Richard P. Allan of the University of Reading in England and Brian J. Soden at the University of Miami in the online journal *Science Express*. The new paper was important "because it uses observations to demonstrate the sensitivity of extreme rainfall to temperature," said Anthony J. Broccoli, the director of the Center for Environmental Prediction at Rutgers University. "Such changes in extreme rainfall are quite important in my view, as flash flooding is produced by the extreme rain events," Broccoli added. "In the U.S., flooding is a greater cause of death than lightning or tornadoes, and presumably poses similar risks elsewhere" (Revkin 2008h). The study analyzed two decades' worth of NASA satellite data through several El Niño cycles. Moreover, according Allan and Soden, "The observed amplification of rainfall extremes is found to be larger than that predicted by models, implying that projections of future changes in rainfall extremes in response to anthropogenic global warming may be underestimated" (Allan and Soden 2008, 1484).

Generally, higher temperatures enhance evaporation, with some compensatory cooling when water is available. Increased evaporation also intensifies drought which, to some degree, compounds itself as moisture is depleted, leading "to increased risk of heat waves and wildfires in association with such droughts; because once the soil moisture is depleted then all the heating goes into raising temperatures and wilting plants" (Trenberth 2003, 1212).

In midlatitude mountain areas, wrote Trenberth and colleagues,

> The winter snow pack forms a vital resource, not only for skiers, but also as a freshwater resource in the spring and summer as the snow melts. Yet warming makes for a shorter snow season with more precipitation falling as rain rather than snow, earlier snowmelt of the snow that does exist, and greater evaporation and ablation. These factors all contribute to diminished snow pack. In the summer of 2002, in the western parts of the United States, exceptionally low snow pack and subsequent low soil moisture likely contributed substantially to the widespread intense drought because of the importance of recycling [in the hydrological cycle]. Could this be a sign of the future? (Trenberth et al. 2003, 1212)

Scientists tend to distrust anecdotal stories, seeking, instead, consistent evidence that strongly supports a given idea or hypothesis. Thus, while the hydrological cycle may seem to be changing, precipitation measurements that support such an idea have been difficult to assemble on a worldwide basis. "In many parts of the world," according to one scientific source, "We still cannot reliably measure true precipitation.... [D]ue to rain gauge undercatch ... precipitation is believed to be underestimated by 10 to 15 percent" (Potter and Colman 2003, 144). Ground-level precipitation is only rarely measured over the oceans that cover two-thirds of the Earth, and satellite estimates do not measure local variability. Thus, "proof" of global hydrological-cycle changes is elusive.

Models of Extreme Weather

According to many climate models, the Earth will experience increasing numbers of heat waves, prolonged drought, intense rainstorms, and other extreme weather provoked by rising average temperatures by the end of the twenty-first century. Researchers used simulations created on supercomputers from nine different climate models for the periods 1980–1999 and 2080–2099. Each model simulated the 2080–2099 interval three times, varying the extent to which greenhouse gases accumulate in the atmosphere to account for uncertainty about how fast society may act to reduce greenhouse-gas emissions ("Scientists Predict" 2006). The study appeared in the December, 2006 edition of *Climatic Change.*

In all scenarios, the models agree that the number of extremely warm nights and the length of heat waves will increase significantly around the world. Most areas above 40 degrees north latitude will experience many more days with heavy precipitation, including the northern United States, Canada, Russia, and most of Europe. Dry spells also could lengthen across the western United States, southern Europe, eastern Brazil, and other areas. Growing seasons could increase across most of North America and Eurasia ("Scientists Predict" 2006).

"It's the extremes, not the averages, that cause the most damage to society and to many ecosystems," said lead author Claudia Tebaldi, a scientist with the National Center for Atmospheric Research (NCAR). "We now have the first model-based consensus on how the risk of dangerous heat waves, intense rains, and other kinds of extreme weather will change in the next century" ("Scientists Predict" 2006).

Increasing Variability of Precipitation

Despite problems of measurement, evidence of increasing variability in rain and snow has been increasing. U.S. government scientists have measured a rise in downpour-style storms in the United States during the last century. In addition, snow levels are rising in many mountain ranges. During the last 50 years, winter precipitation in California's Sierra Nevada has been falling more often as rain (increasing flood risks) instead of snow, which supplies farmers and city dwellers alike as it melts in the spring (Revkin 2002b, A-10). Atmospheric moisture increases more rapidly than temperature. Over the United States and Europe, atmospheric moisture increased 10 to 20 percent from 1980 to 2000. "That's why you see the impact of global warming mostly in intense storms and flooding like we have seen in Europe," Kevin Trenberth, a scientist with NCAR told London's *Financial Times* (Cookson and Griffith 2002, 6).

The frequency and intensity of heavy rainfall events during monsoon storms in central India have increased during the past 50 years as the climate there has warmed, wrote B. N. Goswami and colleagues in *Science.*

> Against a backdrop of rising global surface temperature, the stability of the Indian monsoon rainfall over the past century has been a puzzle. By using a daily rainfall data set, we show significant rising

> trends in the frequency and the magnitude of extreme rain events and a significant decreasing trend in the frequency of moderate events over central India during the monsoon seasons from 1951 to 2000. The seasonal mean rainfall does not show a significant trend, because the contribution from increasing heavy events is offset by decreasing moderate events. A substantial increase in hazards related to heavy rain is expected over central India in the future. (Goswami et al. 2006, 1442)

While they agreed that "anthropogenic changes in atmospheric composition are expected to cause climate changes (especially as enhancement of the hydrological cycle) leading to enhanced flood risk," a team of scientists wrote in *Nature* during 2003 that flooding rains in Europe had not, to date, exceeded natural variations (e.g. "a clear increase in flood occurrence rate"), despite the acute flooding in summer 2002 (Mudelsee et al. 2003, 166). Perhaps more notable than Europe's floods themselves was the fact that they were followed, the next summer, by withering drought and record heat—a definite signature of the deluge-to-drought cycle. Such evidence occurs in winter as well as summer. In late January and early February 2004, for example, Omaha received its usual seasonal snowfall (about 30 inches) in less than two weeks, in the midst of a multiple-year drought. At the same time, some areas of western Nebraska saw nary a flurry.

Frequency of Floods May Rise

Writing in *Nature* about an increasing risk of floods in a changing climate, P. C. D. Milly said, "We find that the frequency of great floods increased substantially during the twentieth century. The recent emergence of a statistically significant positive trend in risk of great floods is consistent with results from the climate model, and the model suggests that the trend will continue" (Milly et al. 2002, 514–515). An increasing risk of flooding in Britain and northern Europe during the twenty-first century was quantified in the January 31, 2002, edition of *Nature* by Tim Palmer, of the European Centre for Medium-Range Weather Forecasts, in Reading, England, and Jouni Raisanen, of the Swedish Meteorological and Hydrological Institute, in Norrkoping, Sweden.

The team analyzed the forecasts of 19 climate models to produce an "ensemble forecast." This study revealed that the probability of very wet summers in the Asian monsoon region will rise (Highfield 2002, 8). This was the first time that such a probability forecast of weather extremes caused by climate change had been assembled, although its methodology has been used for weather and seasonal forecasts. "Our results suggest that the probability of such extreme precipitation events is already on the increase," said Palmer. But it was "extremely difficult to verify a small increase in a statistic about extreme seasonal weather, especially over a small area like the U.K." (Highfield 2002, 8). Palmer and Ralsanen wrote:

> We estimate that the probability of total boreal winter precipitation exceeding two standard deviations above normal will increase by a factor of five over parts of the United Kingdom over the next 100 years. We find similar increases in probability for the Asian monsoon region in boreal summer, with implications for flooding in Bangladesh. (Palmer and Ralsanen 2002, 512)

According to a report by the France-based World Water Council, released during 2002,

> The economic toll of floods, droughts and other weather-related disasters has increased almost tenfold in the last four decades, a devastating pattern that must be halted with more aggressive efforts to mitigate the damage. The same report said that increasingly rapid and extreme climate changes point to a future of intensified natural disasters that will result in more human and economic misery in many parts of the world unless action is taken. (Greenaway 2003, A-5)

"Most countries aren't ready to deal adequately with the severe natural disasters that we get now, a situation that will become much worse," said William Cosgrove, vice president of the council (Gardiner 2003).

Floods Affect 100 Million People a Year

The World Water Council report compiled statistics indicating that between 1971 and 1995 floods affected more than 1.5 billion people worldwide, or 100 million people a year. An estimated 318,000 were killed and more than 18 million left homeless. The economic costs of these disasters rose to an estimated U.S. $300 billion in the 1990s from about U.S. $35 billion in the 1960s (Greenaway 2003, A-5).

Global warming is causing changes in weather patterns as growing numbers of people migrate to vulnerable areas, increasing costs of individual weather events, said Cosgrove. "The forecast is that it's going to continue to get worse unless we start to take actions to mitigate global warming," he added (Gardiner 2003). Scientists cited by the World Water Council expect that climate changes during the twenty-first century will lead to shorter and more intense rainy seasons in some areas, as well as longer, more intense droughts in others, endangering some crops and species and causing a drop in global food production (Gardiner 2003).

Severe summer floods in Europe during 2002 may be an indicator of an emerging pattern, according to Jens H. and Ole B. Christensen, who modeled precipitation patterns in Europe under types of warming conditions that may be prominent in the area by 2070 to 2100. "Our results," they wrote in *Nature*, "indicate that episodes of severe flooding may become more frequent, despite a general trend toward drier summer conditions" (Christensen and Christensen 2003, 805).

Deluges are not consistent, however; in 2003, Europe experienced heat and drought of a generational order. Farmers whose crops drowned in 2002 watched them wither and die in 2003. As much as 80 percent of the grain crop died in eastern Germany, the site of some of 2002's worst floods ("Drought, Excessive Heat" 2003, 3-A). In other words, the trend toward drought or deluge will intensify as warming intensifies the hydrological cycle. A warming atmosphere will contain more water vapor, which will provide "further potential for latent-heat release during the buildup of low-pressure systems, thereby possibly both intensifying the systems and making more water available for precipitation" (Christensen and Christensen 2003, 805).

In a similar vein, Wilhelm May and colleagues expect that

> In Southern Europe, the reduction of precipitation throughout the year will cause serious problems for the water supply in general and for agriculture in particular, which is very vulnerable due to its dependency on irrigation. The elongation of droughts in this region will worsen the living conditions, in particular in summer in combination with the warmer temperatures.... The dryness in southern Europe is accompanied by an intensification of heavy precipitation events. (May, Voss, and Roeckner 2002, 27)

May and colleagues also anticipate more heavy precipitation events farther north in Europe, such as in Scandinavia, where precipitation amounts are expected to increase as climate warms.

FURTHER READING

Allan, Richard P., and Brian J. Soden. "Atmospheric Warming and the Amplification of Precipitation Extremes." *Science* 321 (September 12, 2008):1481–1484.

Christensen, Jens H., and Ole B. Christensen. "Severe Summertime Flooding in Europe." *Nature* 421 (February 20, 2003):805.

Cookson, Clive, and Victoria Griffith. "Blame for Flooding May Be Misplaced: Climate Change Global Warming May Not Be the Reason for Recent Heavy Rainfall in Europe and Asia." *London Financial Times*, August 15, 2002, 6.

"Drought, Excessive Heat Ruining Harvests in Western Europe." Associated Press in Daytona Beach, Florida, *News-Journal*, August 5, 2003, 3-A.

Gardiner, Beth. "Report: Extreme Weather on the Rise, Likely to Get Worse." Associated Press Worldstream, International News, London. February 27, 2003. (LEXIS)

Goswami, B. N.,V. Venugopal, D. Sengupta, M. S. Madhusoodanan, and Prince K. Xavier. "Increasing Trend of Extreme Rain Events Over India in a Warming Environment." *Science* 314 (December 1, 2006):1442–1445.

Greenaway, Norma. "Disaster Toll from Weather up Tenfold: Droughts, Floods Need More Damage Control, Report Says." *Edmonton Journal*, February 28, 2003, A-5.

Highfield, Roger. "Winter Floods 'Five Times More Likely.'" *London Daily Telegraph*, January 31, 2002, 8.

Hume, Stephen. "A Risk We Can't Afford: The Summer of Fire and the Winter of the Deluge Should Prove to the Naysayers That If We Wait too Long to React to Climate Change We'll Be in Grave Peril." *Vancouver Sun*, October 23, 2003, A-13.

May, Wilhelm, Reinhard Voss, and Erich Roeckner. "Changes in the Mean and Extremes of the Hydrological Cycle in Europe under Enhanced Greenhouse Gas Conditions in a Global Time-slice Experiment." In *Climatic Change: Implications for the Hydrological Cycle and for Water Management*, Martin Beniston, 1–30. Dordrecht, Germany: Kluwer Academic Publishers, 2002.

Milly, P. C. D., R. T. Wetherald, K. A. Dunne, and T. L. Delworth. "Increasing Risk of Great Floods in a Changing Climate." *Nature* 415 (January 30, 2002):514–517.

Mudelsee, Mandred, Michael Borngen, Gerd Tetzlaff, and Uwe Grunewald. "No Upward Trends in the Occurrence of Extreme Floods in Central Europe." *Nature* 425 (September 11, 2003):166–169.

Palmer, T. N., and J. Ralsanen. "Quantifying the Risk of Extreme Seasonal Precipitation Events in a

Changing Climate." *Nature* 415 (January 30, 2002): 512–514.
Potter, Thomas D., and Bradley R. Colman, eds. *Handbook of Weather, Climate, and Water: Dynamics, Climate, Physical Meteorology, Weather Systems, and Measurements.* Hoboken, NJ: Wiley-Interscience, 2003.
Revkin, Andrew C. "Forecast for a Warmer World: Deluge and Drought." *New York Times*, August 28, 2002b, A-10.
Revkin, Andrew C. "Tropical Warming Tied to Flooding Rains." *New York Times*, August 8, 2008h. http://www.nytimes.com/2008/08/08/science/earth/08rain.html.
"Scientists Predict Future of Weather Extremes." Environment News Service, October 20, 2006.
Trenberth, Kevin E., Aiguo Dai, Roy M. Rassmussen, and David B. Parsons. "The Changing Character of Precipitation." *Bulletin of the American Meteorological Society* 84, no. 9(September 2003):1205–1217.

Drought in Western North America

Mortality rates have risen substantially even in previously healthy conifer forests of the western United States averaging more than 200 years of age, doubling during two to three decades as new trees often fail to replace those that die. Drought provoked by rising temperatures is a major reason for rising death rates of pine, fir, hemlock, and others, according to a study released early in 2009. The increasing mortality of trees also reduces the ability of the forests to absorb carbon dioxide (Van Mantgem et al., 2009, 52; Pennisi 2009, 447).

"Summers are getting longer," said Nathan L. Stephenson of the U.S. Geological Survey (USGS), a co-author of the study with Phillip van Mantgem, also of USGS. "Trees are under more drought stress" (Navarro 2009). The recent warming in the West "has contributed to widespread hydrologic changes, such as a declining fraction of precipitation falling as snow, declining water snow pack content, earlier spring snowmelt and runoff, and a consequent lengthening of the summer drought," the scientists wrote (Eilperin 2009, A-8). "It's very likely that mortality rates will continue to rise," said Stephenson, adding that the death of older trees is rapidly exceeding the growth of new ones, analogous to a human community where the deaths of old people surpass the number of babies being born. "If you saw that going on in your hometown, you'd be concerned" (Eilperin 2009, A-8).

Pushed by early warmth, the springs of 2002, 2003, and 2004 on the Great Plains exploded very quickly, compared even with the season's usual frantic pace—tulips opened one day, and died the next. Peonies reached for the sky so quickly one could almost see their green fingers unfold. Near McCook, Nebraska, farmers during these years said that, lacking intense rainfall in the spring, wheat without irrigation would die. As if to accentuate that possibility, on the evening of April 16, 2002, the hot air was chased across a 200-mile-wide swath of central and western Nebraska by a black cloud of dust driven by 60 mile-an-hour winds. The dust cloud was compared to the dust-bowl days of the 1930s, as it reduced visibility to near zero and caused several traffic accidents. On May 22, another series of dust storms in central Nebraska provoked a 10-vehicle chain-reaction accident that killed two people. With visibility down to 20 feet in the dust, rescue workers wore masks and goggles.

The Colorado River by 2004 was carrying only half as much water compared to the Dust Bowl drought of the 1930s. Utah's Lake Powell—the second biggest man-made lake in the United States, had lost nearly 60 percent of its water by 2004. Assuming present trends, the lake may lose its ability to generate electricity in three years' time (Lean 2004a, 20). After nine years of drought that caused Lake Powell's water level to fall more than 100 feet, heavy snows during 2006–2007 contributed to a recouping of 50 feet (Clark 2008, 1-D).

Tree-ring studies of the last 1,200 years indicate that the Medieval Warm Period, a time of unusual warmth in parts of the world, was punctuated by droughts that were longer and more intense than the one afflicting the U.S. West between 1999 and 2008. "Whether increased warmth in the future is due to natural variables or greenhouse [gases], it doesn't matter," said Edward R. Cook of Columbia University's Lamont-Doherty Earth Observatory (Cook et al. 2004, 1015–1018).

"If the world continues to warm, one has to worry we could be going into a period of increased drought in the western U.S. I'm not predicting that, [but] the data suggests that we need to be concerned about this" (Boxall 2004, A-17). "Compared to the earlier mega-droughts that are reconstructed to have occurred around A.D. 936, 1034, 1150 and 1253, the current drought does not stand out as an extreme event because it has not yet lasted nearly as long," the authors wrote. "This is a disquieting result because future droughts in the West of similar duration to those seen prior to A.D. 1300 would be disastrous," said Cook (Boxall 2004, A-17; Cook et al. 2004, 1015–1018). "If we are just at the beginning of dramatic warming ... we can simply expect larger, more severe fires," said Grant A. Meyer, a co-author of a study published in *Nature* ("New Research Links" 2004, 11-A).

Irrigation water may play out just as intensifying drought arrives on the Great Plains. The Ogallala Aquifer supplies farms and ranches from Nebraska to Texas with 20 billion gallons of water more each *day* than is being replenished. People in the area are just now realizing that the water that nourishes their way of life is a finite, and rapidly diminishing, resource. This knowledge has been helped along in recent years by intense drought that has rivaled the worst years of the 1930s Dust Bowl.

By 2006, several years of intense drought and demands from irrigation had caused the Ogallala Aquifer to decline 30 feet in six years in parts of southwest Nebraska. "The rate of decline is unprecedented," said Mark Burbach of the University of Nebraska-Lincoln School of Natural Resources (Hendee 2006, A-1). Elsewhere across Nebraska, groundwater had declined 5 to 25 feet in the same period. Groundwater in parts of southwestern Nebraska declined more than 550 feet in 50 years, after the advent of large-scale irrigated agriculture.

Writing in the *Handbook of Weather, Climate, and Water: Atmospheric Chemistry, Hydrology, and Societal Impacts* (2003), hydrological scientist Donald A. Wilhite said:

> Projected changes in climate because of increased concentrations of carbon dioxide and other atmospheric trace gases suggest a possible increase in the frequency and intensity of severe drought in the Great Plains region. In a region where the incidence of drought is already high, any increase in drought frequency will place even greater pressure on the region's already limited water supplies. (Wilhite, in Potter 2003, 756)

"Megadrought" in the U.S. West

During 2006–2007, large areas of southern California experienced the driest autumn and winter the region had seen in at least 112 years, according to a Seasonal Assessment released by the National Weather Service Climate Prediction Center (CPC) in Camp Springs. The assessment through June 2007 found that much of the region was already dry at what would nominally be only the beginning of the usual nearly rainless period in spring and summer. Drought conditions also could expand into Nevada, Utah, western Colorado, and western New Mexico, the CPC reported.

The United States has experienced a switch to a long-term megadrought encompassing most of the western states and portions of the Midwest. Such megadroughts have occurred in the Earth's history, usually with warmer climate. Global warming causes a relaxation of the tropospheric overturning circulation (Soden and Held 2006) with resulting intensification of hot, dry conditions in subtropical regions such as the U.S. Southwest and the Mediterranean. There is already evidence for an increased tendency toward warmer drier conditions in those regions in conjunction with global warming of 0.6°C during the past three decades.

If global warming continues to increase, a danger exists that decreasing winter snowpack, intensifying summer dry conditions, and increasing forest fires may reduce vegetation cover and regional soil-biosphere water-holding capacity. If these conditions reach sufficient intensity and geographic scale, they may become self-perpetuating, and we will have suddenly entered a long-term megadrought in the western United States. Weather would continue to fluctuate from year to year, but water supplies would be much more limited than in prior decades and dust storms may become frequent. "We cannot say what level of global warming is needed to cause such a megadrought, but the likelihood increases with increase of greenhouse gases and global warming" (Hansen 2006b, 31–32).

In Canada's province of Alberta, experts have warned that the area is headed for a massive drought worse than the dust-bowl conditions of the 1930s, in large part because of dwindling supplies from mountain snowpack, coupled with a rising number of people and livestock using water. "I see a disaster shaping up in Alberta and it's not a question of if, it's a question of when," said University of Alberta ecologist David Schindler, a water researcher and long-time critic of Alberta's water policies. "There is going to be a major drought coinciding with global warming and record numbers of people and livestock on the landscape. It's something that gives me nightmares" (Semmens 2003, A-14). Glacier cover in the Canadian Rockies is nearing its lowest point in at least 10,000 years. Statistics Canada reported that 1,300 glaciers in the country have lost between 25 and 75 percent of their mass since 1850. Most losses have been recorded in the last 50 years. The report said that most of losses can be attributed to global warming (Paraskevas 2003, A-8)

Schindler cited research by several Alberta scientists indicating that the average temperature for western Canada is the highest it has been in at least 10,000 years, and it is expected to continue climbing. Such conditions will lead to continued recession of glaciers in the Rocky Mountains, less rain and snowfall, as well as increasing evaporation of water. With some of Alberta's rivers already running up to 80 percent lower than 100 years ago, Schindler said Alberta is overdue for a 10-year drought that tends to hit the province every 100 to 150 years (Semmens 2003, A-14).

Water Levels at Lake Mead

Lake Mead, the vast reservoir for the Colorado River water that sustains the fast-growing cities of Phoenix and Las Vegas, could lose water faster than previously thought and run dry within 13 years, according to a study by scientists at the Scripps Institute of Oceanography.

With weather patterns in a warming world favoring a drier American West, a study by scientists at Scripps indicated that Lake Mead, which spans the border of Nevada and Arizona, could run so low by 2013 that water pumps would become useless. The study has become a center of controversy between scientists at Scripps and other researchers at the U.S. Bureau of Reclamation who assert that its climate models are too crude to forecast the future water level of a single large lake.

The Scripps study found that Lake Mead's water supply has a 50 percent chance of becoming unusable by 2021 if the demand for water remains at present levels and if global warming trends conform to midrange models. Researchers Tim P. Barnett and David W. Pierce of Scripps said that even with an occasional snowy winter (such as 2007–2008) demand for the lake's water exceeds the mount added each year by runoff. "We were really sort of stunned," Barnett said. "We didn't expect such a big problem basically right on our front doorstep. We thought there'd be more time." He added, "You think of what the implications are, and it's pretty scary" (Barringer 2008a).

Other research has found that the Colorado River watershed, of which Lake Mead is a part, has had a long-standing tendency toward drought that makes the last century look unusually wet. Climate models indicate that a warmer climate favors persistent drought in this area.

Desiccating Winds and Storm Tracks

Research conducted at The University of Arizona in Tucson indicated that hotter, drier springs in the U.S. Southwest are caused at least in part by desiccating winds aggravated by human-abetted climate change. The winds are part of upper-level wind patterns that since the 1970s have contributed to a northward shift in the upper-level winds that steer storms, most often in late winter. Fewer storms have been bringing rain and snow to southern California, Nevada, Arizona, Utah, western New Mexico, and western Colorado during that season.

"When you pull the storm track north, it takes the storms with it," said Stephanie A. McAfee, a doctoral candidate in the University of Arizona's (UA) Department of Geosciences. "During the period it's raining less, it also tends to be warmer than it used to be," McAfee said. "We're starting to see the impacts of climate change in the late winter and early spring, particularly in the Southwest. It's a season-specific kind of drought" ("Drier, Warmer" 2008). Drier, warmer conditions earlier in the year affect snowpack, hydrological processes and water resources, McAfee said. Other researchers, including the UA's Laboratory of Tree-Ring Research Director Tom Swetnam, have linked warmer, drier springs to more and larger forest fires ("Drier, Warmer" 2008).

The study associated a general poleward movement of storm-steering westerly winds to changes in the U.S. West's winter storm pattern for the first time. This change has been provoked by the atmospheric effects of global warming and depleted stratospheric ozone. Climate models suggest that similar movements will continue as the atmosphere warms.

Eroding Snowpacks and Urban Water Shortages

Confirming many anecdotal observations, a study of data collected over the past 50 years (published online January 31, 2008, by the journal *Science*) reported that sharp declines in mountain snowpacks are being caused primarily by human-induced global warming, As temperatures warm, mountains receive more rain and less snow, which alters the hydrological cycle in many ways that are detrimental to nature and human activity. Snows have been melting earlier, with more rivers running dry in the summer.

The study examined natural variability of temperatures and precipitation and the effect of sun-spot cycles, as well as worldwide volcanic activity and human-induced greenhouse-gas releases. "We've known for decades that the hydrology of the West is changing, but for much of that time people said it was because of Mother Nature and that she would return to the old patterns in the future," said lead author Barnett. "But we have found very clearly that global warming has done it, that it is the mechanism that explains the change and that things will be getting worse" (Kaufman 2008b).

Gradual annual melting of the snowpack is vital for the generation of electricity, as well as for agriculture and drinking water, and to provide enough water for salmon to reach their breeding grounds. More abundant rain tends to run off too quickly to be useful, especially during dry summers, and may cause floods. "Our results are not good news for those living in the western United States," the researchers wrote, adding that the changes may make "modifications to the water infrastructure of the western U.S. a virtual necessity" (Kaufman 2008b).

The pattern is not linear; some years (such as 2007–2008) are still notable for heavy snows. This study indicated, however, that since 1950, snowpack water content on April 1 has declined, on the average, in eight of the nine mountainous areas, from 10 percent in the Colorado Rockies to 40 percent in the Oregon Cascades. The southern Sierra Nevada range was alone in not showing a decline.

The model's results were very robust, with chances that its findings are incorrect between 1 in 100 and 1 in 1,000. With human-generated greenhouse-gas levels rising, its conclusions will probably become more certain. "Given the amount of carbon in the air and the trends for future releases, we have to expect that conditions will get progressively worse for some time, no matter what we do now," Barnett said. His team included researchers from the Lawrence Livermore National Laboratory, the University of Washington, and the National Institute for Environmental Studies in Japan (Kaufman 2008b).

Jonathan Overpeck, a climate scientist at the University of Arizona in Tucson, said the new study "closes the circle" in terms of understanding what is happening to the climate of the West. "Almost all of the models we've seen in recent years show the area becoming warmer and more arid due to climate change, but the question was always whether we could believe them," he said. "Now someone has done the statistical analysis to connect the dots so they can say with real confidence that this is happening because of greenhouse gases" (Kaufman 2008b).

FURTHER READING

Barringer, Felicity. "Lake Mead Could Be Within a Few Years of Going Dry, Study Finds." *New York Times*, February 13, 2008a. http://www.nytimes.com/2008/02/13/us/13mead.html.

Boxall, Bettina. "Epic Droughts Possible, Study Says: Tree Ring Records Suggest That if Past Is Prologue, Global Warming Could Trigger Much Longer Dry Spells than the One Now in West, Scientists Say." *Los Angeles Times*, October 8, 2004, A-17.

Clark, Jayne. "Lake Powell on the Rise." *USA Today*, May 9, 2008, 1-D, 2-D.

Cook, Edward R., Connie A. Woodhouse, C. Mark Eakin, David M. Meko, and David W. Stahle. "Long-Term Aridity Changes in the Western United States." *Science* 306 (November 5, 2004):1015–1018.

"Drier, Warmer Springs in U.S. Southwest Stem from Human-caused Changes in Winds." NASA Earth Observatory, August 19, 2008, http://earthobservatory.nasa.gov/Newsroom/MediaAlerts/2008/2008081927359.html.

Eilperin, Juliet. "Study Ties Tree Deaths to Change in Climate." *Washington Post*, January 23, 2009, A-8. http://www.washingtonpost.com/wp-dyn/content/article/2009/01/22/AR2009012202473_pf.html.

Hansen, James E. "Declaration of James E. Hansen." *Green Mountain Chrysler-Plymouth-Dodge-Jeep, et al., Plaintiffs v. Thomas W. Torti, Secretary of the Vermont Agency of Natural Resources, et al., Defendants.* Case Nos. 2:05-CV-302 and 2:05-CV-304, Consolidated. United States District Court for the District of Vermont. August 14, 2006b. http://www.giss.nasa.gov/~dcain/recent_papers_proofs/vermont_14aug20061_textwfigs.pdf.

Hendee, David. "Peril Is Seen to State's Water Table." *Omaha World-Herald*, October 22, 2006, A-1, A-2.

Kaufman, Marc. "Decline in Snowpack Is Blamed On Warming Water Supplies In West Affected." *Washington Post*, February 1, 2008b, A-1. http://www.washingtonpost.com/wp-dyn/content/article/2008/01/31/AR2008013101868_pf.html.

Kunzig, Robert. "Drying of the West." *National Geographic*, February 2008, 90–113.

Lean, Geoffrey. "Worst U.S. Drought in 500 Years Fuels Raging California Wildfires." *London Independent*, July 25, 2004a, 20.

Navarro, Mireya. "Environment Blamed in Western Tree Deaths." *New York Times*, January 23, 2009. http://www.nytimes.com/2009/01/23/us/23trees.html.

"New Research Links Global Warming to Wildfires across the West." *Los Angeles Times* in *Omaha World-Herald*, November 5, 2004, 11-A.

Paraskevas, Joe. "Glaciers in the Canadian Rockies Shrinking to Their Lowest Level in 10,000 Years." *National Post* (Canada), December 4, 2003, A-8.

Pennisi, Elizabeth. "Western U.S. Forests Suffer Death by Degrees." *Science* 323 (January 23, 2009):447.

Semmens, Grady. "Ecologists See Disaster in Dwindling Water Supply." *Calgary Herald*, November 27, 2003, A-14.

Soden, Brian J., and Isaac M. Held. "An Assessment of Climate Feedbacks in Coupled Ocean-Atmosphere Models." *Journal of Climate* 19 (2006):3354–3360.

Van Mantgem, Phillip J., Nathan L. Stephenson, John C. Byrne, Lori D. Daniels, Jerry F. Franklin, Peter Z. Fulé, Mark E. Harmon, Andrew J. Larson, Jeremy M. Smith, Alan H. Taylor, and Thomas T. Veblen. "Widespread Increase of Tree Mortality Rates in the Western United States." *Science* 323 (January 23, 2009):521–524.

Wentz, Frank J., Lucrezia Ricciardulli, Kyle Hilburn, and Carl Mears "How Much More Rain Will Global Warming Bring?" *Science* 317 (July 2007):233–235.

Wilhite, Donald A. "Drought in the U.S. Great Plains." In *Handbook of Weather, Climate, and Water: Atmospheric Chemistry, Hydrology, and Societal Impacts*, ed. Thomas D. Potter and Bradley R. Colman, 743–758. Hoboken, NJ: Wiley Interscience, 2003.

E

Economics of Addressing Global Warming, The

If global warming is not addressed, the world economy's production could fall significantly, perhaps by as much as $20 trillion by the end of the twenty-first century, according to figures compiled by some economists. Such a figure would represent 6 to 8 percent of global economic output in 2100. These figures do not include costs of biodiversity loss or of unpredictable events such as extreme weather.

"The climate system has enormous momentum, as does the economic system that emits so much carbon dioxide," said co-author Frank Ackerman, an economist with Tufts University's Global Development and the Environment Institute. "We have to start turning off greenhouse gas emissions now in order to avoid catastrophe in decades to come." Limiting temperature rise to 2°C could avoid $12 trillion in damage annually and cost a quarter of that amount, according to this study. The report "Climate Change the Costs of Inaction" was compiled by economists at Tufts University's Global Development and Environment Institute ("Failure" 2006).

The Costs of Neglect

Costs of neglecting global warming would include a number of costly environmental impacts, such as decreased crop yields in the developing world, as well as impacts of increasingly intense droughts and water shortages, and nearly complete loss of coral reefs, the spread of tropical diseases such as malaria, and the impending extinction of several arctic species, including the polar bear.

"This report demonstrates that climate change will not only be an environmental and social disaster: it will also be an economic catastrophe, especially if global temperatures are allowed to increase by more than two degrees centigrade," said Elizabeth Blast of Friends of the Earth, which commissioned the report ("Failure" 2006). A 3°C temperature rise would compound damage to agriculture around the world, according to the report, including collapse of the Amazon Valley's ecosystem and loss of boreal and alpine ecosystems.

"Setting a real price on carbon emissions is the single most important policy step to take," said Robert N. Stavins, director of the environmental economics program at Harvard University. "Pricing is the way you get both the short-term gains through efficiency and the longer-term gains from investments in research and switching to cleaner fuels" (Lohr 2006).

How much money will be required to avert catastrophic climate change? A consensus estimated averages about 1 percent of world economic activity annually for 50 years. As with all cost forecasts so far into the future, these are broad estimates. In 2006, 1 percent of the U.S. economic activity was more than $120 billion a year, or $400 per person, roughly the amount of money being spent at that time on the Iraq and Afghanistan wars (Lohr 2006). "There's no easy way around the fact that if global warming is a serious risk, there will be serious costs," said W. David Montgomery, an economist at Charles River Associates (Lohr 2006).

The Stern Review

The Stern Review on the Economics of Climate Change, which has been called "the most comprehensive review ever carried out on the economics of climate change," warns that "global warming could inflict worldwide disruption as great as that caused by the two World Wars and the Great Depression ("Failure" 2006). *The Stern Review*, named after its principal author, Sir Nicolas Stern, head of the British Government Economic Service and a former chief economist at the World Bank, estimated that ignoring warming's impacts could cost US$9 trillion. "Hundreds of millions of people could suffer hunger, water shortages and coastal flooding as the world warms," the report predicted ("Failure" 2006).

"The task is urgent," Stern warned. "Delaying action, even by a decade or two, will take us into dangerous territory. We must not let this window of opportunity close" ("Failure" 2006). Stern's report calculated that "tackling climate change would cost 20 times less than doing nothing." The report estimates the cost of business as usual (or doing nothing) at 5 to 20 percent of global gross domestic product (GDP) by the year 2100 (Giles 2006b, 6). While the coal and oil industries have concentrated on the cost of change, the Stern report says that *not* confronting the problem will, in the long run, be much worse for the world economy. Its own calculations peg the cost of stabilizing greenhouse-gas levels at twice preindustrial levels would cost only 1 percent of global GDP. Thus, the report supports restrictions on carbon emissions and the development of alternative energy sources.

"If we take no action to control emissions, each metric ton of CO_2 that we emit now is causing damage worth at least $85—but these costs are not included when investors and consumers make decisions about how to spend their money," according to the *Stern Review.* "Emerging schemes that allow people to trade reductions in CO_2 have demonstrated that there are many opportunities to cut emissions for less than $25 a ton," the review stated. "According to one measure, the benefits over time of actions to shift the world onto a low-carbon path could be in the order of $2.5 trillion each year" ("Failure" 2006).

"The Greatest Market Failure"

The second half of the *Stern Review* describes the policy challenges involved in moving to a low-carbon economy worldwide. In Stern's opinion, lack of action to counter climate change is "the greatest market failure the world has seen" ("Failure" 2006). Stern believes that three elements of policy are required for an effective response. "First, we must establish a carbon price via tax, trade and regulation—without this price there is no incentive to decarbonize," he said ("Failure" 2007). Stern continued:

> Second, we must promote technology: through research and development. Further, private sector investors need confidence that there will be markets for their products: that is why deployment policy also makes sense.... And third we must deal with market failure—for example problems in property and capital markets inhibit investments for energy efficiency. Further, the sticks and carrots of incentives, rightly emphasized by we economists, need to be supported by information. And still further, greater understanding of the issues can itself change the behavior of individuals and firms. ("Failure" 2006)

While climate-change contrarians complain that dealing with global warming will draw money away from other economic activities, the *Stern Review* anticipated that markets for low-carbon technologies will be worth at least $500 billion, and perhaps much more, by 2050. Stern sees tackling climate change as a pro-growth strategy, believing that ignoring it will ultimately undermine economic growth ("Failure" 2006).

Lack of action is not an option, according to Stern, who was supported by British Prime Minister Tony Blair. "The consequences are stark, for our planet and for the people who live on it, threatening the basic elements of life—access to water, food production, health and our environment" ("Failure" 2006).

Blair said that a temperature rise of 2° to 3°C would portend the following:

- Disappearing glaciers that will significantly reduce water supply to over a billion people.
- Rising sea levels that could lead to 200 million people being displaced.
- Declining crops yields that will lead to famine and death particularly in Africa.
- The spread of diseases such as malaria.
- Extinction of up to 40 percent of Earth's animal species. ("Failure" 2006)

Right-wing critics attacked the *Stern Review's* models, especially its assumption that world population will reach 15 billion, a figure that critics regarded as unlikely. Richard Tol, an economist at Princeton University, asserted that Stern's use of existing data was unduly pessimistic. Tol said that Stern had underplayed efforts at adaptation (Giles 2006b, 6). Roger Pielke, Jr., a climate change policy expert at the University of Colorado, said that Stern had "cherry-picked" the literature (Giles 2006, 6).

FURTHER READING

"Failure to Manage Global Warming Would Cripple World Economy." Environment News Service,

October 30, 2006. http://www.ens-newswire.com/ens/oct2006/2006-10-30-06.asp.

Giles, Jim. "How Much Will It Cost to Save the World?" *Nature* 444 (November 2, 2006b):6–7.

Lohr, Steve. "The Energy Challenge: "The Cost of an Overheated Planet," *New York Times*, December 12, 2006. http://www.nytimes.com/2006/12/12/business/worldbusiness/12warm.html.

The Stern Review. www.sternreview.org.uk. Accessed September 26, 2008.

Ecotourism

Taking a jet from New York to Jakarta via Tokyo to live in a pup tent and watch colorful birds in Borneo does the Earth no favors. Heeding *USA Today's* appeal to visit "Ten Great Endangered Places to See While You Still Can" (Frank 2008, 3-D) will do nothing but speed the demise of the various glaciers, polar-bear habitats, and coral reefs that the story pitches as ecotourism hot-spots.

Riding a bicycle a few miles to watch birds is *real* ecotourism, but not the kind you'll see advertised in a travel magazine, because it makes no one money, except the neighborhood bike-repair shop. Likewise, no market in carbon offsets exists for the eco-hero who rides a bicycle to a local levy to catch a few garden-variety fish. Very little money changes hands, and there is no carbon to offset, although our hero is engaging in recreation that produces virtually no carbon dioxide. Carbon dioxide and methane do not care whether you feel guilty about making more of them, whether you are a contrarian who doesn't think the greenhouse effect exists, or an ecotourist in Patagonia on a smog-belching tour boat agonizing over how to offset long-distance airfare while watching glaciers calve into the sea.

Pitching the Apocalypse

Even as the glaciers melt, commission-conscious travel agents under the banner of ecotourism advise clients to hurry up and see them before they are gone. In 2007, Betchart Expeditions, Inc., of Cupertino, California, offered a 12-day tour to see the "Warming Island," off Greenland, which had recently emerged from melting ice, "a compelling indicator of the rapid speed of global warming" (Naik 2007b, A-12). The cost of the full tour was $5,000 to $7,000, plus airfare. In the Greenland coastal village of Illuissat (population 5,000), about 35,000 tourists arrived, most of them on cruise ships, during 2007, up from 10,000 about five years previously. Many came to witness the receding Jacobshavin Glacier, which had lost nine miles in five years.

Two decades ago, winter temperatures routinely fell to −40°F; by 2007, −15°F was the usual winter's low. The harbor, which used to freeze solid, now remains liquid all year, allowing fisherman to pull halibut of the water at all seasons, depleting stocks (Naik 2007b, A-12). One day during the summer of 2005, residents of Pangnirtung, a tiny hamlet on the east coast of Baffin Island near receding glaciers, were greeted by a surprise: a 400-foot European cruise ship, which had dropped anchor unannounced and sent several hundred tourists ashore in small boats. Pangnirtung may be remote and tiny, but it has enough tourists to have been put on the map by Kentucky Fried Chicken.

Ken Shapiro, editor in chief of *TravelAge West*, with a subscription that circulates mainly among travel agents, has named this genre of travel "The Tourism of Doom." Entire travel agencies (such as Quark Expeditions, which specializes in Arctic tours) have grown up to cater to doom tourists. Quark found itself oversubscribed in 2007 and doubled its capacity for 2008 (Salkin 2007). These trips are marketed as environmentally aware and ecosensitive. Some arctic tourists receive certificates of participation and Climate Change Challenge Mission patches, some of the quaintest ecocapitalism on our rapidly warming planet.

"From the tropics to the ice fields, doom is big business," reported Allen Salkin in the *New York Times*.

> Quark Expeditions, a leader in arctic travel, doubled capacity for its 2008 season of trips to the northern and southernmost reaches of the planet. Travel agents report clients are increasingly requesting trips to see the melting glaciers of Patagonia, the threatened coral of the Great Barrier Reef, and the eroding atolls of the Maldives. (Salkin 2007)

These trips are usually marketed as "environmentally aware" and "ecosensitive" experiences during which travelers observe and commiserate over the effects of global warming. In addition to airline trips, this kind of so-called ecotourism requires boats and often new hotels (because they are conducted in remote locations) and, dare we forget, souvenir shops. Ecotourism is little more than a marketing pitch to dress up the same old carbon-intensive product.

The erosion of Mount Kilimanjaro's snows has become a marketing ploy in the adventure-travel industry that takes about 30,000 climbers a year up the mountain's flanks on six different routes at prices ranging (in 2008) from about $4,000 to $6,000 each. Guide services and outfitters tout climate change's effect on the mountain's snowfields in their advertising and on their Web sites. One climb leader, Justin Merle, said of the typical adventure-travel article: "It's like—See Kili before the snow is gone. That's almost a catchphrase" (Modie 2008).

The same pitch has been getting some exercise in the eroding glaciers of southernmost Argentina. Dominique Browning, formerly editor of *House and Garden*, visited Los Glaciares National Park, Argentina, describing his ecotour on the *New York Times* op-ed page. Glacier tourism has become so popular in the area that cinder-block hotels and vacation condos have been "springing up everywhere.... The shops are full of souvenirs, their windows plastered with posters of glaciers and ads for boat tours. Finishing touches are being put on a large casino" (Browning 2008). The new developments have laced formerly pristine views with plastic bags and cigarette butts.

Guilt-Free Travel Deals?

In November 2007, *Travel & Leisure* magazine published a "responsible travel" issue and listed on its cover "13 Guilt-Free Travel Deals," number five being an Inkaterra Rain Forest package. For $497 a person, it included a three-night stay in a cabana on stilts, an excursion to the hotel's private ecological reserve, a boat trip to a native farm, and a 30-minute massage at the hotel spa" (Salkin 2007). Inkaterra advertised helicopter ecotours from Cuzco, Peru, to Machu Picchu, which many visitors reach only after long hikes. Environmental authorities in Peru took a look at the carbon footprint of this ecotour and, after a few flights during May 2007, shut it down (Higgins 2007, 6).

"This is all a ruse, said John Stetson, a spokesman for the Will Steger Foundation, an environmental education organization in Minnesota. "Ecotourism is more of a term for the marketer," he said. "Many people want to do what's right, so when something is marketed as the right thing, they tend to do that." But, he says, traveling by jet to see the icebergs contributes to global warming, which makes the icebergs melt faster. "It's hard to fault somebody who wants to see something before it disappears, but it's unfortunate that in their pursuit of doing that, they contribute to the problem," he said (Salkin 2007).

Abercrombie & Kent, a luxury travel company, has been offering "mission trips" to environmentally sensitive locales. For its Antarctica mission, the 22 participants, who paid $6,190 each for a 13-day tour, gave an additional $500 each to Friends of Conservation (Salkin 2007).

Damage from Ecotourism

Steve McCrea, editor of the *Eco-Tourist Journal*, calls air travel "ecotourism's hidden pollution" (McCrea 1996). Tourists who take the utmost ecological care when they visit exotic locales rarely give a second thought to the greenhouse gases that they generate while reaching their destinations. According to McCrea,

> One ton of carbon dioxide enters the atmosphere for every 4,000 miles that the typical eco-tourist flies. A round trip from New York to San Jose, Costa Rica (the world's leading eco-tourist destination) is 4,200 miles, so the typical eco-tourist generates roughly 2,100 pounds of carbon dioxide by traveling there to sleep in the rainforest for a week. (McCrea 1996)

To balance carbon dioxide generated by their air travel, McCrea suggested that ecotourists plant three trees for every 4,000 miles, to compensate not only for the carbon dioxide, but also for other greenhouse gases created by the combustion of jet fuel. Given the damage to the atmosphere from aviation, however, three trees per 4,000 miles seems hardly even a symbolic remedy.

The same goes for long-distance tourism, generally. "So as we are flying into the Alps for our ski holiday we are contributing to their destruction," wrote one European author. "Our honeymoon flight to the Maldives is slowly sinking it under rising sea levels and destroying coral through bleaching associated with global warming; and finally our safari flight to Africa is contributing to drought, famine and disease. It's not an appetizing thought, is it?" (Francis 2006).

FURTHER READING

Browning, Dominique. "The Melting Point." *New York Times*, February 2, 2008. http://www.nytimes.com/2008/02/02/opinion/02browning.html.

Francis, Justin. Responsibletravel.com. United Kingdom. Accessed April 30, 2006. http://www.responsibletravel.com/Copy/Copy101993.htm.

Frank, Peter. "Ten Great Endangered Places to See While You Still Can." *USA Today*, April 18, 2008, 3-D.
Higgins, Michelle. "Machu Picchu, Without Roughing It." *New York Times*, Travel, August 12, 2006, 6.
McCrea, Steve. "Air Travel: Ecotourism's Hidden Pollution." *San Diego Earth Times*, August 1996. http://www.sdearthtimes.com/et0896/et0896s13.html.
Modie, Neil. "Mount Kilimanjaro: On Africa's Roof, Still Crowned With Snow." *New York Times*, January 20, 2008. http://travel.nytimes.com/2008/01/20/travel/20Explorer.html.
Naik, Gautam. "Arctic Becomes Tourism Hot Spot, but Is It Cool?" *Wall Street Journal*, September 24, 2007b, A-1, A-12.
Salkin, Allen. "Before It Disappears." *New York Times*, December 16, 2007. http://www.nytimes.com/2007/12/16/fashion/16disappear.html.

Electric Car

One of Henry Ford's fantasy cars was electric, but such designs were muscled aside by the development of powerful internal combustion engines during the 1930s. Reacting to oil embargoes and gasoline-price rises during the 1970s, Ford Motor again developed workable designs for small, efficient, electric cars, but shelved them again. The problem is less technological than a matter of driver preference and corporate profitability.

Electric cars may be coming back, however. In January 2007 General Motors (GM) rolled out the hybrid Chevrolet Volt, a concept car that can run 40 miles on electricity alone with a six-hour nighttime charge and, running as a hybrid, gets 150 miles per gallon of gasoline. For commuting trips under 20 miles each way, it can run on a charge from a standard 110-volt garage outlet. Electricity must be generated, of course, and these days that usually involves the burning of fossil fuels.

By mid-2007, GM had committed to hiring 400 technical experts to work on fuel-saving technology. One of their goals was to produce the Volt commercially within three to four years.

At the new, green GM, Lawrence Burns, vice president for research, development, and global planning, told the *Wall Street Journal*, "We have to have people think we are part of the solution, not part of the problem." The Volt, said Burns, is an effort to show consumers that "We get it" on climate change (Boudette 2007, A-8).

A safe, affordable 100,000-mile lithium-ion battery (less powerful versions are used in cell phones and laptops) is still in the future. California's Air Resources Board calculates that lithium-ion packs would cost $3,000 to $4,000 in mass production, cheap enough to be feasible (Ulrich 2007). The Chevy Volt's gasoline engine charges batteries, to a range of 640 miles. In our time, no purely electric vehicle with four seats and the ability to reach highway speeds has been mass produced. Existing electric cars have been ridiculed as "glorified golf carts" (Ulrich 2007).

245 Miles a Gallon

Other companies are upstaging GM. Early in 2008, the Swedish government announced a joint effort by that nation's automakers and power companies to develop electric cars, hybrid vehicles that can be charged with electricity directly from a wall socket. The government is investing 240 Swedish kroner (about US$35 million). The project is a collaboration between Saab Automobile, Volvo Car Corporation, Vattenfall AB, and ETC AB. The project is being carried out with the support of the Swedish Energy Agency.

One very small electric car, the two-seat Zenn (short for "Zero Emission, No Noise"), has been marketed for urban driving at 25 to 35 miles an hour. It gets up to 245 miles a gallon when it uses any gasoline at all. The Zenn began selling at a Berkeley, California, vehicle dealership called "Green Motors" in 2006; it can run 30 to 50 miles on an eight-hour charge, and plugs into any electrical outlet. The Zenn is based on the European "Microcar," but it is made in Canada and has been selling for two decades for about US$15,000. The Zenn weighs 1,200 pounds, half of which is batteries (Schmit 2007, 3-B).

With all the ecohype over electric cars, a few things have gone unsaid. One is that electricity has to come from somewhere. Today, most of it is generated by fossil fuels, half of it by coal, which is a dirtier greenhouse source per unit of energy than oil. In addition, electric cars lack driving range, interior room, and the get-up-and go performance of gas-powered automobiles. Costs, at $50,000 to $100,000, are prohibitive for all but showcase uses, although mass production would reduce sticker shock.

FURTHER READING

Boudette, Neal E. "Shifting Gears, G.M. Now Sees Green." *Wall Street Journal*, May 29, 2007, A-8.
Schmit, Julie. "Small Electric Car Sales Could Get a Jolt." *USA Today*, December 11, 2007, 3-B.
Ulrich, Lawrence. "They're Electric, but Can They Be Fantastic?" *New York Times*, September 23, 2007. http://www.nytimes.com/2007/09/23/automobiles/23AUTO.html.

Energy, Full-Cost Pricing

As early as 1991, a National Academy of Sciences report recommended a "study in detail [of] the 'full social cost pricing' of energy, with a goal of gradually introducing such a system" (National Academy 1991, 73). Such a system would price energy to include not only the costs of production and distribution, but environmental costs as well. The study admitted that "such a policy would not be easy to design or implement" (National Academy 1991, 73). Such a change would involve a full-scale restructuring of the accounting rules in the capitalistic marketplace, which defines "cost" only in terms of financial (not environmental) assets.

Stephen H. Schneider explained full-cost pricing of energy under the rubric of "integrated assessment."

> Integrated assessment is an attempt to merge economy and ecology with the inclusion of externalities and the analysis of end-to-end costs. In other words, the price of a lump of coal isn't simply extraction, storage and transport, but health consequences of mining and burning, as well as the whole range of potential environmental alterations from the production and use of energy. The cost of a car isn't simply the materials, labor and profit, but should also include disposal costs, and tailpipe emissions.... Integrated assessment will fall short on several grounds ... because of the technical difficulty [of] evaluating the whole range of costs and benefits of our activities and because, ultimately, the single unit of comparison ... is monetary currency. (Schneider 2000a)

Thomas R. Casten endorsed an end to annual federal government spending of $4 billion on subsidies for producers of fossil fuels in the United States, mainly from tax credits related to depletion allowances. Such a change could take place under present accounting rules and assumptions. In the longer term, several observers, including Al Gore, propose that "the definition of gross national product (GNP) should be changed to include environmental costs and benefits," and that "the definition of productivity should be changed to reflect calculations of environmental improvement or decline." Gore calls such definitions "eco-nomics" (Gore 1992, 346).

The Sierra Club asserted that subsidies for fossil-fuel exploration should be ended. Instead, according to the Sierra Club, emissions of pollution should be taxed, as government increases research and development expenditures for solar and other renewable energy technologies. According to a Sierra Club study, the cost of generating wind energy fell 85 percent between 1981 and the middle 1990s. Improvements in photovoltaic-cell technology also have reduced the per-unit cost of generating solar power. Despite the growing competitiveness of alternative fuels, Ross Gelbspan argues that 90 percent of U.S. energy subsidies and incentives still go to fossil fuels, compared with 10 percent for energy alternatives that may help slow the rise in atmospheric greenhouse-gas levels (Gelbspan 1997b, 96–97).

FURTHER READING

Gelbspan, Ross. *The Heat Is On: The High Stakes Battle Over Earth's Threatened Climate.* Reading, MA: Addison-Wesley Publishing Co., 1997b.

Gore, Albert, Jr. *Earth in the Balance: Ecology and the Human Spirit.* Boston: Houghton Mifflin Co., 1992.

National Academy of Sciences. *Policy Implications of Greenhouse Warming.* Washington, DC: National Academy Press, 1991.

Schneider, Stephen H. "No Therapy for the Earth: When Personal Denial Goes Global." In *Nature, Environment, and Me: Explorations of Self in a Deteriorating World,* ed. Michael Aleksiuk and Thomas Nelson. Montreal: McGill-Queens University Press, 2000a.

Energy Generation: Paradigm Changes

In one hundred years, students of history may remark at the nature of the fears that stalled responses to climate change early in the twenty-first century. Skeptics of global warming kept change at bay, it may be remarked, by appealing to most people's fear of change that might erode their comfort and employment security, all of which were wedded psychologically to the massive burning of fossil fuels. A necessary change in our energy base may have been stalled, they may conclude, beyond the point at which climate change forced attention, comprehension, and action.

Technological change always generates fear of unemployment. Paradoxically, such changes also always generate economic activity. A change in our basic energy paradigm during the twenty-first century will not cause the ruination of our economic base, as some "skeptics" of climate change believe, any more than the coming of the railroads in the nineteenth century ruined an economy in which the horse was the major land-based vehicle of transportation. The advent of mass automobile ownership early in the twentieth century propelled economic growth, as did the transformation of information gathering and handling via computers in the recent past. The same developments also put out of work blacksmiths, keepers of hand-drawn accounting ledgers, and anyone who repaired manual typesetters.

As of 2008, the federal government of the United States (which, as a nation, produces more than one-fifth of the world's greenhouse gases) was sitting out the next worldwide energy revolution. The United States was being led (if that was the word) by a set of minds still set to the clock of the early twentieth-century fossil-fuel boom. The out-going G. W. Bush administration not only refused to endorse the Kyoto Protocol, but also has (with a few exceptions, such as its endorsement of hydrogen and hybrid-fueled automobiles) failed to take seriously the coming revolution in the technology of energy production and use. In a century, George Bush's bust may sit in a greenhouse-gas museum, not far from a model of an antique internal combustion engine. A plaque may mention his family's intimate ties to the oil industry as a factor in his refusal to think outside that particular "box."

As temperatures rose, energy policy in the United States under the Bush administration generally ignored atmospheric physics. Gas mileage for U.S. internal combustion engines has, in fact, declined during the last two decades; gains in energy efficiency have been more than offset by increases in vehicle size, notably through sports-utility vehicles. And now comes a mass advertising campaign aimed at security-minded United States citizens for the biggest gas-guzzler of all: the fortress-like Hummer. Present-day automotive marketing may seem quaint in a hundred years. By the end of this century, perhaps sooner than that, the internal combustion engine and the oil (and natural-gas) burning furnace will become museum pieces. They will be as antique as a horse and buggy is today. Such change will be beneficial and necessary.

We are overdue for an energy system paradigm shift. Limited supplies of oil and their location in the volatile Middle East argue for new sources, along with accelerating climate change from greenhouse gases accumulating in the atmosphere. According to an editorial in *Business Week*:

> A national policy that cuts fossil-fuel consumption converges with a geopolitical policy of reducing energy dependence on Middle East oil. Reducing carbon dioxide emissions is no longer just a "green" thing. It makes business and foreign policy sense, as well.... In the end, the only real solution may be new energy technologies. There has been little innovation in energy since the internal combustion engine was invented in the 1860s and Thomas Edison built his first commercial electric generating plant in 1882. ("How to Combat" 2004, 108)

Before the end of this century, the urgency of global warming will become manifest to everyone. Solutions to our fossil-fuel dilemma—solar, wind, hydrogen, and others—will evolve during this century. Within our century, necessity will compel invention. Other technologies may develop that have not, as yet, even broached the realm of present-day science fiction, any more than digitized computers had in the days of the Wright Brothers a hundred years ago. We will take this journey because the changing climate, along with our own innate curiosity and creativity, will compel a changing energy paradigm.

Such change will not take place at once. A paradigm change in basic energy technology may require the better part of a century, or longer. Several technologies will evolve together. Oil-based fuels will continue to be used for purposes that require it. (Air transport comes to mind, although engineers already are working on ways to make jet engines more efficient.)

Technological and Political Solutions

Chances of forestalling seriously debilitating climate change revolve around humankind's ability to forge political and technological solutions. One will not work without the other. Technological changes range from the very prosaic (such as mileage improvements on existing gasoline-burning automobiles, changes in building codes, and painting building roofs white) to the exotic, including the invention of microorganisms that eat carbon dioxide and the generation of microwaves from the moon. In between are the makings of a shift in the energy paradigm by the end of the century from fossil fuels to renewable, nonpolluting sources such as solar and wind power. By the end of this century, the internal combustion engine may be as much of an antique piece as a horse and buggy is today.

S. Pacala and R. Socolow, writing in *Science*, have asserted that, using existing technology, "Humanity already possesses the fundamental scientific, technical, and industrial know-how to solve the carbon and climate problem for the next half-century" (Pacala and Socolow 2004, 968). By "solve," they mean that the tools are at hand to meet global energy needs without doubling preindustrial levels of carbon dioxide.

Their "stabilization strategy" involves intense attention to improved automotive fuel economy, reduced reliance on cars, more efficient building construction, improved power-plant efficiency, substitution of natural gas for coal, storage of carbon captured in power plants as well as hydrogen and synthetic fuel plants, more use of nuclear power, development of wind and photovoltaic (solar) energy sources, creation of hydrogen from renewable sources, and more intense use of biofuels, such as ethanol. The strategy advocates more intense management of natural resources, including reductions in deforestation and aggressive management of agricultural soils through such measures as conservation tillage—drilling seeds into soil without plowing (Pacala and Socolow 2004, 969–971).

Future Energy Sources

Martin I. Hoffert and colleagues believe that "Mid-century primary-power transmissions could be several times what we now derive from fossil fuels (about 10 to the 13th power watts), even with improvements in energy efficiency" (Hoffert et al. 2002, 981). Hoffert and colleagues' survey of future energy sources included terrestrial solar and wind energy, solar power satellites, biomass, nuclear fission, nuclear fusion, fission-fusion hybrids, and fossil fuels from which carbon has been sequestered. To this mix they added efficiency improvements, hydrogen production, storage and transport, superconducting global electric grids, and geo-engineering (Hoffert et al. 2002, 981). This effort will challenge the creative resources of scientists as much as the Apollo project that put men on the Moon. In addition, the scientists, said, the use of fossil fuels must decline, a matter of political decision-making that exceeds the reach of science.

Research must begin immediately to build energy infrastructure that is "climate-neutral." Without such action, the atmosphere's concentration of greenhouse gases will double from preindustrial levels by the end of the twenty-first century, the scientists said. "A broad range of intensive research and development is urgently needed to produce technological options that can allow both climate stabilization and economic development," the group said (Revkin 2002c, A-6). The 18 researchers called for intensive new efforts to improve existing technologies and develop others, such as fusion reactors or space-based solar power plants. The researchers provided only the broadest estimate of how much such research effort will cost. "The cost probably will total tens of billions of dollars in both government and private funds" (Revkin 2002c, A-6).

Most existing energy technologies, "have severe deficiencies," the scientists said. Solar panels, new nuclear power options, windmills, filters for fossil-fuel emissions, and other options are either inadequate or require vastly more research and development than is currently planned in the United States or elsewhere (Revkin 2002c, A-6). One author of the analysis, Dr. Haroon S. Kheshgi, a chemical engineer with ExxonMobil, said that "climate change is a serious risk" requiring a shift away from fossil fuels. "You need a quantum jump in technology," he said. "What we're talking about here is a 50- to 100-year time scale" (Revkin 2002c, A-6).

Solving Technological Problems

Hoffert, a New York University physics professor, said he is convinced that technological problems will be solved on a scientific and engineering level. At the same time, he is worried whether the public and its elected officials will realize the urgency of the task (Revkin 2002c, A-6). Several of the study's authors said that they saw few signs that major industrial nations were ready to engage in an ambitious quest for climate-neutral energy. Richard L. Schmalensee, a climate-policy expert and the dean of the Massachusetts Institute of Technology Sloan School of Management, said the issue of climate change remained too complex and contentious to generate the requisite focus. "There is no substitute for political will," he said (Revkin 2002c, A-6).

Some environmental advocates criticized this study's emphasis on still-distant technologies, asserting that this focus could distract from the need to do what is possible at present to reduce greenhouse-gas emissions. "Techno-fixes are pipe dreams in many cases," said Kert Davies, research director for Greenpeace, which has been conducting a broad campaign against Exxon-Mobil. "The real solution," he said, "is cutting the use of fossil fuels by any means necessary" (Revkin 2002c, A-6). Given a sufficient sense of urgency, however, Hoffert believed that technology will solve the problem once a large proportion of the world's people and their leaders realize the seriousness of global warming. "We started World War II with biplanes," he has said. "And seven years later we had jets" (Revkin 2002d, A-8).

A Wide Variety of Solutions

A wide variety of solutions are being pursued around the world, of which the following are only a few examples. Some changes involve localities. Already, several U.S. states are taking actions to limit carbon-dioxide emissions despite a lack of support from the U.S. federal government. Building code changes have been enacted. Wind-power incentives have been enacted even in Bush's home state of Texas, where some oil fields now host wind turbines.

Wind turbines and photovoltaic solar cells are becoming more efficient and competitive. Improvements in farming technology are reducing emissions. Deep-sea sequestration of carbon dioxide is proceeding in experimental form, but with concerns about this technology's effects on ocean biota. Tokyo, where an intense urban heat island has intensified the effects of general warming, has proposed a gigantic ocean-water cooling grid. Britain and other countries are considering carbon taxes.

J. Craig Venter, the maverick scientist who compiled a human genetic map with private money, has decided to tap a $100 million research endowment he has created from his stock holdings to scour the world's deep ocean trenches for bacteria that might be able to convert carbon dioxide to solid form using very little sunlight or other energy. Failing that, Venter has proposed to synthesize such organisms via genetic engineering. He would like to invent two synthetic microorganisms, one to consume carbon dioxide and turn it into raw materials, including the kinds of organic chemicals that are now made from oil and natural gas. The other microorganism would generate hydrogen fuel from water and sunshine.

The coming energy revolution will engender economic growth and become an engine of wealth creation for those who realize the opportunities that it offers. Denmark, for example, is making every family a shareholder in a burgeoning wind-power industry. The United Kingdom is making plans to reduce its greenhouse-gas emissions 50 percent in 50 years. The British program begins to address the position of the Intergovernmental Panel on Climate Change (IPCC) that emissions will have to fall 60 to 70 percent by century's end to avoid significant warming of the lower atmosphere caused by human activities. The Kyoto Protocol, with its reductions of 5 to 15 percent (depending on the country) is barely earnest money compared with the required paradigm change that will reconstruct the system by which most of the world's people obtain and use energy.

Solutions will combine scientific achievement and political change. We will end this century with a new energy system, one that acknowledges nature and works with its needs and cycles. Economic development will become congruent with the requirements of sustaining nature. Coming generations will be able to mitigate the effects of greenhouse gases without the increase in poverty so feared by "skeptics." Within decades, a new energy paradigm will be enriching us, and securing a future that works with the requirements of nature, not against it.

FURTHER READING

Hoffert, Martin I., Ken Caldeira, Gregory Benford, David R. Criswell, Christopher Green, Howard Herzog, Atul K. Jain, Haroon S. Kheshgi, Klaus S. Lackner, John S. Lewis, H. Douglas Lightfoot, Wallace Manheimer, John C. Mankins, Michael E. Mauel, L. John Perkins, Michael E. Schlesinger, Tyler Volk, and Tom M. L. Wigley. "Advanced Technology Paths to Global Climate Stability: Energy for a Greenhouse Planet." *Science* 298 (November 1, 2002):981–987.

"How to Combat Global Warming: In the End, the Only Real Solution May Be New Energy Technologies." *Business Week*, August 16, 2004, 108.

Pacala, S., and R. Socolow. "Stabilization Wedges: Solving the Climate Problem for the Next 50 Years with Current Technologies." *Science* 305 (August 13, 2004):968–972.

Revkin, Andrew C. "Scientists Say a Quest for Clean Energy Must Begin Now." *New York Times*, November 1, 2002c, A-6.

Revkin, Andrew C. "Climate Talks Shift Focus to How to Deal with Changes." *New York Times*, November 3, 2002d, A-8.

Energy Use and Efficiency

According to British Petroleum's *Statistical Review of World Energy* (2006), global use of energy has doubled since the 1970s (Hillman, Fawcett, and Rajan 2007, 38–40). While demand for fossil fuels continues to rise (most significantly in India and China, with large populations and rapid industrialization), alternative forms of energy are slowly gaining market share. Rising costs of fossil fuels, coupled with falling costs for ethanol, wind power, and other renewable forms could propel such non-fossil forms of energy to 25 percent of U.S. energy demand by 2025 at little additional cost, according to a report prepared by the Rand Corp. In 2005, renewables contributed 6 percent of U.S. energy, half of that from hydroelectric (Fialka 2006b, A-10).

Energy Use: Global Survey

An average baby born in the United States of America during the year 2008 will consume more than 50 times the natural resources (including fossil-fuel energy) of an average baby born in Bangladesh. The wealthiest one-quarter of the world's population consumes 80 percent of its aggregate energy resources. According to Anita Gordon and David Suzuki, "The average North American uses the energy equivalent of 10 tons of coal a year. Bangladeshis use less than 100 kilograms (220 pounds)" (Gordon 1991, 106). In other words, the average North American (in the United States and Canada) contributes as much greenhouse effluent to the atmosphere as 60 Angolans or about 20 residents of India.

The type of coal-powered industrialization that began in Britain more than two centuries ago is now taking place in China and India. The major ecological difference is that England was inhabited by roughly 20 million people at the dawn of its Industrial Revolution. China's population at the dawn of the twenty-first century was 1.3 to 1.5 billion, and India's was nearing 1 billion.

China surpassed the United States as the world's largest burner of coal during the late 1980s; by 1990, China was responsible for 10 percent of the world's carbon-dioxide emissions. In 2006, China became the world's largest national source of greenhouse gases, passing the United States as it industrialized, mainly with coal-fired electricity.

U.S. Energy Efficiency Compared with Europe

Energy efficiency has been improving in the United States. In 1973, roughly 1.5 barrels of oil were required to produce $1,000 worth of gross domestic product (in constant dollars), according to the Department of Energy. By 2004, that figure had been halved (Fialka 2006a, A-4). The *rate* of improvement has been slowing since thee 1970s, however.

Energy efficiency in the United States faces an uphill battle, because demand increases as populations rise, while the number of power-consuming appliances in each home increases as well. In 1978, for example, 56 percent of U.S. homes had air conditioning. In 2001, 77 percent did. In addition, as temperatures rise, so too does peak demand for electricity, especially air conditioning. U.S. electricity for industries is relatively inexpensive, at US$0.061 per kilowatt hour (in 2006), less than half the rate in many European countries (such at Britain or Denmark) and one-quarter the cost in Italy (Abboud and Biers 2007, R-4).

While U.S. energy efficiency has increased, the average western European still uses half as much energy as the average American. European cities are older and more compact, while sprawling U.S. urban areas have been made over in the service of the automobile. Europeans have embraced policies ranging from gas taxes to emissions caps and other ways to keep carbon-dioxide production down, including a lack of the mortgage subsidies that encourage large homes in the United States. Congestion charges also have been started for drivers in London and Stockholm (Mallaby 2006).

The United States is only starting to consider a carbon tax, an idea that has a decade-long record in many European countries. According to a report by Sebastian Mallaby in the *Washington Post,* Norway's carbon taxes have compelled companies drilling for natural gas in the North Sea to capture the carbon dioxide they release and inject it back under the seafloor. "Because the United States lacks a carbon tax," wrote Mallaby,

> Natural gas drillers in this country have less incentive to do that. Meanwhile the Energy Department reports that there are plans to build 150 new coal-fired power plants in the United States, enough to generate power for 93 million homes. But because government hasn't created intelligent incentives, most of these new generators won't be fitted with technology to capture and store the carbon. (Mallaby 2006)

Saving Energy in Existing Infrastructure

The American Council for an Energy-efficient Economy estimates that 35 to 40 percent of the electricity consumed in the United States could be saved through existing technology and rigorous conservation measures. Electricity consumed by lighting could be cut nearly in half through the adoption of compact fluorescent light bulbs. These bulbs consume one-sixth the energy used by "standard" bulbs. Singapore, the Philippines,

Indonesia, Malaysia, and Thailand reduced their energy consumption by more than 20 percent in a decade (1985–1995) by implementing more efficient building codes (Gelbspan 1997b, 116).

Some proposed changes are more behavioral than technological—such things as building more bicycle paths and urging workers to "telecommute" when possible. Other observers advocate dietary measures, such as avoiding beef in favor of plant-based foods. Cattle are responsible for 72 percent of methane released by livestock in the United States (Lovins 1991, 378). Roughly 10 pounds of plant foods are required to produce one pound of beef, making it an energy-intensive food source.

Electricity-generating industries, automobiles, the home, and others have been cased for ways to save energy. Many observers believe that at least 50 percent of present energy use could be eliminated through changes available with existing technology. Only about one-third of the energy consumed by a typical coal-burning power plant in the United States actually drives the steam generators that produce electricity, for example. The other two-thirds is lost as waste heat from smokestacks and cooling towers. Increasing attention is being paid to "cogeneration" strategies that capture more of that waste heat for actual energy production. Industry in the United States made 25 percent more efficient use of energy in 1986 than in 1973.

Existing power plants offer many opportunities for conservation: "For a pulverized coal plant, 90 percent of the carbon dioxide can be recovered using a chemical absorption process to clean up flue gases" (Hendriks 1989, 125). In such a process, according to Hendriks, of the Netherlands' University of Utrecht, "The carbon dioxide is recovered by leading the flue gas through a solution containing the absorber" (Hendriks 1989, 127). The cost of removing 90 percent of the carbon dioxide is roughly a 25 percent loss in generation efficiency and roughly a 35 percent increase in the cost of electricity, according to Hendriks.

Basic Conservation Measures

A National Academy of Sciences list of mitigation strategies starts with the most prosaic—a reduction of air-conditioning usage and the urban "heat island" effect through a crash program to paint half the roofs in urban areas white. Next, the report recommends another crash program to replace incandescent lighting with compact fluorescent bulbs. Energy-efficiency technology would then be upgraded in residential and commercial water heaters, commercial lighting, commercial cooking, refrigeration, and appliances. Homes would be heavily insulated to conserve fuel consumed by space-heating and air-conditioning. Industrial energy efficiency would be upgraded through the use of cogeneration, fuel switching, and new technology. (National Academy of Sciences 1991, 54–58)

A study by the Illinois Environmental Protection Agency found that a number of basic energy-saving measures (including heating-system retrofits, improved thermostats, better efficiency in hot-water generation, more insulation, installation of storm windows, and lighting retrofits) could reduce energy use in an average home by 22 percent. The same study found that similar efficiencies in public buildings could reduce their natural-gas consumption by 20 percent, and electricity consumption by 13 percent.

Lovins and Lovins asserted that mitigating global warming will not drastically curtail American lifestyles.

> Nothing could be further from the truth. The fuel-saving technologies that can stabilize global climate while saving money actually provide unchanged services: showers as hot and tingly as now, beer as cold, rooms as brightly lit ... homes as cozy in the winter and cool in the summer.... The quality of these and other services can often be not just sustained but substantially improved by substituting superior engineering for brute force. (Lovins and Lovins 1991, 433)

In Japan, by the beginning of 1999, control of greenhouse-gas emissions had reached the local (or prefecture) level. On April 1, Japan enacted a stringent law to combat global warming. The law is meant to bring Japan into compliance with the 1997 Kyoto Protocol on climate change. Each municipality has been instructed to submit its own plan to bring down greenhouse-gas emissions.

The city of Koga, in Ibaraki Prefecture, during July 1998 began to actively promote the bicycle as a means of daily transport. Two months earlier, the city hosted a conference among local governments from throughout Japan. Participants traveled between conference sites by bicycle. The city published a pamphlet offering hints on how to best use bicycles and purchased

20 bikes for official use. Cyclists have been given the right-of-way on some roads in the city.

Corporations have been adopting conservation measures. For example, Wal-Mart has centralized energy management for its 4,000 stores (which include Sam's Club outlets) at its Bentonville, Arkansas, headquarters to the point at which if a manager wants to lower the air-conditioning setting anywhere, the request must be justified at headquarters, which employs a team of 100 specialists. If a freezer door has been left open at any store more than 45 minutes, headquarters sends a query. The monitoring team has been credited with saving the company several million dollars a year in energy costs (Abboud and Biers 2007, R-4). The energy monitoring team also has put considerable effort into changing distribution networks so that items (especially food) are ordered as close as possible to a given point of sale. Wal-Mart is developing wind energy and waterless urinals in some stores as demonstration projects, with an eye toward reducing energy use 20 percent in existing stores by 2012. Use of light-emitting diode lighting in refrigerated displays may cut energy use by 80 percent (Abboud and Biers 2007, R-4).

Thinking Globally, Acting Locally

During 2000, the City of Oakland, California, approved a plan to buy alternative power for all of its municipal needs. "It leads us in the direction of reducing global warming, stimulating new industry, and sets the pace for the national government," said Oakland Mayor Jerry Brown, a former California governor and presidential candidate (Boxall 2000, 3). Oakland thus joined Santa Monica, which in 1999 became the first city in California to purchase green power for all its municipal needs. Palmdale soon followed and several other communities were preparing to do the same during 2000, including Santa Barbara, San Jose, and Santa Cruz. By the year 2000, one-eighth of the power consumed in California came from non-fossil-fueled sources, which include wind, geothermal, solar, methane, and hydropower from small dams. Large hydroelectric plants are not considered environmentally friendly.

Even as climate scientists rang alarm bells regarding global warming during the 1990s, the suburban rings of most major urban areas in the United States continued to grow, intensifying the reliance of many residents on their automobiles, which often were being driven longer distances to and from work and other engagements, raising greenhouse-gas emissions. A number of observers have recommended denser urban development that is more amenable to mass transit and walking. A group of scientists led by S. T. Boyle graphed gasoline consumption against urban density. Not surprisingly, the greatest energy consumption occurred in the least dense, newer urban areas in the United States (e.g., Houston, Denver, Los Angeles) that expanded rapidly after the automobile diffused early in the twentieth century. The densest (low-consumption) urban areas were in Europe and Asia, with large areas of urban infrastructure predating the automobile, such as London, Hamburg, and Tokyo (Boyle 1990, 240).

Energy Conservation: U.S. States Take Action

Several U.S. states and cities, no longer waiting for the George W. Bush administration to seize the initiative against global warming, have begun taking steps to reduce greenhouse-gas emissions. By 2005, 166 U.S. cities had agreed to meet the goals of the Kyoto Protocol, which the Bush administration had rejected (*Harper's* Index 2005, 11). By mid-2007, 540 cities in the United States, from Boston, Massachusetts, to Portland, Oregon, had pledged to meet Kyoto Protocol standards, as local officials sped ahead of federal policy on global warming. A nationwide poll released in April showed a third of Americans called global warming the world's single largest environmental problem—double the number a year earlier, according a *Washington Post-ABC News*-Stanford University survey. Greg Nickels, Seattle's mayor, started the U.S. Mayors Climate Protection Agreement in 2005. Two years later, it included 522 cities with a population of 65 million.

State and local governments are legislating broader use of energy-saving devices, more energy-efficient building standards, cleaner-burning power plants and more investment in such renewable energy sources as wind and solar power. By 2002, more than half the states in the United States had adopted voluntary or mandatory programs for reducing carbon-dioxide emissions, according to the Pew Center on Global Climate Change. Fifteen states, including President George W. Bush's home state of Texas, have enacted legislation requiring utilities to increase their use of renewable energy sources such as wind power or biomass in generating some of their electricity (Pianin 2002c, A-3).

California, for example, had enacted regulations to reduce car and truck emissions by 2006 that, if they survive a challenge expected from the auto industry, could be a model for New York, New Jersey, and other northeastern states. Texas has legally required that 3 to 4 percent of its electricity come from renewable energy sources, notably wind power, by 2010. Massachusetts had become the first state to legally limit power-plant emissions of carbon dioxide as part of rules controlling pollution by six major coal- and oil-burning facilities in that state. At about the same time, New Hampshire enacted emission controls for that state's three aging coal-fired electricity-generating plants. Nebraska became the state first to enact legislation to require rotation of crop planting, increasing the number of plants and trees that absorb carbon dioxide.

FURTHER READING

Abboud, Leila, and John Biers. "Business Goes on an Energy Diet." *Wall Street Journal,* August 27, 2007, R-1, R-4.

Boyle, S. T., W. Fulkerson, R. Klingholz, I. M. Mintzer, G. I. Pearman, G. Oinchera, J. Reilly, F. Staib, R. J. Swart, and C.-J. Winter. "Group Report: What Are the Economic Costs, Benefits, and Technical Feasibility of Various Options Available to Reduce Greenhouse Potential per Unit of Energy Service?" In *Limiting Greenhouse Effects: Controlling Carbon-dioxide Emissions* ed. G. I. Pearman 229–260. Report of the Dahlem Workshop on Limiting the Greenhouse Effect, Berlin, September 9–14, 1990. New York: John Wiley & Sons, 1991.

Boxall, Bettina. "Oakland Switches to 'Green" Power." *Los Angeles Times,* June 29, 2000, 3.

Fialka, John J. "Energy Independence: A Dry Hole?" *Wall Street Journal,* July 5, 2006a, A-4.

Fialka, John J. "Renewable Fuels May Provide 25 Percent of U.S. Energy by 2025." *Wall Street Journal,* November 13, 2006b, A-10.

Gelbspan, Ross. *The Heat Is On: The High Stakes Battle over Earth's Threatened Climate.* Reading, MA: Addison-Wesley Publishing Co., 1997b.

Gordon, Anita, and David Suzuki. *It's a Matter of Survival.* Cambridge, MA: Harvard University Press, 1991.

Harper's Index, August 2005, 11.

Hendriks, C. A., K. Blok, and W. C. Turkenburg. "The Recovery of Carbon Dioxide from Power Plants." In *Climate and Energy: The Feasibility of Controlling CO2 Emissions,* ed. P. A. Okken, R. J. Swart, and S. Zwerver, 125–142. Dordrecht, Germany: Kluwer Academic Publishers, 1989.

Hillman, Mayer, Tina Fawcett, and Sudhir Chella Rajan. *The Suicidal Planet: How to Prevent Global Climate Catastrophe.* New York: St. Martin's Press/Thomas Dunne Books, 2007.

Lovins, A. B., and L. H. Lovins. "Least-cost Climatic Stabilization." In *Limiting Greenhouse Effects: Controlling Carbon-dioxide Emissions,* ed. G. I. Pearman, 351–442. *Report of the Dahlem Workshop on Limiting the Greenhouse Effect,* Berlin, September 9–14, 1990. New York: John Wiley & Sons, 1991.

Mallaby, Sebastian. "A Dated Carbon Approach." *Washington Post,* July 10, 2006, A-17. http://www.washingtonpost.com/wp-dyn/content/article/2006/07/09/AR2006070900537_pf.html.

National Academy of Sciences. *Policy Implications of Greenhouse Warming.* Washington, DC: National Academy Press, 1991.

Pianin, Eric. "On Global Warming, States Act Locally: At Odds with Bush's Rejection of Mandatory Cuts, Governors and Legislatures Enact Curbs on Greenhouse Gases." *Washington Post,* November 11, 2002c, A-3.

Stokstad, Erik. "California Sets Goals for Cutting Greenhouse Gases." *Science* 308 (June 10, 2005):1530.

Ethanol: Brazil

Ethanol can be made through the fermentation of many natural substances, but sugarcane offers advantages over others, including corn. For each unit of energy expended to turn cane into ethanol, 8.3 times as much energy is created, compared with a maximum of 1.3 times for corn, according to scientists at the Center for Sugarcane Technology and other Brazilian research institutes. "There's no reason why we shouldn't be able to improve that ratio to 10 to 1," said Suani Teixeira Coelho, director of the National Center for Biomass at the University of São Paulo. "It's no miracle. Our energy balance is so favorable not just because we have high yields, but also because we don't use any fossil fuels to process the cane, which is not the case with corn" (Rohter 2006). Sugarcane is generally more economical than oil with the per-barrel price under $30.

By 2007, 40 percent of Brazil's transportation fuel was sugarcane ethanol. Three-quarters of new cars sold in Brazil were flex fuel (compared with 10 percent in the United States), and most fueling stations offered E85 fuel, compared with 1 percent in the United States.

Brazil, using ethanol from sugarcane, became energy self-sufficient in 2006, even as demand for fuel grew. Many gas stations in Brazil have two sets of pumps, marked A for alcohol and G for gasoline. "Renewable fuel has been a fantastic solution for us," Brazil's minister of agriculture, Roberto Rodrigues, said in a recent interview in São Paulo, the capital of São Paulo State, which accounts for 60 percent of sugar production in Brazil. "And it offers a way out of the fossil-fuel trap for others as well" (Rohter 2006). Use of ethanol in Brazil accelerated after 2003 following the introduction of flex-fuel engines, designed to run on ethanol, gasoline, or any mixture of the two. Gasoline sold in Brazil contains about 25 percent alcohol. By 2006, more than 70 percent of the 1.1 million automobiles sold in Brazil had flex-fuel engines (Rohter 2006).

In the past, the residue remaining when sugarcane stalks were compressed to squeeze out juice were discarded. Today, Brazilian sugar mills use that residue to generate the electricity to process cane into ethanol, and use other by-products to fertilize the fields where cane is planted. Some mills are now producing so much electricity that they sell their excess to the national grid. In addition, Brazilian scientists, with money from São Paulo State, have mapped the sugarcane genome. That opens the prospect of planting genetically modified sugar, if the government allows, that could be made into ethanol even more efficiently (Rohter 2006).

FURTHER READING

Rohter, Larry. "With Big Boost From Sugar Cane, Brazil Is Satisfying Its Fuel Needs." *New York Times*, April 10, 2006. http://www.nytimes.com/2006/04/10/world/americas/10brazil.html.

Extinctions and Climate Change

We are in the midst of one of the Earth's most intense, rapid, and pervasive mass extinctions, which has placed in harm's way many flora and fauna that humankind does not eat or keep as pets. The Earth has experienced mass extinctions before, but all of them have resulted from natural causes. Global warming is one product of humankind's increasing dominance of the Earth that is devastating the native habits of many animals and plants, driving many to extinction. Compared with past mass extinctions, which were driven by natural catastrophes such as meteor strikes or large-scale volcanism, the present-day human-driven wave of extinctions has been occurring with frightening speed. Given projected rises in temperature during decades to come, the flora and fauna of our home planet thus far have seen only their initial travails.

Albert K. Bates wrote: "Sixty-five million years ago, 60 to 80 percent of the world's species disappeared in a cataclysmic mass extinction, possibly caused by an asteroid's impact with Earth. Human population, not an asteroid, will cut the remaining number of species in half again, in just the next few years" (Bates and Project Plenty 1990, 137). Another observer, Robert L. Peters, factors climate change into a similar picture of the Earth's biological future:

> Habitat destruction in conjunction with climate change sets the stage for an even larger wave of extinction than previously imagined, based on consideration of human encroachment alone. Small, remnant populations of most species, surrounded by cities, roads, reservoirs, and farmland, would have little chance of reaching new habitat if climate change makes the old unsuitable. Few animals or plants would be able to cross Los Angeles on the way to the promised land. (Peters 1989, 91)

Global warming could play a role in the destruction or fundamental alteration of a third of Earth's plant and animal habitats within a century, bringing extinction to thousands of species, according to a study by Great Britain's World Wide Fund for Nature, an affiliate of the World Wildlife Fund. The report, *Global Warming and Terrestrial Bio-diversity Decline*, was written by Jay Malcolm, professor of forestry at Toronto University, and Adam Markham, the executive director of Clean Air/Cool Planet. The study found that the most vulnerable plant and animal species will be in Arctic and mountain areas, where as many as 20 percent could be driven to extinction. In the north of Canada, Russia, and Scandinavia, where warming was predicted to be most rapid, up to 70 percent of habitat could be lost, according to this study.

Pests and weedy species would fare best, the report said, and "[i]f past fastest rates of migration are a good proxy for what can be attained in a warming world, then radical reductions in greenhouse gas emissions are urgently required to reduce the threat of bio-diversity loss" (Clover 2000, 9). To adapt and survive the expected rate of warming during the next century, the report said that plants may need to move 10 times more quickly than they did when recolonizing previously glaciated land at the end of the last ice age. Few plant species can move at a rate of one kilometer per year, the speed that will be required in many parts of the world.

David Quammen, an environmental writer, believes that the Earth during the next century will become a "planet of weeds," where human domination provokes extinction of most undomesticated plants and animals. Quammen expects that the flora and fauna of the Earth eventually will include food crops, animals raised to be eaten or petted, and a few stubborn weed species that will benefit from a harsher, hotter world. By 2050, Quammen believes that deforestation will cause half the world's wild birds and two-thirds of other wild-animal species to become extinct (Quammen 1998, 61, 69). "Wildlife," he writes, "[w]ill consist of pigeons, coyotes, rats, roaches, house sparrows, crows, and feral dogs." (Quammen 1998, 67). Human beings—"remarkably widespread, prolific, and adaptable," are "the consummate weed" (Quammen 1998, 68).

Mass Extinctions within a Century?

In the first study of its kind, researchers in a range of habitats—including northern Britain, the wet tropics of northeastern Australia, and the Mexican desert—said early in 2004 that given "midrange" climate change scenarios for 2050, they anticipate that 15 to 37 percent of the species in their sample of regions (covering 20 percent of Earth's surface) would be "committed to extinction" (Thomas et al. 2004, 145). The number of extinctions is expected to vary with the severity of warming. The study used United Nations projections that world average temperatures will rise 2.5° to 10.4°F by 2100. "We're not talking about the occasional extinction—we're talking about 1.25 million species. It's a massive number," the authors of this study wrote (Gugliotta 2004, A-1).

This study, described in *Nature*, marked the first time that scientists have produced a global analysis with concrete estimates of the effect of climate change on many various animal and plant habitats. Thomas led a 19-member international team that surveyed habitat decline for 1,103 plant and animal species in Europe; Queensland, Australia; Mexico's Chihuahua Desert; the Brazilian Amazon; and the Cape Floristic Region at South Africa's southern tip (Gugliotta 2004, A-1).

According to the researchers, climate change during the past 30 years already has produced many shifts in the distribution and abundance of plants and animals. Climate change thus has become a major driver of biodiversity change. The survey team used one of ecology's few ironclad laws: the species-area relationship, first postulated by Charles Darwin in his *Origin of Species* (1859), which holds that a smaller habitable area will host a smaller number of viable species. They then projected habitat changes based on various warming scenarios. With a temperature rise of 0.8° to 1.7°C, they anticipated an 18 percent extinction rate; at 1.8° to 2°C, they projected 24 percent, and at more than 2°C, 35 percent (Pounds and Puschendorf 2004, 108).

This study emphasized that examining possible extinctions solely in light of global warming probably understates their potential scope because, in the real world, extinctions also may be caused by such factors as landscape modification, species invasions, and pollution. By projecting effects of rising temperatures alone, the researchers realized that they may be ignoring effects of changes in precipitation on habitat. Declines of amphibians in Costa Rica have been traced to reduced cloudiness in mountainous areas that probably are related to warming, for example.

The authors of this study considered a range of possibilities based on the ability of each species to move to a more congenial habitat to escape warming. If all species were able to move, or "disperse," the study said, only 15 percent would be irrevocably headed for extinction by 2050. If no species were able to move, the extinction rate could rise as high as 37 percent (Gugliotta 2004, A-1). The scientists concluded that "[t]hese estimates show the importance of rapid implementation of technologies to decrease greenhouse-gas emissions and strategies for carbon sequestration" (Thomas et al. 2004, 145).

The survey's findings were disputed by skeptics, such as William O'Keefe, president of the George C. Marshall Institute, who said that the research "ignored species' ability to adapt to higher temperatures" and assumed that technologies will not arise to reduce emissions (Gugliotta 2004, A-1). Some animals and plants are adapting to warming—to a point, as they move upward in elevation or toward the poles in direction, until something blocks their way. In the European Alps, for example, some plant species have been migrating upward by one to four meters each decade (Grabherr, Gottfried, and Pauli 1994, 448). Across Europe, the growing season in controlled mixed-species gardens lengthened by 10.8 days a year from 1959 to 1993 (Menzel and Fabian 1999, 659). In Europe and North America, many migratory birds now arrive earlier in the spring and depart later in the autumn. Butterflies, beetles, dragonflies, and other species are now found farther north, where conditions previously were too cold for them to survive.

What Happened 251 Million Years Ago?

As anticipation of mass extinctions abetted by global warming has become more common, scientists have sought paleoclimatic parallels. According to some scientists, the worst mass extinction in the history of the planet could be replicated in as little as a century if global warming continues at the pace forecast by the Intergovernmental Panel on Climate Change (IPCC). Researchers at England's Bristol University have estimated that a 6°C increase in global temperatures was enough to play a role in the annihilation of up to 95 percent of species that were alive on Earth at the end of the Permian period 251 million years ago, roughly the same amount of warming expected by the IPCC, if levels of greenhouse gases in the atmosphere continue to rise at present rates (Reynolds 2003, 6).

The wave of mass extinction at the end of the Permian period probably was caused by a series of very large volcanic eruptions that triggered a runaway greenhouse effect that nearly extinguished life on Earth. Conditions in what geologists have termed this "postapocalyptic greenhouse" were so severe that 100 million years passed before species diversity returned to former levels. Michael Benton, head of earth sciences at Bristol University, commented: "The end-Permian crisis nearly marked the end of life. It's estimated that fewer than one in ten species survived. Geologists are only now coming to appreciate the severity of this global catastrophe and to understand how and why so many species died out so quickly" (Reynolds 2003, 6).

The Permian heat wave was felt first and most intensely in the tropical latitudes; loss of species diversity spread from there. Reduction of vegetation, soil erosion, and the effects of increasing rainfall wiped out the lush, diverse habitats of the tropics, which today would lead to the loss of animals such as hippos, elephants, and all of the primates, according to Benton (Reynolds 2003, 6). He added:

> The end-Permian extinction event is a good model for what might happen in the future because it was fairly non-specific. The sequence of what happened then is different from today because then the carbon dioxide came from massive volcanic eruptions, whereas today it is coming from industrial activity. However, it doesn't matter where this gas comes from; the fact is that if it is pumped into the atmosphere in high volumes, then that gives us the greenhouse effect and leads to the warming with all the other consequences. (Reynolds 2003, 6)

According to a theory first advanced by Anthony Hallam and Paul Wignall (1997), the volcanic eruptions 251 million years ago provoked a number of biotic feedbacks that accelerated global warming of about 6°C. In his book *When Life Nearly Died: The Greatest Mass Extinction of All Time* (2003), Benton sketched how the warming (which was accompanied by anoxia) may have fed upon itself:

> The end-Permian runaway greenhouse may have been simple. Release of carbon dioxide from the eruption of the Siberian Traps [volcanoes] led to a rise in global temperatures of 6°C or so. Cool polar regions became warm and frozen tundra became unfrozen. The melting might have penetrated to the frozen gas hydrate reservoirs located around the polar oceans, and massive volumes of methane may have burst to the surface of the oceans in huge bubbles. This further input of carbon into the atmosphere caused more warming, which could have melted further gas hydrate reservoirs. So the process went on, running faster and faster. The natural systems that normally reduce carbon dioxide levels could not operate, and eventually the system spiraled out of control, with the biggest crash in the history of life. (Benton 2003, 276–277)

Greg Retallack, an expert on ancient soils at the University of Oregon has speculated that the same methane "belch" was of such a magnitude that it caused mass extinction via oxygen starvation of land animals. Bob Berner of Yale University has calculated that a cascade of effects on wetlands and coral reefs may have reduced oxygen levels in the atmosphere from 35 percent to only 12 percent in 20,000 years. Marine life may have suffocated in the oxygen-poor water ("Suffocation" 2003). One animal, the meter-long reptile *Lystrosaurus*, survived because it had evolved to live in burrows, where oxygen levels were low and carbon-dioxide levels high. According to a report by the New Scientist News Service, "It had developed a barrel chest, thick ribs, enlarged lungs, a muscular diaphragm and short internal nostrils to get the oxygen it needed" ("Suffocation" 2003).

According to Chris Lavers, writing in *Why Elephants Have Big Ears* (2000), a spike of worldwide warming contributed to this mass extinction in part because all of the Earth's continents at the time were combined into one land mass (Lavers 2000, 231). Warming of tropical regions at this time has been estimated at about 11°F, with larger rises near the poles that tended to create a generally warm atmosphere planet-wide, "a flattening of the temperature difference between the poles and the equator" (Lavers 2000, 232). Lavers suspected that this condition drastically slowed or shut down ocean mixing, killing many sea creatures. "Unstirred," wrote Lavers, "the oceans begin to stagnate. Deep waters gradually lost oxygen, and species began to vanish" (Lavers 2000, 233).

What caused this spike in temperatures? The prime suspect, at least in the beginning, is coal-bearing deposits in the southern reaches of Pangaea (the Earth's single land mass), which were oxidized after they were lifted by tectonic activity, releasing large volumes of carbon dioxide when the volcanoes erupted. The level of greenhouse gases in the atmosphere thus increased because of the most concentrated bout of volcanic activity on Earth during the last 600 million years. "This injection of volcanic CO_2," wrote Lavers, "was probably the decisive event that ultimately tipped the biosphere into the new era of the Mesozoic" (Lavers 2000, 235).

A Methane "Burp" and Noah's Ark

Gregory Ryskin, a geologist at Northwestern University, has asserted that a smaller-scale methane "burp" may help to explain the Biblical flood navigated, in *The Bible*, by Noah's Ark. The Biblical flood, according to Ryskin, may be attributable to a methane "burp" from Europe's stagnant Black Sea. Some geological evidence suggests such an event of this type 7,000 to 8,000 years ago. According to an account by Tom Clarke published in *Nature* Online (Clarke 2003; Ryskin 2003, 737), "Ryskin contends that methane from bacterial decay or from frozen methane hydrates in deep oceans began to be released. Under the enormous pressure from water above, the gas dissolved in the water at the bottom of the ocean and was trapped there as its concentration grew." A single disturbance, according to this account,

> [A] small meteorite impact or even a fast-moving mammal, could then have brought the gas-saturated water closer to the surface. Here it would have bubbled out of solution under the reduced pressure. Thereafter the process would have been unstoppable: a huge overturning of the water layers would have released a vast belch of methane. (Clarke 2003)

FURTHER READING

Bates, Albert K. and Project Plenty. *Climate in Crisis: the Greenhouse Effect and What We Can Do.* Summertown, TN: The Book Publishing Co., 1990.

Benton, Michael J. *When Life Nearly Died: The Greatest Mass Extinction of All Time.* London: Thames and Hudson, 2003.

Clarke, Tom. "Boiling Seas Linked to Mass Extinction: Methane Belches May Have Catastrophic Consequences." *Nature* Online, August 22, 2003. http://info.nature.com/cgi-bin24/DM/y/eLod0BfHSK0Ch0DYy0AL.

Clover, Charles. "Thousands of Species 'Threatened by Warming.'" *London Daily Telegraph*, August 31, 2000, 9.

Grabherr, G., M. Gottfried, and H. Pauli. "Climate Effects on Mountain Plants." *Nature* 339 (1994): 448–451.

Gugliotta, Guy. "Warming May Threaten 37 Percent of Species by 2050." *Washington Post*, January 8, 2004, A-1. http://www.washingtonpost.com/wp-dyn/articles/A63153-2004Jan7.html.

Hallam, Anthony, and Paul Wignall. *Mass Extinctions and Their Aftermath.* Oxford: Oxford University Press, 1997.

Lavers, Chris. *Why Elephants Have Big Ears.* New York: St. Martin's Press, 2000.

Menzel, A., and P. Fabian. "Growing Season Extended in Europe." *Nature* 397 (1999):659-662.

Peters, Robert L. "Effects of Global Warming on Biological Diversity." In *The Challenge of Global Warming*, ed. Edwin Abrahamson, 82–95. Washington, D C: Island Press, 1989.

Quammen, David. "Planet of Weeds: Tallying the Losses of Earth's Animals and Plants." *Harpers*, October 1998, 57–69.

Pounds, J. Alan, and Robert Puschendorf. "Clouded Futures." *Nature* 427 (January 8, 2004):107–108.

Reynolds, James. "Earth Is Heading for Mass Extinction in Just a Century." *The Scotsman*, June 18, 2003, 6.

Ryskin, G. "Methane-driven Oceanic Eruptions and Mass Extinctions." *Geology* 31 (2003):737–740.

Thomas, Chris D., Alison Cameron, Rhys E. Green, Michael Bakkenes, Linda J. Beaumont, Yvonne C. Collingham, Barend F. N. Erasmus, Marinez Ferreira de Siqueira, Alan Grainger, Lee Hannah, Lesley Hughes, Brian Huntley, Albert S. van Jaarsveld, Guy F. Midgley, Lera Miles, Miguel A. Ortega-Huerta, A. Townsend Peterson, Oliver L. Phillips, Stephen A. Williams. "Extinction Risk from Climate Change." *Nature* 427 (January 8, 2004):145–148.

"Suffocation Suspected for Greatest Mass Extinction." NewScientist.com, September 9, 2003. http://www.newscientist.com/news/news.jsp?id=ns99994138.

Extremes of Heat and Cold around the World

Weather is always variable to some degree, but extremes have been taking place with a frequency that has often been unknown to human memory or records. While weather has been generally warmer worldwide, reports of heat waves have been punctuated by occasional record cold and deep snowfall during unusual times of the year. Extremes of precipitation have become more numerous as well, as temperatures rise.

Extremes in Europe and North America

Europe, heretofore usually a temperate place with climate shaped by the ocean, has undergone notable extremes in recent years, including heat waves, droughts, deluges, and melting glaciers in the Alps. Such extremes in the Alps could become routine fare there, according to a 2004 report by the European Environment Agency (http://reports.eea.eu.int/climate-report-2-2004/en). The report said that rising temperatures could eliminate three-quarters of Alpine glaciers by 2050 and bring repeats of Europe's mammoth floods of 2002 and the heat wave of 2003. Global warming has been evident for years, but the problem is becoming acute, Jacqueline McGlade, executive director of the Copenhagen-based agency, told the Associated Press. "What is new is the speed of change," she said (Olsen 2004). Ice melt, for example, reduced the mass of Alpine glaciers by one-tenth in 2003 alone, the report said.

Mikhail Koslov and Natalia G. Berlina analyzed records from a Lapland reserve on the Kola Peninsula of Russia during 1930 and 1998. The researchers found "a decline in the length of the snow-free and ice-free periods by 15 to 20 days due to both delayed spring and advanced autumn/winter" (Koslov and Berlina 2002, 387). Emissions of sulfur dioxide from a nearby industrial plant may have contributed to the cooling, which was associated with a snowfall increase of more than 40 percent during the same period.

On May 23, 2002, Cut Bank, Montana, experienced enough snow and wind to pile drifts four feet high, and, a day later, wet snow fell in Denver, Colorado. In Canada, the same spring was the fifth coldest in half a century, breaking a string of five years during which every season had been warmer than average (Spears 2002, A-8). At the same time, the growing season in Albuquerque, New Mexico, was reported to have lengthened by two weeks between 1931 and 2002, to an average of 204 days, according to the Albuquerque Office of the National Weather Service, which attributed the change to the urban heat-island effect and global warming (Fleck 2002, A-1).

Berlin, Germany, experienced its coldest October on record in 2002, with earlier-than-usual snows. Omaha, Nebraska, also experienced a notably cool October 2002, with early snow and a monthly average temperature 6.6°F below average. On May 6, 2003, residents of St. Albert, Alberta (near Edmonton), looked out their windows at a foot of snow. "How long is this global warming going to take," complained St. Albert resident Tim Enger in the *Edmonton Journal*. "I'm waiting" (Enger 2003, A-19).

May through August 2004, was the coldest on record in Winnipeg, Manitoba, since records began in 1872. On the roster of cold surprises in a generally warming world, residents of Galveston and Corpus Christi, Texas, will long remember their white Christmas of 2004. Between those two cities (and a few miles inland), Victoria, Texas, was buried under 10 inches of wet snow, its first white Christmas in 86 years. Brownsville, Texas, received 1.5 inches of snow the same day, the first measurable snow there since 1895. A Brownsville snowball sold for $95 on e-Bay ("How About" 2005, C-16).

Extremes in South Asia

A weeks-long heat wave in southern India killed at least 1,000 people during mid-May 2002 in Andhra Pradesh State. Temperatures in the shade rose as high as 124° degrees F, killing birds in trees and making many homes uninhabitable. By May 22, the death toll in Andhra Pradesh was the highest in any Indian heat wave—until the next year. Most were elderly and poor, but people who worked outside, such as farm laborers and rickshaw drivers, also died in large numbers.

During January 2003, extreme cold moved into some of the same areas of India that had experienced searing heat the previous May. The cold wave killed at least 1,800 people (mainly from exposure) in Bangladesh and India, where many people sleep outside on the ground with no protective clothing, as well as in Nepal. At about the same time, Vietnam lost a third of its rice crop to unusual cold. During the summer of 2003 more than 1,400 people died of heat stress as temperatures rose as high as 122°F in India during three weeks in late May and early June, before the annual monsoon began. The same area also suffered its worst drought in at least 20 years. An analysis by Uday Shankar De, a geophysicist and former research chief at the Indian meteorological Department in Pune, suggested that the number of hot, dry days in India during May and June increased steadily over the previous two decades.

De and colleagues attributed the trend to global climate change. "The increase in extreme events, such as the heat wave, is linked to global warming," asserted J. Srinivsasan, a professor of atmospheric sciences at the Indian Institute of Science in Bangalore (Jayaraman 2003, 673). India's heat wave of 2003 intensified as hot, dry air flowed eastward from deserts in Iran; by May 2003, according to satellite data supplied by the U.S. Oceanic and Atmospheric Administration, two-thirds of vegetation in India was under "severe" or "extreme" drought conditions (Jayaraman 2003, 673).

FURTHER READING

Enger, Tim. 'If This Is Spring, How Come I'm Still Shoveling Snow?" Letter to the Editor, *Edmonton Journal*, May 8, 2003, A-19.

Fleck, John. "Jack Frost's Nip Arrives a Bit Later." *Albuquerque Journal*, October 26, 2002, A-1.

"How about That Weather? The Answer Is Blowing in the Wind, Rain, Snow ..." *Washington Post*, January 26, 2005, C-16.

Jayaraman, K. S. "Monsoon Rains Start to Ease India's Drought." *Nature* 423 (June 12, 2003):673.

Koslov, Mikhail, and Natalia G. Berlina. "Decline in Length of the Summer Season on the Kola Peninsula, Russia." *Climatic Change* 54 (September, 2002):387–398.

Olsen, Jan M. "Europe Is Warned of Changing Climate." Associated Press, August 19, 2004. (LEXIS)

Spears, Tom. "Cold Spring Bucks the Trend." *Ottawa Citizen*, June 7, 2002, A-8.

F

Fall Colors and Warming

New England, long noted for its sharp fall colors, has found them muted in recent years because of rising temperatures. Falling temperatures in autumn are part of the natural cycle that halts production of chlorophyll in plants and causes leaves to change color (sustained freezing early in the season kills leaves before they can change color, however). Shorter days cause plants to form a layer at the base of their stems that cuts off water and nutrients; cold nights (close to or just above freezing) accelerates this process.

Warmer temperatures also encourage fungi that kill the leaves before they change color, especially among the red and sugar maples that are the most colorful fall trees in New England. The turning of the leaves is such a tourist draw in Vermont ($3.4 million spent during 2005) that state tourism officials have been denying that any problem exists. In the meantime, the Associated Press quoted other Vermonters as saying that the leaves not only are blander most years, but that the peak season, such as it now, occurs about 10 days to two weeks later than prior years ("Climate Blamed" 2007, 3-A).

Many trees go from the deep green of late summer to dull brown of late fall without the usual intervening blight yellows and oranges that used to occur in October. "It's nothing like it used to be," said University of Vermont plant biologist Tom Vogelmann ("Climate Blamed" 2007, 3-A).

FURTHER READING

"Climate Blamed for Blah Foliage." Associated Press in *Omaha World-Herald,* October 29, 2007, 3-A.

Farming Technology Improvements

Rattan Lal, a professor of soil science at Ohio State University, has asserted that the atmosphere's load of carbon dioxide could be reduced markedly by several relatively simple changes in farming technology. Contributions of farming to carbon dioxide in the atmosphere have been increasing with rising populations. Carbon dioxide is added to the atmosphere via plowing, so Lal believes that reducing the depth of furrows would significantly reduce the amount of carbon dioxide introduced into the atmosphere by agriculture.

During mid-2003, the U.S. Department of Agriculture (USDA) announced plans to give incentives to farmers for management practices that keep carbon in the soil. For the first time, the USDA began to factor reduction of greenhouse-gas emissions into soil-conservation programs by giving priority to farmers who reduce emissions of carbon dioxide, methane, and nitrous oxides. Such programs represented $3.9 billion in federal spending during the 2003–2004 fiscal year (Clayton 2003, D-1). Farmers were encouraged to use no or low-tillage methods, as well as crop rotation, buffer strips, and other practices that reduce greenhouse-gas emissions as well as soil erosion. Such practices were expected to retain 12 million tons of greenhouse gases by 2012 (Clayton 2003, D-1).

Farming with an eye toward carbon sequestration utilizes soil restoration and woodland regeneration, no-till farming, cover crops, nutrient management, manuring and sludge application, improved grazing, water conservation, efficient irrigation, agroforestry practices, and growth of energy crops on spare lands. Intensive use of such practices, according to one estimate, could "offset fossil-fuel emissions by 0.4 to 1.2 gigatons of carbon per year, or 5 to 15 percent of the global fossil-fuel emissions" (Lai 2004, 1623).

Tim O'Riordan, of the Zuckerman Institute for Connective Environmental Research said:

> We have to put sustainable development at the heart of businesses such as fish farming and

agriculture. We need agricultural stewardship schemes that have incentives for farmers to produce according to sustainable principles, which in turn will deliver healthy soil, water and wild life. This, in turn, should offer jobs in recreation and education for eco-care. We also need the involvement of the local community to ensure that all acts of stewardship have neighborhood understanding and support. (Urquhart and Gilchrist 2002, 9)

FURTHER READING

Clayton, Chris. "U.S.D.A. Will Offer Incentives for Conserving Carbon in Soil." *Omaha World-Herald*, June 7, 2003, D-1, D-2.

Lai, R. "Soil Carbon Sequestration Impacts on Global Climate Change and Food Security." *Science* 304 (June 11, 2004):1623–1627.

Urquhart, Frank, and Jim Gilchrist. "Air Travel to Blame as Well." *The Scotsman*, October 8, 2002. (LEXIS)

Feedback Loops

Many climate scientists believe that the middle of the twenty-first century will witness dramatic acceleration in global warming. In part, this acceleration will result from exhaustion of various natural "sinks" that have been absorbing greenhouse gases. At about the same time, various feedback loops are expected to accelerate natural increases in atmospheric greenhouse-gas levels and, consequently, worldwide temperatures. These include several natural processes that add greenhouse gases to the atmosphere, such as melting permafrost in the Arctic and, in the far future, possible gasification of solid methane deposits (clathrates) in the oceans, as well as an increasingly dark Arctic Ocean that will absorb more heat as the ice cap melts.

In each of these cases, human-provoked warming caused by an overload of greenhouse gases in the atmosphere is expected to aggravate the natural feedback loops like a bank account drawing an environmentally dangerous form of compound interest. Evidence is accumulating that these processes already have begun. The danger, according to many people who are familiar with the paleoclimatic record, is that once this journey has begun in earnest, any return trip may become a matter of many centuries as well as copious human and natural suffering.

In addition to possible increases in greenhouse gases from gasifying permafrost, other feedback mechanisms are expected to enhance the effects of greenhouse warming during the twenty-first century. Two of the most important are increasing amounts of water vapor (itself a greenhouse gas), as well as changes in the planet's albedo, or reflectivity, as formerly ice-covered land and sea surfaces are replaced by darker-colored bare land and open water.

Already, during 2007, methane was bubbling up through melting permafrost—Inuit hunters now sometimes light impromptu fires to warm themselves with it (Funk 2007, 54). "We are taking risks with a system we don't understand that is absolutely loaded with carbon," said Steven Kallick, a Seattle-based expert on the boreal forests for the Pew Charitable Trusts. "The impact could be enormous" (Struck 2007a, A-1). "With permafrost, it may take longer for change to get moving. But it may keep moving, even if we get our emissions under control," said Antoni Lewkowicz, a professor of geography at the University of Ottawa. "It's like a big boulder. Once you get it moving, it won't stop" (Struck 2007a, A-1).

The Earth's Changing Radiative Balance

During the last 400,000 years, the Earth has usually been nearly in energy balance with space. Thus, calculated temperature change was simply a product of the forcing and equilibrium (i.e., long-term) climate sensitivity. This assumption is no longer valid in modern times, as climate forcings are changing so quickly, especially in the past few decades, that the climate system has not come to equilibrium with today's climate forcing. Thus, the observed temperature rise in the past century is only part of that expected for the current level of atmospheric composition. The additional amount of global warming "in the pipeline" is probably ~0.5°C or ~1°F (Hansen 2006b, 10).

"If warming approaches the range of 2° to 3°C, large-scale disastrous climate impacts for humans as well as for other inhabitants of the planet will be virtually certain, according to James E. Hansen, director of the Goddard Institute for Space Studies. "An important point to note is that the tripwire between keeping global warming less than 1°C, as opposed to having a warming that approaches the range 2–3°C, may depend upon a relatively small difference in human-made direct forcings," Hansen has written.

> The reason for that conclusion is that keeping global warming less than 1°C requires having both a moderate limit on CO_2 and a reduction of non-

Widespread fires fill the skies of eastern Siberia with smoke on June 30, 2008. (NASA)

> CO_2 forcings. However, if the warming becomes greater than 1°C, it may be come impractical to achieve and maintain the presumed reductions of the non-CO_2 forcings, in part because positive GHG [greenhouse-gas] feedbacks from tundra or other terrestrial sources may become significant in the warmer climate. (Hansen 2006b, 15–16)

Sir John Houghton, one of the world's leading experts on global warming, told the *London Independent*: "We are getting almost to the point of irreversible meltdown, and will pass it soon if we are not careful" (Lean 2004b, 8). A report by the National Academy of Sciences prepared during 2001 at the request of the George W. Bush administration estimated that in coming years only 40 percent of potential warming will be caused directly by increasing levels of greenhouse gases. The other 60 percent may be caused by feedback mechanisms. "Together," according to this report, "these two feedbacks amplify the simulated climate response to the greenhouse gas forcing by a factor of 2.5" (Victor 2004, 146). The same summary pointed out that this factor is an estimate and that different climate models forecast varying ranges of feedback responses, producing differing estimates of future warming.

The Endurance of Feedbacks

Feedbacks will produce adverse conditions for decades to centuries after carbon dioxide and methane levels in the atmosphere stop rising. The amount of time that temperatures and sea levels may continue to rise after carbon-dioxide levels stabilize is not known with any degree of certainty, however. One scientific observer believes that "[t]he rise in mean global SAT [surface average temperature] and the world ocean level (due to thermal expansion) may ... continue for several centuries after the stabilization of the carbon-dioxide concentration [in the atmosphere], due to the gigantic thermal inertia of the oceans." The same observer also noted that "[t]he response of ice sheets to earlier climate changes may continue for several centuries after the climate stabilizes" (Kondratyev, Krapivin, and Varotsos 2004, 45).

Geophysical evidence suggests that the Earth has suffered bouts of severe warming in the distant past from natural causes that were intensified by release of the planet's stores of greenhouse gases. Considerable scientific inquiry is now aimed at estimating just how much human-provoked warming might cause the process to reach a "runaway" state in which the feedbacks take control and force warming out of control.

Frozen bogs in Siberia contain an estimated 70 billion tons of methane. If the bogs become driyer as they warm, according to one observer, "the methane will oxidize and the emissions will be primarily CO_2. But if the bogs stay wet, as they have been recently, the methane will escape directly into the atmosphere.... with twenty times the heat-trapping power of carbon dioxide" (Romm 2007, 69). About 600 million tons of methane are emitted each year from human and natural sources, so if even a small fraction of the 70 billion tons of methane in the Siberian bogs is emitted, warming will accelerate dramatically.

Another example of feedback loops that will reinforce warming temperatures (and probably drought) occurs in the tropics. A rainforest or jungle creates its own weather. In our time, however, human beings seeking food and shelter compete with nature. Roughly 20 percent of the Amazon rainforest has been destroyed, and approximately another 20 percent has been damaged by logging to an extent that sufficient sunlight can reach the forest floor to cause significant drying. According to some studies,

> Models suggest that when 50 percent of the forest is destroyed (the year 2050 would be a reasonable estimate for that, at present rates) we will reach a tipping point at which drought and heat will combine to at the rain forest and its canopy, reducing local rainfall and further accelerating the drought and local temperature rise, ultimately causing the release into the atmosphere of huge amounts of carbon currently locked in Amazon soils and vegetation, another fearsome feedback loop at work. (Romm 2007, 72)

How Much Warming Takes Us to the "Tipping Point"?

In the Arctic, September is the peak ice-melt season. During 2007, polar ice retreated to record-low levels, exposing dark ocean water to soak up more of the sun's heat, in an albedo (reflectivity) cycle that reinforces itself. As is often the case, the provocations are both natural and human. Greenhouse gases emitted by human transport and industry are part of the story. The rest involves persistent high pressure over the Arctic that allowed more sunshine than usual, along with winds from the south, drawing up warm, ice-eroding air.

The real news of global warming, however, is not how warm it is today, because today's carbon emissions do not give us tomorrow's temperature. Through a complex set of feedbacks ("thermal inertia" to scientists), we will feel today's emissions in our faces roughly half a century from now. In the oceans, the feedback loop is longer, probably a century and a half, maybe two. In 1960, human activity pumped about two billion metric tons of carbon into the world's atmosphere. In 2006, that total was roughly eight billion, and rising, a 400 percent increase in 46 years, food for thought for anyone considering that temperatures in 2006, because of thermal inertia, were reflecting the greenhouse-gas load of roughly 1960 (Volk 2008, 62).

The real debate is not over how much the oceans may rise from melting ice by the end of this century (one to three feet, perhaps), but how much melting will be "in the pipeline" by that time. Hansen estimates that thermal inertia by the year 2100 may guarantee a 25-meter sea-level rise within about two centuries after that. Melting most of the world's mountain glaciers, as well as a large part of the Greenland and West Antarctic, ice sheets would add up to 25 meters.

Researchers have reported that Earth's ancient stores of peat are gasifying into the atmosphere at an accelerating rate that is adding significantly to the atmosphere's overload of greenhouse gases. Given the fact that one-third of the Earth's carbon is stored in far northern latitudes (mainly in tundra and boreal forests), the speed with which warming of the ecosystem releases this carbon dioxide to the atmosphere is vitally important to forecasts of global warming's speed and effects. The amount of carbon stored in arctic ecosystems also includes two-thirds of the amount presently found in the atmosphere. Its release into the atmosphere will depend on the pace of temperature rise—and the Arctic, according to several sources, has been the most rapidly warming region of the Earth.

Surprises in the Carbon Cycle

In cold water, methane clathrates form crystal structures that are somewhat similar to water ice. Warming temperatures could destabilize the clathrates, and release some of their stored methane. Roughly 10 trillion tons of methane is trapped under pressure in crystal structures in permafrost or on the edges of the oceans' continental shelves, "the Earth's largest fossil-fuel reservoir," according to Gerald Dickens, a geologist at James Cook University in Townsville, Australia (Pearce 1998). The greenhouse potential of all the methane stored in clathrates on the continental shelves and in permafrost worldwide is roughly equal to that of all the world's coal reserves (Cline 1992, 34).

Atmospheric scientist Roger Revelle has estimated that, with a 3°C rise in global average temperature, methane emissions from clathrates would increase half a gigaton per year worldwide. Over a century, this rate could be enough to double the amount of methane in the atmosphere. Add to this another 12 gigatons of methane that could be released by clathrates liberated from ocean bottoms under the Arctic Ocean once the ice cap now covering them melts. "It is possible," wrote Jonathan Weiner, "that [this] … feedback effect is already underway and the rise in Earth temperatures in the last hundred years has already sprung many gigatons of methane from their molecular prisons at the bottom of the sea" (Weiner 1990, 118).

Warming temperatures change the behavior of the Earth's hydrological cycle. Warmer ocean water removes less carbon dioxide from the atmosphere than cooler water, so the warming of oceans may feed upon itself in coming years. Water vapor is also a potent absorber of heat in the atmosphere. It has been estimated that a doubling of carbon dioxide in the atmosphere would increase its water content by about 30 percent, raising temperatures an additional 1.4°C (Hansen 1981, 957). Many models project a rise in cloudiness, and attendant atmospheric moisture, in a warmer, more humid world. George M. Woodwell raises the possibility of a rapid surge in global warming beyond any possibility of human control:

> The possibility exists that the warming will proceed to the point where biotic releases from the warming will exceed in magnitude those controlled directly by human activity. If so, the warming will be beyond control by any steps now considered reasonable. We don't know how far we are from that point because we do not know sufficient detail about the circulation of carbon dioxide among the pools of the carbon cycle. We are not going to be able to resolve those questions definitely soon. Meanwhile, the concentration of heat-trapping gases in the atmosphere rises. (Woodwell 1990, 130)

Given Woodwell's expectations, the peoples of the Earth in the year 2000 are approaching a point of no return with regard to feedback loops. Deforestation is accelerating around the world because of growing populations and levels of material affluence. Use of fossil fuels, which has increased at an annual rate of roughly 5 percent during most of this century, shows no signs of stabilizing, much less falling by half in the next 30 years. China, alone, projects burning enough fossil fuel (mainly coal) by 2025 to account for about half the present consumption of fossil fuels by everyone on Earth (Leggett 1990, 27).

When Will Feedbacks Take Control?

How much time remains before critical feedbacks lock into place? In January 2005, a world task force of senior politicians, business leaders, and academics warned that the point of no return will be reached within a decade. The report, *Meeting the Climate Challenge*, was the first to place a figure on the length of time remaining before cascading feedbacks caused by human-provoked climate change irretrievably commit the Earth to disastrous changes, including widespread agricultural failure, water shortages and major droughts, increased disease, sea-level rise, and the death of forests (Byers and Snowe 2005, 1).

The report asserted that the tipping point will occur as the average world temperature increases 2°C above the average prevailing in 1750, before the Industrial Revolution began. By 2005, temperatures already had risen an average of 0.8°C.

The report also asserted that the tipping point will occur as atmospheric concentration of carbon dioxide passes 400 parts per million (ppm). With the level at 379 ppm in 2004, rising at 2 ppm a year, that threshold was about a decade away in 2005. The report was assembled by the Institute for Public Policy Research in the United Kingdom, the Center for American Progress in the United States, and the Australia Institute. The group's chief scientific adviser was Rakendra Pachauri, chairman of the United Nations Intergovernmental Panel on Climate Change (McCarthy 2005, 1).

The report concluded:

> Above the 2-degree level, the risks of abrupt, accelerated, or runaway climate change also increase. The possibilities include reaching climatic tipping points leading, for example, to the loss of the West Antarctic and Greenland ice sheets (which, between them, could raise sea level more than 10 meters over the space of a few centuries), the shutdown of the thermohaline ocean circulation (and, with it, the Gulf Stream), and the transformation of the planet's forests and soils from a net sink of carbon to a net source of carbon. (McCarthy 2005, 1)

The science of warming feedbacks is not settled knowledge. In some areas, such as the role of organic decomposition in soils under warmer conditions, a robust debate continues. Knorr and colleagues have written in *Nature* that "the sensitivity of soil carbon to warming is a major uncertainty in projections of carbon-dioxide concentration and climate" (Knorr et al. 2005, 298). Their findings indicate that "the long-term positive feedback of soil decomposition in a warming world may be even stronger than predicted by global models" (Knorr et al. 2005, 298). In the meantime, the very idea that organic soil decomposition is sensitive to temperature at all has been challenged by other scientists (Giardina and Ryan 2000, 858–861). For climate models, the solution of this debate is no small matter, because soils contain twice as much carbon as the atmosphere (Powlson 2005, 204), so the rate at which warming may accelerate exchange from one to the other is, and will continue to be, an important factor as other feedbacks add more greenhouse gases to the air that sustains us.

Human Political Inertia

"The biggest lag is in the political system," said geoscientist Michael Oppenheimer of Princeton University. At least two decades have passed discussing the seriousness of the threat, he said, and, as he sees it, another 20 years may pass before a worldwide program that is up to the task is in place. In the meantime, the window of time before feedbacks take control narrows. "We can't really afford to do a 'wait and learn' policy," Oppenheimer said. "The most important question is: when do we commit to [contain global warming to] 2°[C]. Really, there isn't a lot of headroom left. We better get cracking." The current pace, said Roger Pielke, Jr. of the University of Colorado at Boulder, "Isn't going to do it" (Kerr 2007e, 1231). *See also:* Methane Burp

FURTHER READING

Byers, Stephen, and Olympia Snowe, co-chairs. International Climate Change Task Force. *Meeting the Climate Challenge: Recommendations of the International Climate Change Task Force.* London: Institute for Public Policy Research, January 2005.

Cline, William R. *The Economics of Global Warming.* Washington, DC: Institute for International Economics, 1992.

Funk, McKenzie. "Cold Rush: The Coming Fight for the Melting North." *Harper's Magazine*, September 2007, 56–55.

Giardina, C., and M. Ryan. "Evidence That Decomposition Rates of Organic Carbon in Mineral Soil Do Not Vary with Temperature." *Nature* 404 (2000): 858–861.

Hansen, James E. "Declaration of James E. Hansen." *Green Mountain Chrysler-Plymouth-Dodge-Jeep, et al., Plaintiffs v. Thomas W. Torti, Secretary of the Vermont Agency of Natural Resources, et al., Defendants.* Case Nos. 2:05-CV-302 and 2:05-CV-304, Consolidated. United States District Court for the District of Vermont. August 14, 2006b. http://www.giss.nasa.gov/~dcain/recent_papers_proofs/vermont_14aug20061_textwfigs.pdf.

Hansen, J., D. Johnson, A. Lacis, S, Lebendeff, D. Rind, and G. Russell. "Climate Impact of Increasing Atmospheric Carbon Dioxide." *Science* 213 (1981):957–956.

Hansen, James E. A. Lacis, D. Rind, G. Russell, P. Stone, I Fung, R. Ruedy, and J. Lerner. "Climate Sensitivity: Analysis of Feedback Mechanisms." *Geophysical Monograph* 29, no. 5, edited by Maurice Ewing (1984):130–163.

Kerr, Richard. "How Urgent Is Climate Change?" *Science* 318 (November 23, 2007e):1230–1231.

Knorr, W., I. C. Prentice, J. I. House, and E. A. Holland. "Long-term Sensitivity of Soil Carbon Turnover to Warming." *Nature* 433 (January 20, 2005): 298–301.

Kondratyev, Kirill, Vladimir F. Krapivin, and Costas A. Varotsos. *Global Carbon Cycle and Climate Change.* Berlin, Germany: Springer/Praxis, 2004.

Lean, Geoffrey. "Global Warming Will Redraw Map of the World." *London Independent,* November 7, 2004b, 8.

Leggett, Jeremy, ed. *Global Warming: The Greenpeace Report.* New York: Oxford University Press, 1990.

McCarthy, Michael. "Countdown to Global Catastrophe." *London Independent,* January 24, 2005, 1.

National Academy of Sciences. *Policy Implications of Greenhouse Warming.* Washington, DC: National Academy Press, 1991.

Pearce, Fred. "Nature Plants Doomsday Devices." *The Guardian* (England), November 25, 1998. http://go2.guardian.co.uk/science/912000568-disast.html.

Powlson, David. "Will Soil Amplify Climate Change?" *Nature* 433 (January 20, 2005):204–205.

Romm, Joseph. *Hell and High Water: Global Warming—the Solution and the Politics—and What We Should Do.* New York: William Morrow, 2007.

Struck, Doug. "In Far North, Peril and Promise: Great Forests Hold Fateful Role in Climate Change." *Washington Post,* February 22, 2007a, A-1. http://www.washingtonpost.com/wp-dyn/content/article/2007/02/21/AR2007022102095_pf.html.

Victor, David G. *Climate Change: Debating America's Policy Options.* New York: Council on Foreign Relations, 2004.

Volk, Tyler, *CO2 Rising: The World's Greatest Environmental Challenge.* Cambridge, MA: MIT Press, 2008.

Weiner, Jonathan. *The Next One Hundred Years: Shaping the Fate of Our Living Earth.* New York: Bantam Books, 1990.

Woodwell, George M. "The Effects of Global Warming." In *Global Warming: The Greenpeace Report,* ed. Jeremy Leggett, 116–132. New York: Oxford University Press, 1990.

Woodwell, G. M., F. T. MacKenzie, R. A. Houghton, M. Apps, E. Gorham, and E. Davidson. "Biotic Feedbacks in the Warming of the Earth." *Climatic Change* 40 (1998):495–518.

First-Flowering Dates, England

A study of the first-flowering date of 385 British plants found that about 200 species flowered an average of 15 days earlier in the 1990s than in previous decades. The study, by Alastair Fitter of York University and his father, Richard, a distinguished naturalist, claimed to have found "the strongest biological signal yet of climate change. The Fitters have been keeping track of blooming dates for half a century, long before global warming was a widespread concern. The senior Fitter was 89 years old in 2002. Fitter senior, who lives in Cambridgeshire, is one of Britain's best-known naturalists, the author of dozens of books on flowers and birds. He began his record after moving to Chinnor in Oxfordshire in 1953, when his son was five years old.

A. H. Fitter and R. S. R. Fitter reported in *Science*:

> The average first-flowering date of 385 British plant species has advanced by 4.5 days during the past decade compared with the previous four decades; 16 percent of species flowered significantly earlier in the 1990s than previously, with an average advancement of 15 days in a decade. Ten species flowered significantly later in the 1990s than previously. These data reveal the strongest biological signal yet of climatic change. (Fitter and Fitter 2002, 1689)

The Fitters also raised the possibility that climate change could alter evolution, as different plants move into synchronous flowering patterns, raising the chances that they will breed natural hybrids (Fitter and Fitter 2002, 1691).

Spring-flowering species were the most responsive to the change, with white dead nettle flowering 55 days earlier during the 1990s than during four previous decades. From 1954 to 1990, its average first flowering date was March 18, but from 1991 to 2000, the plant bloomed on or about January 23. Ivy-leafed toadflax flowered five weeks earlier, hornbeam four weeks earlier as did hairy bittercress. Lesser periwinkle flowered 25 days earlier and lesser celandine three weeks earlier. The opium poppy flowered 20 days earlier during the 1990s than its 1954 to 1990 average, according to the Fitters (Meek 2002a, 6). "Not everything is flowering earlier—otherwise you would just say that the beginning of spring has moved forward eight weeks," Fitter said (Clover 2002b, 14).

"I did an analysis of his data eight or nine years ago, up to 1990, thinking there might be some sign of climate warming, but there wasn't," he said. "When we came to look at the last 10 years, out of the data jumped this incredible signal that it's really happening fast, and it's no coincidence that the 1990s were the warmest decade on record" (Meek 2002a, 6).

FURTHER READING

Clover, Charles. "Global Warming Is Driving Fish North.'" *London Daily Telegraph,* May 31, 2002b, 14.

Fitter, A. H., and R. S. R. Fitter. "Rapid Changes in Flowering Time in British Plants." *Science* 296 (May 31, 2002):1689–1691.

Meek, James. "Wildflowers Study Gives Clear Evidence of Global Warming." *London Guardian*, May 31, 2002a, 6.

Fisheries and Warming

Declines of Cod, Salmon, and Trout

Oceans waters generally warmed by an average of 0.06°C between 1960 and 2000. Surface water, where many fish consumed by humans live, has warmed more quickly, by as much as 0.31°C. The North Atlantic Ocean in particular has been warming more quickly than any other oceanic region, with a temperature increase of 0.5°C during the past 20 years, as warming accelerates (Connor 2002c, 12). Some deeper waters have not escaped temperature rises. Stebbing, the lead author of a study published in the *Journal of the Marine Biological Association,* pointed out that the shelf-edge current, which flows from Spain to the British Shetland Islands at a depth of between 200 to 600 meters (656 to 2,000 feet), warmed by 2°C between 1972 and 1992. He pointed out that because this current runs north at a speed of about 35 miles a day, it is the most likely route for some tropical and semitropical species that have been observed in British waters (Connor 2002c, 12).

According to the Intergovernmental Panel on Climate Change, global warming may cause some fisheries to collapse, while others expand. As Robert Gough has commented:

> Climate changes can exacerbate the effects of over-fishing at a time of inherent instability in world fisheries. In addition, over-fishing creates an increased imbalance in the age composition of a stock, and may reduce the resiliency of the population. Further, changes in ocean currents may result in changes in fish population location and abundance and the loss of certain fish populations. (Gough 1999, 47)

"As the world warms, the only way for wildlife species to live in the temperature they prefer is to move their ranges slowly poleward," said Stebbing (Connor 2002c, 12). He explained:

> Fish are good indicators of temperature change because they are unable to regulate their temperature independently of the surrounding water. They therefore swim to keep themselves in waters of their preferred temperature range. Not only are changes in fish distribution likely to reflect temperature increases, but the arrival of new fish species are well monitored by fishermen, as well as scientists. (Connor 2002c, 12)

A number of U.S. Environmental Protection Agency reports issued as early as 1995 anticipated that several cool-water fish species, especially trout, will diminish as waters warm. This study projects that between 8 and 10 U.S. states could lose all of their cool-water sports fisheries within 50 to 60 years. Between 11 and 16 other states could lose half of their cool-water sports fisheries, according to the study (U.S. Environmental Protection Agency 1995, ix).

Cod Declines 90 Percent in 30 Years

Cod, a cold-water fish similar in some ways to salmon, has been declining as habitats warm. In mid-July 2000, the World Wildlife Fund (WWF) placed North Sea Cod, a staple of British fish and chips (as well as a $60 million annual fishery), on its endangered-species list. Cod stocks in the North Sea have been in nearly continuous decline since the early 1970s; they fell below safe biological limits after 1984 (Smith 2001, 7).

Populations of North Sea Cod have declined 90 percent in 30 years, according to the WWF (Brown 2000a, 3). The cod have been overfished, and their population is falling because they do not breed well in warmer water. The North Sea in the year 2000 was as much as 3°C warmer than it was in 1970 (Brown 2000a, 3). During the 1930s, British fishermen harvested roughly 300,000 tons of cod annually; by 1999, the catch was down to 80,000 tons ("Fished" 2000, 18). Spawning cod in the North Sea fell from about 277,000 English tons during the early 1970s to 54,700 English tons in 2001. Global warming and predation (overfishing) by humans and seals are major factors in the decline of the cod fishery. Another problem advanced by fishermen is dredging and laying of pipes (Smith 2001, 7).

Research by Gregory Beaugrand and colleagues supported the idea that cod are declining in the North Sea not only because of overfishing, but also because larval cod feed on plankton, which have declined due at least in part to rising temperatures. They wrote that "variability in temperature affects larval cod survival; [we conclude] that rising temperature since the mid-1980s has modified the plankton ecosystem in way that reduces the survival of young cod" (Beaugrand et al. 2003, 661). Such observations have been supported by North Sea fishermen, who have been dragging rolls of silk behind their boats for 70 years to monitor the density of plankton populations. Given the plankton's decline, even a complete ban on cod fishing is unlikely to restore the fishery. Cod have declined not only in population, but also in size. The peak of plankton abundance now occurs later in the year, after cod larvae experience their greatest need for them. This mismatch means that fewer larval cod develop into adults.

Dangers to Salmon and Trout

Rising water temperatures may drive trout and salmon from many U.S. waterways, according to a study compiled by Defenders of Wildlife and the Natural Resources Defense Council. The 2002 study, which included eight species of salmon and trout, suggested that the cold-water habitat required by these species could shrink by more than 40 percent during the twenty-first century, given a "business-as-usual" emissions of greenhouse gases. Salmon and trout, both cold-water species, are sensitive to the temperature of their aquatic habitats. In many areas, these fish are already living at the margin of their temperature tolerances, meaning that even modest warming could render a stream uninhabitable ("Warming Streams" 2002). Habitats for some of these species could shrink as much as 17 percent by 2030, 34 percent by 2060, and 42 percent by 2090 ("Warming Streams" 2002). This analysis includes four species of trout (brook, cutthroat, rainbow, and brown) and four species of salmon (pink, coho, chinook, and chum).

Researchers looked at air and water temperature data from more than 2,000 sites across the United States ("Warming Streams" 2002). Anticipated increases in water temperatures vary by location, averaging 0.7° to 1.4°F by 2030, 1.3° to 3.2°F by 2060, and 2.2° to 4.9°F by 2090, depending on future levels of greenhouse gases in the atmosphere ("Warming Streams" 2002). In addition to warming waters, wild trout and salmon populations also are under pressure from habitat loss due to human infrastructure development, competition with hatchery fish, invasive exotic species, and other reasons. "Now we must add climate change to the list of challenges they face," said Mark Shaffer, senior vice president for programs at Defenders of Wildlife. "If we don't address the cumulative impact of all these factors, we will see more of these populations switching from a recreational resource to being listed as threatened or endangered" ("Warming Streams" 2002).

A WWF report issued during 2001 asserted that stocks of wild Atlantic salmon have been cut in half in two decades because of global warming and infections spread by hatchery-bred fish. The WWF report said that wild salmon already had disappeared from 309 of their 2,000 usual breeding areas around the world. Elizabeth Leighton, senior policy officer for the WWF in Great Britain, said, "When a river loses its salmon stock, that population is gone forever. The miracle of [salmon] returning to its spawning ground cannot be repeated" (Milmo and Nash 2001, 11).

The WWF report noted that, as of 2001, 90 percent of healthy wild Atlantic salmon populations returning to Europe could be found in only three countries: Norway, Iceland, and the Irish Republic. Salmon have nearly disappeared from Germany, Switzerland, the Netherlands, and Belgium. Wild salmon are nearly extinct in Estonia, Portugal, Poland, the United States, and the warmer parts of Canada. Wild Atlantic salmon had nearly vanished by 2002 from many rivers in Scotland. Wild stocks were extinct in the rivers Shieldaig, Garvan, Attadale, and Sguord on the west coast of Scotland, with only a handful left in another 10 rivers. The genetic integrity of the few remaining wild fish in Scotland's rivers had been damaged by interbreeding between farmed and wild fish. A wide range of factors was thought to be affecting salmon stocks, including overfishing at sea and global warming. James Butler, director of the Spey Fishery Board, said, "In 14 rivers we found that stocks ranged from no fish to just a handful" (Cramb 2002, 7).

A marine parasite that began spreading among Yukon River Chinook salmon in 2002 may be traced in part to warmer stream waters that allow it to thrive. The parasite has ruined many of the fish, making them inedible, with a fruity odor and ruined meat. The illness, caused by a common microorganism that targets ocean fish, was detected in about 35 percent of king salmon sampled in 2002 and 2003, said Richard Kocan, a fish pathologist overseeing a study for the federal Office of Subsistence Management. That's a significant increase over the number of infected fish found in 1999, 2000, and 2001 (O'Harra 2004, A-1). Kocan said that some Yukon Chinook now spend June and July migrating upstream through water warmed to 59° or higher, temperatures that allow the parasite to spread faster and kill its hosts more quickly. "Some of these fish are swimming within a few degrees of lethal temperatures for healthy salmon, and we know many of them are infected," he said (O'Harra 2004, A-1).

The parasite was first noticed in a few salmon during the mid-1980s, but it grew much worse during the late 1990s. The culprit is a common protozoan called *ichthyophonus*, which probably enters the fish through its food supply, then spreads into organs and flesh. The pathogen has been detected in salmon in the Kuskokwim and Taku rivers, but according to Kocan's report, it is not clear whether it is precisely the same species found in the Yukon Chinook. The parasite also was found in a few Yukon Burbot, raising the possibility that it is now inside the freshwater system. The fish, however, are not toxic and can be eaten safely. "They taste bad, they smell bad and they look bad, but if you're starving, go ahead and eat them," Kocan said (O'Harra 2004, A-1). The parasite inhibits breeding, however, killing salmon before they spawn. An estimated 60 percent of the fish sick with the disease do not make it to spawning grounds.

Warming Limits Fishes' Oxygen Intake

The viviparous eelpout (*Zoarces viviparous*) needs more oxygen at higher temperatures. Because warming water in the Wadden Sea (near the Netherlands) carries less oxygen, the fish there are becoming smaller and scarcer (Portner and Knust 2007, 95–97).

> A mismatch between the demand for oxygen and the capacity of oxygen supply to tissues is the first mechanism to restrict whole-animal tolerance to thermal extremes. We show in the eelpout, *Zoarces viviparus*, a bioindicator fish species for environmental monitoring from North and Baltic Seas[,] that thermally limited oxygen delivery closely matches environmental temperatures beyond which growth performance and abundance decrease. Decrements in aerobic performance in warming seas will thus be the first process to cause extinction or relocation to cooler waters. (Portner and Knust 2007, 95)

Warm summers dramatically reduced populations of the fish the following year. Population declines long before temperature rises threaten the fish as individuals.

Warming Affects Sea Creatures

Roughly 300 surviving North Atlantic right whales may be threatened by a decline in their main food source, plankton, as a result of shifts in ocean circulation. At the same time, the common dolphin (*Delphinus delphis*), a warm-water species, has been increasing its range, while the cold-water range of the white-beaked dolphin (*Lagenorhynchus albirostris*) is shrinking. Predators are following their prey as species of fish change their latitude or depth in response to a warming climate. Exotic southern fish species such as red mullet, anchovy, sardine, and poor cod are now being found in the North Sea. Fish species are unable to regulate their body temperature, and their distribution and abundance are temperature dependent ("Climate Change Dislocates" 2006).

As the North Sea has grown warmer, the Pacific oyster (*Crassostrea gigas*), which was brought to Europe for commercial reasons, has been able to breed in the wild and is now displacing native oysters in the Wadden Sea. Previously, these oysters were unable to survive outside artificial pens ("Climate Change Dislocates" 2006).

In 1984, the jumbo squid (a smaller species that is related to the giant squid) migrated as far north in the eastern Pacific as Point Conception on the California coast. By 2005, some of these squid were caught by fishermen off the southern coast of Alaska. The range of marlins has expanded several miles northward, to the coast of Washington State. A 2003 metaanalysis of 1,700 bird species (mostly birds and some sea animals) found that over several decades their ranges were expanding an average of 600 meters northward per year (Kintisch 2006, 776).

Fish populations in many areas probably will be affected by interruptions in their breeding cycles caused by salt-water intrusion into estuaries as seas rise. Roughly 70 percent of the world's fish use shoreline waters in their breeding cycles. Gough wrote, "Fish production will thus suffer when such nursery habitats are lost" (Gough 1999, 47). Gough warned that many species in the sea and on the land that have adapted to specific environments may not be able to adapt to the speed of coming temperature changes: "Surviving species may succumb to predatory pressure and competition from more exotic species better adapted to the new conditions" (Gough 1999, 48).

FURTHER READING

Beaugrand, Gregory, Keith M. Brander, J. Alistair Lindley, Sami Souissi, and Philip C. Reid. "Plankton Effect on Cod Recruitment in the North Sea." *Nature* 426 (December 11, 2003):661–664.

Brown, Paul. "Overfishing and Global Warming Land Cod on Endangered List." *London Guardian*, July 20, 2000a, 3.

"Climate Change Dislocates Migratory Animals, Birds." Environment News Service, November 17, 2006. http://www.ens-newswire.com.

Connor, Steve. "Strangers in the Seas: Exotic Marine Species are Turning Up Unexpectedly in the Cold Waters of the North Atlantic." *London Independent*, August 5, 2002c, 12–13.

Cramb, Auslan. "Highland River Salmon 'On Verge of Extinction.'" *London Daily Telegraph*, July 15, 2002, 7.

"Fished to the Point of Ruin, North Sea Cod Stocks so Low as to Spell Disaster." *Glasgow Herald* (Scotland), November 7, 2000, 18.

Gough, Robert. "Stress on Stress: Global Warming and Aquatic Resource Depletion." *Native Americas* 16, no. 3/4 (Fall/Winter 1999):46–48. http://nativeamericas.aip.cornell.edu.

Kintisch, El. "As the Seas Warm." Science 313 (August 11, 2006):776–779.

Milmo, Cahal, and Elizabeth Nash. "Fish Farms Push Atlantic Salmon Towards Extinction." *London Independent*, June 1, 2001, 11.

O'Harra, Doug. "Marine Parasite Infects Yukon River King Salmon: Fish Are Left Inedible, Scientists Study Overall Impacts." *Anchorage Daily News*, January 28, 2004. A-1.

Portner, Hans O., and Rainer Knust. "Climate Change Affects Marine Fishes through the Oxygen Limitation of Thermal Tolerance." *Science* 315 (January 5, 2007):95–97.

Smith, Craig S. "One Hundred and Fifty Nations Start Groundwork for Global Warming Policies." *New York Times*, January 18, 2001, 7.

U.S. Environmental Protection Agency. *Ecological Impacts from Climate Change: An Economic Analysis of Freshwater Recreational Fishing*. Washington, DC: EPA, April 1995.

"Warming Streams Could Wipe Out Salmon, Trout." Environment News Service, May 22, 2002. http://ens-news.com/ens/may2002/2002L-05-22-06.html.

Fish Kills, New York Lakes

All along New York's Lake Ontario shoreline, and in many inland lakes and streams, fish died during the summer of 2002 at record rates. The deaths were attributed in part to consistently warm temperatures. "Every day, we're getting calls from all the Finger Lakes, small ponds, a few major rivers and Lake Ontario," said Bill Abraham, the regional fisheries manager for the State Department of Environmental Conservation in Avon. "In my 30 years, I've never seen [a fish kill] so widespread" (Iven 2002). Thousands of fish have washed ashore in eastern Lake Ontario during late June and early July, several lakeside residents said.

High temperatures, which frequently climbed into the upper 80s and low 90s F, have been the main cause of fish deaths on inland bodies of water, Abraham said. According to a report in the *Syracuse Post-Standard*, "the warmer temperatures have caused an explosion in the growth of algae and rooted plants. The plants flood their water environments with oxygen during the day. During the night, however, they absorb the oxygen, choking the fish to death," Abraham explained (Iven 2002).

Overpopulation is also a factor. Most of the dead fish are alewives, also known as sawbellies or mooneyes. For the past three years (before 2002), Lake Ontario produced particularly large populations of these fish, which predators such as salmon and trout feed on. The large population has thinned the supply of the alewife's preferred food, a small shrimp-like creature. Thermal upwelling also plays a role. Winds can push the warm water away from the shore, leaving cold water to replace it. The drastic temperature shift can kill all fish varieties (Iven 2002).

FURTHER READING

Iven, Chris. "Heat Strangles Fish." *Syracuse Post-Standard*, July 9, 2002. http://www.syracuse.com/news/poststandard/index.ssf?/base/news-0/10262073067431.xml.

Flora and Fauna: Worldwide Survey

Relatively modest worldwide warming experienced during the last century (1° to 2°F in most areas, outside of polar regions) already has driven some animals and plants toward cooler environments. In a century or more, substantial warming of a type not seen on Earth for millions of years could provoke mass extinctions of animals with no place left to migrate. Thus, the scientists who are studying previous extinctions are usually not doing so out of pure curiosity, but out of concern that humankind may be setting up a similar situation. Where the climactic rubber meets the zoological road, humankind is already pushing many animals toward extinction, with climate change only one of several provocations.

Not all plants and animals will be harmed by a rapidly warming environment. Unfortunately, however, most of these are species that

The pika. (Courtesy of Getty Images/PhotoDisc)

humankind finds obnoxious, such as poison ivy, ragweed, jellyfish, and ticks.

"Humans are influencing climate through increasing greenhouse gas emissions, and the warming world is causing impacts on physical and biological systems attributable at the global scale," said lead author Cynthia Rosenzweig, a scientist at the NASA Goddard Institute for Space Studies and the Columbia Center for Climate Systems Research. In 1970, Rosenzweig and researchers from 10 other institutions around the world analyzed data from published papers on 829 physical systems and roughly 28,800 plant and animal systems. In physical systems, 95 percent of observed changes are consistent with warming trends. Among living creatures inhabiting such systems, 90 percent of changes are consistent with warming, the researchers found. The study, "Attributing Physical and Biological Impacts to Anthropogenic Climate Change," appeared in the May 15, 2008, issue of *Nature* (Rosenzweig et al. 2008, 353–357).

Australia's nocturnal white lemuroid possum (*Hemibelideus lemuroids*), a furry white marsupial that lives on a few mountaintops in far northern tropical Queensland, may be the first animal to fall victim to a warming climate. The last member of the species was observed in 2005, after an intense heat wave. Four to five hours of exposure at 86°F (30°C) will kill the white lemuroid possum.

Many Species Will Have Nowhere to Go

A 100-year survey indicates that optimal elevations for growth of plant species in European temperate forests have shifted upward by about 30 meters per decade (Lenoir et al. 2008). Evidence of these changes includes earlier first-flowering dates, later freezes, the movement of butterflies, and the spread of scorpions. Many amphibians, for example, have found themselves threatened not only by warming habitats, but also by spreading use of various toxic chemicals and others human-caused insults.

As the pace of warming increases, reports of its effects have become more numerous. Among the most notable effects have been increasing losses of pine forests in western North America to beetles and other pests. Entire regions have been killed from Utah to Alaska. Warmth accelerates the beetles' reproductive cycles. Some species have become more susceptible to infectious diseases. Warming has affected bird migrations; some have starved as migration patterns change and the location of their food sources has not. Some animals migrate upslope on warming mountains until they run out of altitude. The character of tropical forests has been changing as well, as certain species adapt to warming and others die.

Some climate scientists stress the radical nature of climate changes to come. "The frightening thing about our analysis is that it shows that climate change will take us into uncharted waters," said Steve Jackson, a professor in the University of Wyoming Department of Botany who specializes in the ecological consequences of climate change. Jackson conducted the study with University of Wisconsin professor of atmospheric and oceanic sciences John Kutzbach and geographer Jack Williams. "Many species will have no place to go," Jackson said. Physical restrictions on species may amplify the effects of local climate changes. The more relevant question, Williams said, becomes not just whether a given climate still exists, but "will a species be able to keep up with its climatic zone? Most species can't migrate around the world" ("Global Warming Predicted" 2007).

The speed with which temperatures are expected to rise in the next century may be so rapid, according to Brian Huntley, that "[t]here is little likelihood that many organisms, and trees in particular, will be able to migrate fast enough to remain in equilibrium with future climate. For many organisms the situation may be made considerably worse by patterns of human land use and their already depleted populations" (Huntley 1990, 144).

Plants and Animals Face Several Threats

Global warming is one among several human provocations that have been contributing to large-scale alterations of flora and fauna that the human economy places outside its ambit—that is, "wild" populations. The growth of human populations around the world—along with attendant pollution and loss of habitat—have, during the twentieth and twenty-first centuries, set the stage for one of a handful of mass extinction episodes in the geophysical history of the Earth.

"The biotic response to 30 years of enhanced global warming [1970–2000] has become perceptible and substantial," wrote Gian-Reto Walther, who has published several scientific "metastudies," which assess hundreds of specific articles evaluating the response of flora and fauna to changing climatic conditions (Walther 2003, 177). Walther's surveys range the world, describing, for example, increasing growth of palms and broadleaf evergreen trees in Switzerland (Walther 2003, 129–139), the migration of frost-sensitive tropical plants upward along mountainsides near Hong Kong, and the poleward shift of holly in Scandinavia (Walther 2003, 173).

Scientists have examined periods in the distant past in which spikes in worldwide temperatures from natural causes led to widespread extinctions of flora and fauna. These episodes are not being studied as academic exercises, but as cautionary lessons in what can happen when the Earth's climate heats suddenly, as is expected via global warming during the twenty-first century. Along this road, other scientists have been exploring how "enhanced" (i.e., unusually elevated) levels of carbon dioxide and other greenhouse gases affect plant growth and reproduction. Effects on the behavior and reproduction of many animal species, including human beings, already were becoming evident at the beginning of the twenty-first century. What follows is a survey of a world in flux, with the major changes in the lives of Earth's plant and animals still to come.

Significant Changes in Flora and Fauna

Even with a global temperature rise of only 0.6°C during the twentieth century, a small fraction of the warming expected in years to come, significant alterations in flora and fauna have been noticed around the Earth, according to one of the most detailed ecological studies of climate change. An international team of scientists working across a range of disciplines found "a major imprint on wildlife" during the twentieth century (Connor 2002b, 11).

Writing in *Nature*, Walther and colleagues said:

> There is now ample evidence of the ecological impacts of recent climate change, from polar terrestrial to tropical marine environments. The response of both flora and fauna span an array of ecosystems and organizational hierarchies, from the species to the community levels. Despite continued uncertainty as to community and ecosystem trajectories under global change, our review exposes a coherent pattern of ecological change across systems. Although we are only at an early stage in the projected trends of global warming, ecological

> responses to recent climate change are already clearly visible. (Walther 2003, 389)

Walther, an ecologist at the University of Hanover in Germany, and lead author of the study, said that

> We want to emphasize that climate-change impacts are not something we expect for the future but something that is already happening. We are convinced of that. If you have so many studies from so many regions with so many different species involved and all pointing to the same direction of warmer temperatures, then to me it's quite convincing. (Connor 2002b, 11)

The study included work by British specialists in amphibian breeding cycles and Antarctic ecology, German experts on bird migration, and Australian marine biologists studying changes to coral reefs in tropical oceans.

Walther continued:

> It's the first time that researchers from various disciplines have come together to compare their own work. We made comparisons between and across various ecosystems and we compared different species to look for common traits—and they all point to the same direction of warmer temperatures. "We know that the global average temperature has increased by 0.6°C. For many people, this may sound very minor, but we are quite surprised that this minor change has had so many impacts already on natural ecosystems" (Connor 2002b, 11).

Scientists also have found that typical springtime activities, such as arrival and breeding of migrant birds or the first appearance of butterflies and plants, have occurred progressively earlier during the past 40 years. Meanwhile, some warm-weather species and diseases have extended their ranges. "There is much evidence that a steady rise in annual temperatures has been associated with expanding mosquito-borne disease in the highlands of Asia, East Africa and Latin America," the scientists said in this study (Connor 2002b, 11).

To cite a few examples among many: mosses and other plants have begun to grow in parts of the Antarctic that previously were considered too cold for such life. Many coral reefs around the world also have undergone mass "bleaching" on at least six occasions since 1979, all of which are related to warmer sea temperatures. Changes in wind patterns over the Bering Sea and their interaction with local patterns of ocean circulation have affected the distribution of walleye pollock, an important "forage species" for other fish and sea mammals (Connor 2002b, 11). Similar changes in ocean circulation around the Antarctic Peninsula have influenced the breeding range of the krill, an important shrimp-like animal that constitutes the base of the Antarctic food web.

In Britain, warmer winters have caused newts to breed earlier, bringing them into contact with the eggs and young of the common frog at a point at which they are most vulnerable to predators, said Trevor Beebee, professor of molecular biology at the University of Sussex and a co-author of the study (Connor 2002b, 11). "Newts and other amphibians with a protracted breeding season are responding to climate change whereas frogs and toads which usually breed earlier are not," Beebee said. "You can relate these changes to temperature changes observed over the same period" (Connor 2002b, 11). On three continents, a low-latitude, natural genetic variant of the fruit fly (*Drosophila subobscura*) is increasingly found at higher latitudes, paralleling climate warming over the past 25 years (Balany et al. 2006, 1773).

Common changes observed by this team of scientists included earlier spring breeding and first singing of birds, arrival of migrating birds, appearance of butterflies, spawning in amphibians, as well as earlier shooting and flowering of plants. In general, according to this report, spring activities have occurred progressively earlier since the 1960s, at a rate that is easily observable within a single human life span (Walther 2003, 389).

Species Move toward the Poles

Other scientific studies indicate that species are generally moving toward the poles—northward in the Northern Hemisphere and (with a smaller number of examples) southward below the equator. Two studies involving several thousand plant and animal species around the world, from plankton to polar bears, provide ample evidence that climate change is reshaping animal and plant habitats at an accelerating rate. These studies suggest that habitat change is already well under way. To cite one very notable example, a manatee swam past Chelsea Pier on New York City's Hudson River during 2007 (Funk 2007, 47). In mid-October 2008, another manatee was sighted in a Cape Cod harbor near Dennis,

Massachusetts, wandering in water that was turning chilly with the season. It died while being transported to Florida ("Lost Manatee" 2008). Yellow-jacket wasps have been seen for the first time in the far northern Canadian Arctic in places where the Inuit language contains no words to describe them.

"There is a very strong signal from across all regions of the world that the globe is warming," said Terry L. Root a biologist at Stanford University's Center for Environmental Science and Policy, who headed one of the research teams. "Thermometers can tell us that the Earth is warming, but the plants and animals are telling us that global warming is already having a discernible impact. People who don't believe it should take their heads out of the sand and look around" (Toner 2003, 1-A).

Root and colleagues, in their metaanalysis of 143 studies, came to the following conclusion:

> More than 80 percent of the species that show changes are shifting in the direction expected on the basis of known physiological constraints of species. Consequently, the balance of evidence from these studies strongly suggests that a significant impact of global warming is already discernable in animal and plant populations. The synergism of rapid temperature rise and other stresses, in particular habitat destruction, could easily disrupt the connectedness among species and lead to a reformulation of species communities, reflecting differential changes in species, and to numerous extirpations and possibly extinctions. (Root et al. 2003, 57)

Root's team, and another, headed by Camille Parmesan, a biologist at the University of Texas at Austin, looked at hundreds of existing studies of more than 2,000 plants and animals—from shrimp, crabs, and barnacles off the Pacific coast of California to cardinals that nest in Wisconsin—to see whether they could spot the "fingerprint" of changing global temperatures (Toner 2003, 1-A). Camille Parmesan and Gary Yohe calculated that range shifts toward the poles had averaged 6.1 kilometers per decade, with an advance in the beginning of spring by 2.3 days per decade (Parmesan and Yohe 2003, 37).

According to Parmesan and Yohe's analysis, some birds and butterflies had shifted as many as 600 miles northward. Grasses, trees, and other species that lack mobility have moved shorter distances. In a finding that resembled other studies, they found that springtime behavior was occurring earlier, in this case by more than two days per decade. A study of more than 21,000 swallow nests in North America, for example, indicated that the species was laying its eggs nine days earlier in the year 2000 than in 1960 (Toner 2003, 1-A). The Root and Parmesan studies both indicated that the largest changes in temperature as well as habitat movement were taking place in the polar regions, where temperature increases have been greatest.

As climate warms, mollusks and other species, such as barnacles, mussels, some snails, and whelks, probably will migrate from the Pacific Ocean to the Atlantic through the Bering Strait. Geerat J. Vermeij and Peter D. Roopnarine anticipated in *Science* that receding ice cover in the Arctic by 2050 could recreate conditions similar to those of the Pliocene, roughly 3.5 million years ago, when sea life, finding phytoplankton abundant in parts of the Arctic, regularly migrated from ocean to ocean (Vermeij and Roopnarine 2008, 780–781).

Species Migrate to Higher Altitudes

In mountains, various species have responded to warming by moving up in elevation, seeking habitats that were similar to areas they formerly had occupied at lower altitude. In some areas, such the Great Smoky Mountains, the mountain peaks had become shrinking "islands" of cool-weather species (Toner 2003, 1-A).

Warming in southern California has contributed to die-off and rapid migration to higher altitudes by 9 of the 10 most widespread plants, from desert shrubs to white firs. Within the past 30 years, most of these plants moved to elevations 200 feet above their previous growth ranges (Chang 2008). The study may be the first to relate warming's effects to specific widespread plants' migrations in altitude.

Anne E. Kelly of the Center for Environmental Analysis, Centers of Research Excellence in Science and Technology, California State University, Los Angeles, joined Michael Goulden, Department of Earth System Science, University of California at Irvine, in a survey of plant cover comparing 1977 and 2006–2007 along a 2,314-meter elevation gradient in southern California's Santa Rosa Mountains. Southern California's climate has warmed as precipitation variability has increased (and amount of snow has decreased) during the period. They found that the average elevation of the dominant plant species had risen by about 65 meters between 1977 and 2007.

"The plant death was striking, and occurred in most species," said Goulden. "The occurrence of plant death was obvious to everyone living in that area" (Chang 2008). "The speed (of the plant movement) is alarming," said ecologist Travis Huxman of the University of Arizona in Tucson, who did not participate in the study. "It means that we'll likely see vegetation shift a lot faster than we might think" (Chang 2008). Prolonged drought (which may be related to warming) also may play a part in the plants' migrations. "Drought certainly stands out as a real likely explanation. It is an extremely severe event" that could wipe out plants at lower elevations, said U.S. Geological Survey research scientist Jon Keeley. Keeley's study, read with others, provides striking evidence of plant migration (Chang 2008).

"The most remarkable thing is that we have seen so many changes in so many parts of the world from a relatively small increase in temperature," Root told Mike Toner of the *Atlanta Journal-Constitution.* "When you consider that some people are predicting warming that would be 10 times greater by the end of this century, it's spooky to think about what the consequences might be" (Toner 2003, 1-A).

Examples of seasonal and geographic habit change due to warming temperatures abound, according to a news report describing the work by Parmesan and Root. For example, the North American tree swallow is among the bird species beginning springtime activities earlier than historically recorded. Field biologists who kept track of roughly 21,000 tree-swallow nests in the United States and Canada over 40 years concluded that the average egg-laying date has advanced by nine days. Studies in Colorado found that marmots are ending their hibernation about three weeks sooner than during the late 1970s. Measurements taken in Alaska also revealed that the growth of white spruce trees has been stunted in recent years, another expected consequence of a rapidly warming climate (Toner 2003, 1-A).

Commercial crops are moving as well. Reports during 2007 noted that killing frosts during prime blueberry growing season had declined in Quebec by half since the 1950s, With continued warming, Quebec's blueberry industry is poised to overtake growers in Maine, the only other place in the world where the berries are grown commercially.

Higher-elevation animals, such as the Rocky Mountain pika, have found themselves without sustaining habitats as warming ascends mountainsides. Some sea birds have starved as warming waters kill their food sources. Every animal has its own ecological adaptations related to temperature. Warmth alters the sex ratio of Loggerhead sea turtles, for example, to the point at which genders are seriously out of balance, threatening survival of the species. Some animals already have adapted to warmth; red squirrels, for example, have adapted their breeding season; insects find new ranges; and bark beetles, whose spread has been notable in Alaska, have been devouring evergreen forests throughout western North America.

Small Mammals Migrate Upward to Escape the Heat

Many small mammal populations in Yosemite National Park, including shrews, mice, and ground squirrels have migrated upward in altitude during the last 90 years, reacting to a warming climate, according to study leader Craig Moritz, University of California–Berkeley professor of integrative biology and director of the campus's Museum of Vertebrate Zoology. The animals are reacting to a 3°C increase in nighttime low temperatures in the central Sierra Nevada, which also has plated a role in recession of glaciers, including Lyell Glacier, which has lost half of its mass in 100 years ("Warming in Yosemite" 2008; Moritz et al. 2008, 261–264). Moritz and colleagues compared their findings with work done early in the twentieth century by Joseph Grinnell, a zoology professor at the University of California–Berkeley.

Some Arctic Species Colonize New Lands Quickly

Arctic islands have been colonized repeatedly by plant species from distant sources, suggesting that plant ranges can shift rapidly northward in response to global warming. Some arctic plant species can move swiftly to colonize new land when weather warms—even on widely dispersed islands, according to a study of the Svalbard Islands off Norway's coast that traces plant movement through the 20,000 years following the last ice age. This research, conducted by scientists from the University of Oslo indicated that plants may be able to move more quickly than earlier thought as they adapt to climate change.

Inger Greve Alsos, the study's lead author, was quoted by Andrew Revkin in the *New York Times* as saying that

> [N]atural adaptability in the plants might be tested if the projections for rapid Arctic warming from

> the United Nations Intergovernmental Panel on Climate Change come to pass. She also cautioned that the evidence for resilience and long-distance mobility in Arctic plants could be the exception, not the rule. (Revkin 2007g)

Arctic plants may be adept at establishing new habitats because of rapid climate changes in the area, she said.

Root, who has been involved with many studies concluding that plants and animals are measurably feeling the effects of human-driven warming, described the Svalbard research as "great news." She explained, "The large number of documented changes has created quite a concern about the fate of many species." The new study, she said, shows that "some Arctic plants, and hopefully vegetation in other areas, apparently are able to respond in a manner that compensates for the rapid warming" (Alsos et al. 2007, 1606–1609; Revkin 2007g).

Alsos and colleagues commented:

> The ability of species to track their ecological niche after climate change is a major source of uncertainty in predicting their future distribution. By analyzing DNA fingerprinting (amplified fragment-length polymorphism) of nine plant species, we show that long-distance colonization of a remote arctic archipelago, Svalbard, has occurred repeatedly and from several source regions. Propagules are likely carried by wind and drifting sea ice. The genetic effect of restricted colonization was strongly correlated with the temperature requirements of the species, indicating that establishment limits distribution more than dispersal. Thus, it may be appropriate to assume unlimited dispersal when predicting long-term range shifts in the Arctic. (Alsos et al. 2007, 1606)

In the study, Norwegian and French scientists studied more than 4,000 DNA (deoxyribonucleic acid) samples from nine flowering plant species on Svalbard, a group of islands between the Scandinavian mainland and the North Pole. They found that seeds or plant fragments were arriving from Russia, Greenland, or other Arctic regions, often on the wind, driftwood, or dirt on floating ice and bird droppings.

Hardiness Zones Move Northward

The National Arbor Day Foundation during June 2006 released an updated version of the U.S. Department of Agriculture's Hardiness Zone Map, which indicates the lowest winter temperatures across the United States at which various species survive. The Arbor Day map uses data from the National Oceanic and Atmospheric Administration, indicating that most of the United States is one (in some cases two) zones warmer than in 1990. Each zone indicates a rise or fall of about 10°F of extreme low winter temperature. While the National Weather Service uses 30-year averages to adjust its averages, services that cater to gardeners have been cutting that period roughly in half because winter temperatures have been rising so rapidly.

Tree planters, who needed an accurate guide immediately, prompted the Arbor Day Foundation to revise its maps on a shorter timetable, said Woodrow L. Nelson, chief spokesman for the foundation. By 2007, John Denti, the greenhouse manager at the University of North Carolina Botanical Gardens in Charlotte, was tending 50 varieties of palm trees in his yard, as an experiment. "I surely believe in global warming, so I'm planting a lot of marginal stuff and hoping it gets warm enough," Denti said (Dewan 2007).

In recent years, subtropical plants have been sprouting in Atlanta. At the Habersham Gardens nursery, according to a report in the *New York Times*, "a spiky-leafed, sultry coastal oleander has been thriving in a giant urn. We never expected it to come back every year," said Cheryl Aldrich, the assistant manager, guiding a visitor on a tour of plants that would once have needed coddling to survive here, such as the eucalyptus, and angel trumpets. "We've been able to overwinter plants you didn't have a prayer with before" (Dewan 2007). Tara Dillard, an Atlanta landscape designer and garden writer, said she now steers clients away from long-time favorites. "I'm writing a column about rhododendrons right now," Dillard said. "And I think my conclusion is going to have to be not to plant rhododendrons. We have heated out of the rhododendron zone" (Dewan 2007).

Weeds and Other Invasive Species

Horticulturists caution that increasing warmth has two sides. The same mildness that invites a wider variety of plants also brings more insects. Higher levels of carbon dioxide also tend to favor weeds and other invasive species over cultivated plants. Poison ivy becomes more toxic, for example, as ragweed dumps more pollen (and its pollen becomes more potent). Kudzu, a bane of southern forests and waterways that grows so quickly one can almost watch it change, is spreading northward. (Some scientists have proposed that kudzu be cultivated for biofuel.)

Lewis Ziska of the U.S. Department of Agriculture says that a warming atmosphere doubled the per plant production of ragweed (a plant that aggravates human allergies) during the twentieth century. With a carbon-dioxide level of 280 parts per million (ppm), typical of the year 1900, ragweed plants produced an average of 5.5 grams of pollen each. With a level of 370 ppm, a level reflective of the year 2000, the same plants produce an average of 10 grams of pollen. At 600 ppm, Ziska says that ragweed pollen production will double again, to about 20 grams, based on tests in plots with controlled atmosphere.

Some states' signature flora are declining as temperatures warm, including Kansas sunflowers and Mississippi magnolias, which now survive in Omaha, Nebraska. By the end of the twenty-first century, the climate will no longer favor the official state tree or flower in 28 states, according to *The Gardener's Guide to Global Warming*, a report released by the National Wildlife Federation (Dewan 2007). In coming years, climate change could push the growing range of Ohio's buckeye tree out of the state that claims it as a football icon, into rival Michigan. A group ("Save the Buckeye") has been formed to warn against this peril. The state tree of Ohio bears a brown nut with a light circular "eye," which resembles the eye of a buck deer.

Growing Seasons Expanding

Flowers now sometimes bloom and wither earlier, before the spring events meant to observe them. The annual Atlanta Dogwood Festival is now routinely held after the flowers have passed. The same has been true of the Cherry Blossom festival in Washington, D.C. Television commentators during former president Gerald Ford's funeral in Washington, D.C., on December 30, 2006, noted the extraordinary warmth of the season. Bulbs were blooming, some said. The *Washington Post* reported that "In private gardens and public parks, a freakishly mild winter has brought forth a cornucopia of blossoms, from the camellia bushes to the first of the daffodils" (Higgins 2006, B-1).

The length of the autumn fruiting season for fungi in forest soil has increased for the past five decades, in parallel with temperature and rainfall increases in the United Kingdom. Average first-fruiting date of 315 species is earlier, while last fruiting date is later. Fruiting of *mycorrhizal* species that associate with both deciduous and coniferous trees is delayed in deciduous, but not in coniferous, forests. Many species are now fruiting twice a year, indicating increased *mycelial* activity and possibly greater decay rates in ecosystems (Gange et al. 2007, 71).

In parts of Michigan, the climate has warmed enough to accommodate southern magnolia trees, said Arbor Day Foundation spokesman Woodrow L. Nelson. Arizona cypress, another southerly species, suddenly seems a better fit for some sections of the Northeast, he said. "I mean, who would have thought that an Arizona cypress would be a choice for someone in New Jersey?" asked Nelson (Fahrenthold 2006, A-1).

The goya, a bitter Japanese melon that looks like a gnarly cucumber, until recently had been grown mainly on Okinawa. By the year 2000, however, with warming weather, farmers were cultivating the melon as far north as some parts of Honshu. The *unshu-mikan*, similar to a mandarin orange, has been threatened as warm weather has prompted trees to leak sap even in winter, an invitation to pests. The harvest of yellowtail, or *kan-buri*, a cold-water fish, has been declining rapidly in Japanese waters, while harvests of sardines, which have an affinity for warm water, have been increasing ("Bitter Pill" 2003).

Henry David Thoreau's First-flowering Dates

Henry David Thoreau, who died in 1862, at the beginning of the Industrial Revolution in the United States, left behind records of first-flowering dates for about 500 species that are being used a century and a half later as indicators of climate change. Between 1851 and 1858, Thoreau noted the first-flowering dates as part if his notes for an anticipated book on the seasons of Concord, Massachusetts. The records have been used by researchers at Boston and Harvard Universities to determine that many first flowering dates have advanced by about a week during a century and a half.

During that time, the average temperature in Concord has risen about 4.3°F (Rice 2008, 9-D). Biological diversity in the Concord area also has been reduced markedly. The analysis was carried out by Richard B. Primack, a conservation biologist at Boston University, and Abraham J. Miller-Rushing, his graduate student at the time, with

Charles C. Davis, an evolutionary biologist at Harvard. The research published in the *Proceedings of the National Academy of Sciences* found that by 2004 through 2006, 27 percent of species documented by Thoreau are now extinct in the area. Another 36 percent are close to vanishing.

"It's targeting certain branches in the tree of life," said Davis. "They happen to be our most charismatic species—orchids, mints, gentians, lilies, iris." Of 21 orchid species observed by Thoreau in Concord, "we could only find 7," Primack said (Dean 2008; Willis et al. 2008).

Canadian Seal Harvest Falls

Warming and thinning ice is reducing Canada's annual seal harvest that has sparked conflict for years between hunters and animal-rights activists, as several unusually warm winters drown thousands of ice-dwelling seals. As a result, Canadian authorities in early 2007 reduced quotas on the harp seal hunt by about 20 percent.

"We don't know if it's weather or climate. But we have seen a trend in the ice conditions in the last four or five years," said Phil Jenkins, a spokesman for the Department of Fisheries and Oceans. "The pups can't swim for very long. They need stable ice. If the ice deteriorates underneath them, they drown" (Struck 2007c). Rebecca Aldworth, an activist with the Humane Society of the United States, flew over the area this week. "We should have seen vast ice fields, but we saw only a few floating ice pans," she said. "We should have seen thousands of seal pups, but we just saw a few" (Struck 2007c).

"Assisted Migration"

Some scientists have suggested that humanity adapt to climate change by speeding up animal migrations with human assistance. Under this survival strategy, biologists choose a species, which is then moved hundreds of miles to a new, cooler habitat. Assisted migration has provoked mixed reactions from conservation biologists, who realize that it involves risks. The alternative for many species may be extinction, however. When Jason McLachlan, a Notre Dame biologist, gives talks on global warming and extinction, he has noted that "someone will say, 'It's not a problem, since we can just FedEx them to anywhere they need to go'" (Zimmer 2007). With many thousands of species facing extinction, conservation biologists practicing "assisted migration" will face painful decisions. Species in polar regions that have adapted to cold climates may have no alternative habitats. Conservation biologists deciding where to transfer species may not have all the information they need, and may make mistakes, a common result of humanity's hubris. *See also:* Extinctions, and Climate Change

FURTHER READING

"30 Years of Global Warming Has Altered the Planet." Environment News Service, May 16, 2008. http://www.ens-newswire.com/ens/may2008/2008-05-14-01.asp.

Alsos, Inger Greve, Pernille Bronken Eidesen, Dorothee Ehrich, Inger Skrede, Kristine Westergaard, Gro Hilde Jacobsen, Jon Y. Landvik, Pierre Taberlet, and Christian Brochmann. "Frequent Long-Distance Plant Colonization in the Changing Arctic." *Science* 316 (June 15, 2007):1606–1609.

Balany, Joan, Josep M. Oller, Raymond B. Huey, George W. Gilchrist, and Luis Serra. "Global Genetic Change Tracks Global Climate Warming in *Drosophila Subobscura*." *Science* 313 (September 22, 2006):1773–1775.

"Bitter Pill: The Northward Spread of the Okinawan Goya, Warm-weather." Asahi News Service (Japan), January 29, 2003. (LEXIS)

Chang, Alicia. "Plants Move Up Mountain as Temps Rise, Study Shows." Associated Press, August 11, 2008. (LEXIS)

Connor, Steve. "World's Wildlife Shows Effects of Global Warming." *London Independent*, March 28, 2002b, 11.

Dean, Cornelia. "Thoreau Is Rediscovered as a Climatologist." *New York Times*, October 28, 2008. http://www.nytimes.com/2008/10/28/science/earth/28wald.html.

Dewan, Shaila. "Feeling Warmth, Subtropical Plants Move North." *New York Times*, May 3, 2007. http://www.nytimes.com/2007/05/03/science/03flowers.html.

Fahrenthold, David A. "Washington Warming to Southern Plants." *Washington Post*, December 20, 2006, A-1., http://www.washingtonpost.com/wp-dyn/content/article/2006/12/19/AR2006121901769_pf.html.

Funk, McKenzie. "Cold Rush: The Coming Fight for the Melting North." *Harper's Magazine*, September 2007, 56–55.

Gange, A. C., E. G. Gange, T. H. Sparks, and L. Boddy. "Rapid and Recent Changes in Fungal Fruiting Patterns." *Science* 316 (April 6, 2007):71.

"Global Warming Predicted to Create Novel Climates." Environment News Service, March 27, 2007. http://www.ens-newswire.com/ens/mar2007/2007-03-27-03.asp.

Higgins, Adrian. "Kept from Hibernation by a Lingering Warmth." *Washington Post*, January 5, 2006, B-1. http://www.washingtonpost.com/wp-dyn/content/article/2007/01/04/AR2007010401973_pf.html.

Huntley, Brian. "Lessons from the Climates of the Past." In *Global Warming: The Greenpeace Report*, ed. Jeremy Leggett, 133–148. New York: Oxford University Press, 1990.

Kelly, Anne E., and Michael L. Goulden. "Rapid Shifts in Plant Distribution with Recent Climate Change." *Proceedings of the National Academy of Sciences* 105, no. 33 (August 19, 2008):11823–11826.

Lenoir, J., J. C. Gègout, P. A. Marquet, P. de Ruffray, and H. Brisse. "A Significant Upward Shift in Plant Species Optimum Elevation during the 20th Century." *Science* 320 (June 27, 2008):1768–1771.

"Lost Manatee Rescued in Mass. Dies on Trip to Fla." Associated Press, October 2, 2008. (LEXIS)

Moritz, Craig, James L. Patton, Chris J. Conroy, Juan L. Parra, Gary C. White, and Steven R. Beissinger. "Impact of a Century of Climate Change on Small-Mammal Communities in Yosemite National Park, USA." *Science* (October 10, 2008):261–264.

Parmesan, Camille, and Gary Yohe. "A Globally Coherent Fingerprint of Climate Change Impacts across Natural Systems." *Nature* 421 (January 2, 2003):37–42.

Revkin, Andrew C. "Many Arctic Plants Have Adjusted to Big Climate Changes, Study Finds." *New York Times*, June 15, 2007g. http://www.nytimes.com/2007/06/15/science/15arctic.html.

Rice, Doyle. "Warming Affecting Walden, Yellowstone." *USA Today*, October 28, 2008, 9-D.

Root, Terry L., Jeff T. Price, Kimberly L, Hall, Stephen H. Schneider, Cynthia Rosenzweig, and J. Alan Pounds. "Fingerprints of Global Warming on Wild Animals and Plants." *Nature* 421 (January 2, 2003):57–60.

Rosenzweig, Cynthia, David Karoly, Marta Vicarelli, Peter Neofotis, Qigang Wu, Gino Casassa, Annette Menzel, Terry L. Root, Nicole Estrella, Bernard Seguin, Piotr Tryjanowski, Chunzhen Liu, Samuel Rawlins, and Anton Imeson. "Attributing Physical and Biological Impacts to Anthropogenic Climate Change." *Nature* 453 (May 15, 2008):353–357.

Struck, Doug. "Warming Thins Herd for Canada's Seal Hunt: Pups Drown in Melting Ice, Government Reduces Quotas." *Washington Post*, April 4, 2007c, A-8. http://www.washingtonpost.com/wp-dyn/content/article/2007/04/03/AR2007040301754_pf.html.

Toner, Mike. "Warming Rearranges Life in Wild." *Atlanta Journal and Constitution*, January 2, 2003, 1-A.

Vermeij, Geerat J., and Peter D. Roopnarine "The Coming Arctic Invasion." *Science* 321 (August 8, 2008):780–781.

Walther, Gian-Reto. "Plants in a Warmer World." *Perspectives in Plant Ecology, Evolution, and Systematics* 6, no. 3 (2003):169–185.

"Warming in Yosemite National Park Sends Small Mammals Packing to Higher, Cooler Elevations." NASA Earth Observatory, October 9, 2008. http://earthobservatory.nasa.gov/Newsroom/MediaAlerts/2008/2008100927660.html.

Willis, Charles G., Brad Ruhfel, Richard B. Primack, and Charles C. Davis. "Phylogenetic Patterns of Species Loss in Thoreau's Woods Are Driven by Climate Change." *Proceedings of the National Academy of Sciences*, October 27, 2008, doi: 10.1073/pnas.0806446105.

Zimmer, Carl. "A Radical Step to Preserve a Species: Assisted Migration." *New York Times*, January 23, 2007. http://www.nytimes.com/2007/01/23/science/23migrate.html.

Food: The Low Carbon Diet

Food production in the United States today requires six calories of energy, on average, to produce one calorie of food. Much of this energy is consumed not only in transport, but also in energy-intensive cultivation of factory farms, as well as many people's preference for energy-intensive types of food, such as meats (most notably beef) (Hillman, Fawcett, and Rajan 2007, 60).

Far-Fetched Food

Food items sold in U.S. grocery stores travel an average of 1,500 miles to reach consumers, according to David Pimentel, professor of ecology and agricultural science at Cornell University. The food industry burns almost one-fifth of all the petroleum consumed in the United States, about as much as automobiles, according to Michael Pollan's *Omnivore's Dilemma*. The average American eats more than 200 pounds of red meat, poultry, and fish per year, an increase of 23 pounds compared with 1970.

A walk through an average U.S. supermarket is a tour of the world, a cornucopia of food choices. The carbon-conscious sometimes take pride in eating low on the food chain, substituting vegetables and fruits for meat. How carbon conscious, however, are vegetables that come from Chile, Nicaragua, or New Zealand when they are sold out of season? In that case, most of their carbon calories come in the form of transportation, most of it powered by oil. The choices are beguiling. How long has it been since we could indulge our taste for watermelon only during a few summer months?

A tour of a typical U.S. supermarket also reveals that modern transport and various free-trade agreements (the North American Free Trade Agreement [NAFTA]) and others) have rendered distance nearly irrelevant to our food supply. The idea of "seasonal" produce has nearly lost its meaning. During winter, in the United States, many fruits and vegetables are now routinely imported from the Southern

Hemisphere. Geography (and energy expenditure) has become so irrelevant in today's American grocery stores that some salmon are raised in the Pacific Northwest, shipped to China to be cut (where labor is cheap), and shipped back to North America to be sold.

Latino and Asian immigration in the United States has given rise to all manner of imports (packaged goods, mainly) from these countries—Mexico southward, China, South Korea, Vietnam, India, and others. A typical city of a half-million people (Omaha, Nebraska, in this case) includes at least a half-dozen East Asian grocery stores, with goods from China and nearby countries, and at least two specializing in goods from India and Bangladesh. Another sells groceries from the Philippines and Vietnam. Main-line grocery stores now sell extensive offerings from East Asia and Latin America. Omaha has three Spanish-language newspapers that carry advertising for groceries. Many of these goods are shipped from thousands of miles away, usually utilizing oil-based fuels. (The Web site Sustainable Table, www.sustainabletable.org, addresses food and energy issues.)

The energy intensity of food is a worldwide phenomenon. Cod caught off Norway is shipped to China, carved into filets, and shipped back to Norway for sale. (The cost in fuel, not counting greenhouse-gas emissions, is less than the difference in labor.) Argentine lemons fill supermarket shelves in Spain, as lemons grown nearby rot on the ground. Half of Europe's peas in 2008 were being grown in Kenya.

The World Meets at the Supermarket

Cruising the aisles of most supermarkets in the United States yields thousands of food choices from dozens of countries. A few examples of distances to market are calculated in air miles from New York City. The first figure below is miles; the second is kilometers. For surface miles via ocean transport, add 10 percent. Some of these goods probably have been shipped by air, but most have reached the stores via surface, ships, trucks, and trains, which involve longer distances but is much more energy efficient. These are only a few examples of what is available.

- Alaska: salmon (Anchorage: 3,371/5,425)
- Belgium: Banana nectar (3,600/5,840)
- Chile: grapes and other winter fruits (Santiago: 5,107/8,219)
- China: beer (Beijing: 6,843/11,012)
- Costa Rica: pineapples and bananas (San Jose: 2,509/4,039)
- Ecuador: cut flowers (Quito: 2,834/4,560)
- Germany: beer (Frankfurt: 3,858/6,208)
- Honduras: shrimp (Tegucigalpa 2,004/3,225)
- Hong Kong, China: Jasmine rice (8,059/12,968)
- Italy: premium pasta (Rome: 4,283/6,891)
- Kenya: tilapia (fish), shrimp (Nairobi: 7,358/11,842)
- Mexico: mangoes and papayas, watermelon (all seasons), cantaloupe, winter strawberries (Mexico City: 2,090/3,363)
- Netherlands: flower seeds (Amsterdam: 3,652/5,877)
- Nicaragua: winter watermelon, shrimp (Managua: 2,310/3,719)
- Peru: winter asparagus (Lima: 3,640/5,857)
- Philippines: mangoes (Manila: 8,509/13,693)
- Russia: caviar (Moscow: 4,668/7,511)
- Spain: olives (Madrid: 3,591/5,779)
- United Kingdom: Coleman's mustard (London: 3,463/5,772)
- Venezuela: winter watermelon (Caracas: 2,149/3,459)

Within the United States, many fruits and vegetables are shipped nationwide from California (Los Angeles to New York City: 2,451 miles/3,944 kilometers), at what would be international distances in many other countries. In summer, however, produce is more localized. The area around Omaha produces some of the best watermelon on Earth, but for only two to three months (August through early October).

Some U.S. food chains that specialize in organic foods, such as Whole Foods Market, among others, make an effort to purchase locally grown fruits and vegetables. Sometimes, local grocery chains also purchase locally in season in areas with extensive agricultural bases (corn in Nebraska and Iowa, for example). Of course, as in Britain, the United States has a lot of central distribution that trumps geography in all sorts of ridiculous ways.

An Emphasis on Local Food

Countering this trend, more urban residents are raising food on vacant lots. By 2007, 2,000 community gardens were operating in U.S. urban areas. Omaha, Seattle, New York City, Boston, and many other urban U.S. locales also host farmers' markets in season. In Omaha, at least, some of the best-tasting produce in the world comes from small fruit and vegetables stands that pop up around the city every summer and fall.

Some farmers near U.S. urban areas also offer "organic box" plans, by which they supply produce on contract. Food is sometimes low in quality, as some farmers save their best and most salable produce for the open market where they do not have a captive clientele. Thus, a good idea in theory may not work well in practice. Many U.S. states have organic food certification groups, such as the Maine Organic Farmers and Gardeners Association (MOFGA), formed in 1971; another is the California Certified Organic Farmers (CCOF). For more information, see the National Sustainable Agriculture Information Service (http://attra.ncat.org).

The Carbon Footprint of Meat

The world meat industry produces 18 percent of the world's greenhouse-gas emissions, more than transportation. Nitrous oxide in manure (warming effect: 296 times greater than that of CO^2) and methane from animal flatulence (23 times greater) mean that "a 16-oz. T-bone is like a Hummer on a plate" (Will 2007, A-27).

Many people are now going vegetarian as part of a personal climate-change strategy. Billions of chickens, cattle, turkeys, pigs, and cows being raised in factory farms produce methane from digestion and feces. The raising of animals for human consumption is the single-largest source of methane emissions in the United States, and a methane molecule is 23 times as effective at retaining heat as one molecule of carbon dioxide.

Animal agriculture is also a major source of carbon dioxide. Production of animal protein requires about 10 times the fossil-fuel input (producing 10 times the carbon dioxide) compared with production of edible plants. The feeding, killing, and processing of meat is enormously energy intensive.

Taking such things into account, a study by University of Chicago professors Gidon Eshel and Pamela Martinargues demonstrated that a completely vegetarian diet (avoiding eggs and dairy products as well as meat, fish, and fowl) can remove as much carbon dioxide and methane from the air (or, in some cases, more) as driving a hybrid car (compared with one with an internal combustion engine). They calculated that the difference between a meat-centered diet and vegetarianism had the same effect on greenhouse-gas production as switching from an sport-utility vehicle to a standard sedan.

Some overtly "green" businesses are not. Consider Ben & Jerry's ice cream, invented by eco-talking Vermont hippies, a gallon of which

> requires electricity-guzzling refrigeration from manufacture to table, as well as four gallons of milk from cows that, at the same time, produce eight gallons of manure and eight gallons' worth of methane flatulence. The cows eat grain and hay cultivated with tractor fuel, chemical fertilizers, herbicides and insecticides, transported by trains and trucks. (Will 2007, A-27)

The Green Restaurant Movement

Customers at American Flatbread, a pizzeria in the Washington, D.C., suburb of Ashburn, Virginia, drink beer that is brewed nearby. Leaves of their iced tea are grown and packaged on a local farm. Buyers acquire scallions, spinach, cheese, and corn within a few miles. On the restaurant's back wall, a hand-painted map of Loudoun County shows the farms and dairies that the restaurant uses to assemble its menu. Its wood-fueled oven is made of Virginia red clay (Lazo 2007, A-1).

Several other restaurants in the same area have found that using local foods is tasty, popular, and profitable, suiting an area of mixed suburbs and farms that have been struggling to survive urban sprawl. Local farms supply local groceries (including burgeoning organic food markets) as well as restaurants with corn, melons, tomatoes, peppers, squash, eggplant, cucumbers and even lamb. American Flatbread's frozen pizzas also are sold at local grocery stores.

At the Habana Outpost in Brooklyn, a solar-panel awning shades the outdoor dining area, providing the energy for the first solar-powered restaurant in New York City. A bicycle crank provides the power that blends fruit smoothies. "I wanted to make being environmentally responsible fun and friendly, without being didactic," said the eatery's owner, Sean Meenan (Stukin 2007, 51). Cell phones and computers are solar powered. Toilets use rainwater, kitchen waste is composted, and utensils and drinking cups are made of cornstarch. Meenan, who also owns two other eco-eateries in New York City, also buys his organic food as close to home as possible.

Interest in green restaurants has initiated formation of a Green Restaurant Association (GRA) in the United States. The 17-year-old group had certified 300 eating establishments by 2007. Certification is earned by buying locally, avoiding

Styrofoam, and making recycling a priority, among other things, such as using environmentally friendly cleaners, avoiding plastic, using energy-efficient appliances, and sending used frying oil to be used as biodiesel fuel. The GRA's publicity points out that the one million restaurants in the United States consume more electricity than any other type of retail business. The average restaurant turns out 50,000 pounds of garbage (most of it organic) and uses 300,000 gallons of water per year (Stukin 2007, 52).

FURTHER READING

Knoblauch, Jessica A. "Have It Your (the Sustainable) Way." *Environmental Journalism* (Spring 2007):28–30, 46.

Lazo, Alejandro. "A Shorter Link between the Farm and Dinner Plate: Some Restaurants, Grocers Prefer Food Grown Locally." *Washington Post*, July 29, 2007, A-1. http://www.washingtonpost.com/wp-dyn/content/article/2007/07/28/AR2007072801255.html.

Stukin, Stacie. "The Lean, Green Kitchen." *Vegetarian Times*, September 2007, 51–54.

Weise, Elizabeth. "Warming Climate Makes Gardeners' Map Out-of-Date." *USA Today*, April 24, 2008, A-1.

Will, George F. "Fuzzy Climate Math." *Washington Post*, April 12, 2007, A-27. http://www.washingtonpost.com/wp-dyn/content/article/2007/04/11/AR2007041102109_pf.html.

Food Web and Warming in Antarctica: Phytoplankton to Penguins

Global warming could wipe out thousands of Antarctic animal species in the next 100 years, the British Antarctic Survey asserted during 2002. An anticipated temperature rise of 2°C, a fraction of what the Intergovernmental Panel on Climate Change (IPCC) has forecast by the end of the twenty-first century, would be enough to threaten large numbers of fragile invertebrates with extinction, said Professor Lloyd Peck from the British Antarctic Survey. These include exotic creatures found nowhere else on Earth, "such as sea spiders the size of dinner plates, isopods—relatives of the woodlouse, and fluorescent sea gooseberries as big as rugby balls" (Von Radowitz 2002). Peck said, "We are talking about thousands of species, not four or five. It's not a mite on the end of the nose of an elk somewhere" (Von Radowitz 2002).

Laboratory experiments have indicated that these animals and many others cannot survive even small variations in the temperatures of their habitats. Several thousand species of small animals, including mollusks and worms, are likely to die, including 750 species of sand flea alone. In time, fish populations and larger species such as penguins, seals, and whales further up the food chain could be affected, said Peck. He said it was impossible at this stage to guess how severe the consequences might be (Von Radowitz 2002). With a temperature rise of 2° to 3°C, many Antarctic species' movements became sluggish as less oxygen reaches their tissues. With their ability to swim and feed compromised, their survival is threatened.

Krill's Importance in the Food Web

By late 2004, research reported in *Nature* indicated that the density of krill in the Antarctic marine food chain had fallen by 80 percent since the 1970s, creating food shortages that were endangering larger animals and birds, such as whales, seals, penguins, and albatrosses, especially in the vicinity of the Antarctic Peninsula. Angus Atkinson of the British Antarctic Survey, who led the research, said, "This is the first time that we have understood the full scale of this decline. Krill feed on the algae found under the surface of the sea-ice, which acts as a kind of nursery" (Atkinson et al. 2004, 100–103; Henderson 2004c).

The collapse of ice shelves along some of Antarctica's shores changes the ecology of the nearby ocean, with attendant effects on wildlife. According to a report by the Environment News Service, "The new icebergs have changed the Antarctic ecosystem … blocking sunlight needed for growth of the microscopic plants called phytoplankton that form the underpinning of the entire food web." "They are a primary food source for miniscule shrimp-like krill, which in turn are consumed by fish, seals, whales, and penguins. Ice shelf B-15 broke into smaller pieces that prevented the usual movement of sea ice out of the region," said Kevin Arrigo, assistant professor of geophysics at Stanford University. Phytoplankton require open water and sunlight to reproduce, so higher-than-usual amounts of pack ice cause declines in plankton productivity ("Breakaway Bergs" 2002).

Penguins' Populations Fall

Populations of Adelie Penguins on the Antarctic Peninsula are falling as their surroundings warm. About 1985, the Biscoe region of the Antarctic Peninsula was home to about 2,800 breeding pairs of Adelie Penguins. By the year 2000, that number had declined to about 1,000. On nearby islands, the number of breeding pairs has dropped from 32,000 to 11,000 in 30 years. In some cases, melting ice has provided evaporative surface for increasing snowfall, which inhibits the birds' nesting. "The Adelies are the canaries in the coal mine of climate change in the Antarctic," said ecologist Bill Fraser (Montaigne 2004, 36, 39, 47). If warming continues, penguins may abandon much of their home 900-mile-long promontory altogether. The archetypal "tuxedoed" species prefer a cold climate even more so than other penguins (Lean 2002a, 9).

Warming has caused problems for penguins in the Ross Sea. Large icebergs have blocked the way between their breeding colonies and feeding areas. As a result, the penguins are being forced to walk an extra 30 miles (at a one-mile per hour waddle) to get food. Thousands of penguins have died during these treks. Thousands of Emperor Penguin chicks drowned near Britain's Halley base after ice broke up earlier than usual, before they had learned to swim (Lean 2002a, 9). The penguins cannot fly, and so have trouble changing habitat as conditions evolve.

According to a report by Geoffrey Lean in the *London Independent*,

> [The penguins] are feeling the heat most strongly on the Antarctic Peninsula, which juts out from the polar landmass towards South America. Studies of air temperatures around the world over the past half-century suggest that this is one of the three areas on the planet—along with north-western North America and part of Siberia—warming up fastest. The British Antarctic Survey says flowering plants have spread rapidly in the area, glaciers are retreating, and seven huge ice sheets have melted away. (Lean 2002a, 9)

Professor Steven Emslie of the University of North Carolina believes that, if warming continues, the penguins "would continue to decline in the peninsula, and may completely abandon much of it" (Lean 2002a, 9). Studies of fossilized remains that he has conducted near Britain's Rothera base show that the number of the penguins has declined sharply during warmer periods in the past. Researchers for the U.S. National Science Foundation said that one colony of Adelies at Cape Royds will "fail totally." Scientists at the Scripps Institute of Oceanography added that a colony of emperor penguins at Cape Crozier also has failed to raise any chicks (Lean 2002a, 9).

The Ross Sea is home to 25 percent of the world's Emperor Penguins and 30 percent of Adelie Penguins. "We know for certain that penguins suffered breeding losses because of the icebergs in this region," said Arrigo. "There was a lot less food nearby for penguins to get to, so they had to go much farther to feed," he added. "In doing so, they left their nests exposed for longer periods of time than they normally would. That made them vulnerable to predators such as the skua, a large gull that feeds on chicks and the eggs. So penguin breeding success was much lower last year [2001]" ("Breakaway Bergs" 2002). "Now the penguins have another obstacle they have to get around," Arrigo said. "Not only do they have to go farther to find food, but they have to swim around this enormously large iceberg that has found its way in their path. Some rookeries have been abandoned altogether" ("Breakaway Bergs" 2002).

Declining Krill and Penguins' Reproduction Rates

The increasing difficulty of the penguins' lives has cut their reproduction rates. Penguin chick numbers have fallen to about 10 percent of the usual number at Cape Bird and 2 percent of the usual number at Cape Royds and Cape Crozier. At the southernmost colony of Adelie Penguins at Cape Royds, only about 500 nests were established in areas where penguins usually maintain 3,000 to 3,500 active nests, imperiling the entire colony's survival (English 2003). By 2003, the penguins required four days to complete foraging trips for food, sapping time and energy used previously for reproduction. The U.S. Coast Guard icebreaker *Polar Sea*, which supplies the McMurdo base, broke a path through nearby ice to improve the penguins' access to the sea (English 2003).

Climate change may be a factor in the declining numbers of penguins, seals, and albatrosses near the island of South Georgia in the western Antarctic. Melting Antarctic ice also may be eliminating krill, one of the oceans' major food sources. British scientists have said that krill is finding it harder to graze on algae beneath retreating ice shelves, making life more difficult for the animals that feed on krill (Reid and Croxall 2001). Krill populations are declining in the Northern Hemisphere as well. They fell 70 percent during 10 years in the Estuary and Gulf of St. Lawrence, according to new research by scientists with the Maurice Lamontagne Institute, a marine science center associated with Canada's Department of Fisheries and Oceans. A probable cause, the scientists said, is global warming, and the risk is a reduction in the number of whales and fishes in these waters.

Keith Reid and John Croxall of the British Antarctic Survey in Cambridge surveyed the numbers and breeding success of Antarctic Fur Seals (*Arctocephalus gazella*), Macaroni Penguins (*Eudyptes chrysolophus*), Gentoo Penguins (*Pygoscelis papua*), and Black-browed Albatrosses (*Thalassarche melanophrys*). Since the 1980s, the researchers found that the relationship between these species and their favorite prey, Antarctic Krill (*Euphausia superba*), has changed. "Krill graze on algae beneath sea ice," Reid explained, so retreating sea ice in the region may have reduced the krill's ability to reproduce (Loeb 1997). The amount of krill in the diet of seals and penguins closely matches their ability to bear healthy offspring.

Writing in *Science*, however, D. G. Ainley and colleagues took issue with assertions by Croxall and others that populations of Adelie Penguins are declining along the Antarctic coast because of sea-ice retreat. This is true "only along the Northwestern Antarctic Peninsula (about 5 percent of the Antarctic Coast) in the last 50 years," they asserted. Elsewhere, Ainley and colleagues wrote, "This species has been increasing" (Ainley et al. 2003, 429). Furthermore, Ainley and colleagues asserted that the penguins depend on krill only in the summer.

Late in 2000 and early in 2001, during their summer breeding season, thousands of Magellanic Penguins washed up dead on the beaches north of Punta Tombo, in the far south of Argentina. According to a report in the *New York Times*, "Many birds abandoned their nests, leaving chicks to starve. Among the survivors, many were in bad shape, having difficulty finding the fish they needed to sustain themselves" (Yoon 2001a, F-1). Since 1987, the number of Magellanic Penguins at Punta Tombo has declined by 30 percent (Yoon 2001a, F-1).

Researchers have reported that Magellanic penguins near Punta Tombo have been declining in numbers steadily for more than 10 years. Penguin populations are declining around the world, and global warming seems to be a major cause. Ten of the world's 17 penguin species already have been listed as threatened or endangered (Yoon 2001a, F-1). Overfishing and oil spills also threaten penguins in some areas. Warming water depletes the penguins' food sources and causes many species to quit breeding, as large numbers of birds that would be reproducing die of starvation.

El Niño Episodes and Penguin Populations

"If we get a series of intense El Niños, they're going to disappear," said Patricia Majluf, a conservation biologist at the Wildlife Conservation Society, of the colony of Humboldt Penguins she studies, whose numbers also have been declining. "We lost half during one bad El Niño and these are very slow breeding birds" (Yoon 2001a, F-1). El Niño conditions cause marked warming of many oceanic areas in which penguins live and breed. Warmer water also encourages the growth of toxic algae that constitute killing "red tides" in some of the same waters.

As the oceans have warmed during the last half of the twentieth century, winter sea ice no longer extends north of the South Shetland Islands. Pack ice now forms in the area only occasionally, about one or two of every six to eight years. The pack ice contains frozen diatoms, food for young krill. Without the usual ice pack, the penguins go hungry or starve. By 2001, ice had failed to form in many spawning areas six winters in a row, the life span of most krill. Given one more iceless winter, many of the aging krill will die, blowing a very large hole in the Adelies Penguins' food chain.

David Pole Evans, a Falklands farmer, was the first to notice that something was amiss when he saw penguins "just standing around, not looking very fit or healthy" in April 2002. A few weeks later, he found thousands of dead penguins on the shore of Saunders Island and in the surrounding waters (Finch 2002, 12). Evans told the British Broadcasting Corporation that he estimated as many as 9,000 Rockhopper Penguins and 1,000 Gentoo Penguins, which are native to the South Atlantic region, had died within a short time.

Postmortem examinations failed to produce any firm reasons for the penguins' deaths. Nick Huin, the scientific officer from Falklands Conservation, a charity, said, "We're worried, because we don't know exactly why it's happening" (Finch 2002, 12). Tracking devices have been attached to 10 penguins in an attempt to solve the mystery. But Mike Bingham, a researcher who works with the International Penguin Conservation Work Group, based on the islands, said there was "no doubt" that the penguins are dying from starvation (Finch 2002, 12).

The local penguins have been molting a month later than usual. When they molt, they are no longer waterproof and, unable to survive in the ocean, come ashore. Once removed from their food sources, the penguins need to have enough fat reserves to survive on land. One possible explanation for the deaths is overfishing. Another, ironically, is a drop in the sea temperature caused by the melting of Antarctic glaciers as a result of global warming. The cooler waters have caused these penguins' food source, squid, to remain in Argentine waters rather than move on the current to the Falklands (Finch 2002, 12).

Warming in the Southern Ocean during the last quarter of the twentieth century has cut populations of emperor penguins living on the Antarctic coast in half, "because of a decrease in adult survival during the late 1970s," an unusually warm period with reduced coverage of sea ice (Barbraud et al. 2001, 185). A sudden temperature rise during the late 1970s and early 1980s coincided with a sharp drop in the survival rates of adult birds, according to Christopher Barbraud and Henri Wirmerskirch, writing in *Nature.*

Their research compared the size of a colony of penguins at Dumont d'Urville Station in Terre Adelie, Antarctica, with regional weather records over 50 years. Average winter temperatures at the site rose from −17.3°C in the early 1970s to −14.7°C by the late 1990s, reducing the amount of local sea ice. According to a report in the *London Daily Telegraph,* during the same period, penguin numbers at the colony dropped to about 3,000. The death rate was higher for males than females. Although the population has since stabilized, there are concerns that other colonies may have suffered similar declines (Derbyshire 2001a, 6). "In years with high sea-surface temperatures emperor penguins probably have difficulties in finding food, which could increase mortality," the scientists reported (Derbyshire 2001a, 6). Reduced coverage of sea ice also is associated with a lower numbers of krill.

FURTHER READING

Ainley, D. G., G. Ballard, S. D. Emslie, W. R. Fraser, P. R. Wilson, E. J. Woehler, J. P. Croxall, P. N. Trathan, and E. J. Murphy. "Adélie Penguins and Environmental Change." Letter to the Editor, *Science* 300 (April 18, 2003):429.

Atkinson, Angus, Volker Siegel, Evgeny Pakhomov, and Peter Rothery. "Long-term Decline in Krill Stock and Increase in Salps within the Southern Ocean." *Nature* 432 (November 4, 2004):100–103.

Barbraud, Christopher, and Henri Wirmerskirch. "Emperor Penguins and Climate Change." *Nature* 411 (May 10, 2001):183–186.

"Breakaway Bergs Disrupt Antarctic Ecosystem." Environmental News Service, May 9, 2002. http://ens-news.com/ens/may2002/2002L-05-09-01.html.

Derbyshire, David. "'Heatwave' in the Antarctic Halves Penguin Colony." *London Daily Telegraph,* May 10, 2001a, 6.

English, Philip. "Ross Island Penguins Struggling." *New Zealand Herald,* January 9, 2003, n.p. (LEXIS)

Finch, Gavin. "Falklands Penguins Dying in Thousands." *London Independent,* June 19, 2002, 12.

Henderson, Mark. "Southern Krill Decline Threatens Whales, Seals." *London Times* in *Calgary Herald,* November 4, 2004c, A-11.

Lean, Geoffrey. "Antarctic Becomes Too Hot for the Penguins: Decline of 'Dinner Jacket' Species Is a Warning to the World." *London Independent,* February 3, 2002a, 9.

Loeb, V. "Effects of Sea-ice Extent and Krill or Salp Dominance on the Antarctic Food Web." *Nature* 387 (1997):897–900.

Montaigne, Fen. "The Heat Is On: Ecosigns." *National Geographic,* September 2004, 34–55.

Reid, K., and J. P. Croxall. "Environmental Response of Upper Trophic-level Predators Reveals a System Change in an Antarctic Marine Ecosystem." *Proceedings of the Royal Society of London* B268 (2001):377–384.

Von Radowitz, John. "Antarctic Wildlife 'at Risk from Global Warming.'" Press Association News, September 9, 2002. (LEXIS)

Yoon, Carol Kaesuk. "Penguins in Trouble Worldwide." *New York Times,* June 26, 2001a, F-1.

Forest Fires as Feedback Mechanism

Forest fires have become a major example of a feedback loop that may accelerate global warming. According to some research, under some circumstances, wildfires may emit more carbon dioxide to the atmosphere than humankind's contribution. Moreover, many present-day computer simulations of climate change usually do not take fires' contributions into account, and thus may underestimate future greenhouse-gas levels and temperature rises.

Widespread wildfires during the summer of 2002 changed areas of the U.S. West from a carbon sink to a net carbon source, as drought also stunted tree growth, according to computer modeling studies of fires in Colorado conducted by a team of researchers from Colorado State University, the U.S. Geological Survey, and the National Center for Atmospheric Research (NCAR). "We're using the western United States as a case study area where climate and land use are interacting in several interesting ways," said NCAR senior scientist David Schimel. Western lands, particularly evergreen forests, represent about half of all U.S. carbon storage, he said ("Wildfires Add Carbon" 2002). "More carbon is freed from storage during droughts, not only because more dry vegetation burns, but also because plants deprived of water grow slower, absorbing and storing less carbon in their tissues" ("Wildfires add Carbon" 2002).

The boreal forests of the midlatitudes can function as carbon source or sink; when forests are healthy and growing, they take in carbon, but in years of drought or widespread fire, they release it. Warming seems to coincide with events that release carbon, such as increasing wildfires. The number of wildfires in Canada doubled during the 1980s and 1990s from the previous two decades, and some models indicate that they will double again early in the twenty-first century, said Brian Amiro, head of the department of soil science at the University of Manitoba (Struck 2007a, A-1).

Peat Wildfires in Indonesia

Another important case study has been provided by Indonesian fires that polluted air over Southeast Asia during the El Niño years of 1997 and 1998. Roughly 60,000 kilometers of peat swamps, an area twice that of Belgium, dried up and burned in Indonesia during 1997 (Richardson 2002, 5). Susan Page, with Britain's University of Leicester, together with colleagues in England, Germany, and Indonesia, analyzed satellite photos and data gathered on the ground to estimate how much of the fire area's living vegetation and peat deposits burned (Cowen 2002b, 7; Page et al. 2002, 61–65).

Robert Cowen of the *Christian Science Monitor* sketched the situation:

> In Indonesia, nature and human activity had prepared a massive subsurface fuel reservoir. Tropical forests built up thick peat deposits as vegetation died and decayed over many centuries. Forest clearance and drainage for logging and farming have tended to dry the peat. Drought due to the 1997 El Niño was all that was needed to make the circumstances right for a sustained conflagration when forest-clearing fires were lit that year. (Cowen 2002b, 7)

Page and her colleagues explained the difficulty of calculating exactly how much carbon dioxide the fires emitted, but the totals were massive, especially when one adds to Indonesia's fires the many others that have burned around the globe, notably during 2002 in North America during an intense drought. The work of Page and colleagues has major implications for climate-change modeling because, as they wrote in *Nature*:

> Tropical peat lands are one of the largest near-surface reserves of terrestrial organic carbon, and hence their stability has important implications for climate change. In their natural state, lowland tropical peat lands support a luxuriant growth of peat swamp forest overlying peat deposits up to 20 meters thick. Persistent environmental change—in particular, drainage and forest clearing—threatens their stability, and makes them susceptible to fire. This was demonstrated by the occurrence of widespread fires throughout the forested peat lands of Indonesia during the 1997 El Niño event. (Page et al. 2002, 61)

In Indonesia, layers of peat as thick as 20 meters (66 feet) cover an area of about 180,000 square kilometers (112,000 square miles) in Kalimantan (Borneo), Sumatra, and Papua New Guinea (formerly Irian Jaya) (Richardson 2002, 5). Page and colleagues used satellite images of a 2.5 million hectare study area in Central Kalimantan from before and after the 1997 fires. According to their estimates, about 32 percent of the area had burned, of which peat land accounted for 91.5 percent. An estimated 0.19 to 0.23 gigatons of carbon were released to the atmosphere through peat combustion, with an additional 0.05 gigaton released from burning of the overlying vegetation. Extrapolating these estimates to Indonesia as a whole, the researchers estimated that between 0.81 and 2.57 gigatons of carbon were released to the atmosphere in 1997 as a result of burning peat and vegetation in Indonesia (Page et al. 2002, 61). According to the researchers, "This is equivalent to [between] 13 [and] 40 percent of the mean annual global carbon emissions from fossil fuels," which contributed measurably to the largest annual increase in atmospheric carbon-dioxide concentration detected since records began in 1957 (Page et al. 2002, 61).

Jack Rieley of the University of Nottingham, United Kingdom, also believes that burning peat in Borneo is a major factor in rapidly rising atmospheric carbon-dioxide levels. As farmers continue to clear the forests by burning, the bogs catch fire and release carbon for months afterward. A biologist from Borneo told *New Scientist* late in 2004 that the fires have now returned, after an earlier peak during an El Niño–provoked drought during 1998. "During October [2004], the atmosphere around Palangka Raya has been covered in thick smoke, with visibility down to 100 meters. The schools have been shut and flights cancelled," said Suwido Limin from the University of Palangka Raya in the Indonesian province of Central Kalimantan (Pearce 2004).

The fires in Indonesia had other environmental effects as well. Iron fertilization of Indian Ocean waters resulting from the massive wildfires may have played a crucial role in producing a red tide of historic proportions that severely damaged coral reefs, according to Nerilie J. Abram and colleagues, writing in *Science.* Their findings "highlight tropical wildfires as an escalating threat to coastal marine ecosystems" (Abram et al. 2003, 952).

Indonesia was only the worst of several wildfire sites during summer 1998.

> [A] ring of fire circled the globe as rainforests burned: 2,150 square miles in Central America, 1,500 square miles in Mexico in the worst drought in 70 years; 20,000 square milers in Brazil, an area about the size of Slovakia. Fires also engulfed huge tracts of Siberian Russia, Canada, Kenya, Rwanda, Tanzania, Senegal, the Congo, and Florida in the United States. (Brandenburg and Paxson 1999, 241)

Human Populations Aggravates Fire Dangers

The spread of human populations is aggravating fire dangers around the world. The fires that ravaged much of Indonesia during 1997 and 1998 were caused, in part, by drought-provoked El Niño conditions. They were intensified, however, by fires set by peasants who were hired by local speculators as they opened forest land for farming, grazing, and other forms of development. The fires were illegal under Indonesian law when they were set in protected areas—but not if they could be blamed on El Niño, a natural condition. At least 29 companies later were indicted for setting illegal fires in Indonesia's rainforests (Glantz 2003, 196–197).

Page and colleagues pointed out that, "In Indonesia, peat land fires are mostly anthropogenic, started by local (indigenous) and immigrant farmers as part of small-scale land clearance activities, and also, on a much larger scale, by private companies and government agencies as the principal tool for clearing forest before establishing crops" (Page et al. 2002, 61). During the unusually long El Niño dry season of 1997, many of these "managed" fires spread out of control, "consuming not only the surface vegetation but also the underlying peat and tree roots, contributing to the dense haze that blanketed a large part of southeast Asia and causing both severe deterioration in air quality and health problems" (Page et al. 2002, 61).

Commenting on Page's study in *Nature,* David Schimel and David Baker of the National Center for Atmospheric Research in Boulder, Colorado, noted that two other independent studies of atmospheric carbon-dioxide concentrations during that time period support the conclusion that the fires were a major contributor to atmospheric carbon-dioxide levels. Schimel and Baker explained that computer climate simulations assume that processes that emit carbon dioxide and remove it from the atmosphere operate smoothly and continuously (Cowen 2002b, 7; Schimel and Baker 2002, 29–30). Episodic events such as wildfires play havoc with such simulations.

At present, no climate modeler knows exactly how to forecast catastrophic events that occur in small areas that release carbon dioxide, which had been locked away in peat or other carbon and methane reservoirs, on a world scale. Such events "can evidently have a huge impact on the global carbon balance," Schimel and Baker wrote (Cowen 2002b, 7). During 1997, the growth rate of carbon dioxide in the atmosphere was double the usual rate, reaching its highest level on record to that time, in large part because of these peat fires. Most of the carbon injected into the atmosphere during the Indonesian fires resulted from burning of peat rather than combustion of trees (Schimel and Baker 2002, 29). Indonesia's peat is unusually dense and high in carbon content. Today's satellites can detect land-use patterns on a 50-kilometer resolution; it is estimated that a 30-meter resolution will be required to factor effects such as Indonesia's peat fires into global climate models. This is important because relatively small-scale events "can have an appreciable effect on the carbon cycle" (Schimel and Baker 2002, 29, 30).

FURTHER READING

Abram, Nerilie J., Michael K. Gagan, Malcolm T. McCulloch, John Chappell, and Wahyoe S. Hantoro. "Coral Reef Death During the 1997 Indian Ocean Dipole Linked to Indonesian Wildfires." *Science* 301 (August 15, 2003):952–955.

Brandenburg, John E., and Monica Rix Paxson. *Dead Mars, Dying Earth.* Freedom, CA: The Crossing Press, 1999.

Cowen, Robert C. "One Large, Overlooked Factor in Global Warming: Tropical Forest Fires." *Christian Science Monitor,* November 7, 2002b, 14.

Glantz, Michael H. *Climate Affairs: A Primer.* Washington, DC: Island Press, 2003.

Page, Susan E., Florian Siegert, John O. Rieley, Hans-Dieter V. Boehm, Adi Jaya, and Suwido Limin. "The Amount of Carbon Released from Peat and Forest Fires in Indonesia during 1997." *Nature* 420 (November 7, 2002):61–65.

Pearce, Fred. "Massive Peat Burn Is Speeding Climate Change." NewScientist.com, November 3, 2004. http://www.newscientist.com/news/news.jsp?id=ns99996613.

Richardson, Michael. "Indonesian Peat Fires Stoke Rise of Pollution." *International Herald-Tribune,* December 13, 2002, 5.

Schimel, David, and David Baker. "Carbon Cycle: The Wildlife Factor." *Nature* 420 (November 7, 2002): 29–30.

Struck, Doug. "In Far North, Peril and Promise: Great Forests Hold Fateful Role in Climate Change." *Washington Post,* February 22, 2007a, A-1. http://www.washingtonpost.com/wp-dyn/content/article/2007/02/21/AR2007022102095_pf.html.

"Wildfires Add Carbon to the Atmosphere." Environment News Service, December 9, 2002. http://ens-news.com/ens/dec2002/2002-12-09-09.asp.

Forests May Accelerate Warming

Planting forests in temperate regions such as the United States and Europe may not counter global warming and could actually accelerate worldwide warming. By contrast, trees planted in tropical rainforests could indeed slow global warming. Forests have low reflectivity and absorb more heat than many other surfaces, negating some of their absorption of carbon dioxide. One study indicated that worldwide replacement of existing vegetation by trees could lead to a global warming of 2.4°F (1.3°C) by the year 2100. By contrast (and contrary to many popular assumptions, as well as the Kyoto Protocol), global replacement of forests with grassland led to cooling of about 0.7°F (0.38°C) (Gibbard et al. 2005; "No Climate Benefit" 2006).

This research indicated that while tropical forests cool their surroundings by evaporating a great deal of water, northern forests warm their surroundings by absorbing sunlight while shedding relatively little moisture. One simulation covering much of the forested Northern Hemisphere (above 20 degrees latitude) resulted in a temperature rise of more than 6°F. Covering the entire planet's land mass with trees led to a more modest increase of about 2°F. ("Temperate Forests" 2005).

"Although it was previously known that trees could have an overall warming effect in the boreal forests north of 50 degrees, this is the first study to show that temperate forests could lead to net global warming," said Livermore's Seran Gibbard, lead author of the study ("No Climate Benefit" 2006).

Although the Kyoto Protocol, which took effect in February 2005, allows the offset of emissions by planting trees to increase the volume of greenhouse gases removed from the atmosphere, it does not distinguish between forests planted in temperate regions and those planted in tropical latitudes, since this research did not exist when it was negotiated.

"I like forests," said study co-author Ken Caldeira of the Carnegie Institution. "They provide good habitats for plants and animals, and tropical forests are good for climate, so we should be particularly careful to preserve them," Caldeira said. "But in terms of climate change, we should focus our efforts on things that can really make a difference, like energy efficiency and developing new sources of clean energy" ("No Climate Benefit" 2006).

Increasing Danger to Forests

Scientists from the Center for International Forestry Research (CIFOR) in Bogor said in 2008 that measures to reduce vulnerability of forests that supply many millions of people with food and shelter are required in the face of rapid climate change, as they face "an unprecedented combination of flooding, drought, wildfire, and other effects of a warming climate over at least the next 100 years" ("Climate Change Could Destroy" 2008).

"The first is to buffer ecosystems against climate-related disturbances like improving fire management to reduce the risk of uncontrolled wildfires or the control of invasive species," said Bruno Locatelli, a CIFOR scientist. In plantations, this means selecting species that are best adapted to climate change. "The second would help forests to evolve towards new states better suited to the altered climate," he said. "In this way we evolve with the changing climate rather than resist it" ("Climate Change Could Destroy" 2008). In addition to protecting indigenous species and peoples' livelihoods, the scientists pointed out that forests that die off because of climate change eject large quantities of additional greenhouse gases.

FURTHER READING

"Climate Change Could Destroy Vast Forests, Report Warns." Environment News Service, November 28, 2008. http://www.ens-newswire.com/ens/nov2008/2008-11-28-01.asp.

Gibbard, Seran, Kenneth Caldeira, Govindasamy Bala, Thomas J. Phillips, and Michael Wickett. "Climate Effects of Global Land Cover Change." *Geophysical Research Letters* 23 (December 8, 2005), L23705, doi: 10.1029/2005GL024550.

"Greening the Planet." December 8, 1998. http://climatechangedebate.com/archive/12-08_12-15_1998.txt.

"No Climate Benefit Gained by Planting Temperate Forests." Environmental News Service, December 12, 2006. http://www.ens-newswire.com.

"Temperate Forests Could Worsen Global Warming." Press Release, Carnegie Institution, December 6, 2005.

Fossil Fuels

For the past two centuries, at an accelerating rate, the basic composition of the Earth's atmosphere has been materially altered by the fossil-fuel effluvia of machine culture. Human-induced warming of the Earth's climate is emerging as one of the major scientific, social, and economic issues of the twenty-first century, as the effects of climate change become evident in everyday life in locations as varied as small island nations of the Pacific Ocean and the shores of the Arctic Ocean.

Consumption of fossil fuels has been accelerating as human numbers rise around the world, and as countries with large populations (such as China and India) build industry, highways, and electrical-generating infrastructure. Half the fossil-fuel energy consumed since the start of the industrial age has been during the last 20 years (Flannery 2005, 167). Even as awareness of global warming's perils became more apparent, fossil-fuel use continued to increase. The world's inhabitants consumed about 66.6 million barrels of oil a day in 1990, and 83 million in 2007; demands for oil in the United States grew 22 percent during that period. During the same time, demand in China grew almost 200 percent (Friedman 2007, A-27).

The carbon-dioxide content of Earth's atmosphere continued to race upward in 2007 at a rate of 2.9 percent, faster than any of the experts' worst predictions, to a total of 8.47 gigatons, or billions of metric tons, according to the Australia-based Global Carbon Project, an international consortium of scientists that tracks emissions. This output is at the very high end of scenarios outlined by the Intergovernmental Panel on Climate Change (IPCC) and could translate into a global temperature rise of more than 11°F by the end of the century, according to the panel's estimates (Eilperin 2008c). Major contributors included China, India, and Brazil, which have doubled their carbon emissions in less than 20 years. Total carbon emissions from industrial nations as a whole have risen only slightly since 1990.

During September 2008, two scientists with the Scripps Institution of Oceanography at the University of California at San Diego published research in the *Proceedings of the National Academy of Sciences.* Their results indicated that if greenhouse-gas emissions had stopped completely as of 2005, the world's average temperature still would increase by 2.4°C (4.3°F) by the end of the twenty-first century. Richard Moss, vice president and managing director for climate change at the World Wildlife Fund, said the new carbon figures and research show that "we're already locked into more warming than we thought" (Eilperin 2008c; Ramanathan and Feng 2008, 14,245–14,250).

Carbon Emissions: Part of Everyday Life

The emission of carbon dioxide and other greenhouse gases is built into our everyday lives—our modes of transportation, production, and consumption—to the extent that they are enabled by the combustion of fossil fuels. Roughly 80 percent of human industrial activity on Earth is fueled by energy that produces carbon dioxide (and, oftentimes, other greenhouse gases as well) when it is burned. Industrial processes produce waste heat in addition to greenhouse gases, and sometimes, manufactured goods (such as automobiles) also produce waste heat and greenhouse gases as they are operated.

The fossil fuels being burned to power today's machine culture were created by natural processes during the past 600 million years. The geologic record suggests that the richest deposits came into being during times when the Earth was much warmer, and its atmosphere was much richer in carbon dioxide than today. Carbon-dioxide levels were as much as seven times higher than today during the epoch of the dinosaurs, 65 million to 200 million years ago. (Schneider 1989, 21).

During much of this period, the Earth had no permanent ice at all, not even at the poles. The temperature of the Earth also was markedly warmer, and sea levels were about 300 feet above those of the present day. Stephen H. Schneider estimated that if today's Antarctic ice cap were to melt, world sea levels would rise by about 230 feet (Schneider 1989, 162–163). During the glaciations that have characterized the last several million years, carbon-dioxide levels have risen and fallen with the temperature. At the end of the last great ice age, more than 10,000 years ago, carbon-dioxide levels were about 60 percent of today's levels. Such a level, slightly under 200 parts per million (ppm), is close to the lower limit for successful photosynthesis in many green plants.

Ecologist Barry Commoner wrote:

> The amounts of these fuels burned to provide society with energy represent the carbon captured by photosynthesis over millions of years. So, by burning them ... we have returned carbon dioxide to the atmosphere thousands of times faster than the rate at which it was removed by the early tropical forests. (Commoner 1990, 6)

"We are re-injecting our fossil carbon legacy into the atmosphere at an incredibly accelerated pace," wrote Schneider. According to Schneider, human activity is changing the temperature of the atmosphere at between 10 and 100 times as quickly as natural processes at the end of the last ice age (Schneider 1989, 28).

> CO_2 is dumped into the atmosphere at a much faster rate than it can be withdrawn or absorbed by the oceans or living things in the biosphere. CO_2 buildup in the next few decades to centuries could well be one of the principal controlling factors of the near-future climate. (Schneider 1989, 21).

The Abundance and Convenience of Fossil Fuels

In *The Economics of Global Warming* (1992), William R. Cline estimated that 3,500 to 7,000 gigatons of coal could be mined and burned "before the market mechanism chokes off further atmospheric buildup through high prices caused by scarcity of a severely depleted resource" (Cline 1992, 46). Because the total stock of carbon in the atmosphere as of 1990 was roughly 750 gigatons, the market economy is prepared, if unrestrained, to increase carbon-dioxide levels in the atmosphere five- to tenfold before scarcity becomes a factor in the price of coal and other fossil fuels, according to Cline's analysis. There is some climatic irony in the fact that such a level of greenhouse gases in the atmosphere would take the Earth back, roughly, to the atmospheric days of the dinosaurs, a time when temperatures were much higher than today, when the planet had no permanent pack ice, and the dominant animal species (including most of the dinosaurs) were cold blooded.

Following spot oil shortages of the 1970s, the petroleum industry improved its extractive technology to the point at which the supply of oil (and natural gas, often a by-product of oil drilling) will last for centuries, even at increasing rates of consumption. Offshore oil drilling, for example, was restricted to roughly 500 feet below the water's surface during the 1960s. By the mid-1990s, the limit had reached 10,000 feet. As one commentator has written: "'Will we run out of gas?'—a question we began asking during the oil shocks of the 1970s—is now the wrong question. The Earth's supply of carbon-based fuels will last a long time. But if humans burn anywhere near that much carbon, we'll burn up the planet" (Hertsgaard 1999, 111).

If the Earth's entire storehouse of fossil fuels that could be harvested at a cost that the market will bear is burned to provide energy in the fashion of the twentieth century, carbon-dioxide levels in the atmosphere would reach roughly 2800 ppm, or to times the preindustrial level, according to another estimate.

According to Paul Epstein, Harvard University epidemiologist, industrial society's dependence on fossil fuels creates many problems in addition to rising levels of greenhouse gases.

> Extraction damages terrestrial and offshore ecosystems. Drilling in Ecuador, for example, harms the forest and the headwaters to the Amazon. Transport

and the accompanying leaks and spills have enormous impacts on coastal ecosystems. The benzene by-products of refining and impacts on surrounding communities is not healthy. Finally, combustion has multiple hazards. Nitrates in acid precipitation harm forest and lake stocks. Sulfates and hydrocarbons contribute to lung disease, heart disease, and cancer. And now the aggregate of burning fossil fuels and cutting forests (a chief carbon sink) may be destabilizing the climate system itself. Thus, from a health perspective, ending our civilization's addiction to fossil fuels is a healthy "no-regrets" policy. Because they are finite, it is also a necessary step. The only question is, will we change soon enough to reduce the enormous and mounting [climatic] costs of not changing from "business as usual?" (Epstein et al. 1996)

FURTHER READING

Cline, William R. *The Economics of Global Warming.* Washington, DC: Institute for International Economics, 1992.

Commoner, Barry. *Making Peace with the Planet.* New York: Pantheon, 1990.

Eilperin, Juliet. "Carbon Is Building Up in Atmosphere Faster Than Predicted." *Washington Post,* September 26, 2008c, A-2. http://www.washingtonpost.com/wp-dyn/content/article/2008/09/25/AR2008092503989_pf.html.

Epstein, Paul, Georg Grabbher, Tom Karl, Ellen Mosley-Thompson, Kevin Trenberth, and George M. Woodwell. *Current Effects: Global Climate Change. An Ozone Action Roundtable.* Washington, D.C., June 24, 1996. http://www.ozone.org/curreff.html.

Flannery, Tim. *The Weather Makers: How Man Is Changing the Climate and What It Means for Life on Earth.* New York: Atlantic Monthly Press, 2005.

Friedman, Thomas L. "Doha and Dalian." *New York Times,* September 19, 2007, A-27.

Hertsgaard, Mark. "Will We Run Out of Gas? No, We'll Have Plenty of Carbon-based Fuel to See Us through the Next Century. That's the Problem." *TIME,* November 5, 1999, 110–111.

Ramanathan, V. and Y. Feng. "On Avoiding Dangerous Anthropogenic Interference with the Climate System: Formidable Challenges Ahead." *Proceedings of the National Academy of Sciences* 105, no. 38 (September 23, 2008):14,245–14,250.

Schneider, Stephen H. *Global Warming: Are We Entering the Greenhouse Century?* San Francisco: Sierra Club Books, 1989.

Schneider, Stephen H. "No Therapy for the Earth: When Personal Denial Goes Global." In *Nature, Environment, and Me: Explorations of Self in a Deteriorating World,* ed. Michael Aleksiuk and Thomas Nelson. Montreal: McGill-Queens University Press, 2000a.

G

Gaia: The Eradication of Industrial Civilization?

The German philosopher Friedrich Nietzsche once said, "The Earth is a beautiful place, but it has a pox called man" (Browne 1998). Bill McKibben's thesis, in *The End of Nature* (1989) is similar: "We have built a greenhouse, a human creation, where once bloomed a sweet and wild garden" (McKibben 1989, 91). "If industrial civilization is ending nature, it is not utter silliness to talk about ending—or, at least, transforming—industrial civilization," wrote McKibben (1989, 186). The "inertia of affluence, the push of poverty, [and] the soaring population" make McKibben pessimistic about humankind's chances of averting a general ruination of the Earth under greenhouse conditions (McKibben 1989, 204).

John Gribben's *Hothouse Earth* (1990), one of several popular treatments of global warming that appeared shortly after the notably hot summer of 1988, begins with a tribute to the philosophy of the Earth as *Gaia*, a living organism that will restore its ecological balance, even if restoring that balance requires extinguishing the human race. At the beginning of the book, Gribben quotes James Lovelock: "People sometimes have the attitude that '*Gaia* will look after us.' But that's wrong. If the concept means anything at all, Gaia will look after *herself*. And the best way for her to do that might well be to get rid of us." (Gribben 1990, frontispiece).

Gribben believes that what humankind is doing to the Earth's atmosphere will not be undone by human hands. His book ends with a profession that humankind will be unable to forestall the greenhouse effect by its own devices: "There is no prospect at all of bringing a halt to the release of carbon dioxide and other greenhouse gases, and thereby allowing the carbon cycles and the temperature of Gaia to return to normal" (Gribben 1990, 263).

Given rising populations, increasing affluence, and rising levels of greenhouse gases in the atmosphere, it is difficult, early in the twenty-first century, to conceive how the human race will be able to effectively achieve the technological and ideological paradigm shifts necessary to decouple human prosperity from the burning of fossil fuels. It is more difficult to imagine how our squabbling collection of nations and peoples will be able to respond as required to stop (much less reverse) an acceleration toward a warmer, more humid, and more miserable world. That said, a lot of authors have made retrospective fools of themselves by trying to forecast the future, so we can trust in our own fallibility and hope, for the sake of the seventh generation, that our pessimism is at least partially in error.

FURTHER READING

Gribben, John. *Hothouse Earth: The Greenhouse Effect and Gaia.* London: Bantam Press, 1990.

McKibben, Bill. *The End of Nature.* New York: Random House, 1989.

Gelada Baboon

Human encroachment and warming temperatures in the highlands of Ethiopia are eliminating the habitat of the gelada baboon. Primate expert Chadden Hunter said,

> Our research suggests that for each 2°C (3.6°F) degree increase in temperature, the gelada's lower limit for grazing will rise 500 meters (547 yards).... Global temperatures only need to rise a few degrees, and [their habitat] will be, in effect, lifted off the tops of the mountains." (Smucker 2001, 7)

A Swiss study of the Ethiopian highlands showed that the specific grasses on which the baboons feed have receded steadily upward during the past several hundred years. A separate study commissioned by the World Wide Fund for Nature cited the gelada baboon recently as one of several mammals most at risk from the effects of global warming.

American and Ethiopian primate experts agreed that temperature shifts may soon destroy the high-altitude "islands of grass" on which the gelada survives. "Since the gelada survive at the physical limits of the landscape, the likelihood of future global warming raises serious questions over the species' survival," said Jacinta Beehner, an American baboon expert with long experience in Ethiopia (Smucker 2001, 7).

Theropithecus gelada is the last remaining species of its type. Hunter believes that the graminivorous (grass-eating) monkeys may have lost most of their close relatives to past episodes of climate change. "They are the last relic of a great dynasty, barely surviving in the Ethiopian highlands," he said (Smucker 2001, 7). The gelada, which scamper and leap through the highlands in "herds" of 600 or more, maintain a complex social structure that offers "irreplaceable insight into human social behavior," said Hunter (Smucker 2001, 7).

While warming may provoke the baboons' demise within a century, "[t]he most immediate threat to the gelada's survival comes from farming," said Tesfaye Hundessa, manager of the Ethiopian Wildlife Conservation Organization. "They could well be entirely eliminated by humans before the global warming has time to take its toll over the coming decades" (Smucker 2001, 7).

FURTHER READING

Smucker, Philip. "Global Warming Sends Troops of Baboons on the Run: Rising Temperatures and Humans Encroaching on Grasslands Are Endangering the Ethiopian Primates." *Christian Science Monitor,* June 15, 2001, 7.

Geoengineering: Sulfur as Savior?

A tone of desperation is palpable in climate-change science when well-known people seriously propose that filling the stratosphere with sulfur dioxide may be the only way to stop runaway greenhouse warming. Do we really want to pump the stratosphere full of sulfur to shroud the surface from warmth, then live in a perpetual acid mist? Paul J. Crutzen of the Max Planck Institute for Chemistry in Germany, who won a Nobel Prize in 1995 for showing how industrial gases damage the Earth's ozone shield, advanced the idea anew in 2006, citing the "grossly disappointing international political response" to increasing evidence of global warming (Kerr 2006d, 401).

"We should treat these ideas like any other research and get into the mind-set of taking them seriously," said Ralph J. Cicerone, president of the National Academy of Sciences (Broad 2006). Most of these proposals involve geoengineering, large-scale rearranging of the Earth's environment to suit human needs. This idea "should not be taken as a license to go out and pollute," Cicerone said, emphasizing that most scientists believe that reducing greenhouse gases in the atmosphere should be the top priority. He added, however, that "[i]n my opinion, he [Crutzen] has written a brilliant paper" (Broad 2006). Wallace S. Broecker, a geoengineering pioneer at Columbia University, also has proposed spreading tons of sulfur dioxide into the stratosphere, as erupting volcanoes occasionally do. The injections, he calculated in the 1980s, would require a fleet of hundreds of jumbo jets and, as a by-product, would increase acid rain (Broad 2006).

A special issue of the scientific journal *Climatic Change* was devoted to this subject in August 2006. In this issue, Crutzen discussed the theoretical basis of the idea, possible methodologies, and advantages and disadvantages. Several other authors also discussed the history of such proposals, the practical as well as ethical considerations of various approaches, and how best to evaluate them. The authors made it clear that geoengineering climate is a less desirable potential solution to warming than controlling greenhouse emissions, and that only if warming causes sufficiently harmful impacts would geoengineering be the better choice (Editors' Choice 2006, 387).

In a draft of his paper, Crutzen estimated the annual cost of his sulfur proposal at up to $50 billion, or about 5 percent of the world's annual military spending. "Climatic engineering, such as presented here, is the only option available to rapidly reduce temperature rises" if international efforts fail to curb greenhouse gases, Crutzen wrote. "So far," he added, "there is little reason to be optimistic" (Broad 2006). Supporters of this idea contend that any increase in sulfur at the Earth's surface would be small compared with the tons already being emitted from the smokestacks of coal-fueled plants (Broad 2006).

Another proposed solution along the same lines involves the burning of sulfur in ships and power plants to form sulfate aerosols. Mikhail Budyko, a Russian climatologist, has proposed a massive atmospheric infusion of sulfur that would form enough sulfur dioxide to wrap the Earth in radiation-deflecting thin, white clouds within a few months. The net effect, according to Budyko, would be to cool the Earth in a fashion similar to the massive eruption of the volcano Tambora in 1815. The eruption of Tambora ejected enough sulfur into the air to produce, in 1816, an annual cycle known to climate historians as "the year without a summer." Crops across New England and Upstate New York were devastated by frosts that continued into the summer. Mohawk American Indians at Akwesasne, in far northern New York State, reported frosts into June.

The Atmospheric Chemistry of Sulfur

Suspended particulates caused by emissions of sulfur dioxide and some other urban air pollutants (aerosols) increase the net albedo of the Earth, thus usually exerting a cooling influence on planetary temperatures. James E. Hansen has estimated that aerosols cool the climate by about one watt per square meter, "which has substantially offset greenhouse warming" (Hansen, Sato, and Ruedy 1997, 231).

Most sulfur compounds leave the atmosphere within two weeks of their generation, however, while carbon dioxide remains in the air for a century or more. To negate the effects of global warming, the "pollution solution" would require a continuous feed of sulfur into the atmosphere. The resulting precipitation of sulfur-enhanced acidity could have disastrous environmental effects at the ground level.

Mount Pinatubo erupted in 2001, ejecting about 10 million tons of sulfur into the atmosphere, enough to cool the Earth's near-surface atmosphere about 0.5° for a year or two, or roughly the increase attributable to global warming during the previous century. The physical challenge of lifting that much sulfur into the atmosphere year after year would be quite a challenge, especially as increases in greenhouse-gas levels over time require more of it.

Particles Do Not Always Cool the Atmosphere

An increase in atmospheric particulate matter does not always exert a cooling influence, according top some reports. Researchers working with the National Oceanic and Atmospheric Administration (NOAA) have assembled data indicating that periodic increases in atmospheric dust concentrations during the glacial periods of the last 100,000 years may have resulted in significant regional warming, and that this warming may have triggered some of the abrupt climatic changes observed in paleoclimate records.

Jonathan T. Overpeck, working with the Paleoclimatology Program at NOAA's National Geophysical Data Center in Boulder, Colorado, led a team of scientists who, during 1996, conducted global climate model simulations to examine the potential role of tropospheric dust in glacial climates. Comparing "modern dust" with "glacial dust" conditions, they found patterns of regional warming that increased at progressively higher latitudes. The warming was greatest (up to 4.4°C) in regions with dust over areas that were covered with snow and ice ("Abrupt Climate Change" 1996). Under some circumstances "aerosols can reduce cloud cover and thus significantly offset aerosol-induced radiative cooling at the top of the atmosphere on a regional scale" (Ackerman et al. 2000, 1042). Simply put, soot and other such pollution may not mitigate other warming influences on the climate as much as many skeptics claim.

Problems Arising

Aside from creating atmospheric conditions amenable for acid rain, filling the upper atmosphere with sulfur also may deplete stratospheric ozone and reduce overall precipitation, most notably during the African and Indian monsoons, which are crucial for hundreds of millions of subsistence farmers. The "sulfur solution" would reduce pressure to reduce greenhouse emissions, and if it ever stopped, warming would accelerate rapidly (Robock 2008, 1166).

Some scientists object to such a scheme on grounds that it would remove pressure to deal with the problem at its source—that is, to use energy sources other than fossil fuels. "I refuse to go down that road," said biochemist Meinrat Andreae of the Max Planck Institute for Chemistry in Mainz, Germany. "You're papering over the problem so people can keep inflicting damage on the climate system" (Kerr 2006d, 403). "The biggest risk of geo-engineering is that it eliminates pressure to decrease greenhouse gases," agreed Kenneth Caldeira of the Carnegie

Institution Department of Global Ecology at Stanford University (Kerr 2006d). Others argue that such a system could make the Earth dependent on a continuing human-provided sulfur "fix.—Should the supply of sulfur falter, the Earth could heat up quickly within a few years. Some effects of increasing carbon-dioxide levels, such as acidification of the oceans, would not be modified by the sulfur haze.

Other problems quickly arise with the idea of intentionally polluting the atmosphere to counter surface warming. According to one analysis by Tom Wigley, of the National Center for Atmospheric Research, Boulder, Colorado:

> Projected anthropogenic warming and increases in CO_2 concentration present a two-fold threat, both from climate changes and from CO_2 directly through increasing the acidity of the oceans. Future climate change may be reduced through mitigation (reductions in greenhouse gas emissions) or through geo-engineering. Most geo-engineering approaches, however, do not address the problem of increasing ocean acidity. A combined mitigation/geo-engineering strategy could remove this deficiency. Here we consider the deliberate injection of sulfate aerosol precursors into the stratosphere. This action could substantially offset future warming and provide additional time to reduce human dependence on fossil fuels and stabilize CO_2 concentrations cost-effectively at an acceptable level. (Wigley 2006, 452)

The sulfur "solution" is not new. It is a perennial favorite of contrarians and was raised in a 1991 National Academy of Sciences report on mitigation strategies that concluded with consideration of atmospheric modification. These modifications include the placement of mirrored platforms in orbit to reflect sunlight, the use of guns or balloons to add dust to the stratosphere (reflecting sunlight), and the placement of billions of aluminized, hydrogen-filled balloons in the stratosphere, also to reflect incoming solar radiation.

Additional strategies include the use of aircraft to maintain a dust cloud between Earth and sun by making their engines less efficient—in other words, intentional air pollution. Roger P. Angel, an astronomer at the University of Arizona, has proposed a plan to put into orbit small lenses that would bend sunlight away from Earth—trillions of lenses, he now calculates, each about two feet wide, extraordinarily thin and weighing little more than a butterfly (Broad 2006).

Such atmospheric modification probably will remain an intellectual parlor game because the environmental costs of filling the stratosphere with sulfur dioxide (or other pollutants) far outweigh the benefits, even in a warmer and more humid world. The sulfur dioxide would have to be refreshed at least twice a month, as the previous load washed onto the Earth, planetwide, as acid rain. Sulfuric acid also tends to attract chlorine atoms, creating a chemical combination that could assist chlorofluorocarbons (CFCs) in devouring stratospheric ozone.

Some scientists believe that the injection of sulfur into the atmosphere to counteract global warming may threaten the stratospheric ozone shield that protects the Earth and its flora and fauna from ultraviolet radiation. Their model is the sulfur load ejected in 1991 by the eruption of Mount Pinatubo. "We use an empirical relationship between ozone depletion and chlorine activation to estimate how this approach might influence polar ozone," Simone Tilmes and colleagues wrote in *Science.* They continued:

> An injection of sulfur large enough to compensate for surface warming caused by the doubling of atmospheric CO_2 would strongly increase the extent of Arctic ozone depletion during the present century for cold winters and would cause a considerable delay, between 30 and 70 years, in the expected recovery of the Antarctic ozone hole. (Tilmes, Müller, and Salawitch 2008, 1201)

Adding sulfur to the stratosphere would lower temperatures there, promoting the formation of polar stratospheric clouds that provide the surface area to create the chlorine that destroys ozone.

FURTHER READING

"Abrupt Climate Change during Last Glacial Period Could Be Tied to Dust-Induced Global Warming." Press Release, NOAA 96-78, December 4, 1996. http://www.noaa.gov/public-affairs/pr96/dec96/noaa96-78.html.

Ackerman, A. S., O. B. Toon, D. E. Stevens, A. J. Heymsfield, V. Ramanathan, and E. J. Welton. "Reduction of Tropical Cloudiness by Soot." *Science* 288 (May 12, 2000):1042–1047.

Broad, William J. "How to Cool a Planet (Maybe)." *New York Times,* June 27, 2006. http://www.nytimes.com/2006/06/27/science/earth/27cool.html.

Cline, William R. *The Economics of Global Warming.* Washington, DC: Institute for International Economics, 1992.

Editors' Choice. *Science* 314 (October 20, 2006):387.
Hansen, J., M. Sato, and R. Ruedy. "The Missing Climate Forcing." *Philosophical Transactions of the Royal Society of London* B352 (1997):231–240.
Kerr, Richard A. "Pollute the Planet for Climate's Sake? *Science* 314 (October 20, 2006d):401–403.
Robock, Alan. "Whither Geoengineering?" *Science* 320 (May 30, 2008):1166–1167.
Tilmes, Simone, Rolf Müller, and Ross Salawitch. "The Sensitivity of Polar Ozone Depletion to Proposed Geoengineering Schemes." *Science* 320 (May 30, 2008):1201–1204.
Wigley, T. M. L. "A Combined Mitigation/Geoengineering Approach to Climate Stabilization." *Science* 314 (October 20, 2006):452–454.

Geothermal Energy (Iceland and the Philippines)

In Iceland, 90 percent of the country's 290,000 people used geothermal energy to heat their homes in 2007, up from 85 percent in 2003 (Brown 2003, 166). Nearly all new buildings there are constructed with built-in geothermal sources. Iceland's government, working with Shell and Daimler-Chrysler, in 2003 began to convert Reykjavik's city buses from internal combustion to fuel-cell engines, using hydroelectricity to electrolyze water and produce hydrogen.

The next stage is to convert the country's automobiles, then its fishing fleet. These conversions are part of a systemic plan to divorce Iceland's economy from fossil fuels (Brown 2003, 168). By 2006, only 0.1 percent of Iceland's electricity came from fossil fuels, even as the country attracted energy-intensive industries (such as aluminum smelting) drawn by low-cost electricity from renewable sources. Within one generation, Iceland transformed its energy system from a fossil-fuel base to one in which 70 percent of all energy (including transport) comes from renewable sources

Geothermal power by 2008 accounted for about 28 percent of the electricity generated in the Philippines, mainly thanks to program started during an oil shortage in the 1970s. The country has become the world's largest consumer of electricity from geothermal sources, saving billions of dollars on imported oil and coal. The Island of Luzon, especially, has large geothermal resources, such as the Leyte field: "one of nature's most perfectly designed geothermal resources. Located about 1.5 miles underground, it is a great pot of boiling water that covers about 416 square miles" (Harden 2008).

FURTHER READING

Brown, Lester R. *Plan B: Rescuing a Planet under Stress and a Civilization in Trouble.* New York: Earth Policy Institute/W.W. Norton, 2003.
Harden, Blaine. "Filipinos Draw Power from Buried Heat." *Washington Post,* October 4, 2008, A-1. http://www.washingtonpost.com/wp-dyn/content/article/2008/10/03/AR2008100303843_pf.html.

Giant Squid

One species (in addition to jellyfish) that seems to benefit from a warming environment is the giant squid. Squids have reproduced with such success, increasing their physical size and numbers to a point at which, by the end of the twentieth century, on a worldwide scale, they had overtaken humans in terms of total planet-wide biomass, according to some estimates. Human overfishing of other species has favored the squid (reducing competition for food), along with a warmer habitat. Australian scientist George Jackson said he believed that global warming is causing squid to grow larger ("Giant Squid" 2002). While squid may thrive in warmer waters, they face another problem: rising carbon-dioxide levels in the oceans affect their respiration rates, inhibiting their ability to swim.

A report in the Australian science journal *Australasian Science* said that most marine researchers are now in agreement that warming habitats have given cephalopods an advantage not available to any other large sea creature. As a result, they have flourished. Their numbers have been increasing, along with their body size (Benson 2002, 4). In July 2002, according to a report by Agence France Presse, the body of a giant squid weighing about 200 kilograms (440 pounds) washed up on a beach in the Australian state of Tasmania. Within days of that event, hundreds of dead squid washed ashore on the coast of California. George Jackson, with the Institute of Antarctic and Southern Ocean studies in Tasmania, said:

> [S]quid thrived during environmental disasters such as global warming. The animal ate anything in that came their way, bred whenever possible and kept growing.... This trend has been suggested to be due both to the removal of cephalopod predators such as toothed whales and tuna and an increase of cephalopods due to removal of finfish competitors.... The fascinating thing about squid is

that they're short-lived. I haven't found any tropical squid in Australia older than 200 days....

Many of the species have exponential growth, particularly during the juvenile stage so if you increase the water temperature by even a degree it has a tremendous snowballing effect of rapidly increasing their growth rate and their ultimate body size. They get much bigger and they can mature earlier and it just accelerates everything. (Benson 2002, 4)

FURTHER READING

Benson, Simon. "Giant Squid 'Taking over the World.'" *Sydney Daily Telegraph*, July 31, 2002, 4.

"Giant Squid Film Team Makes Spectacular Catch." Agence France Presse, September 14, 2002.

Glacial (Ice Age) Cycle, Prospective End of

Given that we live during an interglacial period, a relatively warm interval between ice ages, absent humanity's contribution to warming climate, the Earth could be expected to eventually cool off and head into a new ice age. This natural cycle, which relied on cyclical variations in atmospheric greenhouse-gas levels, may now have been obliterated by human-made greenhouse gases. A significant fraction of human-made carbon-dioxide emissions will remain in the atmosphere for thousands of years. "Thus," wrote James E. Hansen, director of the NASA Goddard Institute for Space Studies, "[h]umans now control global climate, for better or worse" (Hansen 2006d, 9). Human contributions to greenhouse gases in the atmosphere, if maintained long enough, could interrupt or annul the millions of years of ice age cycles.

Paleoclimatic Precedents

Anthony D. Socci has surveyed epochs of natural cooling and warming in the Earth's atmosphere, with reference to levels of carbon dioxide in the atmosphere roughly 700 million years into the past, during which time he found that the Earth has, at times, been both markedly warmer and colder than today. The range of temperature has remained mainly 15°C warmer or colder than today's levels. According to Socci, "The duration of anthropogenic [human-caused], greenhouse warming may likely exceed 300,000 years, conceivably pre-empting at least three ... global cooling events normally characterized by thickening and spread of continental glaciers" (Socci 1993, 170)

Socci believes that human-induced warming of the Earth's atmosphere is short-circuiting the glacial cycles that have dominated Earth's climate for the last several-score million years (Socci 1993, 170–171). "In other words," Socci commented, "the anthropogenic contribution of carbon dioxide to the atmosphere is in excess of the volume of carbon dioxide which nature calls upon to go from one extreme of climate to the other" (Socci 1993, 171). In raising the atmospheric level of carbon dioxide to about 385 parts per million (ppm), by 2008, from the preindustrial level of 280 ppm in one century, humankind has broken an ages-old natural cycle.

During recent decades, the study of ice cores has fundamentally changed many of science's assumptions about paleoclimate, the climate of the past. Analyzing air bubbles in ice cores of various ages, scientists can "read" a number of things, including the levels of greenhouse trace gases (notably carbon dioxide and methane) during past climatic cycles. Early ice-core work established a relationship between fluctuations in greenhouse-gas levels and temperatures. Because of ice-core records, paleoclimatologists now know that climate can change abruptly as the cycle shifts.

By 1999, a team of scientists led by J. R. Petit, drilling ice cores at Vostok, Antarctica, reached a depth of 3,623 meters and, in so doing, were able to describe the composition of the Earth's atmosphere during the last four glacial cycles, or about 420,000 years. These readings later were extended to more than 800,000 years. The carbon-dioxide level during that period varied in a range of 180 parts per million to 300 parts per million, in a cycle that roughly parallels rises and falls in temperature. The atmospheric level of methane during the same period varied between 320 and 770 parts per billion, in a similar pattern—low during glacial maxima and high during interglacial warm periods. By the year 2000, greenhouse-gas levels (about 385 ppm for carbon dioxide, and 1,700 parts per billion for methane) were much higher than any level reached for as long as humankind has proxy records from ice cores.

The Vostok ice-core investigators have not declared that human-induced warming is short-circuiting the glacial cycle. Their words are more circumspect:

> A striking feature of the Vostok ... record is that the Holocene [the most recent interglacial period],

which has already lasted 11,000 years is, by far, the longest stable warm period recorded in Antarctica during the last 420,000 years.... CO_2 and CH_4 concentrations are strongly correlated with Antarctic temperatures; this is because, overall, our results support the idea that greenhouse gases have contributed significantly to the glacial-interglacial change. This correlation, together with the uniquely elevated concentrations of these gases today, is of relevance with respect to the continuing debate on the future of the Earth's climate. (Petit et al. 1999, 434, 435)

Onset of the Next Ice Age Delayed

Is it possible that ongoing global warming could delay the onset of the next ice age by thousands of years? Belgian researchers raised this issue in the August 23, 2002, issue of *Science.* "We've shown that the input of greenhouse gas could have an impact on the climate 50,000 years in the future," said Marie-France Loutre of the Universite Catholique de Louvain in Belgium, who researched the question with colleague Andre Barger (Berger and Loutre 2002, 1287; Flam 2002).

Princeton climatologist Jorge Sarmiento said that his own work supports Loutre's assertion that increasing levels of carbon dioxide could linger for thousands of years, long enough to influence the climate of the far future. "The warming will certainly launch us into a new interval in terms of climate, far outside what we've seen before," said Duke University climatologist Tom Crowley. He said it's a big enough influence to cause the cycle of ice ages to "skip a beat" (Flam 2002).

Loutre and Berger estimated that human activity will double the concentration of carbon dioxide in the atmosphere over the next century, raising temperatures as much as 10°F. Still, "It could get much worse," said Crowley. "There's a huge reservoir of coal and if people keep burning it, they could more than quadruple the present carbon dioxide concentrations," he said. "I find it hard to believe we will restrain ourselves," he added. "It's really rather startling the changes that people will probably see" (Flam 2002).

"The silliest thing people could say is: We've got an ice age coming, so why are we worrying about global warming?" Sarmiento said. Whether Loutre and Berger's theory is right or not, "[w]e're going to get a lot of global warming before the ice age kicks in" (Flam 2002).

It is possible, over the next several human generations, writes Doug Macdougall in *Frozen Earth: The Once and Future Story of Ice Ages* (2004), that

Warming would be reinforced by the loss of highly reflective ice and snow, and possibly by the decomposition of unstable methane hydrates. The elevated temperatures coupled with the complete loss of continental ice sheets might constitute a threshold-crossing event that would thrust the Earth into a regime from which glaciers could not quickly recover, even with the return of greater [orbital] eccentricity and lower CO_2 levels. Only a few hundred years after Louis Agassiz announced his theory of a global ice age, mankind may inadvertently bring the Pleistocene Ice Age to a premature close, ushering in another long period of ice-free existence for our planet. (Macdougall 2004, 244)

FURTHER READING

Berger, Andre, and Marie-France Loutre. "Climate: An Exceptionally Long Interglacial Ahead?" *Science* 297(August 23, 2002):1287-1288.

Flam, Faye. "It's Hot Now, but Scientists Predict There's an Ice Age Coming." *Philadelphia Inquirer,* August 23, 2002, n.p. (LEXIS)

Hansen, James E. "Declaration of James E. Hansen." *Green Mountain Chrysler-Plymouth-Dodge-Jeep, et al., Plaintiffs v. Thomas W. Torti, Secretary of the Vermont Agency of Natural Resources, et al., Defendants.* Case Nos. 2:05-CV-302 and 2:05-CV-304, Consolidated. United States District Court for the District of Vermont. August 14, 2006d. http://www.giss.nasa.gov/~dcain/recent_papers_proofs/vermont_14aug20061_textwfigs.pdf.

Macdougall, Doug. *Frozen Earth: The Once and Future Story of Ice Ages.* Berkeley: University of California Press, 2004.

Petit, J. R., J. Jouzel, D. Raynaud, N. I. Barkov, J.-M. Barnola, I. Basile, M. Benders, I. Chappellaz, M. Davis, G. Delaygue, M. Delmotte, V. M. Kotlyakov, M. Legrand, V. Y. Lipenkov, C. Lorius, L. Pepin, C. Ritz, E. Saltzman, and M. Stievenard. "Climate and Atmospheric History of the Past 420,000 Years from the Vostok Ice Core, Antarctica." *Nature* 399 (June 3, 1999):429–436.

Socci, Anthony D. "The Climate Continuum: An Overview of Global Warming and Cooling Throughout the History of Planet Earth." In *A Global Warming Forum: Scientific, Economic, and legal Overview,* ed. Richard A. Geyer, 161–207. Boca Raton, Florida: CRC Press, 1993.

Thompson, Lonnie G., Ellen Mosley-Thompson, Henry Brecher, Mary Davis, Blanca León, Don Les, Ping-Nan Lin, Tracy Mashiotta, and Keith Mountain. "Abrupt Tropical Climate Change: Past and Present." *Proceedings of the National Academy of Sciences* 103, no. 28 (July 11, 2006):10,536–10,543.

Glacial Retreat: Comparative Photographs and Survey

Several research centers around the world have been amassing images taken by generations of glaciologists to build a picture of frozen places as the world warms. One archive of before-and-after photographs of Alaskan glaciers has been published by the National Snow and Ice Data Center in Boulder, Colorado, at nsidc.org/data/glacier_photo.

Amanda Phelan of the Sydney *Sunday Telegraph* described a unique European project that uses old photographs, compared with present-day vistas, to document how climate change is affecting glaciers:

> Two photographs, taken only 84 years apart, show how the Blomstrandbreen glacier on a remote island off Norway was reduced from a towering force, dominating the skyline, to little more than a wall of ice rubble. Less than a century on, photographs show the glacier on Svalbard, an island 603 kilometers off Norway, is a pitiful slope—proof that global warming is a threat to our planet, says Greenpeace, which took the second image this year. (Phelan 2002, 47)

German geologists, in association with Greenpeace, used old picture postcards to measure the retreat of glaciers in Europe between 1850 and 2000. A Munich-based team of geological climatologists rummaged through antique shops, markets, state archives, and university libraries for old photos and postcards, collecting 2,500 items. They then hiked to the edges of the glaciers on the photographs and measured changes. They found that in a century and a half, "the giant ice fields shrank about a third in area and lost about half of their volume. In the last 25 years, they have melted even more quickly, losing an additional 20 to 30 percent of their water content" (Hall 2002, 8).

The geologists spent several months hiking through the Alps, sometimes with local climbers. They studied elevation maps, examined satellite images and measured paths and meadows. An observer on the scene described them:

> Equipped with the most modern surveying technology such as gauges measuring ultraviolet light in the one hand and historic hiking maps in the other, the scientists were able to photograph the 60 largest Alpine glaciers pictured on the postcards. (Hall 2002, 8)

"We had to do a lot of climbing and clambering," said the team's project leader, Wolfgang Zaengl. "The postcards were our most invaluable tool," he added (Hall 2002, 8). "We are witnesses to the fastest glacial melting in a thousand years," Zaengl said. "But today we are lucky that we can still see the glaciers. Future generations probably will not" (Hall 2002, 8). Results of this research have been published online at www.gletscherarchiv.de.

FURTHER READING

"As the World Warms: A Glacial Archive That Documents a Melting Landscape." *New York Times*, June 13, 2006. [http://www.nytimes.com/2006/06/13/science/earth/13norw.html]

Hall, Alan. "Postcards a Tell-tale for Icy Retreat." *The Scotsman*, August 3, 2002, 8.

Phelan, Amanda. "Turning up the Heat." *Sydney Sunday Telegraph*, August 18, 2002, 47.

Glacier National Park

Glacier National Park (in northern Montana) lost two-thirds of its ice during roughly a century, raising the possibility that it could be glacier-free within a half-century. The naturalist George Bird Grinnell campaigned for the creation of Glacier National Park more than a century ago, and a 500-acre glacier there was named for him. Today, it has lost two-thirds of its size.

When the 1.4 million-acre Glacier National Park was created early in the twentieth century, it included more than 150 glaciers in rugged crags and valleys along northern Montana's Continental Divide. A hundred years later, only 37 of them remained. By the middle of the twenty-first century, given present trends, the park with "glacier" in its name will have no permanent ice (Toner 2002b, 4-A). The Grinnell Glacier, for example, has been retreating more than 15 feet a year, on the average.

The pace of melting has been accelerating with the advent of the new century. "Even those of us who work on these glaciers are surprised at how quickly they are melting," said Dan Fagre of the U.S. Geological Survey. "At the current rate, we may see the complete disappearance of functioning glaciers in the park within 30 years" (Toner 2002b, 4-A). "Glaciers don't know anything about global warming or its causes, but they are excellent barometers of climate change," Fagre said. "They integrate temperature, solar

Photos of Boulder Glacier, in Montana's Glacier National Park, July 1932 and July 1988. The glacier has completely melted. (AP/Wide World Photos)

radiation and snowfall and express it as a big lump of ice. They don't have a political agenda. They just reflect what's happening" (Toner 2002b, 4-A).

Spring has been arriving earlier in the park; summers are warmer and longer. Most years, more ice melts than the winter snows replace. During the winter, precipitation now falls more often as rain rather than snow, even at relatively high altitudes (Toner 2002b, 4-A).

Weather can still be variable—very much so. For example, after eight years of record warmth and drought, Glacier National Park, in northern Montana, was smothered in snow during the La Niña year of 2007–2008. By early July 2008, the Going-to-the-Sun Road in the park was still snowed in, a record-late date, and tourism was suffering from too much snow (Robbins 2008a, A-9, A-10).

FURTHER READING

Robbins, Jim. "Snow in July Is a Mixed Blessing for the Northern Rockies." *New York Times*, July 2, 2008a, A-9, A-10.

Toner, Mike. "Meltdown in Montana: Scientists Fear Park's Glaciers May Disappear within 30 Years." *Atlanta Journal-Constitution*, June 30, 2002b, 4-A.

Glaciers, Alaska

Fewer than 20 of Alaska's several thousand valley glaciers were advancing after the year 2000. Glacial retreat, thinning, stagnation, or a combination of these changes characterizes all 11 mountain ranges and three island areas that support glaciers in the state, according to U.S. Geological Survey scientist Bruce Molnia ("Alaskan Glaciers" 2001).

"The Earth recently emerged from a global climate event, called the 'Little Ice Age' during which Alaskan glaciers expanded significantly," explained Molnia. The Little Ice Age began to wane in the late nineteenth century. In some areas of Alaska, glacier retreat started during the early eighteenth century, before the beginning of the Industrial Revolution ("Alaskan Glaciers" 2001). "During the twentieth century, most Alaskan glaciers receded and, in some areas, disappeared. But it is important to note that our data do not address whether or not any of these changes are human-induced," said Molnia, who warned against blaming the receding glaciers on any single cause, including human emissions of greenhouse gases ("Alaskan Glaciers" 2001).

Measurements by aircraft using global positioning satellites and laser altimeters show that the 5,000-square-kilometer Malaspina Glacier in Alaska is losing nearly a meter of thickness per year, the equivalent of three cubic kilometers of water. That glacier alone has 10 times the water of all the glaciers in the Alps, which have lost more than half of their ice volume since the 1850s. While climatologists have focused their attention on the great ice sheets of Greenland and Antarctica for signs of melting that could raise sea levels, University of Colorado professor Mark Meier asserted that the impact of smaller-scale glacial melting has been underestimated (Russell 2002, A-4).

Since 1993 Keith Echelmeyer has surveyed the size of 90 glaciers from northern Alaska to the Cascades of Washington. He uses data from satellites and from his own observations, often taken from a single-engine airplane. For reference, Echelmeyer uses U.S. Geological Survey maps surveyed a half-century ago. He has found that 90 percent of the glaciers are losing mass balance, with more summer melting than winter ice accumulation.

"The glaciers in Alaska are giving us a clear picture that indeed something is happening to cause them to thin ... that is climate-related," said Echelmeyer (Monastersky 2001, 31–32). Most of the glaciers are shrinking about a meter a year, but some, such as the Lemon Glacier near Juneau, are losing two to three meters annually. Echelmeyer believes that several of the glaciers he surveys will be gone in 50 to 100 years.

A study by Anthony Arendt and colleagues at the University of Alaska/Fairbanks used airborne laser altimetry to estimate volume changes of 67 glaciers in Alaska between the mid-1950s and mid-1990s. The profiles they developed were compared with contours on U.S. Geological Survey and Canadian topographic maps made from aerial photographs taken in the 1950s to the early 1970s (Pianin 2002b, A-14). According to a report in *Science* (Arendt et al. 2002, 382–386), these glaciers, representing about 20 percent of the glacial area in Alaska and neighboring Canada, have been melting at an average of six feet a year, as some have retreated as much as a few hundred feet annually, a rate that has accelerated during the past seven or eight years.

Examples of Receding Glaciers

Alaska's Columbia Glacier by 2007 was discharging about two cubic miles of ice into Prince William Sound per year, according to Robert Anderson, a University of Colorado-Boulder geology professor who is affiliated with the Institute of Arctic and Alpine Research. The Columbia Glacier has thinned as much as 1,300 feet in places. It has shrunk by about nine miles since 1980 and may recede another nine miles in the next 20 years ("Glaciers" 2007).

In Glacier Bay, Alaska, 95 percent of the ice observed when the area was first mapped during the 1790s had melted by the year 2000. During the ensuing two centuries, the ice that once nearly covered the bay has receded more than 60 miles, and, by 2002, was retreating at about half a mile a year (Toner 2002b, 4-A). The first European explorers to visit the area named it after ice that covered nearly the entire harbor. By the 1880s, steady glacial retreat opened liquid water in a 40-mile-long bay. Today, Glacier Bay extends more than 60 miles ("Alaskan Glaciers" 2001).

Richard Monastersky, writing in *The New Scientist*, described how the Mendenhall Glacier, behind Juneau, Alaska, is shrinking "before the public's eyes," as 300,000 visitors a year witness its retreat, which accelerated to 100 meters a year in 2000 (Monastersky 2001, 31). Since the 1930s, that glacier has lost about a kilometer of its length. Observations taken at the Juneau airport indicate that average temperatures there have risen 1.6°C since 1943 (Monastersky 2001, 32).

Glaciers' Contribution to Sea-Level Rise

Some scientists believe that rapid glacial melting in Alaska could be a harbinger of worldwide sea-level rise. Meier anticipates that the level of the world's oceans will rise between 7 and 11 inches by the end of the twenty-first century, more than twice the level anticipated in 2000 by the Intergovernmental Panel on Climate Change (IPCC) (Russell 2002, A-4). Meier said that the IPCC's previous prediction of a sea-level increase from two inches to four inches by century's end was too low for several reasons, the most important of which was an underestimation of the amount of water he believes will be contributed by the melting of glaciers in the Alps, southern Alaska, and the Patagonian Mountains of South America (Russell 2002, A-4).

Anthony A. Arendt and colleagues calculated that Alaskan glaciers are generating nearly twice the annual volume of melting water as the Greenland Ice Sheet, the largest ice mass in the Northern Hemisphere. According to this study, the Alaskan melt is adding about two-tenths of a millimeter a year to worldwide sea levels. Alaskan ice melt accounts for about 9 percent of the sea-level rise during the last century (Arendt et al. 2002, 382).

"The change we are seeing is more rapid than any climate change that has happened in the last 10 to 20 centuries," said Keith A. Echelmeyer, one of the five researchers who prepared the study (Pianin 2002b, A-14). The scientists did not speculate whether accelerating melting results from human-induced global warming, natural factors, or a combination. Sallie L. Baliunas of the Harvard-Smithsonian Center for Astrophysics in Cambridge, Massachusetts (a long-time global-warming skeptic) contended that Alaskan glacial melting is due to a dramatic but temporary shift in Pacific Ocean warm water and wind patterns that began in 1976. "It doesn't have the fingerprints of enhanced greenhouse gas concentrations," she contended (Pianin 2002b, A-14).

Whatever the cause, "[m]ost glaciers have thinned several hundred feet at low elevation in the last 40 years and about 60 feet at higher elevations," said Keith Echelmayer of the University of Alaska at Fairbanks. The ice cover in the Arctic Ocean itself is shrinking by an area the size of the Netherlands each year (Radford 2002c, 9).

FURTHER READING

"Alaskan Glaciers Retreating." Environment News Service, December 11, 2001. http://ens-news.com/ens/dec2001/2001L-12-11-09.html.

Arendt, Anthony A., Keith A. Echelmeyer, William D. Harrison, Craig S. Lingle, and Virginia B. Valentine. "Rapid Wastage of Alaska Glaciers and Their Contribution to Rising Sea Level." *Science* 297 (July 19, 2002):382–386.

"Glaciers and Ice Caps Quickly Melting Into the Seas." Environment News Service, July 20, 2007. http://www.ens-newswire.com/ens/jul2007/2007-07-20-03.asp.

Monastersky, Richard. "The Long Goodbye: Alaska's Glaciers Appear to Be Disappearing before Our Eyes. Are They a Sign of Things to Come?" *New Scientist*, April 14, 2001, 30–32.

Pianin, Eric. "Study Fuels Worry over Glacial Melting: Research Shows Alaskan Ice Mass Vanishing at Twice Rate Previously Estimated." *Washington Post*, July 19, 2002b, A-14.

Radford, Tim. "85 Percent of Alaskan Glaciers Melting at 'Incredible Rate.'" *London Guardian*, July 19, 2002c, 9.

Russell, Sabin. "Glaciers on Thin Ice: Expert Says Melting to Be Faster than Expected." *San Francisco Chronicle*, February 17, 2002, A-4.

Toner, Mike. "Meltdown in Montana: Scientists Fear Park's Glaciers May Disappear within 30 Years." *Atlanta Journal-Constitution*, June 30, 2002b, 4-A.

Glaciers, Andes

Glaciers in the Andes are melting at record rates, threatening water supplies for cities in the region. Glaciologist Lonnie Thompson, who has achieved global recognition for studying ice cores to learn about climate change, in 1976 took an ice core going back 1,500 years from Quelccaya in the Andes. In 1991, he returned to find that annual accumulation of snow had stopped and that the top 20 yards of ice had melted.

The Qori Kalis, the largest glacier in the Quelccaya ice cap, is retreating 500 feet a year and has lost a fifth of its area since 1963 (Pearce 2007b, 181). Thompson has been following the melting of these glaciers for 30 years. At a news conference at an American Association for the Advancement of Science meeting in February 2007, he warned that Peru's Quelccaya had lost about 22 percent of its glacial mass over the past 20 years and was retreating at 200 feet per year ("Eminent Scientists" 2007).

Cuzco, a city of 400,000, has already resorted to periodic water rationing and started pumping from a river 15 miles away. In Peru's capital, Lima, the city's eight million residents face a dry future without major public works. The situation is similar in Asia, where cities in China, India, Nepal, and Bolivia also face drastic water shortages as glaciers vanish (Struck 2006e, A-1). "The poor will suffer. Our children will suffer," said Adolfo Pena, the local representative of the grassroots political movement Peruvians Without Water. "Lima is built on a desert, and in 20 years, there's not going to be water" (Struck 2006e, A-1).

The government in Lima has postponed projects that might help relieve Lima's coming water deficit because they too are expensive for a poor country. Some projects (one would drill a tunnel through the Andes) also may cause earthquakes in an area that is prone to them. Planners remember an earthquake during 1970 that, by one account, "shook loose a wall of ice and rock from the Huascaran mountain in the Andes north of Lima, burying the town of Yungay and killing tens of thousands of people" (Struck 2006e, A-1).

"The repercussions of this are very scary," agreed Tim Barnett, a climate scientist with the Scripps Institution of Oceanography in San Diego. "When the glaciers are gone, they are gone. What does a place like Lima do? Or, in northwest China, there are 300 million people relying on snowmelt for water supply. There's no way to replace it until the next ice age" (Struck 2006e, A-1).

Near Chacaltaya, Bolivia, the lodge at what is billed as the world's highest ski resort at 17,388 feet, with a view of Lake Titicaca on the horizon, faces a snowless future because the glacier that once surrounded the lodge with copious amounts of snow and ice is quickly melting. Chacaltaya's glacier began melting significantly about 1995. According to one account,

> Scientists say that glaciers are increasingly receding throughout the Andes, but that Chacaltaya's melting has been especially quick. More than 80 percent

of the glacier has been lost in 20 years, said Jaime Argollo Bautista, director of the Institute of Geological Investigation at the University of San Andrés, in La Paz. (Romero 2007)

FURTHER READING

"Eminent Scientists Warn of Disastrous, Permanent Global Warming." Environment News Service, February 19, 2007. http://www.ens-newswire.com/ens/feb2007/2007-02-19-03.asp.

Pearce, Fred. *With Speech and Violence: Why Scientists Fear Tipping Points in Climate Change*. Boston: Beacon Press, 2007b.

Romero, Simon. "Bolivia's Only Ski Resort Is Facing a Snowless Future." *New York Times*, February 2, 2007. www.nytimes.com/2007/02/02/world/americas/02bolivia.html.

Struck, Doug. "On the Roof of Peru, Omens in the Ice Retreat of Once-Mighty Glacier Signals Water Crisis, Mirroring Worldwide Trend." *Washington Post*, July 29, 2006e, A-1. http://www.washingtonpost.com/wp-dyn/content/article/2006/07/28/AR2006072801994.html.

Glaciers, Rocky Mountains: Gone in 30 Years

The glaciers of the Rockies, a major source of water for the western half of North America, will be gone in 20 to 30 years, according to David Schindler, a University of Alberta ecologist who holds Canada's top science prize, the $1 million Gerhard Herzberg Gold Medal. Schindler, also a renowned expert on water quality, asserted that the situation is so dire it cannot be reversed:

> This is the bad news. Because we've procrastinated for the past 25 years, we're locked into considerable warming for the next several decades, and I think that will be enough to finish off the glaciers of the Rockies. Right now we are getting more melting than precipitation. That could be reversed with a cooler climate. I think that, looking ahead a century, we could get them back. All we need to do is reverse the balance. (Remington 2002, A-19)

"When those glaciers go, I wonder what we will be doing for municipalities and agriculture on the southern [Canadian] prairies. Those rivers are going be streams at best once climate warming takes its toll on water supply," said Schindler (Remington 2002, A-19). "The sorts of melting phenomena we see are everywhere—the Alps, the Rockies, the mountains of Africa."

Glacial melt water is released slowly, allowing streams and rivers to flow throughout the year, including mid-to-late summer, when water demand reaches its annual peak. Should glaciers disappear, people will have to rely on trapping water when the snow melts in the spring. "I know what people will say: 'Build more water reservoirs and trap more spring flow.' That doesn't bode very well for people downstream," Schindler said (Remington 2002, A-19).

Schindler has been analyzing climate records for Canadian prairie cities. He has found temperature increases ranging between 1° and 4°C since recordkeeping began in the early twentieth century. "These can't be dismissed as urban heat-island effects, as a lot of the debunkers would like to say. The biggest increase of all is Fort Chipewyan (an isolated community in northeastern Alberta). There is certainly no heat island effect in Fort Chipewyan," Schindler said (Remington 2002, A-19).

Ice patches along the Continental Divide in Colorado have disgorged ancient bison horns that have been radiocarbon dated between 2,090 and 2,280 years old, suggesting that, in some cases, ice along the divide has retreated to levels unseen since before the Christian era (Erickson 2004). "Over the last couple of decades, and especially over the last 10 years, we have entered a period of warming and retreat that is as great, or greater, than any we know of since the end of the last ice age" 10,000 years ago, said glaciologist Tad Pfeffer, of Colorado University's Institute of Arctic and Alpine Research. "The Front Range glaciers and snowfields could be gone in a couple of decades" (Erickson 2004).

FURTHER READING

Erickson, Jim. "Going, Going, Gone? Front-range Glaciers Declining: Researchers Point to a Warming World." *Rocky Mountain News*, October 26, 2004, 5-A.

Remington, Robert. "Goodbye to Glaciers: Thanks to Global Warming, Mountains—the World's Water Towers—are Losing Their Ice. As it Disappears, so Does an Irreplaceable Source of Water." *Financial Post* (Canada), September 6, 2002, A-19.

Global Climate Coalition

Sponsors of the Global Climate Coalition (GCC), a major vehicle of the climate contrarians, at one time included many of the world's largest fossil-fuel corporations. Beginning in the late 1990s, however, a number of large companies (among them General Motors, Ford, Daimler-Chrysler, Southern Co., British Petroleum, Royal Dutch Shell, and Texaco) failed to renew their memberships in the GCC (Wadman 2000, 322).

Donald Pearlman founded the Climate Council to represent companies whose businesses could be affected by global climate change. He secured nongovernmental organization (NGO) status for his lobby at the United Nations, as environmentalists crowned him "King of the Carbon Club" (Gelbspan 1997b, 120). Pearlman specialized in coaching delegates from oil-rich countries such as Kuwait, Saudi Arabia, Syria, Iran, China, and Nigeria on how to use delaying tactics in international meetings on strategies to combat global warming. Meetings of the IPCC have been a special target for this sort of activity.

FURTHER READING

Gelbspan, Ross. *The Heat Is On: The High Stakes Battle over Earth's Threatened Climate.* Reading, MA: Addison-Wesley Publishing Co., 1997b.

Wadman, Meredith. "Car Maker Joins Exodus from Anti-Kyoto Coalition." *Nature* 404 (2000):322.

Global Warming: Importance of the Issue

Lord Peter Levene, board chairman of the insurance company Lloyd's of London, has said that terrorism is not the industry's biggest worry, despite the fact that his company was the largest single insurer of the World Trade Center. Levene said that Lloyd's along with other large international insurance companies is bracing for an increase in weather disasters related to global warming. During January 2005, Rajendra Pachauri, chairman of the Intergovernmental Panel on Climate Change (IPCC), said, with regard to global warming, "We are risking the ability of the human race to survive" (Hertsgaard 2005).

Following his assignment as chief weapons inspector in Iraq, Hans Blix said:

> To me the question of the environment is more ominous than that of peace and war. We will have regional conflicts and use of force, but world conflicts I do not believe will happen any longer. But the environment, that is a creeping danger. I'm more worried about global warming than I am of any major military conflict. ("Hans Blix's" 2003, D-2)

Sir John Houghton, co-chair of the IPCC, agreed. "Global warming is already upon us," he said. "The impacts of global warming are such that that I have no hesitation in describing it as a weapon of mass destruction" (Kambayashi 2003, A-17).

The problem is at once very simple and astoundingly complex. Increasing human populations, rising affluence, and continued dependence on energy derived from fossil fuels are at the crux of the issue. The complexity of the problem is illustrated by the degree to which the daily lives of machine-age peoples depend on fossil fuels. This dependence gives rise to an array of local, regional, and national economic interests.

These interests cause tensions between nations attending negotiations to reduce greenhouse-gas emissions. The cacophony of debate illustrates the strength and diversity of established interests that are being assiduously protected. Add to the human elements of the problem the sheer randomness of climate (as well as the amount of time that passes before a given level of greenhouse gases is actually factored into climate), and the problem becomes complex and intractable enough to (thus far) seriously impede any serious, unified effort by humankind to fashion solutions.

Risks "Real and Palpable"

"The risks of global warming are real, palpable, the effects are accumulating daily, and the costs of correcting the trend rise with each day's delay," warned George M. Woodwell, director of the Woods Hole Research Center ("Eco Bridge" n.d.). Dean Edwin Abrahamson, an early leader in the field, commented: "Fossil fuel burning, deforestation, and the release of industrial chemicals are rapidly heating the earth to temperatures not experienced in human memory. Limiting global heating and climatic change is the central environmental challenge of our time" (Abrahamson 1989, xi).

Mark Lynas, an author who has written extensively on global warming, traveled around the world cataloging the impacts of climate change for a book, *High Tide* (Lynas 2004a). According to Lynas,

> This is a global emergency. We are heading for disaster and yet the world is still on fossil fuel autopilot. There needs to be an immediate phase-out of coal, oil and gas, and a phase-in of clean energy sources.... People can no longer ignore this looming catastrophe. (Reynolds 2003, 6)

Author Ross Gelbspan agreed:

> Climate change is not just another issue. It is *the* overriding threat facing human civilization in the twenty-first century, and so far our institutions are doing dangerously little to address it. Americans in particular are still in denial, thanks largely to the efforts of the fossil-fuel industry and its allies in the Bush Administration. But the nation's biggest environmental organizations and opposition politicians have also displayed a disturbing lack of leadership on this crucial challenge. (Gelbspan 2004a, 24)

Weather is the story; climate is the plot. The innate variability of daily weather sometimes masks longer-term trends. Climatic news comes to us in snapshots: Thousands die of heat in Europe. Corals bleach under heat stress, and polar bears go hungry, lacking seasonal ice from which to hunt seals. Long-range trends go largely unnoticed.

Bill McKibben called global warming "[t]he first civilization-scale challenge that humans have yet faced." Newly emerging science (including some that the Bush administration tried to force NASA climatologist James Hansen to suppress) shows that we have underestimated the scale and urgency of the crisis. Everything frozen on Earth is melting fast, for instance, threatening to produce an inhospitable planet in the decades ahead and an unbearable one in the lifetime of those being born. "Political rhetoric needs to reflect the stark fact that this is an emergency" (McKibben 2006, A-25).

A 90 percent Probability of "Potentially Catastrophic" Global Warming

The Earth faces a 90 percent probability of experiencing potentially catastrophic global warming of 1.7° to 4.9°C provoked by anthropogenic greenhouse gases by the end of the twenty-first century, according to probabilities calculated by Tom Wigley of the National Center for Atmospheric Research in Boulder, Colorado, and Sarah Raper of the University of East Anglia in Norwich, Great Britain. Wigley and Raper used computer models to build on the work of the third report of the IPCC, which predicted a global temperature rise between 1.4°C and 5.8°C (Connor 2001b, 15).

As global greenhouse-gas emissions increase with the passage of time, the global "window" for escaping such a catastrophe narrows. The world has only 15 years to start cutting greenhouse-gas emissions if it is to stand a chance of curbing global warming, according to Bert Metz of the IPCC. Metz told an international climate summit in Bonn during July 2001 that even a five-year delay could push the goal of stabilizing carbon dioxide in the atmosphere "beyond reach" (Henderson 2001b). Metz, chairman of the IPCC mitigation committee, said, "The first steps to fight global warming have to be taken imminently if its effects are to be contained" (Henderson 2001b). By 2008, that "window" had narrowed to less than a decade, according to Hansen and others.

Alarm Bells Ringing

Evidence has been accumulating that sustained, human-induced warming of the Earth's lower troposphere has been in progress since about 1980, accelerating during the 1990s. During 1997 and 1998, the global temperature set records for 15 consecutive months; July 1998 averaged 0.6°F higher than July 1997, an enormous increase if maintained year to year. The year 1998 was the warmest of the millennium, topping 1997 by 0.25°F (Christianson 1999, 275). By 2007, 11 of the 12 warmest years on the instrumental record had occurred since 1995.

Alarm bells have been ringing regarding global warming in the scientific community for the better part of two decades. A statement issued in Toronto during June 1988 representing the views of more than 300 policymakers and scientists from 46 countries, the United Nations, and other international organizations warned:

> Humanity is conducting an unintended, uncontrolled, globally pervasive experiment whose ultimate consequences could be second only to nuclear war. The earth's atmosphere is being changed at an unprecedented rate by pollutants resulting from human activities, inefficient and wasteful fossil fuel use and the effects of rapid population growth in many regions. These changes are already having harmful consequences over many parts of the globe. (Abrahamson 1989, 3)

Michael Meacher, speaking as Great Britain's environment minister, has said that "[c]ombating climate change is the greatest challenge of human history" (Brown 1999, 44). If the atmosphere's carbon-dioxide level doubles over preindustrial levels, which is likely (at present rates of increase) before the year 2100, climate models indicate that temperatures may rise 1.9° to 5.2°C (3.4° to 9.4°F) within a century, producing "a climate warmer than any in human history. The consequences of this amount of warming are unknown and could include extremely unpleasant surprises" (National Academy 1991, 2).

Global Warming and Denial

Despite the many warnings of global warming's importance, not everyone is convinced. Denial has proved resilient, along with motivations to profit from the new climate regime. Witness the many proposals to drill for oil in the Arctic as its ice cap melts, and the jockeying between nations bordering the region for sovereignty that bestows ownership of subsurface resources.

George Marshall, a United Kingdom climate activist and co-executive director of the Climate Outreach Information Network, sees a potentially fatal contradiction in government and some industrial leaders' statements about global warming. They assess the reality, then fail to take concrete steps to address the problem, as if voicing concern, by itself, was enough. "We perform the same trick with these psychological blind spots," he wrote.

> We patch over them and we do not see that we have a fundamental weakness in our rationality. The result is a sustained malfunction in our capacity to assess and evaluate risk. Thus we can simultaneously hold the view that this is a countdown to a global catastrophe … without receiving any reliable warning.… We are like people with leprosy who can intellectually accept that putting their hands in a flame can cause damage but are lacking the pain receptors that would trigger the reflex to withdraw [the] hand. (Marshall 2006)

Global Warming's Enduring Impacts

Many scientists agree that a general climatic warming trend will continue for at least the next hundred years even if fossil-fuel consumption is slashed sharply and immediately. Robert Dickinson of the Georgia Institute of Technology's School of Earth and Atmospheric Sciences presented evidence to support this assessment at the annual meeting of the American Association for the Advancement of Science in Boston on February 17, 2002 ("Global Warming Could Persist" 2002).

"Current climate models can indicate the general nature of climate change for the next 100 to 200 years," Dickinson said.

> But the effects of carbon dioxide that have been released into the atmosphere from the burning of fossil fuels last for at least 100 years. That means that any reductions in carbon dioxide that are expected to be possible over this period will not result in … less global warming than we see today for at least a century. ("Global Warming Could Persist" 2002)

The IPCC presented similar scenarios for other greenhouse gases, as it warned that "[t]he stabilization of greenhouse-gas concentrations does not imply that there will be no further climate change. After stabilization is achieved, global mean surface temperature would continue to rise for some centuries and sea level for many centuries" (Bolin et al. 1995).

By the year 2008, levels of carbon dioxide and methane in the Earth's atmosphere had risen to more than 30 percent higher than the upper range of natural levels during warm spells between glaciations. The "Keeling Curve," which charts carbon-dioxide levels over time, now resides on uncharted atmospheric ground as far back in time as human measurement extends. By early in the year 2008, that record, from Antarctic ice cores, extended to 820,000 years before the present. Other proxies extend the record millions of years, indicating that the atmosphere has not witnessed today's greenhouse-gas levels since the Pliocene, two to three million years ago, or even longer. Some estimates range to 20 million years.

The Evolution of Knowledge

To illustrate how quickly base-line scientific knowledge has been evolving in this area, Paul Pearson of the University of Bristol and Martin Palmer of Imperial College, London, reported in the August 17, 2000, edition of *Nature* that they have developed proxy methods for measuring the atmosphere's carbon-dioxide level to 60 million years before the present. The upshot of Pearson and Palmer's studies is that carbon-dioxide levels at the year 2000 were as high as they have been in at least the last 20 million years. According to their records, however, carbon-dioxide levels reached the vicinity of 2,000 parts per million (ppm) during "the late Palaeocene and earliest Eocene periods, from about 60 to 52 million years ago (Pearson and Palmer 2000, 695).

Pearson and Palmer used plankton shells drilled from the seabed to estimate the acidity (and thus the carbon content) of seawater over a span of time back almost to the era of the dinosaurs. By 2100, at present rates of increase, their figures indicated that the carbon-dioxide level of the Earth's atmosphere could match the level last seen in the Eocene, about 50 million years ago. At that time, the Earth had no permanent pack ice and, as characterized by one English newspaper, "London was a steaming mangrove swamp" (Radford 2000a, 9).

As levels of greenhouse gases have risen, concern has been expressed by a number scientists and policymakers that the Earth may be entering a period of rapid, human-induced warming that may damage animal (including human) and plant life. The IPCC's *Second Assessment* indicated that deserts "are likely to become more extreme—in that, with few exceptions, they are projected to become hotter but not significantly wetter" (Bolin et al. 1995). Temperature increases could threaten organisms, such as oceanic corals, which already live near their heat tolerance limits. Large areas of land may become deserts, a process which often takes the land involved to an ecological dead-end:

> Land degradation in arid, semi-arid and dry sub-humid areas resulting from various factors, including climatic variations and human activities, is more likely to become irreversible if the environment becomes drier and the soil becomes further degraded through erosion and compaction. (Bolin et al. 1995)

In aquatic and coastal ecosystems, such as lakes and streams, warming would, according to the IPCC,

> have the greatest biological effects at high latitudes, where biological productivity would increase, and at the low-latitude boundaries of cold-and cool-water species ranges, where extinctions would be greatest. The geographical distribution of wetlands is likely to shift with changes in temperature and precipitation.... Some coastal ecosystems are particularly at risk, including saltwater marshes, mangrove ecosystems, coastal wetlands, sandy beaches, coral reefs, coral atolls and river deltas. (Bolin et al. 1995)

As time has passed, the effects, real and prospective, have become more urgent.

FURTHER READING

Abrahmson, Dean Edwin. *The Challenge of Global Warming*. Washington, DC: Island Press, 1989.

Bolin, Bert, John T. Houghton, Gylvan Meira Filho, Robert T. Watson, M. C. Zinyowera, James Bruce, Hoesung Lee, Bruce Callander, Richard Moss, Erik Haites, Roberto Acosta Moreno, Tariq Banuri, Zhou Dadi, Bronson Gardner, J. Goldemberg, Jean-Charles Hourcade, Michael Jefferson, Jerry Melillo, Irving Mintzer, Richard Odingo, Martin Parry, Martha Perdomo, Cornelia Quennet-Thielen, Pier Vellinga, and Narasimhan Sundararaman. *Intergovernmental Panel on Climate Change. Second Assessment Synthesis of Scientific-Technical Information Relevant to Interpreting Article 2 of the United Nations Framework Convention on Climate Change*. 1995. Approved by the IPCC at its eleventh session, Rome, December 11–15, 1995. http://www.unep.ch/ipcc/pub/sarsyn.htm.

Brown, Paul. "Global Warming: Worse Than We Thought." *World Press Review*, February 1999, 44.

Christianson, Gale E. *Greenhouse: The 200-Year Story of Global Warming*. New York: Walker and Company, 1999.

Connor, Steve. "Catastrophic Climate Change 90 Percent Certain." *London Independent*, July 20, 2001b, 15.

"Eco Bridge: What Can We Do about Global Warming?" n.d. http://www.ecobridge.org/content/g_wdo.htm.

Gelbspan, Ross. "Boiling Point." *The Nation*, August 16, 2004a, 24–27.

"Hans Blix's Greatest Fear." *New York Times*, March 16, 2003, D-2.

Henderson, Mark. ""World Has 15 Years to Stop Global Warming."" *London Times*, July 21, 2001b, n.p. (LEXIS)

Hertsgaard, Mark. "It's Much Too Late to Sweat Global Warming." *San Francisco Chronicle*, February 13, 2005, n.p.

Kambayashi, Takehiko. "World Weather Prompts New Look at Kyoto." *Washington Times*, September 5, 2003, A-17.

Lynas, Mark. *High Tide: News from a Warming World*. London: Flamingo, 2004a.

Marshall, George. "Sleepwalking into Disaster—Are We in a State of Denial about Climate Change? *Perspectives on Climate Change* No. 4. Climate Outreach and Information Network, Oxford, U.K., 2006. http://www.coineet.org.uk.

McKibben, Bill. "Welcome to the Climate Crisis: How to Tell Whether a Candidate Is Serious about Combating Global Warming." *Washington Post*, May 27, 2006, A-25. http://www.washingtonpost.com/wp-dyn/content/article/2006/05/26/AR2006052601549_pf.html.

National Academy of Sciences. *Policy Implications of Greenhouse Warming*. Washington, DC: National Academy Press, 1991.

Pearson, Paul N., and Martin R. Palmer. "Atmospheric Carbon Dioxide Concentrations Over the Past 60 Million Years." *Nature* 406 (August 17, 2000):695–699.

Radford, Tim. "Greenhouse Buildup Worst for 20m[Million] Years." *London Guardian*, August 17, 2000a, 9.

Reynolds, James. "Earth Is Heading for Mass Extinction in Just a Century." *The Scotsman*, June 18, 2003, 6.

Global Warming: Origins as a Political Issue

Rising temperatures and speculation regarding their effects have made global warming a subject of copious and increasing media discourse during recent years. The amount of debate and discourse has increased as rapidly rising temperatures provoke serious attempts at legislating greenhouse-gas limitations by a Democratic Congress elected in 2006 and 2008. Few scientific issues exceed this one in terms of political and economic gravitas. At stake is not only the future of the Earth as a habitable environment, but the ways in which more than six billion people use energy. To an informed observer, the discourse often displays the tone of a political campaign, a horse race, or a religious revival more than a scientific debate.

Origins of the Modern Debate

The modern debate over whether the Earth is warming due to human activity began in policy circles during 1979, after a small number of well-known scientists reported to the Council on Environmental Quality that "[m]an is setting in motion a series of events that seem certain to cause a significant warming of world climates unless mitigating steps are taken immediately" (Pomerance 1989, 260). At the same time, the National Academy of Sciences initiated a study of the greenhouse effect. Also during 1979, in the United States, the President's Council on Environmental Quality mentioned global warming: "The possibility of global climate change induced by an increase of carbon dioxide in the atmosphere is the subject of intense discussion and controversy among scientists" (Anderson 1999).

An alarm regarding global warming was sounded in the British journal *Nature* on May 3, 1979: "The release of carbon dioxide to the atmosphere by the burning of fossil fuels is, conceivably, the most important environmental issue in the world today" (Bernard, 1993, 6). At about the same time, a study conducted by a scientific team chaired by meteorologist Jule Charney estimated that doubling the carbon dioxide level in the atmosphere would raise the average global temperatures by about 3°C, plus or minus 1.5°C. Four years later, the U.S. Environmental Protection Agency released a report, "Can We Delay a Greenhouse Warming?" A National Academy of Sciences report, also issued in 1983, stated that "[w]e do not believe that the evidence at hand about CO_2-induced climate change would support steps to change current fuel-use patterns away from fossil fuels" (Pomerance 1989, 261).

Until the early 1980s, those who argued that infrared forcing had (or would) raise the temperature of the atmosphere near the Earth's surface had a statistical problem. Starting in 1940, until about 1975, average global temperatures actually fell slightly. After 1975, however, temperatures began a steady, accelerating rise, which made global warming a notable political issue by the late 1980s.

Many people began to look at the atmosphere as a global commons requiring protection for survival. Author John J. Nance, in *What Goes Up: The Global Assault on our Atmosphere*, quoted Susan Soloman, an atmospheric chemist:

> I can't go home and dump my garbage in my neighbor's back yard. The police would arrest me in five minutes. But [until they were banned] I could take a tank of chlorofluorocarbons, put it in my backyard, turn it on and let it go into the atmosphere all day long, and no one [could] stop me. Somehow that's very wrong. (Nance 1991, 9)

The potential impact of warming caused by infrared forcing was raised again at a United Nations–sponsored conferences in Villach, Austria, during the mid-1980s. Shortly after these conferences, Senators David Durenberger of Minnesota and Albert Gore of Tennessee called for an international "Year of the Greenhouse" to raise the issue in public consciousness. Gore already had played a role in congressional hearings on the issue in 1982 and 1984, when he was serving in the House of Representatives.

The Time to "Cry Wolf"

As a political issue in the United States, global warming came of age during the notably hot summer of 1988. The year 1988 provided something of a wake-up call in the debate over global warming because it was the warmest year since reliable records had been kept in the middle of the nineteenth century. During 1988, 400 electrical transformers in Los Angeles blew out on a single day as temperatures rose to 110°F. Two thousand daily temperature records were set that year in the United States. Widespread heat and drought caused some crop yields in the U.S. Midwest to fall between 30 and 40 percent. In

Moscow, Russians escaping their hottest summer in a century flocked to rivers and lakes, and drowned in record numbers (Christianson 1999, 197).

During 1988, Colorado Senator Timothy E. Wirth, whose hearings on global warming the previous winter had drawn little attention, played the weather card. He called another hearing, this time during the summer. As it happened, the hearing convened on a particularly hot, humid day in Washington, D.C., during which the high temperature reached a record 101°F. At Wirth's hearing, James E. Hansen, director of NASA's Goddard Institute of Space Studies, testified that the unusually warm temperatures of the 1980s were an early portent of global warming caused by the burning of fossil fuels, and not solely a result of natural variation. Hansen's remarks became front-page news nationwide within hours.

Hansen thus continued a running battle through the 1980s, during the Reagan and Bush administrations, to call the science of global warming to public attention despite repeated threats to the funding of the Goddard Institute. The Office of Management and Budget forced Hansen to censor his findings several times. The pressure was so intense that Hansen sometimes asked to testify as a private citizen rather than as a federal employee (Hansen 1989).

In the meantime, the scientific debate over global warming was intensifying. By the end of 1988, the United Nations General Assembly had approved the creation of the Intergovernmental Panel on Climate Change (IPCC). A year later, Hansen said that it was "time to cry wolf":

> When is the proper time to cry wolf? Must we wait until the prey, in this case the world's environment, is mangled by the wolf's grip? The danger of crying too soon, which much of the scientific community fears, is that a few cool years may discredit the whole issue. But I believe that decision-makers and the man-in-the-street can be educated about natural climate variability.... A greater danger is to wait too long. The climate system has great inertia, so as yet we have realized only a part of the climate change which will be caused by gases we have already added to the atmosphere. Add to this the inertia of the world's energy, economic, and political systems, which will affect any plans to reduce greenhouse gas emissions. Although I am optimistic that we can still avoid the worst-case climate scenarios, the time to cry wolf is here. (Nance 1991, 267–268)

Hansen elaborated:

> I said three things [in 1988]. The first was that I believed the earth was getting warmer and I could say that with 99 percent confidence. The second was that with a high degree of confidence we could associate the warming and the greenhouse effect. The third was that, in our climate model, by the late 1980s and early 1990s, there's already a noticeable increase in the frequency of drought. (Parsons 1995, 7)

Between June 27 and 30, 1988, as the Earth's warmest summer on record (to that time) was getting under way, more than 300 leaders in science, politics, law, and environmental studies gathered in Toronto at the invitation of Canada's government to address problems related to climate change, including prospects of global warming. A scientific consensus was forming around the idea that human activity already was altering the Earth's atmosphere at an unprecedented rate. A consensus statement issued by the Montreal climate conference asserted that "[t]here can be a time lag of the order of decades between the emission of gases into the atmosphere and their full manifestation in atmospheric and biological consequences. Past emissions have already committed planet Earth to a significant warming" (Ferguson 1989, 48).

Rising Political Debate and Temperatures during the 1990s

Debate regarding the greenhouse effect intensified when temperature readings in 1990 eclipsed the record warmth of 1988. Later in the decade, 1995 became the warmest year, followed by 1997, and 1998. During most of these years, the Southern Oscillation (El Niño) weather pattern (called "ENSO" in climate-change literature) was an important factor in world weather. ENSO involves a marked warming of the tropical ocean off the west coast of South America. The pattern occurred more often during the 1990s than at any time for the century and a half during which detailed worldwide weather records have been available.

Some atmospheric scientists have asserted that the ENSO was associated with a gradual warming of the lower atmosphere during most of the twentieth century. In 1996, Kevin E. Trenberth and Timothy J. Hoar of the National Center for Atmospheric Research pointed out that El Niño periods had occurred more frequently in the

1980s and 1990s than during the previous century (Christianson 1999, 224).

During the 1992 presidential campaign in the United States, candidate Bill Clinton criticized the Bush administration's refusal to join in worldwide diplomatic efforts to reduce emissions of greenhouse gases. Clinton's first budget proposed a carbon tax, a measure that was quickly dropped under pressure from Republicans in Congress. In Congress, the carbon tax died in committee. Meanwhile, a Climate Convention signed by 161 countries at the Conference on Environment and Development in Rio de Janeiro, during 1992, contained a directive, in Article 2, favoring stabilization of greenhouse gases "at levels and on a time scale that do not produce unacceptable damage to ecosystems and that allow for sustainable economic development" (Woodwell 1995, v).

At about the same time, several studies (e.g., Karl et al. 1993; Easterling et al. 1997) indicated that daily minimum temperatures had increased more rapidly during much of the twentieth century than daily maxima. Easterling and colleagues reported that between 1950 and the mid-1990s, daily minima increased at a rate of 1.86°C per century, while maxima increased 0.88°C (less than half as much) during the same period.

A study by Henry F. Diaz and Raymond S. Bradley supported climate-model forecasts that daily minimum temperatures will rise more rapidly than maximums in a warmer world. Diaz and Bradley, who studied temperature changes during the twentieth century at high-elevation sites, wrote:

> The signal appears to be more closely related to increases in daily minimum temperatures than changes in the daily maximum. The changes in surface temperature vary spatially, with Europe (particularly Western Europe) and parts of Asia displaying the strongest high-altitude warming during the period of record. (Diaz and Bradley 1997, 253)

A number of legislative bodies in different parts of the world took initiatives to limit greenhouse-gas emissions soon after the memorably hot summer of 1988. During 1989, the Netherlands passed a National Environmental Policy Plan, which required a freeze on carbon-dioxide emissions at 1989–1990 levels. The parliament of Norway decided to limit carbon-dioxide emissions in that country to 1989 levels by the year 2000, with a decline in emissions mandated after that. On June 13, 1990, just before Germany was reunified, the West German Cabinet committed the country to a 25 percent reduction in greenhouse gases, based on 1987 levels, by the year 2005. Later, the unified government stood behind these limits, but added allowances for energy-inefficient industries in what had been East Germany.

Britain's Early Activist Role in Climate-Change Diplomacy

Britain's government has diplomatically nudged the United States on several occasions on the global-warming issue. The British government's involvement in global warming as a political issue dates to the tenure of Tory Prime Minister Margaret Thatcher, who had earned an undergraduate degree in chemistry and served as a teacher of the subject. In 1997, as 160 nations prepared to negotiate the Kyoto Protocol, British Deputy Prime Minister John Prescott traveled to Washington, D.C., to advocate a stronger response to the problem by the United States.

Britain's environmental secretary, John Gummer, called early for an emissions-reduction target of 50 percent, beginning with an end to all subsidies for oil and coal use, as Britain's *Financial Times* pinned the ineffectiveness of the Kyoto Protocol on resistance from established interests in the United States:

> Right now the final protocol promises to be a sad affair given the power of the oil and coal lobby in the United States, which has still failed to meet its obligations under the Rio Treaty.... It is therefore a shame that the U. S. government lacks the courage to admit that while some jobs in some industries may be lost, new jobs will more than compensate. ("Global Warming" 1997)

The George W. Bush White House

In 2001, climate contrarians acquired a new degree of influence with an ally in the White House, George W. Bush. Republicans also assumed control of the House of Representatives and Senate, where the ruling ideology (expressed by Oklahoma Senator James Inhofe) portrayed global warming as an expensive hoax. This point of view soon was being reflected with censorship of government scientists and attempts to rewrite NASA's mission statement. The space agency had come to concentrate on earth sciences. Under James E. Hansen, for example, the NASA

Goddard Institute for Space Studies in New York City had become mainly concerned with global-warming issues.

Unknown to scientists at NASA (who were not consulted), during late January or early February of 2006—just as articles describing how Hansen was resisting White House pressure hit the front pages of the *New York Times* and *Washington Post*—the Bush administration's annual budget reached Congress with an altered version of the agency's mission statement. Gone was the statement that had been used since 2002 ("To understand and protect our home planet; to explore the universe and search for life, to inspire the next generation of explorers … as only NASA can"). The phrase "to understand and protect our home planet" had vanished. The 2002 statement, an extension of NASA's original statement of purpose ("the expansion of human knowledge of the Earth and of phenomena in the atmosphere and space") had been adopted with advice from NASA's 19,000 employees. The new, sans Earth mission was written by fiat. NASA employees did not learn of their new mission statement until summer (Revkin 2006d, A-1, A-10).

Without explicitly saying so, the change seemed aimed squarely at parts of NASA, such as the Goddard Institute for Space Studies, which had sifted its focus over the years to earth sciences. Hansen said that the change might reflect White House eagerness to shift federal resources away from study of global warming. "They're making it clear that they have the authority to make this change, that the president sets the objectives for NASA, and that they prefer that NASA work on something that's not causing them a problem," Hansen told Andrew Revkin of the *New York Times* (2006d, A-10).

FURTHER READING

Anderson, J. W. "The History of Climate Change as a Political Issue." The Weathervane: A Global Forum on Climate Policy Presented by Resources for the Future, August 1999. http://www.weathervane.rff.org/features/feature005.html.

Bernard, Harold W., Jr. *Global Warming: Signs to Watch For.* Bloomington: Indiana University Press, 1993.

Christianson, Gale E. *Greenhouse: The 200-Year Story of Global Warming.* New York: Walker and Company, 1999.

Diaz, Henry F., and Raymond S. Bradley. "Temperature Variations during the Last Century at High-elevation Sites." *Climatic Change* 36 (1997): 253–279.

Easterling, David R., Briony Horton, Phillip D. Jones, Thomas C. Peterson, Thomas R. Karl, David E. Parker, M. James Salinger, Vyacheslav Razuvayev, Neil Plummer, Paul Jamason, and Christopher K. Folland. "Maximum and Minimum Temperature Trends for the Globe." *Science* 277 (1997):364–366.

Ferguson, H. L. "The Changing Atmosphere: Implications for Global Security." In *The Challenge of Global Warming,* ed. Dean Edwin Abrahamson, 48–62. Washington, DC: Island Press, 1989.

"Global Warming." Editorial, *The Financial Times,* March 11, 1997. http://benetton.dkrz.de:3688/homepages/georg/kimo/0254.html.

Hansen, James E. "The Greenhouse, the White House, and Our House." Typescript of a speech at the International Platform Association, Washington, D.C., August 3, 1989.

Karl, T. R., P. D. Jones, R. W. Knight, G. Kukla, N. Plummer, V. Razuvayev, K. P. Gallo, J. Lindsay, R. J. Charlson, and T. C. Peterson. "A New Perspective on Recent Global Warming: Asymmetric Trends of Daily Maximum and Minimum Temperature." *Bulletin of the American Meteorological Society* 74 (1993):1007–1023.

Nance, John J. *What Goes Up: the Global Assault on Our Atmosphere.* New York: William Morrow and Co., 1991.

Parsons, Michael L. *Global Warming: The Truth behind the Myth.* New York: Plenum Press/Insight, 1995.

Pomerance, Rafe. "The Dangers from Climate Warming: A Public Awakening." In *The Challenge of Global Warming,* ed. Edwin Abrahamson, 259–269. Washington, DC: Island Press, 1989.

Revkin, Andrew C. "NASA's Goals Delete Mention of Home Planet." *New York Times,* July 22, 2006d, A-1, A-10.

Woodwell, George M., and Fred T. MacKenzie, eds. *Biotic Feedbacks in the Global Climate System: Will the Warming Feed the Warming?* New York: Oxford University Press, 1995.

Gore, Albert (March 31, 1948–)

Al Gore, U.S. vice president under Bill Clinton who narrowly missed election as president in the hotly contested 2000 election, became a major political figure in the global warming debate. Through his participation in diplomacy and his public advocacy of the issue, Gore became a leading figure worldwide as an interpreter of global-warming science to the public.

Gore's leading role in publicizing global warming was recognized in 2007 with a Nobel Peace Prize that he shared with the Intergovernmental Panel on Climate Change (IPCC). In its formal citation, the Nobel committee called Gore

Former Vice President Al Gore (1948–) smiles during a book-signing event to promote his book *An Inconvenient Truth* at a Tokyo bookstore on January 14, 2007. (AP/Wide World Photos)

"probably the single individual who has done most to create greater world-wide understanding of the measures that need to be adopted" (Gibbs 2007, A-13).

Accepting the Nobel Peace Prize on December 12, 2007, seven years to the day after the U.S. Supreme Court called a halt to a Florida vote recount involving "hanging chads" and sealed George W. Bush's victory as president (even after Gore had won the popular vote), Gore said: "I read my own political obituary in a judgment that seemed to me harsh and mistaken—if not premature. But that unwelcome verdict also brought a precious if painful gift: an opportunity to search for fresh new ways to serve my purpose" (Kolbert 2007d, 43).

Gore authored a book, *Earth in the Balance* (1992), and a well-known documentary film, *An Inconvenient Truth* (2006), which won an Oscar. David Denby, writing in *The New Yorker* said of *An Inconvenient Truth*, "The science is detailed, deep-layered, vivid, and terrifying.... If even half of what Gore says is true, this may the most galvanizing documentary you will see in your lifetime" (Denby 2006, 23). Released in May 2006, *An Inconvenient Truth*, directed by Davis Guggenheim, took in gross receipts of more than $46 million, making it one of the top-grossing documentaries ever made. A companion book by Gore quickly became a bestseller, reaching number one on the *New York Times* list (Broad 2007). The film was credited with raising global warming's salience as a major political issue in the United States.

Late in September 2006, Gore endorsed a popular movement in the United States to seek an immediate freeze in greenhouse gases. He made the endorsement during a speech at the New York University Law School. "Merely engaging in high-minded debates about theoretical future reductions while continuing to steadily increase emissions represents a self-delusional and reckless approach," Gore said. "In some ways, that approach is worse than doing nothing at all, because it lulls the gullible into thinking that something is actually being done, when in fact it is not" (Revkin 2006j).

During 2008, Gore also played a leading role in shaping and financing a $300 million publicity campaign against global warming, which ranked among the most expensive public advocacy campaigns in United States history.

Scientists' Criticisms of Gore

Some scientists have said that Al Gore's film is by turns inaccurate and alarmist. "I don't want to pick on Al Gore," Don J. Easterbrook, an emeritus professor of geology at Western Washington University, told hundreds of experts at the annual meeting of the Geological Society of America. "But there are a lot of inaccuracies in the statements we are seeing, and we have to temper that with real data" (Broad 2007). Gore, in an e-mail exchange about his critics, said his work made "the most important and salient points" about climate change, if not "some nuances and distinctions" scientists might want. "The degree of scientific consensus on global warming has never been stronger," he said, adding, "I am trying to communicate the essence of it in the lay language that I understand" (Broad 2007).

"He's a very polarizing figure in the science community," said Roger A. Pielke Jr., an environmental scientist at the University of Colorado. "Very quickly, these discussions turn from the issue to the person, and become a referendum on Mr. Gore" (Broad 2007). Paul Reiter, a global-warming contrarian who directs the insects and infectious diseases unit of the Pasteur Institute in Paris, faulted Gore for alleging that global warming spreads malaria. "For 12 years, my colleagues and I have protested against the unsubstantiated claims," Reiter wrote in the *International Herald-Tribune.* "We have done the studies and challenged the alarmists, but they continue to ignore the facts" (Broad 2007).

Scientific Support for Gore

Gore does enjoy support among scientists who believe that his science is basically sound. During December 2006, Gore spoke in San Francisco at the American Geophysical Union's annual meeting to an audience of several thousand scientists. "He has credibility in this community," said Tim Killeen, the group's president and director of the National Center for Atmospheric Research, a top group studying climate change. "There's no question he's read a lot and is able to respond in a very effective way" (Broad 2007).

James E. Hansen, director of NASA's Goddard Institute for Space Studies and a frequent adviser to Gore, said, "Al does an exceptionally good job of seeing the forest for the trees," adding that Gore often did so "better than scientists" (Broad 2007). Hansen noted imperfections and technical flaws in Gore's work. For example, Hansen faulted Gore's simplistic association of strengthening hurricanes with warming seawater (many other factors play a role), especially following the devastating storms (including Katrina) during 2005. "We need to be more careful in describing the hurricane story than he is," Hansen said. "On the other hand," Hansen continued, "he has the bottom line right: most storms, at least those driven by the latent heat of vaporization, will tend to be stronger, or have the potential to be stronger, in a warmer climate" (Broad 2007).

Gore defended his work as fundamentally accurate. "Of course," he said, "there will always be questions around the edges of the science, and we have to rely upon the scientific community to continue to ask and to challenge and to answer those questions." He said "not every single adviser" agreed with him on every point, "but we do agree on the fundamentals—that warming is real and caused by humans" (Broad 2007).

Michael Oppenheimer, a professor of geosciences and international affairs at Princeton who advised Gore on the book and movie, said that reasonable scientists disagreed on the malaria issue and other points that the critics had raised. In general, he said, Gore has distinguished himself with integrity. "On balance, he did quite well—a credible and entertaining job on a difficult subject," Dr. Oppenheimer said. "For that, he deserves a lot of credit. If you rake him over the coals, you're going to find people who disagree. But in terms of the big picture, he got it right" (Broad 2007).

An Inconvenient Truth

An Inconvenient Truth was made after Laurie David, a prominent Hollywood environmentalist, saw Gore give a short version of his presentation at an event held just before the premiere of the climate disaster movie *The Day After Tomorrow.* Stunned by the power of Gore's talk, David helped organize presentations in New York and Los Angeles for people involved in the news media, environmental groups, business, and entertainment.

By the time she had done the Los Angeles event, "I realized we had to make a movie out of it," she said. "What's the guy going to do? There are not physically enough hours in the day to travel to every town and city to show this thing" (Revkin 2006c). David helped recruit a team of filmmakers and investors and persuaded Gore to be followed by a film crew as he traveled around the world.

Support for Renewable Energy Sources

Not deterred by the stridency of his critics, Gore has led a popular surge into support of renewable energy sources, especially wind power. Speaking on July 17, 2008, Gore said: "We're borrowing money from China to buy oil from the Persian Gulf to burn it in ways that destroy the planet.... Every bit of that's got to change." He urged a 10-year goal of getting 100 percent of U.S. electricity from renewable sources and clean, rather than carbon-based, fuels. Within weeks, T. Boone Pickens, who earned a billion-dollar fortune in the oil industry, switched his emphasis to wind power and bought an extensive advertising campaign. Pickens also invested heavily in wind-power infrastructure.

Gore said that enough solar energy falls on the surface of the Earth every 40 minutes to meet 100 percent of the entire world's energy needs for a full year. "Tapping just a small portion of this solar energy could provide all of the electricity America uses," he said (Gore 2008). "Today I challenge our nation to commit to producing 100 percent of our electricity from renewable energy and truly clean carbon-free sources within 10 years," Gore said. "This goal is achievable, affordable and transformative. It represents a challenge to all Americans—in every walk of life: to our political leaders, entrepreneurs, innovators, engineers, and to every citizen" (Gore 2008).

Gore developed his proposal in some detail:

> To be sure, reaching the goal of 100 percent renewable and truly clean electricity within 10 years will require us to overcome many obstacles. At present, for example, we do not have a unified national grid that is sufficiently advanced to link the areas where the sun shines and the wind blows to the cities in the East and the West that need the electricity. Our national electric grid is critical infrastructure, as vital to the health and security of our economy as our highways and telecommunication networks. Today, our grids are antiquated, fragile, and vulnerable to cascading failure. Power outages and defects in the current grid system cost US businesses more than $120 billion dollars a year. It has to be upgraded anyway. (Gore 2008)

"The political system, like the environment, is nonlinear," Al Gore said. "In 1941 it was impossible for us to build 1,000 airplanes. In 1942 it was easy. As this pattern becomes ever more clear, there will be a rising public demand for action" (Revkin 2006c). The film concludes with Gore stating that the one element missing in the fight against global warming was political will. In a line that some have interpreted as a hint of electoral ambitions, Gore added, "In America, political will is a renewable resource."

In a cover story for *Entertainment Weekly* (July 21, 2006), Gore said of *An Inconvenient Truth,* "This isn't a political film.... Global warming isn't a political issue. It's about the survival of the planet. Nobody is going to care who won or lost any election when the Earth is uninhabitable" (Svetsky 2006, 32).

FURTHER READING

Broad, William J. "From a Rapt Audience, a Call to Cool the Hype." *New York Times*, March 13, 2007. http://www.nytimes.com/2007/03/13/science/13gore.html.

Denby, David. "Review: *An Inconvenient Truth.*" *The New Yorker*, June 19, 2006, 23.

Dionne, E. J., Jr. "Gore's Energy Oomph." *Washington Post*, July 18, 2008. http://www.washingtonpost.com/wp-dyn/content/article/2008/07/17/AR2008071701840_pf.html.

Gibbs, Walter, and Sarah Lyall. "Gore Shares Peace Prize for Climate Change Work." *New York Times*, October 13, 2007, A-1, A-13.

Gore, Al. "A Generational Challenge to Repower America." July 17, 2008. http://www.wecansolveit.org/content/pages/304/.

Kolbert, Elizabeth. "Testing the Climate." (Talk of the Town) *The New Yorker*, December 24 and 31, 2007d, 43–44.

Revkin, Andrew C. "*An Inconvenient Truth*: Al Gore's Fight against Global Warming. *New York Times*, May 22, 2006c. http://www.nytimes.com/2006/05/22/movies/22gore.html.

Revkin, Andrew C. "Gore Calls for Immediate Freeze on Heat-Trapping Gas Emissions." *New York Times*, September 19, 2006j. http://www.nytimes.com/2006/09/19/washington/19gore.html.

Svetsky, Benjamin. "How Al Gore Tamed Hollywood." *Entertainment Weekly*, July 21, 2006, 26–32.

Gray Whales and El Niño

During 1999 and 2000, following intense El Niño conditions that sharply reduced their food supply, hundreds of dead gray whales floated ashore along the U.S., Canadian, and Mexican west coasts, from Alaska to Puget Sound to San Francisco Bay, and Baja California. According to a report in the *Los Angeles Times*, "The putrid carcasses became such a nuisance in 1999 and 2000 that beach communities took to towing the 35-ton cadavers out to sea or burying them with backhoes. Eskimo whalers reported harpooning 'stinky' whales that appeared to be rotting alive, too smelly even for dogs to eat" (McFarling and Weiss 2002, 1).

The whales, which weigh 35 to 50 tons each, spend their summers in the Bering Strait, gorging on millions of amphipods—crustaceans that live in tubes in the mud and sand on the shallow ocean floor. The whales can eat other foods, but the amphipods make up 95 percent of their diets in the Arctic. During the last 50 years, much of the ice that covers Arctic waters has melted earlier than usual. Lack of ice disrupts the food web. Even though the Arctic has been warming on average, a few locations are colder than usual. In some areas, the late spring ice has been slow to recede, keeping whales from reaching accustomed feeding grounds.

The die-off of whales largely ended after El Niño subsided, but gray whale populations plunged by more than one-third, falling from an estimated peak of 26,600 in 1998 to about 17,400 during the spring of 2002, the lowest in nearly two decades. The population was as many as 118,000 before commercial whaling during the 1800s attacked them.

Populations rose to about 22,000 by 2007, but began to decline again as waters in their habitats continued to warm (Eilperin 2007c, A-12). "That's a jolting decline for a long-lived species," said Ray Highsmith, a professor of marine science at the University of Alaska at Fairbanks and an expert on the main food source for gray whales. "If the numbers are right, there's something seriously wrong" (McFarling and Weiss 2002, 1). The recovering population of whales may have been eating more than nature could provide. At the same time, nature itself, buffeted by global warming and shorter-term climate changes such as El Niño, may have been producing less of the cocktail-shrimp-size seafloor amphipods that are the primary food of gray whales (McFarling and Weiss 2002, 1).

"All of a sudden, in 1999, the bottom fell out. We went from 1,400 calves to 420. Strandings jumped from 35 to 270," said Wayne Perryman, a biologist with the Southwest Fisheries Science Center in La Jolla. "That's not a subtle signal" (McFarling and Weiss 2002, 1). The casualties didn't just include the sick, weak, young, and very old. Many of the dead animals were in the prime of their 50-year life spans.

In the spring of 2000, veterinarian Frances Gulland of Sausalito's Marine Mammal Center conducted full necropsies on three animals and found as many distinct causes of death: viral encephalitis, the biotoxin domoic acid, and parasitic abscesses. "All of those could have initially started as malnutrition," Gulland said. "The real question is, why were they so malnourished? Why did they get whatever caused them to die?" (McFarling and Weiss 2002, 1). Both the living and dead animals were so skinny that their ribs stuck out. Their scrawniness was visible even from aerial photographs.

FURTHER READING

Eilperin, Juliet. "Warming May be Hurting Gray Whales' Recovery." *Washington Post*, September 11, 2007c, A-12. http://www.washingtonpost.com/wp-dyn/content/article/2007/09/10/AR2007091002143_pf.html.

McFarling, Usha Lee, and Kenneth R. Weiss. "A Whale of a Food Shortage." *Los Angeles Times*, June 24, 2002, 1.

Great Barrier Reef, Australia

Australia's 1,200-mile-long Great Barrier Reef is the largest coral-reef system in the world, including more than 2,600 individual reefs and about 300 islands. The Barrier Reef, which has been described as the Amazon jungle of the marine world, has been severely degraded, however, having suffered, during 2002, its worst bleaching event on record ("New Wave" 2002). The destruction of the Great Barrier Reef has been accelerated not only by warming water temperatures, but also by agricultural runoff, including fertilizers, herbicides, and pesticides, which kill corals. Runoff also causes increases in algal blooms, which feed increasing populations of crown-of-thorns starfish that devour corals.

Extensive areas of Australia's Great Barrier Reef showed no signs of recovery a year after extensive bleaching during the summer of 2001–2002. At that time, aerial surveys showed that about 60 percent of the reef's 6,700 square kilometers had been affected. According to Ray Berkelmans, a research scientist at the Australian Institute of Marine Science in Townsville, recovery was "poor to non-existent.... What had been beautiful reef is now acres and acres of dead coral covered with algae" (Roberts 2003, 4). Berkelmans and other scientists expressed concern that another hot summer could damage the reef even more. The Great Barrier Reef has experienced six bleaching events since 1980.

Half of Coral Cover Lost

By 2004, the Great Barrier Reef had lost about half its coral cover, compared with the 1960s. During the 1960s, about 40 percent of the reef was covered with corals; by 2004, the coverage averaged just 20 percent (Williams 2004, 11). Australian scientists said that this decline was attributable entirely to human impact, most notably from rising water temperatures caused by global warming, overfishing, and water pollution. David Bellwood of James Cook University in Townsville, co-author of a major review published in *Nature* (2004, 827–833), said that the loss of coral was not a surprise to scientists who knew of the damage done by three major outbreaks of crown-of-thorns starfish since the 1960s and two large-scale bleaching events in 1998 and 2000. "Data has been accumulating for years on this and we've now gotten around to pulling it all together and looking at the overall pattern," he said (Williams 2004, 11).

Coral bleaching in the Great Barrier Reef Marine Park may be the worst on record, scientists said in late May 2002, after the most comprehensive aerial survey ever conducted. The survey is aimed at helping unravel the implications of global warming for reef management ("Pacific too Hot" 2002). "Our aerial surveys found that nearly 60 percent of the reef area in the marine park was heat stressed to some extent as indicated by bleaching," said Berkelmans. "Until now, the coral bleaching episode in 1998 was the worst on record, but the 2002 event was probably worse because more reef area was affected," said Berkelmans ("Pacific Too Hot" 2002).

The survey by scientists from the Australian Institute of Marine Science, CRC Reef, and the Great Barrier Reef Marine Park Authority examined more than 640 locations from the northern tip to the southern end of the Great Barrier Reef Marine Park using light aircraft. The team also used SCUBA (self-contained underwater breathing apparatus) to confirm results and determine whether corals were likely to recover from bleaching or would die. The aerial surveys indicated that bleaching was worst in the Princess Charlotte Bay region, near the Turtle Island Group, on inshore reefs from Cape Upstart to the Whitsundays and in some reefs in the Sir James Smith Group and the Keppel Island area. Moderate to very high bleaching was observed inshore and offshore from around Cape Flattery to Mackay ("Pacific Too Hot" 2002).

Few Reefs Escape Bleaching

"Our underwater surveys found that few reefs escaped bleaching, but it appears likely that most reefs will recover with only minor death of corals," said Paul Marshall of the Great Barrier Reef Marine Park Authority, who led the underwater surveys. "We did find that some of the most severely bleached reefs were devastated with 50 percent and 90 percent of coral dead at some sites" ("Pacific Too Hot" 2002).

Several toxic pollutants carried from Australian farms by floods have been threatening inshore areas of the Barrier Reef, according to a study by marine experts. The report, by the Great Barrier Reef Marine Park Authority, found pollution levels in floodwaters, or flood plumes, were four times worse than 15 years previously (Freeman 2002, 7). "Of about 750 inshore reefs in the park designated as a world heritage site by the United Nations, 200 were considered at high risk and more than 400 were at risk," said Sheriden Morris, the park authority's water-quality director (Freeman 2002, 7).

This study said that floodwaters occasionally surge onto the reefs from 26 river systems in Queensland, carrying sediment, nutrients, herbicides, and pesticides. Sediment and nutrients damage sea grass beds and create algae growth, while pesticides and herbicides stop the growth of plankton and sea grass, the report said. The study found that concentrations of dissolved nutrients in flood plumes were well above levels known to damage coral-reef ecosystems (Freeman 2002, 7). Imogen Zethoven, World Wildlife Fund Great Barrier Reef campaign manager, said that most of the pollution comes from agricultural land. "You're getting high levels of fertilizer run-off from sugar cane growing properties and horticultural properties.... We've got to find a way of growing cattle, cane and fruit and vegetables where those industries can coexist harmoniously with the tourism and fishing industries," she said (Freeman 2002, 7).

By the summer of 2002, authorities in Australia were issuing individual heat-wave alerts for events that they believed could harm the Great Barrier Reef. Pervasive warming of shallow waters above the reefs was the most immediate concern. In the meantime, Australia's federal government, following the example of the United States under the administration of George W. Bush, was refusing to ratify the Kyoto Protocol, even as heat records were broken, wildfires nipped at the capital of Canberra, and the magnificent Barrier Reef decayed.

"My Grandchildren Won't See a Great Barrier Reef"

Australian Institute of Marine Science senior principal researcher John Veron said during 2002 that the Great Barrier Reef would be so severely degraded in 50 years by global warming and other anthropomorphic insults that "[m]y grandchildren won't see a Great Barrier Reef like I did, that's for sure.... It will be mostly dead." He continued:

> Words don't exist to describe what's just around the corner—coral reefs, more than anything else, are first in line for the effects of global warming. It's too late for a lot of the areas like the Great Barrier Reef and the reefs off Western Australia; I can't see any escape from that conclusion. ("Warming Doom" 2002)

In 50 years, coral bleaching, considered dangerously severe at the beginning of the twenty-first century may be regarded as mild by comparison, Veron said ("Warming Doom" 2002).

In contrast to Veron's bleak assessment, a few other reports asserted that the Great Barrier Reef is improving because of better management. These reports were a distinct minority, however. One such report trumpeted a belief that "[f]ears for the future of Australia's Great Barrier Reef were ... laid to rest with the revelation that it is now one of the world's healthiest coral reefs" ("Great Barrier" 2002, 11).

According to the Australian Institute of Marine Science, this report continued, "Only about six percent of the vast reef ... is now suffering from the phenomenon. The government's Great Barrier Marine Park Authority is credited with the restoration. It worked tirelessly to ensure water quality, protect fish stocks and set up marine sanctuaries" ("Great Barrier" 2002, 11).

The report continued, quoting Wilkinson:

> Reefs, if they are left alone and not stressed, will recover quite rapidly. Our first state-of-the-reef report in 1998 identified mass coral bleaching, which killed off about 16 percent of the world's coral stocks.... The latest report shows there has been recovery but it is in areas that are quarantined from other activity. ("Great Barrier" 2002, 11)

In addition to the destruction of the Great Barrier Reef, warming oceans are playing a major role in the decline of another Australian marine landmark, the giant kelp forests of Tasmania. Two-third of the kelp beds along Tasmania's east coast have died during the last 50 years, probably as a result of a 1.5° to 2°C rise in water temperatures. Kelp generally dies if waters rise above 20°C. Additionally, warm-water species that feed on kelp (such as sea urchins) have been moving in, reducing, according to marine ecologist Craig Johnson of the University of Tasmania, "rich, luxurious, highly diverse seaweed beds to barren, over-grazed 'moonscapes'" ("Kelp Points" 2004, A-8). The kelp forests are a major tourist attraction and habitat for rock lobster and abalone ("Kelp Points" 2004, A-8). *See also:* Coral Reefs

FURTHER READING

Bellwood, D. R., T. P. Hughes, C. Folke, and M. Nystrom. "Confronting the Coral Reef Crisis." *Nature* 429 (June 24, 2004):827–833.

Freeman, James, and Eleanor Cowie. "Pollutants Threaten the Great Barrier Reef." *Glasgow Herald*, January 25, 2002, 7.

"Great Barrier Reef Is Springing Back to Life." *Western Daily Press* (Australia), December 18, 2002, 11.

"Kelp Points to Worrying Sea Change." *Canberra Times*, August 30, 2004, A-8.

"New Wave of Bleaching Hits Coral Reefs Worldwide." Environment News Service, October 29, 2002. http://ens-news.com/ens/oct2002/2002-10-29-19.asp#anchor1.

"Pacific Too Hot for Corals of World's Largest Reef." Environment News Service, May 23, 2002. http://ens-news.com/ens/may2002/2002-05-23-01.asp.

Roberts, Greg. "Great Barrier Grief as Warm-water Bleaching Lingers." *Sydney Morning Herald*, January 20, 2003, 4.

"Warming Doom for Great Barrier Reef." Australian Associated Press in *The Mercury* (Hobart, Australia), February 16, 2002, n.p. (LEXIS)

Williams, Brian. "Reef Down to Half Its Former Self." *Courier Mail* (Queensland, Australia), June 24, 2004, 11.

Great Britain, Weather Conditions and Leadership in Greenhouse Diplomacy

The British government has been among the world's most acutely aware of global warming's potential consequences. In stark contrast to the United States, where the George W. Bush administration was doing its best to edit the problem out of public consciousness at the turn of the millennium, British officialdom sounded sharp and frequent warnings. "In recent years more and more people have accepted that climate change is happening and will affect the lives of our children and grandchildren. I fear we need to start worrying about ourselves as well," said Margaret Beckett, British Environment, Food and Rural Affairs secretary (Clover 2002a, 1).

Also unlike the U.S. federal government, political leaders in Great Britain have long been international leaders in greenhouse diplomacy. The tradition stems from Margaret Thatcher, a conservative politically, but also an astute student of the physical sciences who realized early the stakes of the issue. Even the right wing in Britain advocates political action to counter global warming; British conservatives found George W. Bush's resistance to the scientific record laughable. Britain, a rather small and densely populated island compared with the United States, also has been shaken by large-scale changes in its heretofore usually benign maritime climate.

On October 30, 2006, shortly before he left office, British Prime Minister Tony Blair warned that the Earth is encountering "disastrous" and "irreversible" climate changes. Meanwhile, David Cameron, the popular leader of the opposition Conservative Party, vowed to install a wind-power generator and solar panels at the prime minister's residence if he won the office. Blair and Cameron were rivals, but both shared opposition to global warming. This bond illustrated the unity among leaders across the political spectrum in Britain on the issue (Sullivan 2006, A-18).

Unlike the United States, global warming dominates British political discourse "as leaders take to bicycles and hybrid cars in an effort to be more green than their opponents. The Conservatives recently adopted a tree as the party's logo, raising eyebrows among many in a party that is generally more committed to lowering taxes than planting trees. A political cartoon in the *London Times* newspaper showed several national candidates hugging trees and pointing at each other, saying, "The other two are faking it!" (Sullivan 2006, A-18).

In Great Britain, the London tabloids often feast on fears of weather gone wild. The British government is acutely aware of climate change's perils, a subject of many reports that argue, for example, that sizable parts of London may be abandoned to rising seas within a century. There exists on this island a palatable feeling of climatic assault by sea and atmosphere, as parts of Dover's White Cliffs crumble and some of Scotland's St. Andrews golf course's famous links surrender to the sea.

By October 2004, even Britain's Queen Elizabeth was criticizing Bush's inertia on the subject. The leader of Britain's opposition Conservative Party, Michael Howard, also criticized U.S. President George W. Bush for failing to tackle climate change. One day after the Conservatives' challenge, British Prime Minister Tony Blair pressured the United States and Russia to face up to the "catastrophic consequences" of climate change, as he issued his starkest warning yet about the "alarming and unsustainable" consequences of global warming. He said that within the lifetime of his children—and possibly his own—the impact on the world could be so far-reaching and "irreversible in its destructive power" that it will alter human existence radically (Jones and Clover 2004, 2).

Great Britain's Warming and Wild Weather

In central England, the growing season has lengthened by one month since 1900, with an annual temperature increase of 1°C. Even before Europe's searing summer of 2003, climate change had become an important factor in English political discourse.

The worst storm experienced by England in a decade caused road and rail chaos across the country, killed six people, and left hundreds of millions of pounds worth of damage in its wake on October 30, 2000. Torrential rain and winds up to 90 miles per hour uprooted trees, blocked roads, and cut electricity supplies across southern England and Wales. According to newspaper reports, a tornado ripped through a trailer park in Selsey in West Sussex less than 48 hours after a similar twister had devastated parts of Bognor Regis. In Yorkshire, the first blizzards of the winter coincided with flash floods. English weather recordkeepers said that October's rainfall in East Sussex, one of the driest parts of the country, had been nearly three times its average, at 226 millimeters (nine inches). September also had been exceptionally wet.

Marilyn McKenzie Hedger, head of the U.K. Climate Impacts Program based at Oxford University, said, "These events should be a wake-up call to everyone to discover how we are going to cope with climate change" (Brown 2000b, 1). Michael Meacher, U.K. environment minister, said that while it would be foolish to blame global warming every time extreme weather conditions occur, "[t]he increasing frequency and intensity of extreme climate phenomena suggest that although global warming is certainly not the sole cause, it is very likely to be a major contributory factor" (Brown 2000b, 1). The storm included the lowest barometric pressure on record in the United Kingdom during October, 951 millibars.

The next day, John Prescott, deputy prime minister, said that extreme weather events must now be regarded as usual fare in Britain as global warming takes hold. Railways, power lines, and flood defenses must adapt, he said. Government officials, local authorities, emergency services, and the environment agency were summoned to a meeting in London a day later, where Prescott demanded action, saying, "Our infrastructure should be robust enough, and our preparations rigorous enough, to withstand the kind of weather we have just experienced." Prescott continued:

> We aren't putting the amount of resources and investment in for what we call more extreme conditions, which we must now accept [are] normal. We have to ask ourselves: should our power lines come down every time we have such storms? Should 1,000 trees fall across our railway lines in the southeast? Should we do more to prevent flooding? Are our drainage systems really adequate? (Brown 2000c, 1)

During the fall of 2002, Britain was being battered by intense storminess that was described in the *London Guardian*:

> As big storms go, yesterday morning's [October 27, 2002] was not quite on the scale of October 16, 1987, when some 15 million trees were uprooted in a wild night that changed the face of southern England. But ecologists said yesterday it was a timely reminder that the terrifying weather once assumed to take place only every 250 years is now liable to occur far more frequently. The 1987 storm, which left a devastating trail across 10 counties and killed 18 people, was said to be the greatest in Britain since 1709. British Broadcasting Corporation weatherman Michael Fish had said the night before that "no hurricane" was expected, but wind speeds of over 115 m.p.h. in Norfolk and along the south coast were not far short of those usually seen in the Caribbean and the Pacific. (Vidal 2002e, 3)

Great Britain: "Future Flooding"

During 2004, a panel of 60 British climate-change experts released a government-sponsored report, "Future Flooding," which asserted that the homes of as many as four million Britons may be at risk of inundation by 2050. The report said that the national cost of flooding may rise from $2.6 billion a year about 2000 to $52 billion annually by 2080. Some government officials warned that the government might be forced to consider an "orderly retreat" from London because parts of the 2,000-year-old city are below sea level. Professor Paul Samuels, who is leading a Europe-wide study of flooding, said London could be "mostly gone in the next few centuries" (Melvin 2004, 3-A).

The flooding report said that Britain must create "green corridors" in cities to act as safety valves into which floodwaters can be channeled. It said parts of some urban areas may have to be abandoned, and oil refineries moved inland. Many homes, it warned, may become uninsurable. Samuels, who suggested that the government retreat from London, was working on the premise that the tidal section of the Thames River would rise as much as to three feet in a century, a situation exacerbated by the subsidence of the land on which some of London is built (Melvin 2004, 3-A).

English scientists have considered different scenarios for high, medium, and low emissions of greenhouse gases, predicting the following changes in the British climate by 2080: A rise in average temperature of 2° to 3.5°C, probably with greater warming in the south and east. Generally, the climate could become be like that of Normandy, the Loire, or Bordeaux, varying according to the level of global greenhouse-gas emissions. Hot days in summer will be more frequent, with some above 40°C (104°F) in lowland Britain under the high-emission scenario. According to the highest-emission projections, the United Kingdom's summer rainfall may decrease by 50 percent and winter rainfall may increase by 30 percent. Snowfall will decrease throughout Britain. Scotland may experience 90 percent less snow, according to the highest-emissions scenario. Sea levels could rise by 26 to 86 centimeters, (10 to 34 inches). The probability of extreme storm surges will increase from 1 in 50 years to 9 in 10 years under the high-emissions scenario (Clover 2002a, 1).

After several years of excessive rains, Great Britain by 2003 was experiencing drought. At Hyde Park, London, between February and April 2003, rainfall measured 2.9 inches. During the same period in 2002, the total precipitation was

5.4 inches, and in 2001, concluding the wettest 12 months on record in England and Wales, 10.8 inches of rain fell at the same station during the same period.

Tim Sparks, a research biologist at the Centre for Ecology and Hydrology in Cambridge, said that England's spring of 2002 could be its earliest in 300 years:

> Winter is being shrunk at both ends. We had a very late autumn last year and most of January and February have been mild. Nobody has heard the first cuckoo yet—usually [in] mid-April—but the first frogspawn was observed on December 10, the first primroses as early as October, and the first snowdrops a week before Christmas. (Vidal 2002a, 3)

Britain's Warmest October on Record

As concern about global warming was smothered in the United States by a deluge of terrorism concerns following the September 11, 2001, attacks on New York City and Washington, D.C., Britain experienced its warmest October on record. "Butterflies from America, birds from the Mediterranean, mushrooms afoot and spring flowers in bloom—all part of Britain's warmest-ever October, according to records kept since 1659," said one account (McCarthy 2001, 12). The average temperature for the month is 13.5°C, much higher than the long-term average of 10.6°C. The previous October record of 13°C had been set in 1969. During the warmest spell, in the middle of the month, several places, including London, reached a temperature of more than 25°C.

According to a report in the *London Independent*,

> Wildlife has responded accordingly. Bees, moths, and dragonflies were all to be seen yesterday [October 29, 2001] at the Wetland Centre nature reserve in Barnes, southwest London. Butterflies have been on the wing astonishingly late—meadow browns, holly blues and speckled woods have been prolific and are still visible. On October 21, Martin Warren, of the charity Butterfly Conservation, saw a Silver-studded Blue in the New Forest, the latest recorded [sighting] for the area for more than 70 years. (McCarthy 2001, 12)

An influx of monarch butterflies from North America was blown by storm winds across the Atlantic when they should have been migrating to Mexico.

In parts of Devon, daffodils poked through the soil out of cycle. In the Gordano Valley, Somerset, a scientist affiliated with the charity Plantlife found flowering marsh marigolds, which are usually in bloom from March to June. Plantlife volunteers also reported seeing other spring flowers in bloom, including cow parsley and blue fleabane (McCarthy 2001, 12). Many birds that should have been migrating southward from southern Europe reversed course and flew northward to Britain. A spokesman for the British Meteorological Office said, "October really has been an astonishingly warm month. We can't say this is by itself proof of global warming, but it is certainly another piece of the jigsaw" (McCarthy 2001, 12).

Of the record October warmth in Britain, an observer wrote in the London *Times*:

> Keats would have been as bewildered as the bees that "think warm days will never cease." Our autumn is not his: the mists have been dispersed by hot sun or drowned by tropical downpours, and mellow fruitfulness has been rejuvenated by trees in bud, flowers in bloom and grass still pushing up sturdily across a million lawns. In Kent the moss'd cottage trees may soon no longer bend with apples, as farmers grub up their orchards and plant walnuts, sunflowers and vines more commonly found in the distant oases of Uzbekistan than the Garden of England. No longer are stubble-plains touched with rosy hue; most are now under several feet of water, as glassy floods spread far across the land. (Walnuts 2001)

Monsoon Britain

By 2000, England's summers sometimes were becoming quasi-tropical. According to one newspaper report in on August 11, 2002, "In the last three weeks alone, temperatures above 30°C and brutal storms capable of sending down several normal weeks of rain in a few minutes were recorded in London." Scientists believe such events are merely a taste of things to come. Experts at the Tyndall Centre of Climate Change at the University of East Anglia warn that precipitation is "becoming increasingly erratic as the planet heats up, offering humankind arguably its biggest challenge to date" (Townsend 2002, 15).

In mid-August 2002, a sudden deluge dumped 20 days' worth of rain on London in 30 minutes,

> causing chaos as the antiquated drainage systems and transport infrastructure failed almost instantly and flash floods turned the city's streets into rivers. Glasgow, was still recovering from its own "freak storm" that overwhelmed parts of the sewage

system. The crisis forced scores of families into temporary accommodation and thousands had to boil drinking water as floods contaminated the supply. (Townsend 2002, 15)

On August 10, 2002, seaside towns in North Yorkshire were hit by flash floods, as more than 100 people were evacuated from their homes.

During August 2004, a deluge over London flushed 600,000 tons of sewage into the Thames. Two weeks later, mud and debris carried by another deluge swept into the British coastal town of Boscastle, which lies at the confluence of three rivers, sweeping away homes and cars. The flood was fed by more than six inches of rain (2.5 inches of which fell in one hour) and complicated by an unusually high tide that impeded the water's flow into the sea. During the same series of storms, residents of Knighton reported a shoal of fish falling from the sky during a thunderstorm. August 2004 was among England's wettest in recorded history, with copious flash floods, landslides, lightning strikes, hailstorms, and at least 14 tornadoes.

Occasional wet summers are hardly unknown in England, however. Boscastle itself was flooded in 1847, 1957, and 1958, as well as in 2004. The frequency of Britain's climatic violence is new, however. John Turnpenny, senior research adviser at the Tyndall Centre, said that he expected "monsoon" conditions to become more prevalent, particularly in southern England. That means the violent storms of the early twenty-first century could become commonplace, while heat waves resembling those of 1995 and 2003—during which maximum temperatures remained above 25°C for 17 days in August—"[are] forecast in Britain for two out of every three years by the time that this century ends" (Townsend 2002, 15).

Elaine Jones, a British government climate specialist, said that the Victorian infrastructure of London, particularly its sewers, needs to be improved if it is to maintain its position as one of the world's great cities. "The drains can't take the water away. The result is that you risk sewage in the streets and all those associated public health risks," Jones told the *London Observer* (Townsend 2002, 15).

Palm Trees and Banana Plants in English Gardens

Traditional English gardens have been changing as climate warms. As described in an Associated Press dispatch carried in Canada's *Financial Post*:

The fabled English garden with its velvety green lawn and vivid daffodils, delphiniums and bluebells is under threat from global warming, leading conservation groups said late in 2002. Within the next 50 to 80 years, palm trees, figs and oranges may find themselves more at home in Britain's hotter, drier summers, the National Trust and the Royal Horticultural Society said, releasing a new report on the impact of climate change. *Gardening in the Global Greenhouse: The Impacts of Climate Change on Gardens in the U.K.* was commissioned by the two organizations and the government, as well as water, forestry and botanical organizations. (Woods 2002, S-10)

The Chelsea Flower Show in May 2002 "strongly reflected the trend for Mediterranean-style plants suitable for dry conditions" (Johnson 2002, 5). Climate models for England projected warmer, drier summers and wetter winters. Landscape architects are faced with a paradox of finding plants that can survive hotter, drier summers while building landscapes that can carry off a larger volume of winter floodwaters. Guy Barter, head of the Royal Horticultural Society advisory service, said, "Olive trees, grapes, avocados and even banana plants could all become common garden features. The air could be full of the scent of acacia.... We will also see more gardens with heat-resistant trees, and cacti and yucca. But the problem will be flooding in winter" (Johnson 2002, 5).

The May 2002 edition of the British Meteorological Society's magazine *Weather* reported evidence that the English growing season has been lengthening by an average of a day a year; in 2000 it was the longest on record, at 330 days (leaving a "winter" of 35 days). "If the trend continues, it is possible we'll have a year-round growing season within a generation," said Tim Mitchell of the Tyndall Research Centre (Johnson 2002, 5).

At about the same time, the *London Guardian* reported:

Daffodils [are] blooming near Buntingford in Hertfordshire. Helped by one of mildest winters on record, other spring indicators such as budding blackthorn and the first butterfly have arrived early. Scientists say the trend will continue. The evidence of the past five weeks suggests that this has been one of the warmest starts to a year since records began more than 300 years ago. The signs that spring has not just stirred but has actually sprung up to three weeks earlier than usual are now everywhere. Daffodils have been blooming in Scotland for weeks, toads have been on their ancient migratory marches to breeding ponds well before schedule, hawthorn buds are bursting in hedgerows, and the lesser celandine and other vernal indicators are flowering in woodland. (Vidal 2002a, 3)

Biological and botanical events that heretofore had taken place in March, or even April, during 2002 began in February: "Lambs were 'gamboling around,' the elder leaves and forsythia flowers were out and the horse chestnuts and blackthorn were budding. Tits were nesting, frogs were spawning, and gardeners were mowing lawns that had grown shaggy during the winter" (Vidal 2002a, 3). *See also:* Global Warming: Origins as a Political Issue

FURTHER READING

Brown, Paul. "Global Warming—It's with Us Now: Six Dead AS Storms Bring Chaos Throughout the Country." *London Guardian*, October 31, 2000b, 1.

Brown, Paul. [No headline]. *London Guardian*, November 1, 2000c, 1.

Clover, Charles. "2002 'Warmest for 1,000 Years.'" *London Daily Telegraph*, April 26, 2002a, 1.

Johnson, Andrew. "Climate to Bring New Gardening Revolution: Hot Summers and Wet Winters Could Kill Our Best-loved Plants." *London Independent*, May 12, 2002, 5.

Jones, George, and Charles Clover. "Blair Warns of Climate Catastrophe: 'Shocked' Prime Minister Puts Pressure on U.S. and Russia over Emissions." *London Daily Telegraph*, September 15, 2004, 2.

Laurance, Jeremy. "Climate Change to Kill Thousands, Ministers Warned." *London Independent*, February 9, 2001, 2.

McCarthy, Michael. "Global Warming: Warm Spell Sees Nature Defying the Seasons." *London Independent*, October 30, 2001, 12.

Melvin, Don. "There'll Always Be an England? Study of Global Warming Says Sea Is Winning." *Atlanta Journal-Constitution*, June 5, 2004, 3-A.

Sullivan, Kevin. "In Britain, All Parties Want to Color the Flag Green." *Washington Post*, October 31, 2006, A-18.

Townsend, Mark. "Monsoon Britain: As Storms Bombard Europe, Experts Say That What We Still Call 'Freak' Weather Could Soon Be the Norm." *London Observer*, August 11, 2002, 15.

Vidal, John. "The Darling Buds of February: Daffodils Flower and Frogs Spawn as Spring Gets Earlier and Earlier." *London Guardian*, February 23, 2002a, 3.

Vidal, John. "Better Get Used to It, Say Climate Experts." *London Guardian*, October 28, 2002e, 3.

"Walnuts and Vineyards." London *Times*, October 29, 2001, n.p. (LEXIS)

Woods, Audrey. "English Gardens Disappearing in Global Warmth: Will Be Replaced by Palm Trees." Associated Press in *Financial Post* (Canada), November 20, 2002, S-10.

Great Lakes, North America

A warming climate for inland lakes (notably the Great Lakes of North America) generally will not raise water levels, as in the oceans, but rather decrease water levels, as declining snowfall in the hinterlands deprives the lakes of runoff. Reduced ice cover and higher temperatures also lead to increased evaporation. "A decade ago," wrote Kari Lydersemn in the *Washington Post*, "Chicago winters meant monumental ice hillocks and caves forming along the lakeshore, skirted by interlocking ice sheets like a giant jigsaw puzzle. Today, it is rare to see more than a thin frozen shelf or a few small ice floes sloshing in Lake Michigan below the city's skyline" (Lydersen 2008, A-4).

Lower water levels impede shipping, forcing lighter loads, and require power plants that use lake water to extend cooling pipes. Lower water keeps boaters from using docks and kills animals that once flourished in wetlands. Depending on the size of the ship, every inch of lost draft (the level at which a ship rides in the water) means 50 to 270 tons less cargo (Lydersen 2008, A-4). Low water also increases groundings. Delayed deliveries of raw materials affect factory production.

"We firmly believe the changes we're seeing are impacting fisheries, possibly in a dramatic way," said Jeff Skelding of the National Wildlife Federation. "Disruption of habitat will impede fish species from being able to reproduce" (Lydersen 2008, A-4).

Many Factors Influence Water Levels

Factors other than climatic warming also may be reducing lake levels in the Great Lakes. One is the "bathtub drain" effect caused by dredging in the St. Clair River between Lake Huron and Lake St. Clair, which also influences the level of Lake Erie. Dredging removes water from the upper Great Lakes to the St. Lawrence Seaway, then to the Atlantic Ocean. "Isostatic rebound," a rising of Earth's crust after glaciers have melted, also may be a relatively minor factor. The Great Lakes are remnants of melted glaciers.

As temperatures have warmed, ice fishermen on the Great Lakes have found that they must check weather forecasts before planning their activities. In recent years, the lakes have frozen later than usual—and in some years, ice has not reached thickness required to support the weight of people, their equipment, and their cars. Lake Eire, for example, froze every year between 1953 and 1998, then failed to freeze three of the next nine years, according to the Army Corps of Engineers (Higgins 2007). The next winter, in early January, the lake's water was 41°F, a record high. "This year [2006] we had the second-warmest November and December in history," said Alan Blackburn, a meteorologist with the National Weather Service in Buffalo. "We're seeing more warm years in the recent decade. There's something happening here" (Higgins 2007).

Without seasonal ice cover, millions of liters of water have been evaporating from the Great Lakes. Part of what evaporated fell as lake-effect snow at the eastern end of the lakes, creating something of a paradox: global warming, at least for the time being, was burying some lakefront areas in heaver-than-usual snows. Some of these snows were wildly variable.

Wild Lake-Effect Weather

Having had no snow in November 2001, and only 1.5 inches until the third week of December, residents of Buffalo, New York, greeted Christmas Eve with nearly two feet. During the ensuing week, Buffalo had another nearly three-foot storm; with a few other smaller storms, Buffalo ended the year with its snowiest month on record (about 83 inches), nearly all of which fell in *one* week. How is seven feet of snow in one week evidence of global warming? Lake Eire's water was much warmer than usual, and when cold air moved in suddenly, it set up the most ferocious period of lake-effect snowfall in Buffalo's recorded history.

During October 2006, a weird combination of summer and winter weather swept off the lakes. Douglas M. George, a resident on Oneida Castle, about 30 miles east of Syracuse, recalled:

> A nice clear day here in Oneida—but 100 miles west the folks in Rochester-Buffalo have found out what global warming really means. A truly unique storm struck that region last night with a foot of "thunder snow" coating the ground, trees, roads and power lines. This was very strange as the water temperature of Lake Erie is a relatively warm 60° but a frigid upper air mass from central Canada flowed through over the waters and dumped its load of moisture as soon as it reached land. Most of the trees in the northeast have their leaves so the weight of the snow—one of the wet, heavy kind—has resulted in many cracking and falling down.... Parts of the New York lakeshore near Buffalo got two feet of snow, which quickly turned to mountains of slush. (George 2006)

Another account described the same storm:

> Heavy wet snow started falling about 2 P.M. Thursday. Snow built up on the leaves and branches of trees, bending and breaking even the healthiest of trunks. Heightening the clash of seasons, lightning flashed through the night and loud claps of thunder mingled with the cracks and pops of trees giving way under the weight of the snow. "It sounded like gunshots," said Percy Jackson, 18, who spent the day off from school shoveling sidewalks. The falling trees severed power lines, damaged cars and houses, and blocked countless streets. (Staba 2006)

The storm reached east to the Rochester suburbs, snapping power lines and cutting power to almost 400,000 homes and businesses, 70 percent of the city. Three people died during the storm, two in automobile accidents. The third was struck by a falling tree laden with heavy, wet snow. Don Paul, a meteorologist at WIVB-TV4 in Buffalo, who had worked in the area for more than 20 years, said, "Of all the events I've seen here, this storm involves the most widespread devastation in the most populated area," he said. "It's absolutely an historic storm" (Staba 2006).

According to the National Weather Service, 22.6 inches of snow fell at Buffalo Niagara International Airport during the storm, a record for October. "Our crews were wrapping up their workdays yesterday and the weather forecasters were still talking about rain," said Steve Brady, a spokesman for National Grid, which provides electricity to Buffalo as well as its largest suburbs. "We've had windstorms in the fall and ice storms in the fall, but never a snowstorm like this, this early" (Staba 2006). A similar "thundersnow" struck Omaha at the end of October 1997 as about a foot of snow destroyed many of the city's old maple and ash trees and knocked out power in some parts of the city for a week.

Great Lakes Water Levels Fall

By 2002, the Great Lakes were at their lowest point in 35 years, as various experts said that water levels were likely to drop even more because of unusually warm winter weather. By 2001, cargo ships were being forced to lighten their loads and many boat ramps became inaccessible. On lakes Michigan and Huron, the water levels dropped by more than 100 centimeters, beginning in 1997, and remained 35 centimeters below average into 2002. Lake Superior was more than 15 centimeters below average, Lake Erie was 10 centimeters below, and Lake Ontario was 2.5 centimeters below its average level (Mitchell 2002). During 2003 and 2004, the pattern reversed itself, at least for a time, with snowfall in the area returning to near average. With a La Niña weather pattern in control, the winter of 2007–2008 was very snowy in the Great Lakes watershed.

Even given cyclical weather patterns, by 2050, according to one study, water levels also may fall low enough to render the hydroelectric works of Niagara Falls nearly useless. "It's going to affect everything," said Rich Thomas, chief of water management at the Army Corps of Engineers in Buffalo, where the impact of climate change has been a growing concern (Zremski 2002, A-1).

Temperatures in the Great Lakes watershed are expected to rise, on average, from 3.5° to 9°F during the twenty-first century. "For the Great Lakes region, the next century could bring one of the greatest environmental transformations since the end of the last Ice Age," the U.S. Environmental Protection Agency said in a study on global warming in the Great Lakes (Zremski 2002, A-1). Temperatures in western New York might come to resemble the climate of western Maryland in 2000, according to David Easterling, chief scientist for the National Climatic Data Center (Zremski 2002, A-1).

Jeremy Zremski, writing in the *Buffalo News*, said:

> As temperatures rise, Lake Erie as we know it would be transformed. Like the rest of the Great Lakes, it would start to evaporate, meaning water levels would fall by as much as five feet over the next century. Most scientists expect the bulk of the drop to occur in the next few decades. The remaining water would be warmer and might never freeze during winter. (Zremski 2002, A-1)

Fewer Storms, Greater Intensity

Fed by greater evaporation from the Great Lakes, overall precipitation in the area could increase 10 to 20 percent, according to the U.S. Environmental Protection Agency. "A lot of scientists mention getting fewer storms but of greater intensity," said Helen Domske, a researcher at the University at Buffalo's Great Lakes Program (Zremski 2002, A-1). During the first few decades of warming, lake-effect snowstorms could be more frequent in Buffalo, thanks to Lake Erie's lack of ice. Easterling said, however, that temperatures probably will warm to a point at which lake-effect rain may become more common than snow. As a result, lake-effect snow could decrease by half within a century (Zremski 2002, A-1).

A comparative study of snowfall records in and outside of the Great Lakes region indicated a statistically significant increase in lake-effect snowfall in the region since the 1930s. In areas where the lake effect is small, little change has been observed. Warmer lake waters and decreased ice cover were cited by the researchers as the major reasons for the snowfall increases in lake-effect areas. A team of researchers, led by Colgate Associate Professor of Geography Adam W. Burnett, published the study, "Increasing Great Lake-Effect Snowfall during the Twentieth Century: A Regional Response to Global Warming?" in the November issue of the *Journal of Climate* (Burnett et al. 2003, 3535–3542).

Syracuse, New York, one of the snowiest cities in the United States, experienced four of its heaviest annual snowfalls on record during the 1990s, the warmest decade in the twentieth century. "Recent increases in the water temperature of the Great Lakes are consistent with global warming," said Burnett. "This widens the gap between water temperature and air temperature—the ideal condition for [lake-effect] snowfall" ("Global Warming Means" 2003).

The research team compared snowfall records from 15 weather stations within the Great Lakes region with 10 stations at sites outside of the region. Records dating to 1931 were examined for eight of the lake-effect and six of the non-lake-effect areas. Records for the rest of the sample dated back to 1950. "We found a statistically significant increase in snowfall in the lake-effect region since 1931, but no such increase in the non-lake-effect area during the same period," said Burnett. "This leads us to believe that recent increases in lake-effect snowfall are not the result of changes in regional weather disturbances" ("Global Warming Means" 2003).

Great Lakes water levels have dropped before, and the low levels of the years after 1997 followed a period of relatively high water levels in the lakes, which include 18 percent of the world's freshwater supply. Recent low levels were notable, according to climatic experts, because of the "amount of lowering and the rapidity with which it occurred," as well as the fact that "the primary hydroclimatological driver was high air temperatures [increasing evaporation], not extremely low precipitation" (Assel, Quinn, and Sellinger 2004, 1150).

Great Lakes and New York Lakes Fail to Freeze

The Great Lakes failed to freeze during the winter of 2001–2002. In addition, several lakes in Upstate New York remained liquid for the first time in at least three decades. In his 30 years of studying freeze-thaw cycles of more than 250 lakes in New York State, Kenton Stewart had never before seen some of these lakes remain unfrozen for an entire winter. "The majority of the lakes in the state still froze, but a surprising number that developed ice covers in previous winters, had only a partial skim of ice that winter, or did not freeze at all," said Stewart, professor emeritus of biological sciences at the University at Buffalo ("New York Lakes" 2002). In subsequent winters, however, temperatures cooled and the lakes froze again. Shortening freezing seasons are not restricted to New York lakes, of course. The waters of Lake Mendota, near Madison, Wisconsin, for example, freeze about 40 fewer days today than during 1860 (Glick 2004, 32).

According to a report by the Environment News Service, Stewart said, "Lakes that did not freeze this winter include some that did so during an El Niño year. Those that did freeze did so one to three weeks later than usual" ("New York Lakes" 2002). "One surprising thing about the unusually mild winter is that while it was as mild as some of the strong El Niño events that we've seen, it was not associated with an El Niño event in the Pacific Ocean that can have an atmospheric influence," said Stewart. "It also was not foreseen by the Climate Prediction Center of the National Oceanographic and Atmospheric Agency" ("New York Lakes" 2002).

Among the New York lakes that failed to freeze during 2001–2002 were Irondequoit Bay in Rochester; Hemlock and Canadice lakes, south of Rochester; Cross Lake located west of Syracuse; Onondaga Lake in Syracuse; Otisco Lake located west of Syracuse; Big Green Lake in Green Lake State Park, east of Syracuse; and Ashokan and other water supply reservoirs north of New York City ("New York Lakes" 2002).

FURTHER READING

Assel, Raymond A., Frank H. Quinn, and Cynthia E. Sellinger. "Hyrdoclimatic Factors of the Recent Record Drop in Laurentian Great Lakes Water Levels." *Bulletin of the American Meteorological Society* 85, no. 8 (August 2004):1143–1150.

Burnett, Adam W., Matthew E. Kirby, Henry T. Mullins, and William P. Patterson. "Increasing Great Lake-effect Snowfall during the Twentieth Century: A Regional Response to Global Warming?" *Journal of Climate* 16, no. 21 (November 1, 2003): 3535–3542.

George, Douglas M. Personal communication, October 13, 2006.

Glick, Daniel. "The Heat Is On: Geosigns." *National Geographic*, September 2004, 12–33.

"Global Warming Means More Snow for Great Lakes Region." Ascribe Newswire, November 4, 2003. (LEXIS)

Higgins, Matt. "Warm Temperatures Chill the Ice Fishing Season." *New York Times*, January 10, 2007. http://www.nytimes.com/2007/01/10/sports/othersports/10outdoors.html.

Lydersen, Kari. "Great Lakes' Lower Water Levels Propel a Cascade of Hardships." *Washington Post*, January 27, 2008, A-4. http://www.washingtonpost.com/wp-dyn/content/article/2008/01/26/AR2008012601748_pf.html.

Mitchell, John G. "Down the Drain: The Incredible Shrinking Great Lakes." *National Geographic*, September 2002, 34–51.

"New York Lakes Fail to Freeze." Environment News Service, March 21, 2002. http://ens-news.com/ens/mar2002/2002L-03-21-09.html#anchor1.

Staba, David. "Snowstorm Blankets Buffalo, Killing at Least 3." *New York Times*, October 14, 2006. http://www.nytimes.com/2006/10/14/nyregion/14storm.html.

Zremski, Jeremy. "A Chilling Forecast on Global Warming." *Buffalo News*, August 8, 2002, A-1.

Great Plains: Warming and Drought in the Past

Paleoclimatic precedent exists for the kind of substantial drought that has become common in the U.S. West in recent years. Prevailing winds from the hot, arid southwest during the growing season in the Great Plains of the United States during warm weather 800 to 1,000 years ago spawned a drought that destroyed the Anasazis' civilization in the Southwest. During this period, the prevailing winds shifted from the south, as today (which brings in moisture from the Gulf of Mexico) to the southwest, which is hot and dry.

A team of scientists (Venkataramana Sridhar, David B. Loope, James B. Swinehart, Joseph A. Mason, Robert J. Oglesby, and Clinton M. Rowe) from the University of Nebraska at Lincoln and the University of Wisconsin have discerned a major wind shift by studying the shapes of dunes in the Sand Hills, which were once a sea bottom in north-central Nebraska. During the Medieval Warm Period, the drought killed the dunes' vegetative cover. The winds caused the dunes to migrate, leaving today's patterns, which are again anchored by grasses. As the scientists wrote, "Longitudinal dunes built during the Medieval Warm Period (800 to 1,000 years before the present) record the last major period of sand mobility. These dunes are oriented NW-SE and are composed of cross-strata with bipolar dip directions" (Sridhar et al. 2006, 345).

"Such a westward shift [in prevailing winds] would … greatly reduce the flow of moist air into the central Great Plains, thereby generating severe drought," the scientists wrote in the July 21, 2006, edition of *Science* (Sridhar et al. 2006, 346). While today the "dry line" brings spring and summer thunderstorms that traverse Nebraska from the Rocky Mountains, the scientists wrote that "during the Medieval Warm Period the mean position of the dry line moved much further east, such that the Sand Hills were most often in the dry, hot air with greatly reduced precipitation" (Sridhar et al. 2006, 346).

A similar wind shift coupled with depletion of the Ogallala Aquifer could make much of Nebraska too dry for agriculture in coming decades. Large parts of this aquifer, the largest body of underground water in North America, are being drawn down by feet per year while precipitation recharges it at inches a year.

Such examinations of dune patterns can be used as a proxy for atmospheric circulation that governs weather and climate over periods from decades to centuries long, which leave no records. About half of Sand Hills' annual precipitation falls during May, June, and July, nourishing a grassland ecosystem. During the Dust Bowl years of the 1930s (and to a lesser extent during the 1950s drought), part of this ecosystem broke down, resulting in what the scientists call "isolated blowouts" (Sridhar et al. 2006, 345). Historical accounts indicate that some dune crests lost their grass to drought occasionally even during the generally cooler nineteenth century. These accounts paled, however, beside the sustained drought that accompanied the Medieval Warm Period.

FURTHER READING

Sridhar, Venkataramana, David B. Loope, James B. Swinehart, Joseph A. Mason, Robert J. Oglesby, and Clinton M. Rowe. "Large Wind Shift on the Great Plains during the Medieval Warm Period." *Science* 313 (July 21, 2006):345–347.

Greenhouse Effect, as an Idea

Humanity's impact on climate is usually dated to the beginning of the Industrial Revolution with its combustion of fossil fuels, about 200 years ago. The first carbon-based fuel, however, was firewood, first burned about 500,000 years ago. William F. Ruddiman argued in *Plows, Plagues, and Petroleum: How Humans Took Control of Climate* (2005) that a measurable human carbon (and methane) footprint dates back several thousand years, to the origins of large-scale agriculture and urbanization. Ruddiman wrote that human contributions to increasing temperature have, from time to time, been interrupted by major disease outbreaks that have impeded the pace of deforestation and other activities that create greenhouse gases. Human activity may even have aborted a minor glaciation in northeast Canada long before the first stirrings of the Industrial Revolution, Ruddiman has asserted.

The fossil-fueled Industrial Revolution was born in England. As coal-fired industry (as well as home heating and cooking) filled English skies with acrid smoke, some English homeowners protested to coal's use as a fuel. Others described the horrors of coal mines, into which children as young as six years of age were sent to work. Queen Elizabeth sometimes forbade the burning of coal in London while Parliament was in session. In 1661, John Evelyn wrote a book that complained about the noxious nature of coal smoke.

The coal-burning steam engine was invented by Thomas Newcomen in 1712 and refined into a form that was widely adaptable for industrial processes by James Watt, beginning in 1769. Within a century of industrialism's first stirrings,

John Tyndall. (Library of Congress)

during the 1820s, Jean Baptiste Joseph Fourier, a Frenchman, compared the atmosphere to a greenhouse.

During the 1860s, John Tyndall, an Irishman, developed the idea of an "atmospheric envelope," suggesting that water vapor and carbon dioxide in the atmosphere are responsible for retaining heat radiated from the sun. Tyndall also wrote that climate might warm or cool based on the amount of carbon dioxide and other gases in the atmosphere. Tyndall, who speculated in 1861 that a decline in carbon-dioxide levels could have accounted for the ice ages, was the first person to make quantitative, spectroscopic measurements showing that water vapor and carbon dioxide absorb thermal radiation and therefore trap solar heating in the atmosphere.

Savante Arrhenius and the Role of "Carbonic Acid in the Air"

In 1896, Savante August Arrhenius, a Swedish chemist, published a paper in the *London, Edinburgh, and Dublin Philosophical Magazine and Journal of Science* titled "On the Influence of Carbonic Acid in the Air upon the Temperature of the Ground" (Arrhenius 1896). In his paper, Arrhenius theorized that a rise in the atmospheric level of carbon dioxide could raise the temperature of the air. Arrhenius was not the only person thinking along these lines at the time; Swedish geologist Arvid Hogbom had delivered a lecture on the same idea three years earlier, which Arrhenius incorporated into his article.

Arrhenius was a well-known scientist in his own time, not for his theories describing the greenhouse effect, but for his work in electrical conductivity, for which he was awarded a Nobel Prize in 1903. Later in his life, Arrhenius directed the Nobel Institute in Stockholm.

His work in global-warming theory was not much discussed during Arrhenius' own life. Using available measurements of absorption and transmission by water vapor and carbon dioxide, he developed the first quantitative mathematical model of the Earth's greenhouse effect and obtained results of acceptable accuracy by today's standards for equilibrium climate sensitivity to carbon-dioxide changes.

Arrhenius developed his theory through the use of equations, by which he calculated that a doubling of carbon dioxide in the atmosphere would raise air temperatures about 10°F. Arrhenius thought 3,000 years would have to pass before human-generated carbon-dioxide levels would double, a miscalculation.

Arrhenius applauded the possibility of global warming, telling audiences that a warmer world "would allow all our descendants, even if they only be those of a distant future, to live under a warmer sky and in a less harsh environment than we were granted" (Christianson 1999, 115). During 1908, in his book *Worlds in the Making*, Arrhenius wrote:

> By the influence of the increasing percentage of carbonic acid in the atmosphere, we may hope to enjoy ages with more equable and better climates, especially as regards the colder regions of the Earth, ages when the Earth will bring forth much more abundant crops than at present for the benefit of rapidly propagating mankind. (Christianson 1999, 115)

Arrhenius' ideas were not widely discussed, but they did not completely die during the early twentieth century. Alfred J. Lotka, an American physicist, warned in 1924 that "[e]conomically we are living on our capital; biologically, we are changing radically the complexion of our share in the carbon cycle by throwing into the atmosphere, from coal fires and metallurgical furnaces, ten times as much carbon dioxide as in the process of breathing" (Oppenheimer and Boyle 1990, 1990, 35).

Calculating on the basis of fossil-fuel use in 1920, at the beginning of the automotive age, Lotka ventured that the level of carbon dioxide in the atmosphere would double in 500 years because of human activities, one-sixth of the time period forecast by Arrhenius.

By the late 1930s, the prospect of a warming climate based on rising greenhouse-gas levels in the atmosphere was catching the eye of G. D. Callendar, a British meteorologist. Callendar gathered records from more than 200 weather stations around the world to argue that the Earth had warmed 0.4°C between the 1880s and the 1930s because of carbon-dioxide emissions by industry. Callendar's assertions were met with skepticism by many English scientists at the time; however, he was laying the foundation for modern-day efforts to make more precise measurements of atmospheric trace-gas trends and radiative properties and to design more capable climate models to simulate climate change.

Two decades after Callendar, in 1956, Gilbert Plass, a scientist at Johns Hopkins University in Baltimore, suggested that carbon dioxide was an important climate-control mechanism. He projected that burning of fossil fuels would raise the global temperature 1.1°C (2°F) by the end of the twentieth century, very close to the actual worldwide increase. *See also:* Arrhenius, Savante

FURTHER READING

Arrhenius, Savante. "On the Influence of Carbonic Acid in the Air upon the Temperature of the Ground." *London, Edinburgh, and Dublin Philosophical Magazine and Journal of Science,* 5th series (April 1896):237–276.

Christianson, Gale E. *Greenhouse: The 200-Year Story of Global Warming.* New York: Walker and Company, 1999.

Oppenheimer, Michael, and Robert H. Boyle: *Dead Heat: The Race against the Greenhouse Effect.* New York: Basic Books, 1990.

Ruddiman, William F. *Plows, Plagues, and Petroleum: How Humans Took Control of Climate.* Princeton, NJ: Princeton University Press, 2005.

Greenhouse-Gas Emissions, Worldwide

Carbon dioxide and other greenhouse gases, such as methane, nitrous oxides, and chlorofluorocarbons (CFCs), retain heat in the atmosphere. Since the beginning of the industrial age more than two centuries ago, humankind has been raising their proportion in our air. We are carbon-creating creatures. Throughout the twentieth and twenty-first centuries, we have assembled an increasing array of machines, all of which produce carbon dioxide and other gases that have been changing the atmospheric balance in ways that retain an increasing amount of heat in the atmosphere. In 1910, the average American commanded about 1.5 horsepower worth of mechanized energy; in 1990, largely due to the general acquisition of automobiles, the average person commanded the power (and the greenhouse gas effluent) of 130 horsepower. By 2008, that figure exceeded 180 horsepower.

The Earth's atmosphere is composed of 78.1 percent nitrogen and 20.9 percent oxygen. All the other gases, including those responsible for the greenhouse effect, make up only about 1 percent of the atmosphere. Carbon dioxide (CO_2) is 0.035 percent; methane (CH_4) is 0.00017 percent, and ozone 0.000001–0.000004 percent. To a certain extent, the greenhouse effect is necessary to keep the Earth at a temperature that sustains life as we know it. Without "infrared forcing" (popularly, the greenhouse effect), the average temperature of the Earth would be about 33°C (60°F) colder than today's averages, too cold to sustain the Earth's existing plant and animal life.

The proportion of carbon dioxide in the atmosphere has risen from 280 parts per million (ppm) at the dawn of the industrial age to roughly 385 ppm in 2007. Other greenhouse gases also have risen in similar proportion, or more. During the 1990s, a vivid public debate grew around the world regarding how much warmer the Earth has become, and how much warmer it may become. The debate continues with increasing vibrancy and salience today as greenhouse-gas levels and temperatures continue to rise.

Greenhouse gas levels hit record levels in 2007, according to the World Meteorological Organization. Carbon dioxide reached 383.1 ppm, an increase of 0.5 percent from 2006, according to the World Meteorological Organization report. Concentrations of nitrous oxide also reached record highs in 2007, up 0.25 percent from the year before. Methane levels increased 0.34 percent, exceeding the highest value so far, which was recorded in 2003.

Global emissions of greenhouse gases, in millions of metric tons of carbon, increased from

6.192 million in 1990 to 8.381 in 2006, 35.3 percent in 16 years, according to the Carbon Dioxide Information Analysis Center of the U.S. Department of Energy. By 2007, China was emitting 24 percent of the world's human-generated carbon dioxide, compared with the United States at 21 percent, the European Union at 12 percent, India at 8 percent, and the Russian Federation at 6 percent, according to the Netherlands Environmental Assessment Agency.

Among these influences, human generation of greenhouse gases emerged as the dominant forcing during the twentieth century. By the year 2008, scientists had tested the composition of the atmosphere to roughly 60 million years in the past. The level of carbon dioxide today is believed, according to such measurements, to be as high as it has been in at least 20 million years.

More People, Greater Affluence

Not only are human populations increasing, but these ever-larger numbers of human beings are experiencing greater affluence, overloading the Earth's atmosphere with carbon dioxide, methane, and other trace gases. What we loosely call "the American way," an intoxicating mixture of hedonistic affluence and individual gratification, is also the world's most potent creator of atmospheric carbon dioxide (and other greenhouse gases) in the history of the planet.

Even as international protocols called for freezes or reductions in fossil-fuel use during the 1990s, the burning of carbon-based fuels continued to increase around the world. Nearly every country and region experienced substantial increases in fossil-fuel use: 20 percent in Brazil, 28 percent in India, 40 percent in Indonesia, and 27 percent in China (Ridenour 1998). The only region to experience the type of decline that scientists say will be required to forestall substantial global warming were Russia and Ukraine, whose fossil-fuel use declined 40 to 45 percent during the 1990s because of widespread economic collapse. Several more prosperous nations in Europe also stabilized or slightly reduced their generation of greenhouse gases.

Human life and labor have become more mechanized than even a few decades ago; agriculture, for example, has ceased, for the most part, to be the province of family farmers, as it was a century ago. It has, instead, become industrialized on a large, mass-production scale, with increasing utilization of the fossil fuels and fertilizers that produce greenhouse gases. The industrialization of agriculture has been accompanied by a gradual increase in agricultural-industrial scale that has made many of the small towns of the U.S. Midwest economically obsolete. In Nebraska, for example, the cities of Omaha and Lincoln thrive, with unemployment rates among the lowest in the world, while many small farming towns in its hinterland crumble. Animal protein and cereal crops, such as wheat and corn, are now produced on a factory model, with attendant air, soil, and water pollution.

Greenhouse-Gas Emissions by Country and Region

Energy consumption and greenhouse-gas emissions vary widely by region. North American energy usage per capita is roughly five to six times as high north of the Rio Grande River as in Latin America. Energy consumption per person in the United States is more than twice what it is in Europe, and roughly 15 times the amount in Sub-Saharan Africa. The largest factor in this difference is the pervasiveness in any given area of private automobiles and other motorized vehicles. In 1950, the United States produced 40 percent of the world's industrial carbon dioxide. That proportion fell to 25 percent in 1975, 22 percent in 1988, and 21 percent in 2007. Emissions in the United States were growing during this period, but those of the rest of the world, on balance, were increasing even more rapidly.

Greenhouse-gas emissions in the United States (including carbon dioxide, methane, nitrous oxide, hydrofluorocarbons, perfluorocarbons, and sulfur hexafluoride) declined by 1.1 percent during 2006 from the previous year, according to the U.S. Environmental Protection Agency (EPA). Americans burned fewer fossil fuels, including those to generate electricity. A large part of the difference resulted from a relatively warm winter and cool summer, both of which reduced the need for power and fuel. Rising fuel prices also had an impact on consumption of gasoline and diesel fuel. In some cases, natural gas was substituted for coal in power generation ("U.S. Greenhouse" 2008). Even with these reductions, the EPA report indicated that overall U.S. greenhouse-gas emissions rose 14.7 percent between 1990 and 2006, as the country's economy grew by 59 percent.

The Washington, D.C., metropolitan area, for example, alone produces more carbon dioxide than several European countries, as the region's gridlocked traffic and coal-fired power plants provide an enormous carbon footprint. The Metropolitan Washington Council of Governments compiled the first official assay of the federal capital's carbon emissions—65.6 million metric tons of carbon dioxide were emitted during 2005, more than in all of Hungary, Finland, Sweden, Denmark, or Switzerland, each of which had more people (Fahrenthold 2007, C-1).

Carbon-Dioxide Emissions Accelerate

Marine and atmospheric scientist Mike Raupach, who co-chairs the Global Carbon Project at the Australian Commonwealth Scientific and Industrial Research Organization (CSIRO) told a meeting of scientists in Tasmania that carbon-dioxide emissions into the world's atmosphere are accelerating. "From 2000 to 2005, the growth rate of carbon dioxide emissions was more than 2.5 percent per year, whereas in the 1990s it was less than 1 percent per year," he said ("Growth of Global" 2006).

Paul Fraser, also with CSIRO Marine and Atmospheric Research, said that atmospheric concentrations of carbon dioxide grew by two parts per million in 2005, the fourth year in a row of above-average growth. "To have four years in a row of above average carbon dioxide growth is unprecedented," said Fraser, who is program manager for the CSIRO Measurement, Processes and Remote Sensing Program ("Growth of Global" 2006).

Fraser said that the 30-year record of air collected at the Australian Bureau of Meteorology's observation station at Tasmania's Cape Grim showed growth rates of just over one part per million in the early 1980s, but in recent years, carbon dioxide has increased at almost twice that rate. "The trend over recent years suggests the growth rate is accelerating, signifying that fossil fuels are having an impact on greenhouse gas concentrations in a way we haven't seen in the past," he said ("Growth of Global" 2006).

The Complexity of Atmospheric Chemistry and Dynamics

Determining the net climatic effect of a given mix of greenhouse gases is no simple matter, in part because the mix is so complex. Some gases (an example is carbon dioxide) may continue to increase in the atmosphere, while others (an example during the 1990s was methane) may stabilize or even decline. Other gases (an example being nitrous oxides) may counteract some of the effects of others.

After doubling from preindustrial times to about 1980, the proportion of methane in the atmosphere parted paths with carbon dioxide during the 1990s, as its rate of accumulation slowed markedly, and, during 1992 and 1993, actually declined. According to a report in the British scientific journal *Nature*, methane levels may have declined during 1992 and 1993 for several reasons. Mount Pinatubo's eruption in 1991 caused a cooling of global temperatures; cooler temperatures caused less methane than usual to be released from boreal wetlands.

At the same time, the countries comprising the former Soviet Union continued a decade-long decline in methane emissions from oil and natural gas production because of social and political collapse. In flusher times, the Soviet Union's oil and gas drillers vented immense amounts of methane from their operations. Decreased biomass burning in the tropics also may have contributed to the decline in atmospheric methane. The scientists who disclosed the news that methane levels have stabilized speculate that atmospheric levels will rise slowly again for a few decades, and then stabilize (Dlugokencky et al. 1999, 449). While scientists can measure the level of methane in the atmosphere, they have difficulty accounting for rises or falls in that level because methane sources are many, varied, and often too small to detect in global-scale calculations.

The interactions of various greenhouse gases in the atmosphere are so complex in some cases that scientists are only now beginning to understand them. Mark G. Lawrence and Paul J. Crutzen surveyed the role of nitrogen oxides in the troposphere, or lower atmosphere, finding that they "would be expected to reduce the atmospheric lifetimes of greenhouse gases—such as methane—as well as to increase aerosol production rates and cloud reflectivities, therefore exerting a cooling influence on the climate" (Lawrence 1999, 168). According to Lawrence and his co-authors, human activities account for about half the atmosphere's nitrogen oxides. Their investigation centered on nitrogen-oxide emissions by ships, which eject enough of the chemicals into the air of frequently used shipping lanes to raise nitrogen oxide levels to as much as 100 times natural "background" levels (Lawrence 1999, 168).

The oceans absorb large amounts of carbon dioxide, perhaps half of what humankind produces, but their capacity is not infinite. Models indicate that by roughly 2050, the oceans' capacity as a carbon "sink" (absorber) of carbon dioxide may have been reached. At the surface, the gas dissolves in water. Through complex chemical processes, some of the carbon becomes part of the tissues of marine organisms. Carbon-containing seashells eventually settle to the ocean floor, forming carbonate sediments such as limestone. The vast capacity of the oceans and other carbon sinks is being overwhelmed, however, by the volume of greenhouse gases being poured into the atmosphere as human ingenuity invents new ways to put fossil-fuel-generated energy to work, and as populations and material living standards rise around the world.

Greenhouse Gases Other Than Carbon Dioxide

In addition to carbon dioxide, during the 1990s, human activity was adding about 550 metric tons of methane to the atmosphere each year. Methane's preindustrial range in the atmosphere was 320 to 780 parts per billion (ppb); by 2000, that level had risen to about 1,700 ppb—a steeper rise, in proportional terms, than carbon dioxide. After a pause during the 1990s, the level of methane began to rise again after 2005. Carbon dioxide is 200 times more plentiful in the atmosphere than methane, but a molecule of methane can retain 20 to 25 times as much heat as one molecule of carbon dioxide.

Atmospheric methane is produced by many human activities, from transporting natural gas to raising meat animals, dumping garbage in landfills, and growing rice. Methane contributes about half as much retained heat to the atmosphere as carbon dioxide. The rate of increase in atmospheric methane (1 percent a year) was about twice as rapid as that of carbon dioxide for much of the twentieth century (Jager 1991, 79).

Carbon monoxide shares greenhouse-gas properties with carbon dioxide; its presence in the atmosphere has been rising 0.8 to 1.5 percent a year. Most of the carbon monoxide in the atmosphere is produced by humans, much of it by burning fossil fuels and by deforestation. By the late 1980s, roughly one billion metric tons of human-produced carbon monoxide was being added to the global atmospheric inventory each year.

Add to this mix of trace gases tropospheric ozone, which is produced photochemically in the atmosphere from the oxidation of carbon monoxide, methane, or other hydrocarbons in the presence of nitrogen oxides, which act as catalysts. Tropospheric ozone has been increasing about 1 percent a year. Levels in the air have increased 20 percent to 50 percent during the twentieth century. This type of ozone, which is contributed to the atmosphere by several industrial processes, absorbs infrared radiation and thereby, like carbon dioxide and methane, contributes to infrared forcing.

A few other gases, such as nitrous oxide ("laughing gas") add to infrared forcing as well. Greenhouse gases include the chlorofluorocarbon (CFC) family, made up of several synthetic chemicals that have been implicated in the destruction of the Earth's stratospheric ozone layer. The CFCs, which have been banned by the Montreal Protocol since the late 1980s, also retain heat in the atmosphere. Molecule for molecule, CFCs retain thousands of times as much heat as carbon dioxide.

A molecule of sulfur hexafluoride, one of the chemicals subject to controls under the Kyoto Protocol, is 23,900 times more potent over 100 years than a molecule of carbon dioxide. Most CFC use is being phased out under international protocols, but its effects "will only decrease very slowly next century," writes John T. Houghton (Houghton 1997, 37). Global warming near the Earth's surface may be contributing to the destruction of stratospheric ozone.

Humankind continues to add new greenhouse gases to the atmospheric mix. For example, nitrogen trifluoride (NF_3), a gas used in manufacture of flat-panel television sets, computer displays, microcircuits, and thin-film solar panels is 17,000 times more potent at retaining heat than carbon dioxide. It is also four times more prevalent in the atmosphere than previously estimated, according to research led by geochemistry professor Ray Weissby, head of a team at Scripps Institution of Oceanography at the University of California–San Diego ("The Most Potent" 2008).

The Role of Water Vapor

Water vapor, Earth's most abundant greenhouse gas, is potent enough, at elevated temperatures, to double the climate warming caused by increased levels of carbon dioxide in the atmosphere, according to Andrew Dessler and colleagues from Texas A&M University in College Station, who using data from NASA's Aqua satellite to measure humidity levels in the lowest 10 miles of the atmosphere. "This new data set shows that as surface temperature increases, so does atmospheric humidity," Dessler said. "Dumping greenhouse gases into the atmosphere makes the atmosphere more humid. And since water vapor is itself a greenhouse gas, the increase in humidity amplifies the warming from carbon dioxide" ("Water Vapor" 2008).

This work, published in the American Geophysical Union's *Geophysical Research Letters* combined global observations of shifts in temperature with satellite data to construct a model of the interplay between water vapor, carbon dioxide, and other atmosphere-warming gases. "Everyone agrees that if you add carbon dioxide to the atmosphere, then warming will result," Dessler said. "So the real question is, how much warming?" The answer, according to Dessler and colleagues, can be found by estimating the magnitude of water vapor feedback. Increasing water vapor leads to warmer temperatures, which causes more water vapor to be absorbed into the air. Warming and water absorption increase in a spiraling cycle ("Water Vapor" 2008).

Dessler and colleagues' work indicates that if the Earth warms 1.8°F, increases in water vapor will trap an extra two watts of energy per square meter (about 11 square feet). "This study confirms that what was predicted by the models is really happening in the atmosphere," said Eric Fetzer, an atmospheric scientist who works with Atmospheric Infrared Sounder data from NASA's Aqua satellite at the Jet Propulsion Laboratory in Pasadena, California. "Water vapor is the big player in the atmosphere as far as climate is concerned" (Dessler, Zhang, and Yang 2008; "Water Vapor" 2008).

FURTHER READING

"Australian Judge Blocks Coal Mine on Climate Grounds." Environment News Service, November 29, 2006. http://www.ens-newswire.com/ens/nov2006/2006-11-29-03.asp.

Brown, Paul. "Analysis: Blair Sets out Far-reaching Vision but Where Are the Practical Policies?" *London Guardian*, February 25, 2003, 13.

Dessler, A. E., Z. Zhang, and P. Yang. "Water-vapor Climate Feedback Inferred from Climate Fluctuations, 2003–2008." *Geophysical Research Letters* 35 (2008), doi: 10.1029/2008GL035333.

Dlugokencky, E. J., K. A. Masrie, P. M. Lang, and P. P. Tans. "Continuing Decline in the Growth Rate of the Atmospheric Methane Burden." *Nature* 393 (June 4, 1999):447–450.

Fahrenthold, David A. "D.C. Area Outpaces Nations in Pollution: High Carbon Emission Blamed on Coal Plants." *Washington Post*, September 30, 2007, C-1. http://www.washingtonpost.com/wp-dyn/content/article/2007/09/29/AR2007092900959.html.

"Growth of Global Greenhouse Gas Emissions Accelerating." Environment News Service, November 29, 2006. http://www.ens-newswire.com/ens/nov2006/2006-11-29-02.asp.

Hegeri, Gabriele. "Climate Change: the Past as Guide to the Future." *Nature* 392 (April 23, 1998):758–759.

Houghton, John. *Global Warming: The Complete Briefing*. Cambridge: Cambridge University Press, 1997.

Jager, J., and H. L. Ferguson. *Climate Change: Science, Impacts, and Policy. Proceedings of the Second World Climate Conference*. Cambridge: Cambridge University Press, 1991.

Lawrence, Mark G., and Paul J. Crutzen. "Influence of Nitrous Oxide Emissions from Ships on Tropospheric Photochemistry and Climate." *Nature* 402 (November 11, 1999):167–168.

Ridenour, David. "Hypocrisy in Buenos Aires: Millions of Gallons of Fuel to Be Burned By Those Seeking Curbs on Fuel Use." National Policy Analysis: A Publication of the National Center for Public Policy Research, No. 217, October 1998. http://nationalcenter.org/NPA217.html.

Schwander, Dominique Raynaud, Valèrie Masson-Delmotte, and Jean Jouzel. "Atmospheric Methane and Nitrous Oxide of the Late Pleistocene from Antarctic Ice Cores." *Science* 310 (November 25, 2005): 1317–1321.

"Study of Ancient Air Bubbles Raises Concern about Today's Greenhouse Gases." Associated Press in *Omaha World-Herald*, November 25, 2005, 4-A.

"The Most Potent Unknown Greenhouse Gas Revealed." Environment News Service, October 24, 2008. http://www.ens-newswire.com/ens/oct2008/2008-10-24-01.asp.

"U.S. Greenhouse Gas Emissions Lower in 2006." Environment News Service, April 15, 2008. http://www.ens-newswire.com/ens/apr2008/2008-04-15-01.asp.

"Water Vapor Confirmed as Major Player in Climate Change." NASA Earth Observatory. November 17, 2008. http://earthobservatory.nasa.gov/Newsroom/view.php?id=35952.

Greenland, Ice Melt

The largest mass of ice in the Northern Hemisphere resides atop Greenland, which includes about 10 percent of the world's ice. This ice is being measured and monitored as never before,

Summer snowmelt on Greenland's ice cap, 2005. (Konrad Steffen, CIRES/University of Colorado)

by satellites, aircraft, and dozens of down-swaddled scientists who are braving 30-below-zero temperatures and deadly snow-cloaked crevasses that corrugate the slumping edges of the ice cap (Revkin 2004b).

Large parts of Greenland's northern ice sheet experienced a record number of melting days during the summer of 2008, according to Dr. Marco Tedesco, assistant professor of earth and atmospheric sciences at the City College of New York (CCNY), and colleagues, who based their findings on an analysis of microwave brightness temperature recorded by the Special Sensor Microwave Imager (SSM/I) onboard the F13 satellite. "Having such extreme melting so far north, where it is usually colder than the southern regions is extremely interesting," said Tedesco. "In 2007, the record occurred in southern Greenland, mostly at high elevation areas where in 2008 extreme snowmelt occurred along the northern coast" ("Satellite Data" 2008). Melting in northern Greenland lasted up to 18 days longer than during previous years. Some of the most rapid melting took place near Ellesmere Island, including the Petermann glacier, which lost 29 square kilometers in July.

Greenland's ice is only a fraction of Antarctica's, but it is melting more rapidly, in part because summers are warmer, allowing for more rapid runoff. Greenland's southern tip is no farther north than Juneau or Stockholm. The persistence of the ice cap is due to its mass, the fact that the ice makes its own climate. The ice itself reflects sunlight and heat, and deflects weather systems from the south. The elevation of the ice sheet helps to keep it cold. As it erodes, these advantages diminish (Appenzeller 2007, 68).

Philippe Huybrechts, a glaciologist and ice-sheet modeler at the free University of Brussels, has modeled the behavior of Greenland's ice sheet, finding that, with an anticipated annual temperature increase of 8°C, "the ice sheet would shrink to a small glaciated area far inland and sea level would rise by six meters" (Schiermeier 2004a, 114). If, over the course of several centuries, global warming provokes the melting of Greenland's ice cap, climate models run by several scientists indicate that the "removal of the Greenland ice sheet due to a prolonged climatic warming would be irreversible" (Toniazzo, Gregory, and Huybrechts 2004, 21).

Rapid Changes in Greenland's Climate

Upon close examination of paleoclimatic records, scientists examining ice cores from Greenland have found that Greenland's climate can change very quickly, giving cold comfort to anyone contemplating the stability of worldwide sea levels. The island's climate "flipped" to a different state within one to three years at the onset of the present interglacial period, for example. A northward shift of the Intertropical Convergence Zone may have triggered abrupt shifts in Northern Hemisphere atmospheric circulation, resulting in changes of 2° to 4°C in Greenland temperatures from one year to the next (Steffensen et al. 2008, 654). Other scientists have found evidence that Greenland's surface had been covered with pine forests within a few hundred thousand years of the present, with its ice mass a fraction of today's size.

Ice melt in Greenland has accelerated significantly since 1990, according to a report in the *Journal of Climate* (January 15, 2008) coauthored by Konrad Steffen, director of the Cooperative Institute for Research in Environmental Sciences and professor of Geography at the University of Colorado. A scientific team surveyed the rate of summer melting there between 1958 and 2006, and found that the five largest melting years all had occurred since 1995. The year 1998 was the biggest (109 cubic miles), followed by 2003, 2006, 1995, and 2002. Preliminary data suggest that the melting in 2007 may have exceeded all previous years. "Ice is moving faster into the ocean, and that will add to the sea-level rise," said Steffen ("Ice Sheet" 2008, 18-A).

In Ilulissat (meaning "iceberg"), on the west coast of Greenland, rain fell during December 2007 and January 2008. "Twenty years ago, if I had told the people of Ilulissat that it would rain at Christmas 2007, they would have just laughed at me. Today it is a reality," said Steffen (Friedman 2008b). Melting of the Greenland ice sheet increased 30 percent between 1979 and 2007. By that time, Greenland was losing 200 cubic kilometers of ice per year—from actual melting, as well as from ice sliding into the ocean from outlet glaciers along its edges—which far exceeds the volume of all the ice in the European Alps, he added. "Everything is happening faster than anticipated," he said (Friedman 2008b). Air temperatures on the Greenland ice sheet increased by about 7°F between 1991 and 2008.

"The amount of ice lost by Greenland over the last year [2007] is the equivalent of two times all the ice in the Alps, or a layer of water more than one-half mile deep covering Washington, D.C.," according to Steffen (Kolbert 2007d, 43). The 2007 melt extent on the Greenland ice sheet broke the 2005 summer melt record by 10 percent, making it the largest ever recorded there since satellite measurements began in 1979 (Steffen 2007). Ice is melting most rapidly at the edges of the ice sheet. Although Greenland has been thickening at higher elevations due to increases in snowfall (warmer air holds more moisture, thus more snow), the gain is more than offset by an accelerating mass loss, primarily from rapidly thinning and accelerating outlet glaciers, Steffen said.

Examining ice melt patterns in Greenland from the 1960s to the present, Edward Hanna and colleagues, writing in the *Journal of Climate*, found a strong correlation after 1990 to rapidly rising temperatures. Significant warming in south Greenland, especially, since about 1990 "reflects general Northern Hemisphere and global warming (Hanna et al. 2008, 331). Rapidly rising temperatures especially since 2001 have been reflected in significant increases in runoff from the southern flanks of Greenland's ice sheet. Inland areas of Greenland have cooled slightly at the same time, they found, while experiencing generally heavier snowfalls.

Greenland's Ice "Will Not Re-form"

During the winters from 2003 to 2007, Greenland lost two to three times as much ice in summer melt as it regained during winter snows (Witze 2008, 798). Greenland is a relic of the last ice age, "stranded out of time" (Witze 2008, 798). Even without human-induced global warming, its glaciers would not re-form under present conditions. Ice loss is accelerating irregularly year by year, as well. During the 2007 melting season, with temperatures 4° to 6°C higher than the previous 30 years' average, 500 billion tons of ice vanished, 30 percent more than the previous year and 4 percent more than the previous record in 2005 (Witze 2008, 799). During years of record melting, such as 2005 and 2007, high-pressure systems over Greenland keep storms away, clearing skies, allowing the summer sun to shine for extended hours.

By the summer of 2007, Greenland's ice cap was studded by more than 1,000 shallow meltwater lakes, some as wide as five kilometers, "like Minnesota, except white," wrote Alexandra Witze in *Nature* (2008, 800). Tens of millions of cubic meters of water swirl from these lakes to the base of the ice sheet in a matter of days, opening huge waterfalls where none had previously existed. "It has been only in the last five years that we have realized that—hey—the ice sheet is falling apart," said Ian M. Howat of the University of California–Santa Cruz (Witze 2008, 801).

After summer melt collects in lakes on the ice sheets, the water then finds fissures in the ice (called moulins), which conduct the water to the glacier's base, lubricating its movement toward the sea. The more ice melt, the faster the glacier moves. Sometimes surface lakes disappear down moulins nearly instantly (Appenzeller 2007, 68). Increases in the speed with which glaciers flow to the sea is, itself, evidence of a rapidly warming climate over a longer time period.

In early 2008, Andrew C. Revkin of the *New York Times* sketched this picture of Greenland's eroding ice cap:

> For a lengthening string of warm years, a lacework of blue lakes and rivulets of meltwater have been spreading ever higher on the ice cap. The melting surface darkens, absorbing up to four times as much energy from the sun as un-melted snow, which reflects sunlight. Natural drainpipes called moulins carry water from the surface into the depths, in some places reaching bedrock. The process slightly, but measurably, lubricates and accelerates the grinding passage of ice toward the sea. Most important, many glaciologists say, is the breakup of huge semi-submerged clots of ice where some large Greenland glaciers, particularly along the west coast, squeeze through fjords as they meet the warming ocean. As these passages have cleared, this has sharply accelerated the flow of many of these creeping, corrugated, frozen rivers. (Revkin 2008a)

All of these changes have many glaciologists "a little nervous these days—shell-shocked," said Ted Scambos, the lead scientist at the National Snow and Ice Data Center in Boulder, Colorado, and a veteran of both Greenland and Antarctic studies (Revkin 2008a).

Greenland's Ice Loss Year by Year

Ice loss in Greenland during 2007 and 2008 came on top of extensive loses in previous years. Between 2004 and 2006, Greenland's ice sheet melted two and one-half times faster than during the two previous years. Greenland lost about 164 cubic miles of ice from April 2004 to April 2006, more than the volume of water in Lake Erie ("Greenland's Ice" 2006). "The acceleration rate really took off in 2004," said Isabella Velicogna, a researcher at the University of Colorado's Boulder-based Cooperative Institute for Research in Environmental Sciences. "We think the changes we are seeing are probably a pretty good indicator of the changing climatic conditions in Greenland, particularly in the southern region" ("Greenland's Ice" 2006). Temperatures in southern Greenland, where most of the melting has taken place to date, have risen by about 4.4°F in the past two decades (Velicogna and Wahr 2006b, 329).

The Greenland ice sheet lost between 192 and 258 million tons of ice each year between April 2002 and April 2006 (Murray 2006, 277). Chen and colleagues (2006), who were mapping Arctic sea ice, also picked up ice mass loss in northeast Greenland (Chen et al. 2006). Outlet glaciers that feed ice from central Greenland to the ocean also have been depleting rapidly. This depletion has been most rapid in the southeast, where it approaches 10 meters per year (Murray 2006, 277). Carl Egede Boggild, a professor of snow-and-ice physics at the University Center of Svalbard, said Greenland could be losing more than 80 cubic miles of ice per year (Rudolf 2007).

How Ice Melts

Many climate scientists until recently placed the impacts of melting polar ice sheets in the far future, a matter of centuries. Then they learned that their models were not square with real-world ice melt patterns, which were much faster, and often in unpredictable, nonlinear ways. "When you look at the ice sheet, the models didn't work, which puts us on shaky ground," said Richard Alley, a geosciences professor at Pennsylvania State University (Rudolf 2007).

Scientists now have admitted that they have no accurate way to predict the melting of polar ice. Thus, estimates of how much ice may melt, and how quickly, cover a wide range. In addition, the actual melting takes place perhaps a century or two after a given quantity of greenhouse gases have been consumed and released into the air because of thermal inertia. The major question becomes not how much ice may melt by a given date, but how much more melting is "in the pipeline," guaranteed by the greenhouse gases already burned for which effects have not been fully realized.

In July 2006, researchers floating in a small boat studied melting dynamics of a mile-wide glacial lake in Greenland. Ten days later, in an hour and a half, the lake was sucked through a crack in the ice with a force equaling Niagara Falls. This event was studied by several scientists led by Ian Joughin of the University of Washington and Sarah Das of the Woods Hole

Oceanographic Institution in Woods Hole, Massachusetts. Their description of how ice melts in Greenland illustrates how summertime melting accelerates ice loss. Their results, published in *Science Express* on April 17, 2008, concluded that summer melt is only one factor influencing the erosion of Greenland's ice.

"For years people have said that the increasing length and intensity of the melt season in Greenland could yield an increase in ice discharge," said Joughin, lead author on the paper in *Science Express*. "Greater melt in future summers would cause ice to flow faster toward the coast and draw down more of the ice sheet" ("Researchers Warm" 2008). This study found that the violent draining of surface lakes had a short-lived influence on local movement of the ice sheet. Water is then quickly distributed to the ice sheet's base, which contributes to glaciers' movement toward the sea. "If you're really going to get a lot of ice out of Greenland, that would have to occur through outlet glaciers, but those are not being affected very much by seasonal melt," Joughin said. "The outlet glaciers are more affected by the removal of their shelves and grounded ice in their fjords, which decreases resistance to ice flow" ("Researchers Warm" 2008).

Moving "Melt Zones"

Greenland's coastal ice often melts with a boost from the sea, warm water upwelling from offshore, warming land that has been locked in ice for millennia. "Changes in the ocean eat the ice sheet from underneath," said Das. "Warmer water causes the glaciers to calve and melt back more quickly" (Faris 2008).

Additional global warming of 2° to 3°C under a business-as-usual scenario would be expected to yield about 5°C additional warming over Greenland," said James E. Hansen, director of NASA's Goddard Institute for Space Studies.

> Such a level of additional warming would spread summer melt over practically the entire ice sheet and considerably lengthen the melt season. It is inconceivable that the ice sheet could long withstand such increased melt-water before entering a period of rapid disintegration, but it is very difficult to predict when such a period of large rapid change would begin. (Hansen 2006b, 20)

Scientists have been learning that ice melts in nonlinear ways—and often much more quickly than it accumulates. Greenland's Jakobshavn Glacier (which probably delivered the iceberg that sunk the *Titanic*), at four miles wide and 1,000 feet thick, shunts more ice into the sea than any other, and its pace has been accelerating. In the past decade (to 2006), it has doubled its speed, on average, delivering 11 cubic miles of ice to the sea each year. The glacier's ice tongue, the point at which the glacier meets the ocean, also has retreated four miles since 2000 (Appenzeller 2007, 61).

"We may be very close to the threshold where the Greenland ice cap will melt irreversibly," said Tavi Murray, professor of glaciology at the University of Wales. Tulaczyk added, "The observations that we are seeing now point in that direction" (Lean 2005).

During the last few years, Greenland's "melt zone," where summer warmth turns snow on the edge of the ice cap into slush and ponds of water, has expanded inland, reaching elevations more than a mile high in some places, said Steffen. Some tongues of floating ice, where glaciers protrude into the sea, are thinning rapidly. Measurements during 2004 by Steffen and others on the Petermann Glacier in northern Greenland indicated that more than 150 feet of thickness had melted away under that tongue in one year. "If other ice streams start to react in a similar way," he said, "then we will actually produce much more fresh water" (Revkin 2004b).

W. S. B. Paterson and Niels Reeh presented direct measurements of the changes in Greenland's surface elevation between 1954 and 1995 on a traverse of the north Greenland ice sheet. Writing in *Science*, they said:

> We find only small changes in the eastern part of the transect, except for some thickening of the north ice stream. On the west side, however, the thinning rates of the ice sheet are significantly higher, and the thinning extends to higher elevations than had been anticipated from previous studies. (Paterson and Reeh 2001, 60)

"The higher elevation appears to be stable, but in a lot of areas around the coast the ice is thinning," said Waleed Abdalati, a manager in the Earth Sciences Department of NASA's Goddard Space Flight Center "There is a net loss of ice, particularly in the south" (Brown 2002, A-30).

Modeling by Jonathan Gregory of the Reading, U.K. Centre for Global Atmospheric Modeling

> suggests that as ice is lost, portions of the surface of Greenland's interior will heat and up at lower

> elevations where the air is warmer. Less snowfall and more rain would cause the ice to disappear at a faster rate than it is being replaced, leading in turn to further drops in elevation. (Schiermeier 2004a, 114–115)

Moreover, models suggest that the ice sheet would not reappear even if temperatures cooled in the future. The ice sheet creates its own climate, "depending on itself to exist" (Schiermeier 2004a, 114–115). According to these models, warming of 3°C could initiate eventual melting of the entire ice sheet over 1,000 years or more, raising global sea levels by about seven meters. Gregory and colleagues have written that "concentrations of greenhouse gases probably will have reached levels before the year 2100 that are sufficient to raise the temperature past this warming threshold" (Gregory, Huybrechts, and Raper 2004, 616). Models are complicated by several factors. For example, warmer temperatures initially could increase snowfall, delaying glacial melting. Ice melt also may depress the Gulf Stream, causing cooling over Greenland.

Evidence published in *Geophysical Research Letters* late in 2005 (Howat et al. 2005) described a sudden thinning of Greenland's Helheim Glacier, on the island's east coast. Tulaczyk said that between 2000 and 2005 the glacier retreated four and a half miles. As this glacier has retreated and thinned, effects have spread inland "very fast indeed," said Tulaczyk. "If the 2005 speedup also produces strong thinning, then much of the glacier's main trunk may un-ground, leading to further retreat," the authors wrote (Howat et al. 2005). Because the center of the Greenland ice cap is only 150 miles away, the researchers fear that it, too, may soon be affected (Lean 2005).

Speed of Ice Melt Surprises Glaciologists

The speed at which Greenland's ice has been melting has surprised even veteran glaciologists. Robert Lee Hotz of the *Los Angeles Times* described the reaction of Jay Zwally, who works at NASA's Goddard Space Flight Center in Greenbelt, Maryland, who has been studying the polar regions since 1972:

> Gripping a bottle of Jack Daniel's between his knees, Jay Zwally savored the warmth inside the tiny plane as it flew low across Greenland's biggest and fastest-moving outlet glacier. Mile upon mile of the steep fjord was choked with icy rubble from the glacier's disintegrated leading edge. More than six miles of the Jakobshavn had simply crumbled into open water.
>
> "My God!" Zwally shouted over the hornet whine of the engines.
>
> From satellite sensors and seasons in the field, Zwally, 67, knew the ice sheet below in a way that few could match. Even after a lifetime of study, the raffish NASA glaciologist with a silver dolphin in one pierced ear was dismayed by how quickly the breakup had occurred. (Hotz 2006)

During the late summer of 2008, the margin of the Jakobshavn Glacier retreated inland farther than it has at any time on record, during the past 150 years. Scientists studying proxies said that the glacier is probably at a roughly 5,000-year low point. During the summer of 2008, the northern section of the glacier broke up, eroding three square miles of ice ("Satellite Images" 2008). The Jakobshavn, the island's largest single glacier, lost 36 square miles of surface area between 2001 and 2005.

At the same time, the Petermann Glacier, in northern Greenland, continued to break up. An 11-square-mile (29-square-kilometer) piece of this floating ice mass broke away between July 10 and July 24, 2008. The Petermann Glacier already had lost 33 square miles of floating ice during 2000 and 2001. The entire Petermann Glacier, all of which floats on the sea, covers roughly 500 square miles, 10 by 50 miles.

Jason Box, an associate professor of geography at Ohio State and his colleagues, graduate students Russell Benson and David Decker, working with the Byrd Polar Research Center, have been studying cracks in the glacier that may portend future breakup. "If the Petermann glacier breaks up back to the upstream rift, the loss would be as much as 60 square miles (160 square kilometers)," Box said, representing a loss of one-third of the massive ice field ("Satellite Images" 2008).

Ice Melt in Winter

By 2005, some ice melted in Greenland even during December. "We have never seen that," Steffen said, combing the ice crystals from his beard. "It is significantly warmer now, and it happened quite suddenly. This year, the temperatures were warmer than I have ever experienced" (Hotz 2006).

Melting ice is changing the geography of Greenland, as islands never previously mapped emerge from the melting ice cap. "We are already in a new era of geography," said Arctic explorer Will Steger. "This phenomenon—of an island all of a sudden appearing out of nowhere and the ice melting around it—is a real common phenomenon now." With 27,555 miles of coastline and thousands of fjords, inlets, bays, and straits, Greenland's geography is becoming obsolete almost as soon as new maps are created. Routes once pioneered on a dogsled are routinely paddled in a kayak now; many features, like the Ward Hunt Ice Shelf in Greenland's northwest, have disappeared for good. During August 2006, Steger discovered a new island himself off the coast of Svalbard, "We saw it ourselves up there, just how fast the ice is going," he said (Rudolf 2007).

Scientists who think they are looking at solid ice now often find mazes of tunnels and cracks below the surface, melting ice at work. "We are witnessing enormous changes, and it will take some time before we understand how it happened, although it is clearly a result of warming around the glaciers," said Eric Rignot, a scientist at the California Institute of Technology's Jet Propulsion Laboratory (Vedantam 2006, A-1).

Ice Melt Imperils Hunters

Erosion of Greenland's ice poses practical dangers for indigenous hunters. DeNeen L. Brown of the *Washington Post* described an Inuit hunter's confrontation with glacial ice in Greenland made more dangerous by a warming climate. Aqqaluk Lynge had been chasing a seal. Fear chilled him when the seal dove under the ice and didn't return. Patience when seal hunting is essential, so he waited, but the seal never came back. When an animal begins to act strangely, such as not coming up for breath, something tremendous is happening in nature, wrote Brown.

> The iceberg was at his back. Suddenly it began moving like a monster that was waking up. Lynge … looked up in alarm, knowing that these floating mountains … for all their frozen beauty, are ruthless and deadly. So he decided to get moving … but the engine on his motorboat wouldn't start. Just then he noticed the iceberg moving. If the tip is moving, he knew, it could mean that one end is moving up and the other end is moving down.… A friend in another boat nearby quickly gave him a tow. Soon they were speeding away from the iceberg, not waiting to look back. Behind them they heard it turning. "We looked back and saw the whole iceberg was collapsing, exploding almost," he said. "We were so afraid." Then it flipped, creating a great tidal wave that crashed hard onto nearby shorelines. By then the two men were out of the wave's path. "When we were finally far away, we could breathe normally again. We were looking back and seeing nothing was left. It exploded underneath the surface of the sea." (Brown 2002, A-30)

Economic Benefits from a Warmer Climate

Some of Greenland's 56,000 people, most of whom live on the coasts, near the edge of the ice cap, are reaping benefits from a warming climate in which average winter temperatures rose about 10°F between 1991 and 2007. New pastures are being used to graze sheep and new fields grow potatoes. The hay-growing season has been lengthening in southern coastal Greenland, along with Chinese cabbage, several types of flowers, and turnips. Greenland farmers raised 22,000 lambs for local meat markets during 2006, another market that is thriving on the warming edges of the retreating ice cap. Some of the lamb has been exported to restaurants in Europe.

By 2007, a few food markets in Greenland sold local cauliflower, broccoli, and cabbage for the first time. Eight sheep farmers grew potatoes commercially. Five grew vegetables, and home gardeners harvested a few strawberries. Greenland beer's unique selling point is the purity of its glacier-fed water (native-grown hops may be next). Ewes were having fatter lambs, and more of them during growing season that extended from the middle of May to about September 15, three weeks more than 10 years previously. "Now spring is coming earlier, and you can have earlier lambings and longer grazing periods," said Eenoraq Frederiksen, 68, a sheep farmer whose farm, near Qassiarsuk, is accessible by a harrowing drive across a rudimentary road plowed in the hillside. "Young people now have a lot of possibilities for the future" (Lyall 2007b).

Kim Hoegh-Dam, who believes that warming coastal waters near Greenland will bring cod that have been abandoning the North Sea and other more southerly waters, has raised more than $1 million to buy cod trawlers and three processing plants. "Global warming will increase the cod tremendously and will bring other species up from the south," he said (Struck 2007b, A-1). A government trawler sent to test the cod runs caught 25 tons in one hour, so much that its crew cut its trip short. While some conservations voice concerns about overfishing (a factor in past declining cod catches), others' mouths water at the prospects not only for cod, but other sea creatures that thrive in relatively cold water, such as shrimp.

"The only limiting factor on human endeavor in Greenland is the temperature," Hoegh said, while bouncing on a fast motorboat past icebergs to visit the agriculture station in this country of few roads. Warm the temperature a bit, and new endeavors pop out like lambs from ewes, he believes (Struck 2007b, A-1).

Ilulissat, Greenland's third-largest village with 4,500 people, includes one posh hotel, The Arctic, which was doubling its size during 2008. From the windows of the Arctic, icebergs move so quickly that they appear to have formed and re-formed overnight. "Nobody would have predicted 10 to 15 years ago that Greenland would lose ice that fast," Steffen said. "That revises all of the textbooks" (Henry 2008).

Farther north along Greenland's coast, fishermen work from boats for longer periods as pack ice forms later and melts earlier. In recent winters, pack ice has failed to form on large areas of the coast, allowing fishing by boat year-round for halibut and other species. Daneborg Station on the Wollaston Foreland peninsula in northeastern Greenland reported that during 1997 the water was open for 80 days. Now it stays ice-free for 140 days, said Soren Rysgaard, a researcher for the Greenland Institute of Natural Resources (Struck 2007b, A-1).

A warming climate has prompted changes in traditional ways of life. For example, Doug Struck reported in the *Washington Post* that Ono Fleischer, one of the most renowned dogsledders in Greenland, took a dog team across the huge island in 19 days last spring to marry his companion, Karo Thomsen, in a village in eastern Greenland. "They intended to sled back," wrote Struck, "but a warm rain put a dangerous glaze on the ice cap. They had to give away their 12 dogs and fly back to Ilulissat" (Struck 2007b, A-1). Some villages in northern Greenland had to appeal for emergency food when a lack of ice prevented them from hunting for seals. Sled dogs had to eat donated European dog food instead of seal scraps. "If a seal hunter can't hunt, what is he to do?" asked Alfred Jakobsen, Greenland's minister of the environment (Struck 2007b, A-1).

Business entrepreneurs are knocking on the door in Greenland, looking for more fossil fuels. Four oil companies have applied to explore offshore as mining companies prospect for uranium and gold. Two aluminum companies want to build smelters using glacial meltwater for hydroelectric power. The U.S. Geological Survey estimates that waters off Greenland's northeastern coast may contain as much as 31 billion barrels of oil and gas. More oil may be found on the west coast, enough to tempt Exxon Mobil, Chevron, Canada's Husky Energy and Cairn Energy, and Sweden's PA Resources. Greenlanders in November 2008 approved a self-rule charter that directs mineral royalties to national development. The idea is to leverage the fruits of global warming to wean Greenland off its annual $680 million subsidy from Denmark.

FURTHER READING

Appenzeller, Tim. "The Big Thaw." *National Geographic*, June 2007, 56–71.

Brown, DeNeen L. "Greenland's Glaciers Crumble: Global Warming Melts Polar Ice Cap into Deadly Icebergs." *Washington Post*, October 13, 2002, A-30.

Chen, J. L., C. R. Wilson, and B. D. Tapley. "Satellite Gravity Measurements Confirm Accelerated Melting of Greenland Ice Sheet." *Science* Online, August 10, 2006, doi: 10.1126/science.1129007.

Dyurgerow, Mark B., and Mark F. Meier. "Twentieth-century Climate Change: Evidence from Small Glaciers." *Proceedings of the National Academy of Sciences of the United States of America* 97, no. 4 (February 15, 2000):1351–1354.

Faris, Stephan. "Phenomenon: Ice Free." *New York Times Sunday Magazine*, July 27, 2008. http://www.nytimes.com/2008/07/27/magazine/27wwln-phenom-t.html.

Friedman, Thomas. "Learning to Speak Climate." *New York Times*, August 6, 2008b. http://www.nytimes.com/2008/08/06/opinion/06friedman.html.

"Greenland Ice Sheet Melting Rapidly." Environment News Service, October 20, 2006.

"Greenland's Ice Melt Accelerating." Environment News Service. September 21, 2006. http://www.ens-newswire.com/ens/sep2006/2006-09-21-01.asp.

Gregory, Jonathan M., Philippe Huybrechts, and Sarah C. B. Raper. "Threatened Loss of the Greenland Ice Sheet." *Nature* 428 (April 8, 2004):616.

Hanna, Edward, Philippe Huybrechts, Konrad Steffen, John Cappelen, Russell Huff, Christopher Shuman, Tristram Irvine-Flynn, Stephen Wise, and Michael Griffiths. "Increased Runoff from Melt from the Greenland Ice Sheet: A Response to Global Warming." *Journal of Climate* 21, no. 2 (January 15, 2008):331–341.

Hansen, James E. "Declaration of James E. Hansen." *Green Mountain Chrysler-Plymouth-Dodge-Jeep, et al., Plaintiffs v. Thomas W. Torti, Secretary of the Vermont Agency of Natural Resources, et al.,*

Defendants. Case Nos. 2:05-CV-302 and 2:05-CV-304, Consolidated. United States District Court for the District of Vermont. August 14, 2006b. http://www.giss.nasa.gov/~dcain/recent_papers_proofs/vermont_14aug20061_textwfigs.pdf.

Henry, Tom. "Global Warming Grips Greenland, Leaving Lasting Mark." *Toledo Blade*, October 12, 2008. http://www.toledoblade.com/apps/pbcs.dll/article?AID=2008810109858.

Hotz, Robert Lee. "Greenland's Ice Sheet Is Slip, Sliding Away." *Los Angeles Times*, June 24, 2006, n.p.

Howat, I. M., I. Joughin, S. Tulaczyk, and S. Gogineni, 2005, "Rapid Retreat and Acceleration of Helheim Glacier, East Greenland." *Geophysical Research Letters* 32 (2005), L22502, doi: 10.1029/2005GL024737.

"Ice Sheet of Greenland Melting Away at Faster Pace." Bloomberg News in *Omaha World-Herald*, January 20, 2008, 18-A.

Joughin, Ian, Sarah B. Das, Matt A. King, Ben E. Smith, Ian M. Howat, and Twila Moon. "Seasonal Speedup Along the Western Flank of the Greenland Ice Sheet." *Science Express*, April 17, 2008, 1153288v1, doi: 10.1126/science.1153288.

Kerr, Richard A. "Greenland Ice Slipping Away but Not All That Quickly." *Science* 320 (April 18, 2008c):301.

Kolbert, Elizabeth. "Testing the Climate." (Talk of the Town) *The New Yorker*, December 31, 2007d, 43–44.

Lean, Geoffrey. "The Big Thaw: Global Disaster Will Follow if the Ice Cap on Greenland Melts—Now Scientists Say It Is Vanishing Far Faster Than Even They Expected." *London Independent*, November 20, 2005 (in Common Dreams News Center). http://www.commondreams.org/headlines05/1120-03.htm.

Lyall, Sarah. "Warming Revives Flora and Fauna in Greenland." *New York Times*, October 28, 2007b. http://www.nytimes.com/2007/10/28/world/europe/28greenland.html.

Murray, Tavi. "Greenland's Ice on the Scales." *Nature* 443 (September 21, 2006):277–278.

Paterson, W. S. B., and N. Reeh. "Thinning of the Ice Sheet in Northwest Greenland over the Past Forty Years." *Nature* 414 (November 1, 2001):60–62.

"Researchers Warm Up to Melt's Role in Greenland Ice Loss." NASA Earth Observatory, April 17, 2008. http://earthobservatory.nasa.gov/Newsroom/NasaNews/2008/2008041726522.html.

Revkin, Andrew C. "In Greenland, Ice and Instability." *New York Times*, January 8, 2008a. http://www.nytimes.com/2008/01/08/science/earth/08gree.html.

Rudolf, John Collins. "The Warming of Greenland." *New York Times*, January 16, 2007. http://www.nytimes.com/2007/01/16/science/earth/16gree.html.

"Satellite Data Reveals Extreme Summer Snowmelt in Northern Greenland, CCNY Professor Says." NASA Earth Observatory, October 8, 2008. http://earthobservatory.nasa.gov/Newsroom/MediaAlerts/2008/2008100827658.html.

"Satellite Images Show Continued Breakup of Greenland's Largest Glaciers, Predict Disintegration in Near Future." NASA Earth Observatory, August 20, 2008. http://earthobservatory.nasa.gov/Newsroom/MediaAlerts/2008/2008082027361.html.

Schiermeier, Quirin. "A Rising Tide: The Ice Covering Greenland Holds Enough Water to Raise the Oceans Six Metres—and it's Starting to Melt." *Nature* 428 (March 11, 2004a):114–115.

Steffen, Konrad. "Greenland Melt Accelerating, According to Colorado University-Boulder Study." Press Release, University of Colorado, December 11, 2007. NASA Earth Observatory, December 25, 2007. http://www.news.uiuc.edu/news/07/1210nitrogen.html.

Steffensen, Jørgen Peder, Katrine K. Andersen, Matthias Bigler, Henrik B. Clausen, Dorthe Dahl-Jensen,1 Hubertus Fischer, Kumiko Goto-Azuma, Margareta Hansson, Sigfús J. Johnsen, Jean Jouzel, Valérie Masson-Delmotte, Trevor Popp, Sune O. Rasmussen, Regine Röthlisberger, Urs Ruth, Bernhard Stauffer, Marie-Louise Siggaard-Andersen, Árny E. Sveinbjörnsdóttir, Anders Svensson, and James W. C. White. "High-Resolution Greenland Ice Core Data Show Abrupt Climate Change Happens in Few Years." *Science* 321(August 1, 2008):680–684.

Struck, Doug. "Icy Island Warms to Climate Change: Greenlanders Exploit 'Gifts from Nature' While Facing New Hardships." *Washington Post*, June 7, 2007b, A-1 http://www.washingtonpost.com/wp-dyn/content/article/2007/06/06/AR2007060602783.html?referrer=email.

Toniazzo, T., J. M. Gregory, and P. Huybrechts. "Climatic Impact of a Greenland Deglaciation and its Possible Irreversibility." *Journal of Climate* 17, no. 1 (January 1, 2004):21–33.

Vedantam, Shankar. "Glacier Melt Could Signal Faster Rise in Ocean Levels." *Washington Post*, February 17, 2006, A-1. http://www.washingtonpost.com/wp-dyn/content/article/2006/02/16/AR2006021601292_pf.html.

Velicogna, Isabella, and John Wahr. "Acceleration of Greenland Ice Mass Loss in Spring 2004." *Nature* 443 (September 21, 2006b):329–331.

Witze, Alexandra. "Climate Change: Losing Greenland." *Nature* 452 (April 17, 2008):798–802.

Growing Seasons in Europe and Asia

A network of phenological observers monitors gardens stretching from the north of Norway southeastward to Macedonia, as well as from Valentia Observatory's phenological garden in County Kerry, Ireland, eastward to Finland. The observers used clones of a single sample of each species bred in Germany when the first gardens were established. Phenology, "the timing of seasonal activities of animals and plants" (Walther et al. 2002, 389), has been used to assemble a

40-year record that was analyzed by two German scientists, Annette Menzel and Peter Fabian, from the University of Munich. Observations gathered with this system indicate that spring events, including leaf unfolding, have advanced by about six days during the period, whereas autumn events such as leaf coloring now occur, on average, 4.8 days later. Therefore, the average growing season in Europe has lengthened by 10.8 days since the early 1960s (McWilliams 2001, 26).

These changes can be attributed almost entirely to changes in average temperatures, especially in winter. Other variables, including soil composition, water supply, biological factors, and the general surroundings, have as far as possible been kept identical. The results agree with similar studies carried out in the United States. The U.S. study concluded that during three decades the average first leaf date had moved progressively earlier each year, advancing from around May 4 in 1965 to April 23 by the early 1990s. According to one observer,

> Both studies confirm what we already know—that average winter temperatures in both regions have been rising slightly, but significantly, throughout the period concerned. But more importantly, the results establish phenology as an important tool for the monitoring of global warming. (McWilliams 2001, 26)

During 2001 and 2002, British phenologists gathered information at various locations in the British Isles, "showing that the arrival of spring is no longer constrained to its traditional March slot and the end of autumn now continues well beyond late October" (Hann 2002). The phenologists' studies, taken collectively, indicated that

> higher-than-average temperatures from January to April [of 2002] led to almost every characteristic of this year's spring occurring up to three weeks earlier than in 2001. On average, insects such as bees and butterflies, were three weeks early, while plants flowered two weeks ahead and many birds, including the turtle dove, arrived a week earlier than usual (Hann 2002)

According to these studies, other notable early spring signs included the following:

- the call of the cuckoo—five days earlier
- hazel flowering—23 days earlier
- emergence of snowdrops—seven days earlier
- the first shimmering bluebell carpets—16 days earlier
- alder leafing—13 days earlier
- hawthorn leafing –17 days earlier (Hann 2002)

Changes in the autumn included the following:

- beech leaf coloring—12 days later
- oak leaf coloring—nine days later
- beech leaf fall—12 days later
- oak leaf fall—five days later (Hann 2000)

According to the study, "some people in milder parts of the United Kingdom were condemned to cutting their lawns all year-round" (Hann 2002).

Great Britain's Woodland Trust recruited 13,500 gardeners and wildlife watchers throughout the United Kingdom to monitor the first signs of spring early in 2004. A volunteer in Lochgilphead, Scotland, reported the first frogspawn on February 1, six weeks before its usual expected occurrence. The report also confirmed the earliest recorded sighting of a bumblebee in Scotland, on February 15. Ten more sightings were recorded during the same month. Blackbirds were observed building nests on January 8 near Kircudbright, two or three months ahead of usual. In London, during the 1920s and 1930s, bumblebees were never sighted before the first week of February. In 2003, one was sighted in Isleworth, West London, on December 23. Another was noted on Christmas Eve in Devon, lured out by mild temperatures. During the second week of February 2004, however, many of these "early wakers" were killed by a hard freeze, snow, and ice (Derbyshire 2004, 15).

The phenologists concluded that

> [i]nterconnected and often complex relationships between trees, insects and animals in woods are also affected by early spring arrivals. For example, results from this spring [2002] show that some synchronous relationships between birds, insects and plants could become disturbed. Crucially, for many species, this could have serious implications for their future survival. As ancient woods become even more fragmented and isolated some characteristic plants and animals may face further threats as the climate changes (Hann 2002)

Researchers at Boston University and the American Geophysical Union have described dramatic changes in the timing of both the appearance and fall of leaves (as recorded by weather satellites during the past 21 years), according to data published in the September 16, 2000, issue of the *Journal of Geophysical Research.*

The growing season is now almost 18 days longer in Eurasia, with spring arriving a week early and autumn delayed by 10 days. In North America, the growing season appears to be as much as 12 days longer (Dube 2001).

While a longer growing season may seem like cause for celebration, researcher Ranga Myneni, an associate professor in the department of geography at Boston University, argued that the news is cause for concern. "What is good for plants is not necessarily good for the planet" (Dube 2001).

Wolfgang Lucht and colleagues documented a two-decade-long trend toward earlier spring bud-burst and increased leaf cover in northern latitudes that was briefly interrupted following the 1991 eruption of Mount Pinatubo during 1992 and 1993 (Lucht et al. 2002, 1687). "We conclude," they wrote, "That there has been a greening trend in the high northern latitudes, associated with the gradual lengthening of the growing season … and reduced snow-cover extent" (Lucht et al. 2002, 1688).

FURTHER READING

Derbyshire, David. "Baffled Bumble Bee Lured out Early by Changing Climate." *London Daily Telegraph*, March 12, 2004, 15.

Dube, Francine. "North America's Growing Season 12 Days Longer: What Is Good for Plants Is Not Necessarily Good for the Planet." *National Post* (Canada), September 5, 2001. http://www.nationalpost.com.

Hann, Judith. "Spring Wakes Early, but Will Autumn Lie in Late Again? What Will Tomorrow's World Look Like?" United Kingdom Woodland Trust, 2002. http://www.woodland-trust.org.uk/news/subindex.asp?aid=328.

Lucht, Wolfgang, I. Colin Prentice, Ranga B. Myneni, Stephen Sitch, Pierre Friedlingstein, Wolfgang Cramer, Philippe Bousquet, Wolfgang Buermann, and Benjamin Smith. "Climatic Control of the High-latitude Vegetation Greening Trend and Pinatubo Effect." *Science* 296 (May 31, 2002):1687–1689.

McWilliams, Brendan. "Study of Plants Confirms Global Warming." *Irish Times*, November 1, 2001, 26.

Walther, Gian-Reto, Eric Post, Peter Convey, Annette Menzel, Camille Parmesan, Trevor J.C. Beebee, Jean-Marc Fromentin, Ove Hoegh-Guldberg, and Franz Bairlein. "Ecological Responses to Recent Climate Change." *Nature* 416 (March 28, 2002):389–395.

Gulf of Mexico Coast: Prospective Climate Changes

Global warming will have serious long-term effects on the U.S. Gulf Coast, including more frequent flooding, worse droughts, and a diminishing supply of fresh water, according to a study by a team of 10 scientists with expertise in the region. The study detailed potential consequences during the next 50 to 100 years of higher temperatures and rising sea levels on the Gulf Coastal Plain from the Laguna Madre to the Florida Keys (Freemantle 2001, 32).

The Gulf Coast region is disproportionately susceptible to changes in temperature and sea level because a small increase in temperature in this humid area can result in a drastic rise in the area's July heat index; also, a small rise in sea level poses serious problems in a coastal region that is flat, overdeveloped, and slowly sinking even without sea-level rise and thermal expansion provoked by global warming (Freemantle 2001, 32).

Projected Sea-Level Rises

This study said that sea level may rise between 8 and 20 inches during the twenty-first century. Adding sea-level rise to ground subsidence, coastal residents can expect a net sea-level rise of 15 to 44 inches (Freemantle 2001, 32). At the same time, many scientists' forecasting models agreed that precipitation could become more uneven. Extreme rainfall events, such as the flooding caused in the Houston area in June 2001 by the remnants of Tropical Storm Allison, are expected to occur more frequently (Freemantle 2001, 32).

Joseph Siry, director of the Florida Climate Alliance, called the report "an early-warning advisory about a problem that shows no signs of going away" (Hollingsworth 2001, 1). "The potential impacts are so great [that] we must plan now to avert them," he said. The study suggested that accelerated warming of global climate might lead to a catastrophic rise in local sea level, dramatic weather fluctuations, and an increase in heat-related and insect-borne illness (Hollingsworth 2001, 1).

Higher storm surges probably will worsen the impact of future hurricanes in Florida and other states bordering the Gulf of Mexico. The Gulf Coast climate study warned that, because of rising seas, salt-water intrusion is likely to contaminate more drinking supplies in Texas, Louisiana, Mississippi, Alabama, and Florida. The report, "Confronting Climate Change in the Gulf Coast Region," was sponsored by the Union of Concerned Scientists and the Ecological Society of

America. A more concise report, "Feeling the Heat in Florida," was sponsored by the Natural Resources Defense Council.

Parts of coastal Louisiana and Mississippi could lose as much as one foot of elevation within 10 years according to an analysis by the National Oceanic and Atmospheric Administration (NOAA) National Geodetic Survey. The NOAA researchers have warned that populated areas will face increased dangers from storm surges and flooding caused by the ongoing subsidence of coastal areas along the northern Gulf of Mexico. Coastal wetlands in Louisiana have been disappearing at the rate of 33 football fields per day. The NOAA researchers estimate that at the present rate of subsidence, 15,000 square miles of land along the southern Louisiana coast will subside to sea level or below within the next 70 years ("Coastal Gulf" 2003). Shoreline in this area is sinking due to natural processes as well as the withdrawal of subsurface oil and water. In southern Louisiana, roughly one million acres of coastal marsh were converted to open water between 1940 and 2000, with losses accelerating over time concurrent with a quickening pace of sea-level rise (Inkley et al. 2004, 8, 13). Subsidence adds to the effects of sea-level rise due to rising temperatures. The state of Louisiana lost about 1,900 square miles of coastal marshes, an area the size of Delaware, during the twentieth century. It may lose another 700 square miles by 2050 at present rates of erosion and subsidence. By that time, one-third of the state's coast will have washed into the Gulf of Mexico within a century and a half.

Matt Crenson of the Associated Press sketched a picture of coastal erosion's practical effects:

> A few decades ago Isle de Jean Charles was a patch of high ground in a sea of grassy marsh teeming with catfish and crawdads. Today the small community is a true island, regularly flooded during storms and sometimes even at high tide. In a few years it will be submerged completely. Deme Naquin, 75 years old, remembers paddling a flat-bottomed pirogue to school as a boy. Now he's getting ready to leave the only place he has ever called home. The U.S. government has offered to resettle the island's 270 residents because a new hurricane protection plan leaves them outside the ramparts. (Crenson 2002)

"Another hurricane and the road's going to be gone," said Chad Naquin, Deme's 29-year-old grandson. "It would be hard to leave, but in the long run it would be the best thing" (Crenson 2002). As many as 35 square miles of Louisiana's wetlands sink into the Gulf of Mexico each year. In some places, the coastline has retreated as much as 30 miles.

Effects on Florida

Florida already has been experiencing early effects of global warming in the form of retreating shorelines, dying coral reefs, more wildfires, and saltwater seeping into the freshwater aquifers, according to a peer-reviewed study by scientists from Florida universities (Hollingsworth 2001, 1). Maximum summer temperatures in Florida could increase 3° to 7°F, boosting July's heat index (the combined effect of heat and humidity) by 10° to 25°.

"We're already doomed to having the consequences of global climate change no matter what we do," said Mark Harwell, a professor of marine ecology at the University of Miami. "But we can significantly reduce the consequences if we acknowledge and plan for them now" (Hollingsworth 2001, 1). About 1,000 people move to Florida every day, a pace expected to more than double the population to 34 million in 30 years. "Water supply is critical to this scenario. It's going to be the problem for Florida in the future," said Susanne Moser of the Union of Concerned Scientists (Hollingsworth 2001, 1).

An anticipated sea-level rise of 8 to 20 inches during the twenty-first century could imperil houses that were within 200 feet of Florida shorelines by the year 2100. Rising sea levels are likely to inundate many coastal areas during coming decades. Rising seas will force Floridians to build extensive seawalls to protect their waterfront homes, condominiums, and businesses, the scientists said. The disappearance of the beaches is likely to cost Florida's tourism industry dearly. The state's major real-estate assets are its beaches, now framed by ranks of condominiums. Ricardo Alvarez of the International Hurricane Center at Florida International University, said during a news conference announcing the study results, "we sell good beaches, climate and sunshine" (Pittman 2001, 3-B).

"We're Already Living on the Edge"

"Here, in Louisiana, we're particularly vulnerable to climate change because we're already living on the edge," said Denise Reed, a wetlands scientist at the University of New Orleans and one of the report's authors. "Many of our coastal communities, Dulac, Irish Bayou, Yscloskey, already have water lapping at their roads. Add a foot of water from climate change, and you're facing more than an inconvenience" (Schleifstein 2001, 1).

Most states along the Atlantic and Gulf coasts of the United States have been experiencing problems similar to Louisiana's. "We're not going to be the only ones in the boat," said Al Naomi, a project manager in the New Orleans District of the U.S. Army Corps of Engineers. "We're just in the boat first" (Crenson 2002). Loss of coastal wetlands may devastate Louisiana's $2 billion-a-year fishing industry. Many fish, crabs, shrimp, and other marine animals rely on wetlands as a nursery, where their young feed. By some calculations, according to Crenson's report, Louisiana's wetlands are involved in producing as much as 40 percent of the seafood caught in the United States (Crenson 2002).

Sinking land and rising waters imperil coastal residents as well as wildlife. Even before Hurricane Katrina hit the area in 2005, the Louisiana Red Cross estimated that between 25,000 and 100,000 people could become refugees if a major hurricane hit New Orleans from the southeast, driving the waters of Lake Pontchartrain into the city. Many of these homes already lie below mean sea level, often behind dikes.

According to Crenson, the Red Cross was so pessimistic about southern Louisiana's prospects in the face of a major storm that it closed all of its relief centers south of Interstate 10, which runs across the state from Lake Charles to New Orleans. The Red Cross said it would be foolhardy to operate disaster centers in an area that people should abandon in a hurricane (Crenson 2002). Ivor van Heerden, deputy director of the Louisiana State University Hurricane Center, described a horrific ordeal for anybody riding out a major hurricane in New Orleans. "If you survive the flying debris and your house collapsing," van Heerden said, "then you're going to have to deal with a minimum of 13 feet of water" (Crenson 2002).

Years before Katrina, New Orleans was warned that the weeks or months after a major hurricane could present the city's residents with many problems. New Orleans sits in a bowl ringed by protective levees, so water could stagnate, contaminated with toxic waste from dozens of petrochemical plants that line the Mississippi, as well as human waste and decomposing carcasses (Crenson 2002). "This is a $50 billion disaster," van Heerden said, surveying expensive new housing developments on the shores of Lake Pontchartrain (Crenson 2002). The real thing turned out to be much costlier, and deadlier, than the projections.

FURTHER READING

"Coastal Gulf States Are Sinking." Environment News Service. April 21, 2003. http://ens-news.com/ens/apr2003/2003-04-21-09.asp#anchor4.

Crenson, Matt. "Louisiana Sinking: One State's Environmental Nightmare Could Become Common Problem." Associated Press, August 10, 2002. (LEXIS).

Freemantle, Tony. "Global Warming Likely to Hit Texas: Scientists Say Temperature Rise Will Change Rainfall, Gulf Coast Region." *Houston Chronicle*, October 24, 2001, 32.

Hollingsworth, Jan. "Global Warming Studies Put Heat on State: Tampa Bay Area Labeled Extremely Vulnerable." *Tampa Tribune*, October 24, 2001, 1.

Inkley, D. B., M. G. Anderson, A. R. Blaustein, V. R. Burkett, B. Felzer, B. Griffith, J. Price, and T. L. Root. *Global Climate Change and Wildlife in North America*. Washington, DC: Wildlife Society, 2004. http://www.nwf.org/news.

Pittman, Craig. "Global Warming Report Warns: Seas Will Rise." *St. Petersburg Times*, October 24, 2001, 3-B.

Schleifstein, Mark. "The Gulf Will Rise, Report Predicts: Maybe 44 Inches, Scientists Say." *New Orleans Times-Picayune*, October 24, 2001, 1.

H

Hadley Cells

While warmer air holds more moisture, not everyone will see more precipitation in a globally warmed world. Many deserts already are expanding in a worldwide pattern influenced by atmospheric circulation patterns that meteorologists call "Hadley Cells."

Most deserts range between 20 and 40 degrees north and south latitude. While precipitation patterns are also influenced by other factors (such as ready access, or lack thereof, to ocean-borne moisture), rainfall is strongly influenced by Hadley Cells, which determine whether air generally rises or falls at certain latitudes. Rising air portends instability, low pressure, and storminess; descending air generally provokes high pressure and clear skies. In a warmer world, Hadley Cells expand, which causes deserts to expand, a process that is already evident from news reports around the world

Drought and Deluge

Extreme weather is not a novelty in much of the world, but warming tends to aggravate extremes of both drought and deluge. Reports of an intensifying hydrological cycle have been plentiful outside the United States. In some areas, deluges may punctuate severe droughts. On July 6, 2002, the northeastern edge of what was once called "The Great American Desert," near Ogallala, Nebraska, received as much as 10 inches of rain during a drought, running off the hardened soil, washing out sections of Interstate 80, killing a truck driver, and provoking evacuation of residents. Both approaches of a bridge over the South Platte River were washed out. The rainfall was two to three times the amount that previously had fallen in the area during the entire year of 2002.

Nearly a year later, during the night of June 22, 2003, a stagnant super-cell thunderstorm dumped 12 to 15 inches of rain (half the area's annual average) south and east of Grand Island, Nebraska, an area that also was suffering intense drought at the time. The same storm spawned several tornadoes, killing one person and injuring several others. This storm, which destroyed large parts of Aurora, Nebraska, produced hail that was among the largest ever reported in the United States, as well as a tornado that stood nearly in one place for half an hour, devastating the town of Deshler.

Droughts around the World

Droughts in regions where Hadley Cells favor descending air now span the globe, from Australia, to Spain, Iraq, Afghanistan, parts of China, and the U.S. Southwest, including California. In June 2008, California declared a drought and warned that water rationing might follow. California this year had its driest spring in 88 years. Los Angeles announced plans to use cleansed sewage water to augment supplies ("Hunger 2008").

In China, the Gobi Desert—also within the northern reaches of Hadley Cell range—has been expanding, sending occasional dust storms into Beijing and aggravating air pollution from coal-fired power plants that give that city and others in China the dirtiest air on Earth.

By mid-2008, several thousands of people had been forced out of their homes because of food and water shortages in northern Afghanistan. Large parts of the country received much less than normal rain and snow. Nearly 2,000 families departed the drought-stricken Chemtal district of Balkh province in one week late in May 2008, even after the government donated 37 tons of food and some water. Water and food shortages have forced several families to eat grass, and some died of starvation. Many farm fields were ravaged by locusts as well.

Drought intensified in the Fertile Crescent region of northern Iraq and eastern Syria during the winter of 2007–2008 as sufficient winter rains and snows failed to arrive in the mountains of Turkey, which have fed rivers in the area since the beginnings of human urban civilization. Lack of moisture also limited the irrigation that is crucial for agriculture in the desert ("Drought in Iraq" 2008).

Barcelona, Spain, was so dry by May 2008 that water was being imported, for the first time ever, by ship ("Drought Forces" 2008, A-13). A desalinization plant and pipeline from the Ebro River (to the west) was being planned. Parts of southeastern Spain were turning to desert by 2008, even amid new developments of vacation homes and golf courses. Water has become a valuable commodity and source of copious conflict. This area has experienced cyclical droughts in the past, but this one may be long-lasting. Local aquifers are retreating below the range of pumps as well, as the area's climate comes to resemble that of northern Africa. The Spanish Environment Ministry warns that a third of the country may turn to desert in coming years (Rosenthal 2008d).

FURTHER READING

"Drought Forces Barcelona to Ship in Drinking Water." *Wall Street Journal*, May 14, 2008, A-13.

"Drought in Iraq." NASA Earth Observatory, June 4, 2008. http://earthobservatory.nasa.gov/Newsroom/NewImages/images.php3?img_id=18046

"Hunger, Water Scarcity Displaces Thousands Of Afghans." Reuters in *New York Times*, June 4, 2008. http://www.nytimes.com/reuters/world/international-afghan-displacement.html.

Rosenthal, Elisabeth. "Water Is New Battleground in Drying Spain." *New York Times*, June 3, 2008d, A-1, A-12.

Hansen, James E. (March 29, 1941–)

If the U.S. president's Cabinet included a Secretary of Climate Security, James E. Hansen, who directs NASA's Goddard Institute for Space Sciences in New York City, might be the best nominee. Hansen is too much of a scientist—and not enough of a politician—to accept the job, however. Regarding global warming, he may be the Earth's Paul Revere. Hansen has been an international leader in both the scientific and political worlds since global warming became an important issue during the 1980s.

Climate scientist Dr. James Hansen (1941–) speaks at the Green Apple Festival on April 20, 2008, in Washington, D.C. (Courtesy of Getty Images)

Hansen's long-term forecast: 10 more years of "business as usual" greenhouse-gas emissions and we'll cross the "tipping point." At that point, humanity loses any chance of turning back accelerating feedbacks that lead to runaway heating of the Earth's atmosphere. Mark your calendars; Jim Hansen has been forecasting global greenhouse weather for more than 25 years and his forecasting record is very good, even as he has run rear-guard actions to keep three Republican U.S. presidents backed by fossil-fuel partisans from defunding his laboratory and censoring his statements. His science is impeccable, and his prescience is well known among his peers.

Personally, Hansen is unassuming to a fault, an Iowan with a measured speaking style that some New Yorkers might mistake as a little slow on the draw, until he opens up on global warming. At that point, anyone engaging him in conversation will discover a hard edge, and a wealth of information. Six feet tall, with receding, light-brown hair, Hansen favors plaid shirts that would put him at home on an Iowa farm. His nearly aw-shucks manner seems at odds with his work as a tough-as-nails scientist who has been speaking truth to power for three decades.

Author Mark Bowen wrote in *Thin Ice: Unlocking the Secrets of Climate in the World's Highest Mountains* (2005):

> As testament to [Hansen's] fine character, he never accused his colleagues of pettiness. In the 30 years he has been working on the greenhouse issue, he has been the object of endless scorn and personal attack, yet he has never responded in kind. He has always focused resolutely on the science and the facts and, moreover, gone out of his way to point out the shortcomings in his own arguments. (Bowen 2005, 159)

Hansen enjoys reading fiction with heroes who flout convention and are persecuted for sticking to their principles. He also maintains an extensive e-mail list of people to whom he sends early drafts of his papers, asking for criticism.

Hansen's Early Life

Hansen, the fifth of seven children (he has four older sisters), was born in a farmhouse and raised in Denison, Iowa, about 60 miles northeast of Omaha. Hansen's father, a tenant farmer, moved to Denison when Jim was four years of age, and took up work as a bartender; his mother worked as a waitress. Hansen was known for his intelligence as a young man, even when he was not being overly studious. Hansen, a life-long New York Yankees fan, played second base in the Babe Ruth league.

With a scholarship and money saved from his *Omaha World-Herald* paper route, Hansen attended the University of Iowa and graduated summa cum laude in 1963, majoring in mathematics and physics. At the same school, in physics, Hansen earned a master's degree in astronomy. The University of Iowa was an exciting place to study astronomy; the department had its own satellite, and its chairman was James Van Allen, who discovered the Earth-girdling radiation belts named after him. Hansen decided to specialize in the atmosphere of Venus at a time when scientists were discovering that the planet's super-hothouse atmosphere (with temperatures above 850°F) was 95 percent carbon dioxide. He earned a doctorate in 1967 with a dissertation on Venus. "I was so shy and unconfident that when I had an opportunity to take a course under Prof. Van Allen, I avoided it because I didn't want him to realize how ignorant I was," Hansen told an audience at his *alma mater* in 2004 (Johansen 2006b, 27).

His doctorate completed, Hansen went to work at the NASA Goddard Institute for Space Studies (GISS), which is affiliated with Columbia University in New York City. On leave for a year, 1969–1970, at the Leiden Observatory in the Netherlands, Hansen met his wife, Anniek. During 1976, Hansen was working as principal investigator on the Pioneer Venus Orbiter when a Harvard postdoctoral researcher asked him to help calculate the greenhouse effect of human-generated emissions in Earth's atmosphere. Soon Hansen was captivated by potential global warming on Earth. As he worked at GISS, it also was becoming more involved in Earth studies, a trend that accelerated after his appointment as director in 1981.

Attempts at Censorship

Government censors cross Hansen's path at their peril. He is an expert at turning any attempt to shut him up into international headlines as he insists that NASA's mission is to keep the public informed, not to follow the weathervane of White House spin control.

Hansen was heading GISS in 1981, when, with several colleagues, he was the first to use the term "global warming" in a scientific context. Following an important article in *Science* (Hansen et al. 1981, 957–966), Reagan administration functionaries withdrew GISS funding from the Energy Department.

In 1988, with George H. W. Bush in the White House, Hansen went public with warnings about global warming before the U.S. Senate, on a very hot, humid summer day in Washington, D.C., part of a notably hot summer nationwide. During 1989, Hansen asked,

> When is the proper time to cry wolf? Must we wait until the prey, in this case the world's environment, is mangled by the wolf's grip? … The climate system has great inertia, so as yet we have realized only a part of the climate change which will be caused by gases we have already added to the atmosphere. Add to this the inertia of the world's energy, economic, and political systems, which will affect any plans to reduce greenhouse gas emissions. (Johansen 2006b, 27)

In 2006, the George W. Bush White House assigned Hansen a 24-year-old political operative,

George Deutsch, to manage his media relations. At age 65, Hansen had been waging weather wars since before his "minder" had been born. Several times over many years, government operatives have edited Hansen's warnings. Each time, he has then spoken as a private citizen and provoked a public furor. Deutsch seemed not to know enough history of the weather wars to realize what he was getting into. Soon Hansen was making headlines in the *Washington Post* and the *New York Times* as he compared censorship of science to the tactics of Nazi Germany and the old Soviet Union—maintaining that the Bush administration is fiddling as the planet burns.

Deutsch's purported journalism degree from Texas A&M quickly was exposed as a fraud, and he was forced to resign, meanwhile pouting that he had been a victim of a Democratic Party ambush. Deutsch's sole qualification for the job was his service on the Bush-Cheney reelection campaign. Hansen himself is a political independent, who readily acknowledges that he voted for John Kerry in 2004.

Facing off with an administration known for its fossil-fuel interests, Hansen is scientifically armed and politically dangerous. Following cancellation of a speech in Washington, D.C., because of White House pressure, Hansen called upon Van Allen, who taught at the University of Iowa until his death in 2006. Van Allen arranged a presentation in Iowa City October 26, 2004, during which Hansen—speaking explicitly as a private citizen, having paid his own expenses—said that "[t]his process [of censorship] is in direct opposition to the most fundamental precepts of science.... This, I believe, is a recipe for environmental disaster" (Johansen 2006b, 27).

Two years after Hansen and other NASA employees described a pattern of distortion and suppression of climate science by political appointees, NASA's inspector general agreed with them, concluding that such manipulation was "inconsistent" with the law. The inquiry began in 2006 after a request by 14 U.S. senators.

Taking Earth's Temperature

Every month, Hansen's lab takes the Earth's temperature, monitoring 10,000 temperature gauges around the planet. Hansen anticipates that the Earth won't experience another ice age unless we stop using fossil fuels—and even then it would take several thousand years to restore natural equilibrium to natural cycles that have been disrupted by the burning of fossil fuels. "The human race now controls climate," said Hansen, "For better or for worse" (Johansen 2006b, 27).

In a presentation at the American Geophysical Union annual meeting in San Francisco on December 6, 2005, Hansen addressed a basic question: how much "wiggle room" do the Earth and its inhabitants have before global warming becomes a truly unavoidable disaster? Further warming of more than 1°C "will make the Earth warmer than it has been in a million years," Hansen said.

> "Business-as-usual" scenarios, with fossil fuel CO_2 emissions continuing to increase at about 2 percent a year as in the past decade, yield additional warming of 2° or 3°C this century and imply changes that constitute practically a different planet ... the Earth's climate is nearing, but has not passed, a tipping point, beyond which it will be impossible to avoid climate change with far-ranging undesirable consequences. (Johansen 2006b, 28)

"A different planet," said Hansen, will include

> not only loss of the Arctic as we know it, with all that implies for wildlife and indigenous peoples, but losses on a much vaster scale due to worldwide rising seas. Sea level will increase slowly at first, as losses at the fringes of Greenland and Antarctica due to accelerating ice streams are nearly balanced by increased snowfall and ice sheet thickening in the ice sheet interiors. (Johansen 2006b, 28)

The Dynamics of Melting Ice

As Greenland and West Antarctic ice is softened and lubricated by meltwater and as buttressing ice shelves disappear due to a warming ocean, Hansen continued,

> The balance will tip toward ice loss, thus bringing multiple positive feedbacks into play and causing rapid ice sheet disintegration. The Earth's history suggests that with warming of 2° to 3° degrees C the new equilibrium sea level will include not only most of the ice from Greenland and West Antarctica, but a portion of East Antarctica, raising sea level of the order of 25 meters (80 feet). (Johansen 2006b, 28)

This is not to say that the sea will be 80 feet higher at the end of this century. It *does* indicate that enough warming to raise sea levels that much will be "in the pipeline," with profound implications within 200 years for the world's coastlines. Many coastal cities—Shanghai, New York City, London, and Calcutta are only a few examples—will be in peril. Hansen cannot tell us exactly *when* the toilets will back up at the White House (about 50 feet above sea level), but if we don't cut our fossil-fuel consumption soon, they will. The question is *when.*

Hansen still hopes that "the grim 'business-as usual' climate change" may be avoided by slowing the growth of greenhouse-gas emissions during the first quarter of the present century, requiring "strong policy leadership and international cooperation." Hansen pointedly emphasizes that special interests representing the fossil-fuel industries are in the political saddle, "a roadblock wielding undue influence over policymakers." The special interests, said Hansen, "seek to maintain short-term profits with little regard to either the long-term impact on the planet that will be inherited by our children and grandchildren or the long-term economic well-being of our country" (Johansen 2006b, 28).

"How long have we got?" Hansen asked.

> We have to stabilize emissions of carbon dioxide within a decade, or temperatures will warm by more than 1° [Celsius]. That will be warmer than it has been for half a million years, and many things could become unstoppable. If we are to stop that, we cannot wait for new technologies like capturing emissions from burning coal. We have to act with what we have. This decade, that means focusing on energy efficiency and renewable sources of energy that do not burn carbon. We don't have much time left. (Johansen 2006b, 28)

Ideological Weather at NASA

Unknown to scientists at NASA (who were not consulted), during late January or early February of 2006—just as articles describing how Hansen was resisting White House pressure hit the front pages of the *New York Times* and *Washington Post*—the Bush administration's annual budget reached Congress with an altered version of the agency's Mission Statement. Gone was the statement that had been used since 2002: "To understand and protect our home planet; to explore the universe and search for life, to inspire the next generation of explorers ... as only NASA can." The phrase "to understand and protect our home planet" had vanished. The 2002 statement, an extension of NASA's original statement of purpose ("the expansion of human knowledge of the Earth and of phenomena in the atmosphere and space") had been adopted with advice from NASA's 19,000 employees. The new, sans Earth mission was written by fiat. NASA employees did not learn of their new mission statement until summer (Revkin 2006d, A-1, A-10).

Without explicitly saying so, the change seemed aimed squarely at parts of NASA, such as the GISS, which had sifted its focus over the years to earth sciences. Hansen said that the change might reflect White House eagerness to shift federal resources away from study of global warming. "They're making it clear that they have the authority to make this change, that the president sets the objectives for NASA, and that they prefer that NASA work on something that's nor causing them a problem," Hansen told Andrew Revkin of the *New York Times* (2006d, A-10).

On March 13, Hansen sent a letter to his correspondents observing that the ideological weather improved at NASA after he went public. At the Environmental Protection Agency, however, "where double-speak ('sound science,' 'clear skies,') has achieved a level that would make George Orwell envious, [the situation] is much bleaker, based on the impression that I receive from limited discussion with colleagues there," Hansen wrote. "Unless some new event demands it"—and precedent indicates the battle never ends—Hansen said he would like to avoid whistle-blowing activities in favor of full-time science, "quantifying options for dealing with global warming" (Johansen 2006b, 28). A few weeks after Hansen's letter, NASA issued new rules protecting scientists' rights to communicate their work to the public.

Hansen Battles Climatic Self-Deception Worldwide

During June and July 2008, Hansen traveled to Europe and Japan and found a "sobering degree of self-deception in countries that are among the best-educated on climate change" (Johansen 2008b, 7). Hansen believes that even the most advanced governments have not realized the urgency of the situation and have placed their bets on market-driven solutions such as carbon-emissions (cap-and-trade) solutions. In the meantime, even Germany, Great Britain, and Japan—all countries that he visited—are still building coal-fired power plants without adequate technology to remove polluting carbon dioxide from their effluent.

In Germany, during June, Hansen spoke with Minister of the Environment Sigmar Gabriel. After the meeting, in a diplomatic way, Hansen said that Gabriel and other German officials (who have been avid advocates of curtailing emissions in a general sense) just didn't get the point. "I am grateful for Minister Gabriel's generosity with his time, and I have no doubt about his sincerity in dealing with climate change," Hansen said. "However, we did not come to a common understanding about ... the stark policy implication of the data, [that] I assert, [creates an] urgency for a moratorium on coal-fired power plants.... In effect we agreed to disagree, as we were both trying to be cordial" (Johansen 2008b, 7).

In Great Britain, Hansen was dealt a similar hand by officials who maintained that emissions trading would bring greenhouse-gas emissions back into a safe range. In Japan, the prime minister's representatives declined to accept his letter until they realized that it was all over the Tokyo newspapers. They, too, professed faith in cap-and-trade. Replied Hansen: "Emissions trading is such an unutterably bogus concept that we should toss it on the slagheap right off the bat. It is a shell game that allows corporations to buy the right to pollute" (Johansen 2008b, 7).

Hansen versus Coal-Fired Power

Hansen's letter, which also has been presented to several U.S. state governors, argued that

> If CO_2 emissions from coal were phased out over the period 2010–2030, and if use of unconventional fossil fuels (tar shale, tar sands) remained negligible, atmospheric CO_2 would peak at 400–425 parts per million. In that case, improved forestry and agricultural practices, especially reforestation, could get atmospheric CO_2 back beneath 350 ppm within a century or less. During the overshoot phase we might hope that ocean and ice sheet inertia may keep climate impacts tolerable, avoiding the most disastrous effects. (Johansen 2008b, 7)

"However," continued Hansen,

> If coal use continues or expands (as it is now) CO_2 will be headed to the 500–600 ppm range ... we will hand our children a planet that has entered a long chaotic transient period with climate changes out of their control, as the planet heads inexorably toward an ice-free state. (Hansen 2008c)

Furthermore, Hansen is telling all heads of government, "The danger of carbon caps and percent reduction goals is that they allow self-deception, a pretense that the climate problem is being solved. Unless they are accompanied by phase-out of coal emissions, they have practically no impact on climate change." The European Union has required that new coal plants be "carbon capture ready," but the technology is not yet available. Such a requirement without the technology, said Hansen, is "sobering self-deception." (Hansen 2008c).

Coal-fired plants can be made unnecessary, said Hansen, by investment in an electricity grid that leaks less power (as much as half our power is lost in transmission), along with increasing use of wind and solar power in a crash program resembling the Manhattan Project during World War II.

Climatic Trash Talkers

Jim Hansen often deals with personal attacks. He gave a speech to the National Press Club on June 23, 2008, which, following copious press attention, brought him hundreds of e-mails that ranged from critical to nasty. While a few such messages a day is not uncommon for him, Hansen sensed something organized here.

Most of the e-mailers rehearsed a familiar contrarian mantra, spiced with a generous helping of personal attacks on Hansen himself. Most insisted that the sun is the primary cause of climate change, with a large number convinced that a new ice age is right around the geophysical corner.

Many of the messages asserted that the Earth is entering an ice age, with the temperature in 2008 already having lost all of the warming of the past several decades. Hansen's calculations indicate that the main cause of our brief recent cooling is a La Niña episode in the Pacific Ocean.

By mid-2008, however, the La Niña was ending, and tropical temperatures had returned nearly to average. A new El Niño episode may propel temperatures to or past recent high levels during the next few years. "The low temperatures in the first half of 2008 lead us to estimate that the mean 2008 global temperature will be perhaps in the range about 10th to 15th warmest year in our record," Hansen said.

The contrarians are not having this, as they swift-boat Hansen mercilessly, calling for a new ice age. One e-mail to him virtually shouted:

> What kind of kindergarten fairy land are you living in, Jimmy? ... I've been doing my level best for the

> last 2 years to tell people what a fraud you and your ilk are.... I've also been telling people that you were basically bribed to lie ... and that you don't even have a degree in climatology ... so your biggest enemy is ME, buddy boy ... not the oil companies ... I am completely impartial and have no economic stake in the argument one way or the other ... which gives me a lot more credibility then [sic] you.... So you just keep sounding off like some closet Hitler and spewing your lies ... I will continue to trash your professional reputation and your pretended manhood.... I always keep my word when dealing with incipient [sic] nazis like you.... I will destroy you, Jimmy ... do not doubt it ... the science and the facts are on my side. (Johansen 2008b)

No one ever said that saving the Earth from some members of the human race would be easy.

FURTHER READING

Bowen, Mark. *Thin Ice: Unlocking the Secrets of Climate in the World's Highest Mountains.* New York: Henry Holt, 2005.

Hansen, J., D. Johnson, A. Lacis, S, Lebendeff, D. Rind, and G. Russell. "Climate Impact of Increasing Atmospheric Carbon Dioxide." *Science* 213 (1981):957–966.

Hansen, James E. "Trip Report." August 5, 2008c. http://www.columbia.edu/~jeh1/mailings/20080804_TripReport.pdf.

Johansen, Bruce E. "The Paul Revere of Global Warming." *The Progressive*, August 2006b, 26-28.

Johansen, Bruce E. "Hansen Battles Climatic Self-deception World-wide." *Nebraska Report*, October 2008b, 7.

Revkin, Andrew C. "NASA's Goals Delete Mention of Home Planet." *New York Times*, July 22, 2006d, A-1, A-10.

Hay Fever

Warmer temperatures and shorter winters have been related to a rising incidence of hay fever. The epidemic seems to be worldwide. Hay fever, otherwise known as seasonal allergic rhinitis, is caused by an allergy to pollens and fungal spores. David Adam, writing in the *London Guardian*, summarized the world hay-fever situation:

> They are sneezing in Stockholm, throats are itchy in India and Irish eyes are streaming. From Algeria to Iceland and Hong Kong to Aberdeen, record numbers of people are suffering the misery of hay fever—and it's getting worse. Everything from global warming to air pollution is conspiring to make hay fever the number-one global irritant. The figures are truly remarkable. The number of people rubbing their eyes in British doctor's surgeries [offices] has risen fivefold since the 1950s, and about a quarter of people in the United Kingdom are now believed to be hay-fever sufferers. (Adam 2003, 4)

Similar trends have been seen across Europe, Adam wrote. Cases of hay fever doubled or even tripled in Sweden and Finland during the 1970s and 1980s, while Swiss surveys show about 1 in 10 are affected, an increase from fewer than 1 in 100 when a similar count was made in 1926. About 40 percent of the people in Australia and the United States said they suffered from hay fever or similar allergies by 2002. Areas such as West Africa, where hay fever was once all but unknown, have been reporting it (Adam 2003, 4).

English scientists (as well as street-level observers) have maintained that the allergy season is arriving earlier because of global warming. Many trees and grasses are flowering sooner and for extended periods, creating more of the pollen that is the main trigger of hay fever. In 2002, the hay-fever season began as early as January 30, according to research by Tim Sparks of the Centre for Ecology and Hydrology. The same research also indicated that varieties of common grass were flowering up to 13 days earlier than in 2001. Hazel and birch trees also were causing allergic reactions in many hay-fever sufferers (Chapman 2003, 23).

These findings were part of the world's largest phenological survey by the Centre and the Woodland Trust. By early February 2003, said Sparks, hazel trees were in flower. Jean Emberlin, of the British National Pollen Research Unit, said, "Last year, the grass pollen season was exceptionally long because the weather was wet and warm. The season extended into August instead of ending in July" (Chapman 2003, 23). Young people appear to be worst hit, with 36 percent suffering from it. The figure in the wider population is 15 to 25 percent, with rates doubling since 1965 (Chapman 2003, 23).

In the past, London hay-fever sufferers usually have not stocked up on remedies in earnest until the beginning of May. Pharmacies in 2003 reported a 30 percent increase in sales each week between mid-March and mid-April. Muriel Simmonds, chief executive of the medical-charity Allergy U.K., cited "definite evidence" that hay fever was returning to the British Isles earlier year over year (Galloway and Rhodes 2003, 16).

Simmonds's organization has moved National Allergy Week from June to May because of the earlier onset of hay fever. "This is the earliest I've ever seen it in London," she said (Galloway and Rhodes 2003, 16). *See also:* Asthma

FURTHER READING

Adam, David. "Hatchoooooh! Record Numbers of People are Complaining of Hay Fever." *London Guardian*, June 18, 2003, 4.

Chapman, James. "Early Spring Misery for 12 Million Hay Fever Sufferers." *London Daily Mail*, February 4, 2003, 23.

Galloway, Elaine, and Chloe Rhodes. "Warm Spell Brings Early Start to Hay-fever Misery." *London Evening Standard*, April 14, 2003, 16.

Heat-Island Effect, Urban

Cities tend to absorb heat more quickly than surrounding countryside due to a number of reasons that have little to do with the basic atmospheric physics of global warming. The larger a city, and the more dense its degree of urbanization, the greater the warming.

The "urban heat-island effect" was first identified by a meteorologist, Luke Howard, in 1818. Extra heat is produced in urban areas by a city's many sources of waste heat, from building heating and air conditioning, as well as from motor vehicles, among other sources. Heating also increases when open fields and forests become streets, sidewalks, parking lots, and buildings. The dark colors of city structures, especially the asphalt streets and parking lots that make up as much as 30 percent of many urban surfaces have a very low albedo (reflectivity), so most of the sun's heat energy is absorbed, not reflected.

Cities also warm more rapidly than surrounding countryside because they usually are drier, and have less surface water and plant mass (both of which cool the air through evaporation) than most rural areas. Furthermore, as new housing and businesses spreads from urban areas, some of the cities' urban heat follows with them, spreading in widening suburban circles.

In a compact urban area such as Manhattan Island, the total heat generation of the city can add quite substantially to solar radiation. By one estimate, the heat energy generated by motor vehicles and waste space heat on Manhattan during an average winter day sometimes exceeds that of incoming solar radiation (Weiner 1990, 262).

Robert Balling, who professes skepticism regarding greenhouse warming as a global issue, is something of an expert on the urban heat-island effect in his hometown, Phoenix, Arizona. Since World War II, when many Phoenix residents slept on porches outdoors, average summertime lows in Phoenix have risen above the human comfort zone. Average summer-time lows have risen from 73°F to more than 80°F during the last half of the twentieth century. During the same 50 years, the Phoenix area's human population has increased by nearly 20 times, from roughly 150,000 to 2.8 million. Dale Quattrochi, senior research scientist at NASA's Global Hydrology and Climate Center in Huntsville, Alabama, estimated that Phoenix temperatures probably will increase as much as 15° to 20°F over historic averages during the next several decades (Yozwiak 1998).

Heat Island, Tokyo

According to the *Daily Yomiuri* of Tokyo, temperatures in several Japanese cities averaged between 3.2° and 3.9°C above long-term averages during the winter of 2001–2002. The increase in temperatures was most notable in and around Tokyo. During the early-morning hours (midnight to 5 A.M.), average temperatures in Tokyo have risen by 7.2°F in a century. In 1900, the number of "tropical nights" with minimum temperatures above 77°F was zero to five in an average summer. By the early twenty-first century, the number of such nights reached 30 to 40 during most summers (Brooke 2002, A-3). On July 20, 2004, the temperature in Tokyo hit a record-breaking 39.5°C (103.1°), the hottest temperature recorded there since records began in 1923.

Tokyo winters also have become milder, with night-time temperatures rarely dropping below freezing. Snow in Tokyo is increasingly rare. None at all fell in the city during the winter of 2001–2002. Leaves used to start turning color in the end of November, said Shinsuke Hagiwara, chief researcher of National Institute for Nature Study. "Now they only start in mid-December" (Brooke 2002, A-3). During the spring of 2002, cherry blossoms in Tokyo opened so early that when Prime Minister Junichiro Koizumi held the government's annual cherry blossom viewing party in April, the blossoms had fallen from the trees. A type of mosquito carrying dengue fever, usually found in warmer places, by 2002 had expanded its range to 60 miles north of Tokyo, according to Mutsuo Kobayashi, a medical entomologist (Brooke 2002, A-3).

Rural Japan is awash in anecdotes of unusual warming as well. In rural areas as well as Tokyo, cherry blossoms have been blooming earlier than in the past, leaving people to "hold their blossom-viewing parties under leaves instead of flowers" (Hatsuhisa 2002). The Prefecture of Niigata on the Sea of Japan, two hours north of Tokyo by bullet train, which once was known for heavy "ocean effect" snows carried by cold air from Siberia, reported a scarcity of snow during the winter of 2001–2002. During March 2002, no snow fell there for the first time since weather records had been kept (Hatsuhisa 2002). Many resorts that depend on snowfall (mostly for skiing) were forced to close. During the same month, two-thirds of Japan's weather-observation stations reported their highest temperatures in a statistical record that, in most cases, reaches to 1886.

FURTHER READING

Brooke, James. "'Heat Island' Tokyo Is Global Warming's Vanguard." *New York Times*, August 13, 2002, A-3.
Hatsuhisa, Takashima. "Climate." *Journal of Japanese Trade and Industry*, September 1, 2002, n.p. (LEXIS)
Weiner, Jonathan. *The Next One Hundred Years: Shaping the Fate of Our Living Earth*. New York: Bantam Books, 1990.
Yozwiak, Steve. "'Island' Sizzle: Valley an Increasingly Hot Spot." *Arizona Republic* (Phoenix), September 25, 1998. http://www.sepp.org/reality/arizrepub.html.

Heat Waves

Absent increased use of artificial cooling mechanisms, a major consequence of a warmer climate has been increasing frequency and intensity of heat waves, and consequent heat illnesses.

Heat illness usually begin with the minor discomfort and inconvenience of heat cramps, caused when blood vessels widen to dissipate extra heat. These muscle spasms usually disappear with relief from the heat and a drink of cold water. Heat stress is cumulative, however, even over many days. Increasingly warm nights and lack of relief can lead to heat stress in more serious forms, one of which is heat exhaustion. A person feels dizzy and weak, progressing to nausea and vomiting. As uncomfortable as it may be, heat exhaustion is usually curable with fluids in a cool environment.

Young residents fetch water in recycled plastic cans as they walk on a dry riverbed of Sabarmati River in Ahmadabad, India, May 9, 2006. (AP/Wide World Photos)

The truly dangerous stage of heat stress is heat stroke, which can set in with a deceptive euphoria. A person passes from heat exhaustion to heart stroke as body temperature rises to about 105°F. Once body temperature reaches the level of a medical emergency, lethargy sets in, along with severe headaches and disorientation, delirium, comas, and, in some cases, death from damage to the brain and nervous system (Parker and Shapiro 2008, 32–34).

Surges of Heat in a Longer Summer

Infrared forcing (the greenhouse effect, or global warming) does not cause the Earth to heat evenly; weather is still variable, so a generally warming climate expresses itself in occasional bursts of heat in specific areas, as well as cold episodes from time to time. Heat often occurs earlier in spring and may last longer into the early fall. Different areas of the globe experience record heat in various years as the general warming trend coincides with conditions, such as southerly winds (in the Northern Hemisphere) and persistent high pressure.

As warm seasons expand, some events meant for cooler weather collide with a new climatic reality. Heat is relative to humidity as well, which impedes cooling through evaporation. In Chicago, for example, on October 7, 2007, the annual marathon was held on a very humid day when the temperature rose to the upper 80°s F shortly before noon, almost 25°F above average for the season. The heat, combined with high humidity killed one runner, Chad Schieber, 35, of Midland, Michigan, and sickened several hundred others, forcing cancellation of the race for the first time in its 30-year history.

More than 300 runners left the race course in ambulances with nausea, heart palpitations, and dizziness, all symptoms of heat illness. Fifteen city buses, their air conditioners pumping at maximum, were commandeered as aid stations. Some of the runners had core body temperatures as high as 107°F (Davey 2007b, A-1). The same heat wave set records across the U.S. Midwest and East, accompanied by stifling humidity. Detroit (at 90°F) and Indianapolis (at 91°F) had their latest 90° highs on the instrumental record (since the 1870s). Louisville Kentucky, at 93°F, set an all-time record for October.

Record Highs Worldwide during Summer 2003

In Phoenix, Arizona, on July 14, 2003, the high temperature reached 116°F; the diurnal low that same day was 96°F. Windshields popped in the heat and flip-flops stuck to pavement. In Milan, Italy, at about the same time, an African "sirocco" blew northward from the Sahara, making it "a little dangerous to wear stilettos [spike high-heeled shoes]; they sank half an inch into the melting tar, impaling one's legs to the spot where one's torso pitched forward, head first, like a stone being launched from a sling-shot" (Thurman 2003, 78). One day in June 2003, the temperature hit 40°C (about 104°F) in Milan.

As stilettos were sinking into asphalt in Milan and windshields were popping in Phoenix, much of southern China was experiencing weeks of searing heat, causing drinking-water shortages and massive damage to crops. Hardest-hit Hunan, Jiangxi, Fujian, and Zhejiang provinces experienced temperatures above 40°C day after day.

On August 6, 2003, London, England, experienced its highest temperature in recorded history (since reliable records began about 1770), at 95.7°F (34°C). Four days later, the temperature at Heathrow International Airport peaked at 100.2°F—the first three-digit reading in the climatic history of the British Isles. In Gravesend, Kent, the same day, the temperature reached 100.8°F. British commuter railroads were slowed because of fear that tracks might buckle in the heat. England was hardly alone; during the same week, one location in Germany recorded an all-time record high of 40.8°C (105.4°F). A reporting station in Switzerland reached 41.5°C (106.7°F), and one in Portugal soared to 47.3°C (117.1°F) (McCarthy 2003b, 3). The same week, some areas in southwestern Spain hit 117°F.

The year 2003 turned out to be the warmest in the entire Central England Temperature Record, which dates to 1659. The year's final average temperature of 10.65°C beat the previous records of 1990 (10.63°C) and 1999 (10.62°C). The trend was not restricted to Europe; Albuquerque, New Mexico, for example, set a record-high minima in 2003, at 4°F above long-term averages (Fleck 2003, B-1). Power supplies failed in many urban areas, and sun-seeking tourists from the likes of Ireland and Scandinavia returned home from Italy and Greece having spent their holidays in hotel rooms, with temperature outside near 100°F, too hot to sunbathe.

In France, during the same summer, an estimated 14,800 people died, as many elderly expired in apartments with no air conditioning. During the two weeks when heat was most intense in France, the nation's mortality rate increased by 54 percent (Schar and Jendritzky 2004, 559). A politically charged debate enveloped the country as temperatures exceeded 100°F several times during August of 2003. The nation's health minister resigned. Morgues overflowed with bodies as a quarter of France's 58 nuclear power plants shut down at the height of the heat wave; rivers had become too warm to cool them properly. Paris recorded an overnight low of 78°F one morning, its highest on record. During the same period, more than 1,300 people died of heat-related causes in Portugal. About 7,000 heat-related deaths were recorded in Germany, and nearly 4,200 lives each in both Spain and Italy. More than 2,000 people died in Britain ("Summer Heat Wave" 2003, 2).

One analysis put Europe's death toll at 35,000 or more during the scorching summer of 2003. The Earth Policy Institute, based in Washington, D.C., warned that such deaths are likely to increase, as "even more extreme weather events lie ahead" ("Summer Heat Wave" 2003, 2). Europe's heat wave of summer 2003 was the climatic crown jewel of the hottest decade in the history of the area, which reaches back at least to about 1500 C.E. The average temperature was 2°C (3.6°F) higher than the long-term average.

"Taking into account the uncertainties in our reconstructions, it appears that the summer of 2003 was very likely warmer than any summer back to 1500," said Jurg Luterbacher, who led a 500-year study of European temperatures. Results from regional climate model simulations suggest that about every second summer will be as hot or even hotter than 2003 by the end of the twenty-first century (Luterbacher et al. 2004, 1502). The summer of 2003 in Europe was by far the hottest single season in the 500-year record examined in this study. Luterbacher and colleagues stated with a 95 percent degree of confidence that the "European climate is very likely warmer than that of any time during the past 500 years" (Luterbacher et al. 2004, 1499).

In this study, described in *Science*, Luterbacher and colleagues used temperature data from sources such as ice cores and tree rings to reconstruct the climatic history of Europe during the preceding 500 years. Against this background, the data from the past century, and particularly from its final decades, stand out as extraordinary, and point strongly to human-induced global warming as the key influence (Henderson 2004a, 10). While the summer 2003 heat wave in Europe was a statistical anomaly according to present average temperatures, a team of Swiss scientists found that, given expected increases in greenhouse-gas levels, such events appear increasingly likely in the future (Schar and Jendritzky 2004, 332–336).

Could Europe's Heat of 2003 become Typical?

According to British scientists, Europe's scorching heat wave of 2003 will be considered typical seasonal weather by the middle of the twenty-first century, and may be below average in a century. The British team made a case that the summer of 2003 was Europe's hottest in southern, western, and central Europe in at least five centuries. From the eastern Atlantic to the Black Sea, the mercury was 2.3°C (4.14°F) above average. According to their models, by the 2040s at least one European summer in two will be hotter than in 2003. "By the end of this century, 2003 would be classed as an anomalously cold summer relative to the new climate," the scientists wrote in the December 2, 2004, edition of *Nature* ("Phew" 2004).

According to a team from the Swiss Federal Institute of Technology in Zurich, the European heat wave of summer 2003 was of a severity that could be expected (according to past weather records, at any rate) once every 46,000 years. This indicates how truly paradigm-breaking it was (McGuire 2005, 53).

According to their models, summers in 2100 will average about 6°C (10.8°F) hotter than 2003's averages. "We estimate it is very likely (confidence level more than 90 percent) that human influence has at least doubled the risk of a heat wave exceeding this threshold magnitude," wrote Peter A. Stott and colleagues in *Nature* (Stott, Stone, and Allen 2004, 610). They continued:

> It seems likely that past human influence has more than doubled the risk of European summer temperature as hot as 2003, and with the likelihood of such events projected to increase 100-fold over the next four decades, it is difficult to avoid the conclusion that potentially dangerous anthropogenic interference in the climate system is already underway. (Stott, Stone, and Allen 2004, 613)

Separately, scientists at the French meteorological agency Meteo France said that it expects summer temperatures in France to rise by between 4° and 7°C (7.2° and 12.6°F) by 2100. "By the end of the century, a summer with temperatures as we had in 2003 will be considered a cool summer," said researcher Michel Deque ("Phew" 2004).

As Europe sizzled, U.S. President George W. Bush formed a "century club" composed of people willing to jog with him at his Crawford, Texas, ranch when the temperature exceeded 100°F. After exercising, the joggers then reclined in the air-conditioned comfort of Bush's ranch house—a comfort not available to most of the Europeans who had died in the heat. Skeptics of global warming asserted that the French needed more air conditioning, as environmentalists replied that they had not heretofore required mechanically cooled air to survive. Air conditioning requires copious amounts of energy, often generated by fossil fuels, which eventually will aggravate warming.

Heat and drought during the summer of 2003 reduced grain harvests across Europe, forcing food prices to rise worldwide. During late August 2003, the International Grains Council, an intergovernmental body, reduced its forecast of the world's grain harvest for 2003 by 36 million tons, as a result of "heat and drought, particularly in Europe" (Lean 2003, 4). The damage was most severe in Eastern Europe, which harvested its worst wheat crop in three decades. In Ukraine, the harvest fell from 21 million tons in 2002 to five million in 2003, while Romania has its worst crop on record. Germany was the worst-hit European Union country; some farmers in the southeast regions of that country lost half their grain harvest (Lean 2003, 4). World grain reserves fell sharply, continuing a four-year pattern.

Worldwide Warmth

Elsewhere in the world, long-term evidence of warming was plentiful. According to the Mongolian Ministry of Nature and Environment, temperatures there have risen during 60 years (1942 to 2002) by 1.56°C annually, including 3.61°C during winter. Rising temperatures have come with intensifying drought in many areas ("Climate Warms" 2003).

During 2001, Washington, D.C.'s annual Cherry Blossom Parade was moved up a week to coincide with earlier blooming dates in the area. (In 2003, however, after an unusually cold and snowy winter, the blossoms emerged several days later than usual.) Summer 2002 (June, July, and August) was the hottest on record in the United States, except for 1934 and 1936. On September 9, 2002, the high reached 98°F in Burlington, Vermont, and 97°F in Bangor, Maine, both unusually hot for September. As late as October 1, 2002, Toronto, Ontario, basked in a record high of 29.2°C, which broke the old daily record of 26.9°C, set in 1988. Beaches were open for swimming along Lake Ontario. The following winter was unusually cold and snowy in some of the same areas, however.

Each year becomes notable for heat waves in different locations "If you draw a line from Portland, Maine, to Rutland, Vermont, below that line there's just miserable weather," said Andy Woodcock during 2006. Woodcock, a meteorologist with the National Weather Service's Washington, D.C., office, added "It basically goes from Rutland down to Cuba. Really, the whole East Coast is in an excessive heat warning. This is probably the most widespread as far as the heat goes that I've ever seen" (Zezima 2006). With highs near 100°F over a large area, dew points reached 72 to 76 in the high humidity. The heat index in Albany, New York reached 110°F on August 2. "I've spent every summer here since I was born, and I just can't remember it being this warm," said Stuart Smith, harbormaster in Chatham, on Cape Cod's elbow. "People at the beach all moved their beach chairs down right to the water" (Zezima 2006).

The next summer, a persistent high-pressure system set up over the southeastern United States, giving 100-plus temperatures for weeks at a time during late July into most of August. Atlanta, Georgia, experienced five of its hottest days on record during August 2007. At the same time, a similar pattern took hold over southeastern Europe. By late August, 100-degree temperatures and hot-dry winds played a role (along with arson) in provoking wildfires that ravaged Greece. Two weeks later, torrential rains pelted the same areas, eroding hillsides stripped of their foliage by the fires.

Gerald A. Meehl and Claudia Tebaldi studied heat waves in Chicago during 1995 and Paris during 2003, forecasting that "future heat waves in these areas will become more intense, more frequent, and longer-lasting in the second half of the twenty-first century" (Meehl and Tebaldi 2004, 994). They anticipate that areas suffering severe heat waves will probably experience more intense heat than surrounding areas: "The model show[s] that present-day heat waves over Europe and North America coincide with a specific atmospheric circulation pattern that is intensified by ongoing increases in greenhouse gases" (Meehl and Tebaldi 2004, 994). They concluded that "areas already experiencing strong heat waves (e.g., Southwest, Midwest, and Southeast United States and the Mediterranean region) could experience even more intense heat waves in the future" (Meehl and Tebaldi 2004, 997).

Heat waves often are associated with semistationary high pressure at the surface and aloft that produce clear skies, light winds, warm-air advection, and prolonged hot conditions. Meehl and Tebaldi's model suggests that these conditions will occur more frequently with increasing concentrations of greenhouse gases in the atmosphere (Meehl and Tebaldi 2004, 996).

Heat Waves and Mortality

Historically, heat stress has been the foremost weather-related cause of death in the United

States. During the second half of the twentieth century, however, even as temperatures rose, the rate of heat-related deaths declined dramatically, because of increased use of air conditioning, better medical care, and increased public awareness of heat stress effects. Robert Davis, associate professor of environmental sciences at the University of Virginia and colleagues studied heat-related mortality in 28 major U.S. cities from 1964 though 1998. He found that the heat-related death rate, 41 per million people a year in the 1960s and 1970s, declined to 10.4 per million during the 1990s (Davis et al. 2003).

Laurence S. Kalkstein has estimated that a doubling of the carbon-dioxide level in the atmosphere could increase heat-related mortality to seven times present levels if acclimatization is not factored in. With acclimatization (human adaptation to higher temperatures), the estimated increase in heat-wave mortality estimated by Kalkstein is four times the present rate (Kalkstein 1993, 1397). Kalkstein observed that each urban area has its own "temperature threshold," at which the death rate from heat illnesses rises rapidly. Seattle, for example, has a lower threshold than Dallas. "Mortality rates in warmer cities seemed to be less affected no matter how high the temperature rose," Kalkstein wrote (1993, 1398). He suggested that residents of urban areas in poor countries will find adaptation more difficult because of limited access to air conditioning.

FURTHER READING

"Climate Warms Twice as Fast." British Broadcasting Company Monitoring Asia-Pacific, January 6, 2003. (LEXIS)

Davey, Monica. "Death, Havoc, and Heat Mar Chicago Race." *New York Times*, October 8, 2007b, A-1, A-14.

Davis, Robert E., Paul C. Knappenberger, Patrick J. Michaels, and Wendy M. Novicoff. "Changing Heat-related Mortality in the United States." *Environmental Health Perspectives*, July 23, 2003, doi: 10.1289/ehp.6336.

Fleck, John. "Dry Days, Warm Nights." *Albuquerque Journal*, December 28, 2003, B-1.

Henderson, Mark. "Past Ten Summers Were the Hottest in 500 Years." *London Times*, March 5, 2004a, 10.

Kalkstein, Laurence S. "Direct Impacts in Cities." *Lancet* 342 (December 4, 1993):1397–1400.

Lean, Geoffrey. "Hot Summer Sparks Global Food Crisis." *London Independent*, August 31, 2003, 4.

Luterbacher, Jurg, Daniel Dietrich, Elena Xoplaki, Martin Grosjean, and Heinz Wanner. "European Seasonal and Annual Temperature Variability, Trends, and Extremes since 1500." *Science* 303 (March 5, 2004):1499–1503.

McCarthy, Michael. "The Four Degrees: How Europe's Hottest Summer Shows Global Warming Is Transforming Our World." *London Independent*, December 8, 2003b, 3.

McGuire, Bill. *Surviving Armageddon: Solutions for a Threatened Planet*. New York: Oxford University Press, 2005.

Meehl, Gerald A., and Claudia Tebaldi. "More Intense, More Frequent, and Longer Lasting Heat Waves in the 21st Century." *Science* 305 (August 13, 2004):994–997.

Parker, Cindy L. and Steven M. Shapiro. *Climate Chaos: Your Health at Risk: What You Can Do to Protect Yourself and Your Family*. Westport, CT: Praeger, 2008.

"Phew, What a Scorcher—and It's Going to Get Worse." Agence France Presse, December 1, 2004. (LEXIS)

Schar, Christoph, and Gerd Jendritzky. "Hot News from Summer 2003." *Nature* 432 (December 2, 2004):559–561.

Stott, Peter A., D. A. Stone, and M. R. Allen. "Human Contribution to the European Heatwave of 2003." *Nature* 432 (December 2, 2004):610–613.

"Summer Heat Wave in Europe Killed 35,000." United Press International, October 10, 2003. (LEXIS)

Thurman, Judith. "In Fashion: Broad Stripes and Bright Stars: The Spring-summer Men's Fashion Shows in Milan and Paris." *The New Yorker*, July 28, 2003, 78–82.

Zezima, Katie. "Heat Blankets U.S., Causing Discomfort." *New York Times*, August 3, 2006. http://www.nytimes.com/2006/08/03/us/03swelter.html.

Himalayas, Glacial Retreat

About 90 percent of the Tibetan Plateau's glaciers have retreated during the past century; their rate of decay has increased substantially during the past decade, according to "Global Outlook For Ice and Snow," a report released during 2007 by the United Nations Environment Programme.

The snow pack in this region could shrink a further 43 to 81 percent by 2100, according to projections from Intergovernmental Panel on Climate Change (IPCC) (Gartner 2007a). Thousands of Himalayan glaciers feed several major rivers, partially sustaining one-sixth of the Earth's population, a billion people downstream. Their retreat threatens the region's drinking-water supply, agricultural production, and vulnerability to disease and floods. Glaciers on the Himalayas contain 100 times as much ice as the

Alps and provide more than half the drinking water for 40 percent of the world's population in seven major Asian rivers, including the Ganges and the Indus.

Temperatures in the region have risen by 1°C since the 1950s, causing thousands of glaciers to retreat by an average of 30 meters a year. The worst recorded collapse of one of these dams occurred in 1954, when 300,000 cubic meters of water and rock poured without warning into China in a 40 meter-high flood surge from the Sangwang dam on the Tibet-Nepal border. The city of Gyangze, 120 kilometers away, was destroyed. The dead totaled many thousands (Brown 2002b, 13).

The Indian Space Research Organization has used satellite imaging to measure changes in 466 glaciers, finding more than a 20 percent reduction in their size between 1962 and 2001. Another study found that the Parbati Glacier, among the largest, was retreating by 170 feet a year during the 1990s. Another glacier, the Dokriani, lost an average of 55 feet a year. Temperatures in the northwestern Himalayas have risen by 2.2°C in the last two decades (Sengupta 2007). Roughly 15,000 glaciers bordering India, China, and Nepal—many on the Tibetan Plateau, "The Roof of the World," which supplies the Ganges, Indus, and Brahmaputra rivers—have been melting more quickly than at any time in recorded history, according to research surveying conditions through 2006 published in November 2008 by Lonnie Thompson and colleagues in *Geophysical Research Letters.*

Studying ice melt on the Naimona'nyi glacier, Thompson's team was stunned to learn that all the snow built up since 1944 had melted. "We were very surprised not to find the 1962–1963 horizon, and even more surprised not to find the 1951–1952 signal," Thompson said. In more than 20 years of sampling glaciers all over the world, this was the first time both markers were missing. As more heat is trapped in the atmosphere, said Thompson, it holds more water vapor. This humidity condenses as it rises, releasing heat in mountains. "At the highest elevations, we're seeing something like an average of 0.3°C warming per decade," Thompson said ("Tibetan Glaciers" 2008).

Glaciers Retreat on and Near Mount Everest

In the Himalayas, the Rongbuk Glacier on the north face of Mount Everest retreated between 170 and 270 meters from 1966 to 1997. The glacier from which Sir Edmund Hillary and Tenzing Norgay set out to climb Mount Everest nearly a half-century ago has retreated about three miles up its host mountain. Much of that glacier has turned to meltwater, according to United Nations observers who visited the site (Williams 2002, 2). Roger Payne, one of the observation team's leaders, said, "Back in 1953 when Hillary and Tenzing set off to climb Everest they stepped out of their base camp and straight on to the ice. You would now have to walk for over two hours [from the same site] to get on to the ice" (Williams 2002, 2). According to Jeff Kargel of the U.S. Geological Survey, international coordinator for Global Land Ice Measurements from Space (GLIMS), "Glaciers in the Himalayas are wasting at alarming and accelerating rates, as indicated by comparisons of satellite and historic data, and as shown by the widespread, rapid growth of lakes on the glacier surfaces" ("Glacial Retreat" 2002).

Meltwater from the Himalayas' Imja Glacier, 10 kilometers (6 miles) to the east-southeast of Mount Everest, has created a vast lake held back only by an unstable natural dam composed of boulder debris that once marked the edge of the glacier. A collapse of this dam could send a wall of water up to 100 meters high surging down the valley that is inhabited by the Sherpas who assist climbing expeditions' ascents of Mount Everest. The valley is also the main approach route to the Everest Base Camp. "We know it's going to go shooting down the flood plain, and in a mountainous area, that's where the people live," said Lisa Graumlich, who directs the Big Sky Institute at Montana State University (McFarling 2002b, A-4).

The Dig Tsho Glacial outburst in Nepal during 1985 destroyed a hydroelectric plant, wiped out 14 bridges, and drowned dozens of villagers. The danger is so obvious, Graumlich said, that some Himalayan villages have installed primitive warning systems—basically a system of horns—in an attempt to save lives during the next flood (McFarling 2002b, A-4). "We're just watching [glacial lakes] form in the Himalayas and Peru," said Alton C. Byers, director of research and education for the West Virginia-based Mountain Institute. "All you have to do is release that dam and you'll lose vast amounts of water in seconds" (McFarling 2002b, A-4).

Headwaters of the Ganges

By 2002, the snout of the Himalayan glacier that feeds the mighty Ganga [Ganges River] had developed giant fractures and crevices along a 10-kilometer stretch, indicating massive ice melting. During 15 years of researching such phenomena, Syed Iqbal Hasnain, who heads the Glacier Research Group at Delhi's Jawaharlal Nehru University, had never seen such a rapid deterioration of the frozen massif. He said, "If the rate continues, we could see much of the Gangotri glacier and others in the Himalayas vanish in the next couple of decades" (Chengappa 2002, 40). Hundreds of millions of people who live within the watershed of the Ganges depend on watershed fed by Himalayan glaciers to some degree. Half of India's hydroelectric power is generated from glacial runoff (Lynas 2004a, 238). The Indus River supplies 90 percent of the water used in desert areas of Pakistan from glaciers that have been rapidly losing mass for most of the twentieth century (Lynas 2004a, 238).

More than 40 lakes high in the Himalayas that have formed from rapidly melting glaciers are expected to burst their banks before the year 2010, sending millions of gallons of water and rocks cascading onto towns and villages (Brown 2002b, 13). These lakes are growing larger and more unstable as temperatures rise. As many as 44 such lakes have been identified, 20 in Nepal and 24 in Bhutan. There are thought to be hundreds more such "liquid time bombs" in India, Pakistan, Afghanistan, Tibet, and China (Brown 2002b, 13).

Surendra Shrestha, Asian regional coordinator for the United Nations Environment Programme's early warning division, said, "These 44 could burst their banks with potentially catastrophic results for people and property hundreds of kilometers downstream" (Brown 2002b, 13). Nepal's Tsho Rolpa lake in has grown sixfold in size since the 1950s, and is now 2.6 kilometers long, 500 meters wide, and as deep as 107 meters. A flood from this lake could cause serious damage as far as 108 kilometers downstream in the village of Tribeni, threatening 10,000 lives (Brown 2002b, 13).

FURTHER READING

Brown, Paul. "Scientists Warn of Himalayan Floods: Global Warming Melts Glaciers and Produces Many Unstable Lakes." *London Guardian*, April 17, 2002b, 13.

Chengappa, Raj. "The Monsoon: What's Wrong with the Weather?" *India Today*, August 12, 2002, 40.

Gartner, John. "Climbers Bring Climate Change From Mountaintop to Laptops." Environment News Service, July 31, 2007a. http://www.ens-newswire.com/ens/jul2007/2007-07-31-04.asp

"Glacial Retreat Seen Worldwide." Environment News Service, May 30, 2002. http://ens-news.com/ens/may2002/2002-05-30-09.asp#anchor2; see also http://www.gsfc.nasa.gov/topstory/20020530glaciers.html.

Lynas, Mark. *High Tide: The Truth About Our Climate Crisis*. New York: Picador/St. Martins, 2004a.

McFarling, Usha Lee. "Glacial Melting Takes Human Toll: Avalanche in Russia and Other Disasters Show That Global Warming Is Beginning to Affect Areas Much Closer to Home." *Los Angeles Times*, September 25, 2002b, A-4.

Sengupta, Somini. "Glaciers in Retreat." *New York Times*, July 17, 2007. http://www.nytimes.com/2007/07/17/science/earth/17glacier.html.

"Tibetan Glaciers Melting at Stunning Rate." Discovery.com, November 24, 2008. http://dsc.discovery.com/news/2008/11/24/tibet-glaciers-warming-02.html.

Williams, Frances. "Everest Hit by Effects of Global Warming." *London Financial Times*, June 6, 2002, 2.

Human Health, World Survey

When climate-change scientists and diplomats met in Buenos Aires during 1998, they were greeted by news that mosquitoes carrying dengue fever had invaded more than a third of the homes in Argentina's most populous province, home to 14 million people. The *aedes aegypti* mosquito appeared in Argentina in 1986; within 12 years, it was found in 36 percent of homes in Buenos Aires province, according to Dr. Alfredo Seijo of the Hospital Munoz. "*Aedes aegypti* now exists from the south of the United States to Buenos Aires province and this is obviously due to climatic changes which have taken place in Latin America over the past few years," Seijo told a news conference organized by the World Wildlife Fund at the United Nations climate talks in Argentina (Webb 1998a).

Dengue fever, for which no vaccine exists, had nearly disappeared from the Americas by the 1970s. During the 1980s, however, the disease increased dramatically in South America, infecting more than 300,000 people there by 1995. Also during 1995, Peru and the Amazon Valley were especially hard hit by the area's largest epidemic of yellow fever since 1950, which is carried by the same mosquito that transmits dengue fever. The annual world incidence of dengue fever, which averaged about 100,000 cases between 1981 and 1985, averaged 450,000 cases a year between 1986 and 1990 (Gelbspan 1997b, 149).

Mosquitoes, Termites, and Cockroaches

Dengue fever is one of a number of mosquito-vector diseases that have been increasing their coverage in many areas of the Earth—climbing altitude in the tropics and rising in latitude in temperate zones—as global temperatures have warmed. Rising temperatures and humidity increase the range of many illnesses spread by insects, including mosquitoes, which are warm-weather insects that die at temperatures below a range of 50° to 61°F, depending on species. Dengue, a common disease in tropical regions, is a prolonged, flu-like viral infection that can cause internal bleeding, fever, and sometimes death. Dengue, which is sometimes called "breakbone fever," may be accompanied by headache, rash, and severe joint pain. The World Health Organization lists dengue fever as the tenth deadliest disease worldwide.

During 1995, an explosion of termites, mosquitoes, and cockroaches hit New Orleans, following an unprecedented five years without a killing frost. "Termites are everywhere. The city is totally, completely, inundated with them," said Ed Bordees, a New Orleans health official, who added, "The number of mosquitoes laying eggs has increased tenfold" (Gelbspan 1997b, 15). The situation in New Orleans was aggravated not only by unusual warmth, but also by above-average rainfall totaling about 80 inches the previous year. Some of the 200-year-old oaks along New Orleans' St. Charles Avenue were found to have been eaten alive from the inside by billions of tiny, blind, Formosan termites. The same year, dengue fever spread from Mexico across the border into Texas for the first time since records have been kept. Dengue fever, like malaria, is carried by a mosquito with a range that is defined by temperature. At the same time, Colombia was enduring swarms of mosquitoes and outbreaks of the diseases they carry, including dengue fever and encephalitis, triggered by a record heat wave followed by heavy rains.

Mild winters with a lack of freezing conditions allow many disease-carrying insects to expand their ranges. "Indeed," commented Paul Epstein of Harvard Medical School's Center for Health and the Global Environment,

> Fossil records indicate that when changes in climate occur, insects shift their range far more rapidly than do grasses, shrubs, and forests, and move to more favorable latitudes and elevations hundreds of years before larger animals do. "Beetles," concluded one climatologist, "are better paleo-thermometers than bears." (Epstein 1998)

Global Warming and Disease Vectors

John T. Houghton, author of *Global Warming: The Complete Briefing* (1997), believes that global warming will accelerate the spread of many diseases from the tropics to the middle latitudes. Malaria could increase from its present level, Houghton warned. "Other diseases which are likely to spread for the same reason are yellow fever, dengue fever, and ... viral encephalitis," he wrote (Houghton 1997, 132). After 1980, small outbreaks of locally transmitted malaria occurred in Texas, Georgia, Florida, Michigan, New Jersey, New York, and California, usually during hot, wet spells. Worldwide, according to Epstein, between 1.5 and 3 million people worldwide die of malaria each year, mostly children. Mosquitoes and parasites that carry the disease have evolved immunities to many drugs.

According to Epstein, "If tropical weather is expanding it means that tropical diseases will expand. We're seeing malaria in Houston, Texas" (Glick 1998). Epstein suggested that a resurgence of infectious disease may be one result of global warming. Warming may appear beneficial at first, Epstein said. Initially, some plants benefit from additional warmth and moisture, an earlier spring, and more carbon dioxide and nitrogen in the air. "But," he cautioned, "[w]arming and increased CO_2 can also stimulate microbes and their carriers" (Epstein 1998).

Since 1976, Epstein reported, 30 diseases have emerged that are new to medicine. Old ones, such as drug-resistant tuberculosis, have been given new life by new diseases (such as HIV/AIDS) that compromise the human immune system. By 1998, tuberculosis was claiming three million lives annually around the world. "Malaria, dengue, yellow fever, cholera, and a number of rodent-borne viruses are also appearing with increased frequency," Epstein reported (Epstein 1998). During 1995, mortality from infectious diseases attributed to causes other than HIV/AIDS rose 22 percent above the levels of 15 years earlier in the United States. Adding deaths complicated by HIV and AIDS, deaths from infectious diseases have risen 58 percent in 15 years (Epstein 1998). The Intergovernmental Panel on Climate Change (IPCC) included a chapter on public health in an update of its 1990 assessment that concluded: "climate change is likely to have wide-ranging and mostly adverse impacts on human health, with significant loss of life" (Taubes 1997).

Epstein identified three tendencies in global climate change, and related each to an increasingly virile environment for infectious diseases. The three indicators are as follows:

1. Increased air temperatures at altitudes of two to four miles above the surface in the Southern Hemisphere;
2. A disproportionate rise in minimum temperatures, in either daily or seasonally averaged readings;
3. An increase in extreme weather events, such as droughts and sudden heavy rains. (Epstein 1998)

A Sierra Club study indicated that frequent and lengthy El Niño episodes during the middle and late 1990s provided an indication of how sensitive some diseases can be to changes in climate. The Sierra Club study cited evidence that warming waters in the Pacific Ocean contributed to a severe outbreak of cholera that led to thousands of deaths in Latin American countries during the 1990s. According to health experts quoted by the Sierra Club study, "The current outbreak [of dengue fever], with its proximity to Texas, is at least a reminder of the risks that a warming climate might pose" (Sierra Club 1999). The Sierra Club study concluded: "While it is difficult to prove that any particular outbreak was caused or exacerbated by global warming, such incidents provide a hint of what might occur as global warming escalates" (Sierra Club 1999).

Willem Martens and colleagues, writing in *Climatic Change*, attempted to sketch how a warmer, wetter climate would affect transmission of three vectorborne diseases: malaria, schistosomiasis, and dengue fever. Martens and colleagues anticipated that the periphery of the currently endemic areas "will expand with global warming, with diseases notable at higher elevations in the tropics, an expectation that has been borne out by several observers." Martens and colleagues expected that "[t]he increase in epidemic potential of malaria and dengue transmission may be estimated at 12 to 27 percent and 31 to 47 percent respectively" (Martens, Jetten, and Focks 1997, 145). In contrast, they forecast that the transmission potential of schistosomiasis may decrease 11 to 17 percent.

Another disease that may become more common in the temperate zones of a warmer world is diarrhea, which during the 1990s was killing more than three million children a year worldwide, mainly in the tropics of Asia, Africa, and the Americas. The bacteria that cause diarrhea thrive in warm weather, especially after heavy rainfall.

Warming and Excess Deaths Worldwide

At least 150,000 people die each year as a direct result of global warming, three major United Nations organizations asserted late in 2003. Warming already has been factor in a noticeable increase in malnutrition, as well as outbreaks of diarrhea and malaria, the three largest killers in the poorest countries of the world, according to these reports. Diarmid Campbell-Lendrum, a World Health Organization scientist, said that estimates of deaths were extremely conservative. Furthermore, he said, the number of deaths attributable to rapid warming is expected to double during the coming 30 years. "People may say that this is a small total compared with the totals who die anyway, but these are needless deaths. We must do our best to take preventative measures," he said (Brown 2003d, 19).

The report, produced by the World Health Organization, the U.N. Environment Programme, and the World Meteorological Programme, detailed how increased warmth has intensified the spread of diseases. Diarrhea spread by bacteria, mostly through unclean water and food, develops and diffuses more quickly in warmer temperatures and humidity. Dirty water is the largest killer of children less than five years of age. In Lima, Peru, a six-year study at a clinic set up to treat diarrheal complaints showed a 12 percent increase in cases for every 1°C rise in temperature during cooler months and a 4 percent increase in warmer months (Brown 2003d, 19). Similar results were found in a survey of 18 Pacific islands. The problem is made worse by high rainfall or drought, during which water supplies become contaminated.

Many diseases spread by rats and insects also become more common in warmer weather. Malaria, dengue fever, and Lyme disease are all on the rise. Many threats can be curtailed by dispensing preventive medicine and providing clean water and sanitation. Climate change makes these issues more urgent, the report said. The combined effects of increased warmth and the greater volume of standing water brought by storms create malaria epidemics by providing breeding sites and an accelerated life cycle for mosquitoes. In Africa, where the death toll from malaria is highest, mosquitoes carrying the disease are spreading into mountain areas previously too cool for them to thrive, according to the report.

Ancient Influenza Strains in Old Ice

Ancient strains of influenza that have not afflicted humankind for many centuries could be released by melting ice and migratory birds, according to Professor Scott Rogers, chair of biology at Bowling Green State University. "We've found viral RNA in the ice in Siberia, and it's along the major flight paths of migrating waterfowl," whose pathways take them to North America, Asia, and Australia, and interconnect with other migratory paths to Europe and Africa, Rogers said. The virus that Rogers and his collaborators found resembles a strain that emerged between 1933 and 1938 and again in the 1960s ("Melting Ice" 2006). The research was published in the December 2006 edition of *Journal of Virology.*

These viruses may be preserved in ice and released after human immunity has lapsed. Survivors of the worldwide flu pandemic of 1918 had immunity to the H1N1 strain, for example, but their descendents did not. A recurrence "could take hold as an epidemic," said Rogers, who explained that no one yet knows whether long-frozen viruses retain their ability to infect people. But "we think they can survive a long time" in ice, he said, saying that tomato mosaic virus has been found in 140,000 year old ice in Greenland ("Melting Ice" 2006).

Temperature, Humidity, Air Pollution, and Human Maladies

Aside from insect-vector diseases, warmer, more humid weather may aggravate urban air pollution (especially tropospheric ozone), which is a factor in many human maladies, such as asthma. Pim Martens of the Netherlands' Center for Integrative Studies (and a senior scientist at the University of Maastricht) pointed to a number of studies that indicate that air pollution's effects on human health rise as temperatures increase: "Simultaneous exposure to heat and pollution appears too be more harmful than the sum of the individual effects" (Martens 1999, 535). Allergenic pollens and spores are more readily dispersed during hot, dry summers, Martens said.

Jonathan Patz, an associate professor of environmental studies and population health sciences at the University of Wisconsin at Madison and a lead author of the 2007 IPCC assessment of North America, said the currently projected warming alone probably will mean that, by 2050, the U.S. Northeast will experience 68 percent more "red ozone alert" days, which indicate the air is unhealthy to breathe (Eilperin 2007a, A-5).

Asthma is aggravated by heat (which increases the pollen production of many plants) as well as air pollution. According to the American Lung Association, more than 5,600 people died of asthma in the United States during 1995, a 45.3 percent increase in mortality over 10 years, and a 75 percent increase since 1980. Roughly a third of those cases occurred in children less than 18 years of age. Since 1980, children under age five have experienced a 160 percent increase in asthma (Sierra Club 1999).

Poor and minority children are likely to develop asthma at rising rates in a warming climate combined with air pollution. As climate warms, allergens such as pollen and mold increase, interacting with urban pollutants such as ozone and soot to fuel a growing epidemic of asthma. Asthma among U.S. preschool children, age 3 to 5, rose by 160 percent between 1980 and 1994 ("Findings" 2004). "This is a real wake-up call for people who think global warming is only going to be a problem way off in the future," said Christine Rogers, senior research scientist at the Harvard School of Public Health ("Findings" 2004).

"What happens to asthmatics in the heat and humidity? Well, we can't breathe," said Barbara Mann, an author, teacher, and asthmatic who lives in Toledo, Ohio.

> I'm permanently on three different prescription inhalers that I must use at regular intervals, four times a day—and that's just when I'm well. When I'm sick, there are antibiotics to bust up the hardened mucus in my lungs, along with other regimens of pills to aid the process. When my lungs are irritated by pollutants or by natural "triggers" such as pollens or infected, my air passages swell shut, suffocating me. (Mann 1999)

According to Dr. Joel Schwartz, an epidemiologist at Harvard University, present-day air-pollution concentrations are responsible for 70,000 early deaths per year and more than 100,000 excess hospitalizations for heart and lung disease in the United States. This could increase 10 to 20 percent in the United States as a result of global warming, with significantly greater increases in countries that are more polluted to begin with, according to Schwartz (Sierra Club 1999).

In addition to aggravation of specific diseases, persistent heat and humidity has a general debilitating influence on most warm-blooded animals, including human beings. Farm animals are especially adversely affected when the air temperature remains higher than usual throughout the night. During the latter half of the twentieth century, according to Epstein, night-time minimum temperatures over land areas have risen at a rate of 1.86°C per 100 years, while maximum temperatures have risen at 0.88°C during the same period (Epstein 1998).

Greenhouse-Gas Reduction and Health

Luis Cifuentes and colleagues observed in *Science*: "The same actions that can reduce the long-term buildup of greenhouse gases—reductions in burning of fossil fuels—can also yield powerful, immediate benefits to public health by reducing the adverse effects of local air pollution" (Cifuentes et al. 2001, 1257). Their study cited estimates that 18,700 premature deaths per year could be avoided in the United States by reducing emissions from older coal-fired electricity-generating plants (Cifuentes et al. 2001, 1257). Deaths due to bronchial problems, heart disease, and other ailments would be reduced substantially. Another of the surveyed studies indicated that air pollution from traffic caused more deaths than traffic accidents.

The World Health Organization estimated that in 1995 460,000 avoidable deaths occurred annually worldwide "as a result of suspended particulate matter, largely from outdoor urban exposures" (Cifuentes et al. 2001, 1257). The authors noted that several types of pollution rise with temperatures. According to estimates developed by Cifuentes and colleagues, reducing fossil-fuel air pollution in four of the world's largest cities (New York, Mexico City, São Paulo, and Santiago) could prevent 64,000 premature deaths between 1980 and 2000, reducing rates of infant mortality, asthma, cardiovascular problems, and respiratory ailments.

"The benefits of lowering emissions are immediate" because many of the gases emitted when fuels are burned are also pollutants, said George Thurston, one of the review's authors, who is an associate professor of environmental medicine at the New York University School of Medicine. "Universal studies have shown when air pollution levels go up, you get an increase in the numbers of deaths and hospital admissions, missed days at work and school, and other adverse effects," Thurston said (Surendran 2001, A-20).

Another study reported that alternative transportation policies initiated during the busy 1996 Summer Olympics in Atlanta "not only reduced vehicle exhaust and air pollutants such as ozone by about 30 percent, they also decreased the number of acute asthma attacks by 40 percent and pediatric emergency admissions by about 19 percent (Surendran 2001, A-20).

Health Benefits from Warming?

A climate contrarians' journal, *The World Climate Report,* replied to an editorial in the prominent British medical journal *Lancet* that had asserted that malaria and other mosquito-borne diseases will spread into the temperate zones with warming temperatures. According to the contrarians, malaria's spread has very little to do with temperature or humidity, and more to do with medical technology and availability of air conditioning. The contrarians assert that epidemics of malaria were common in most of the United States before the 1950s. In 1878, according to their argument, about 100,000 people were infected in the United States, and one-quarter of them died.

Pim Martens has written that while the overall impact of global warming on human health is expected to be markedly negative, human beings may experience a few positive outcomes. Some diseases that thrive in cold weather (such as influenza) may find their ranges and effects reduced in a warmer world. The elderly might die less frequently of cardiovascular and pulmonary ailments, which peak during cold weather. "Whether the milder winters could offset the mortality during the summer heat waves is one of the questions that demands further research," Martens wrote (Martens 1999, 535).

Countering the views of Epstein and others, some health researchers contend that global warming will do little to increase the incidence of tropical diseases. "For mosquito-borne diseases such as dengue [fever], yellow fever, and malaria, the assumption that warming will foster the spread of the vector is simplistic," contended Bob Zimmerman, an entomologist with the Pan American Health Organization (PAHO). Zimmerman pointed out that in the Amazon basin, more than 20 species of *Anopheles* mosquitoes can transmit malaria, and each is adapted to a different habitat: "All of these are going to be impacted by rainfall, temperature, and humidity in different ways. There could actually be decreases in malaria in certain regions, depending on what happens" (Taubes 1997).

Virologist Barry Beaty of Colorado State University in Fort Collins, Colorado, agreed with Zimmerman: "You don't have to be a rocket scientist to say we've got a problem," he said. "But global warming is not the current problem. It is a collapse in public-health measures, an increase in drug resistance in parasites, and an increase in pesticide resistance in vector populations. Mosquitoes and parasites are efficiently exploiting these problems" (Taubes 1997).

Countering the majority view that a warmer world will cause malaria to spread, David J. Rogers and Sarah E. Randolph, using their own models, wrote in *Science* that even extreme rises in temperature will not spread the disease. They argued that the spread of malaria is too poorly understood to base a forecast several decades into the future on temperature as a singular variable. For example, the "Dengie marshes" of Essex, in England, a breeding ground for malaria-carrying mosquitoes in the seventeenth century, have dried up, making an increase in temperatures not a factor in relation to malaria's transmission. Malaria is not a new disease in the temperate zones. It was common in the Roman Empire. A British invasion of Holland in 1806 failed to drive out French troops because large numbers of the British became ill with malaria. Malaria was a public health problem in most of the eastern United States during warm, humid summers before medications were developed for it about a century ago.

Paul Reiter, a dengue expert with the Centers for Disease Control and Prevention's Puerto Rico office, argued against the relative importance of climate in human disease by pointing to periods in the past during which malaria and other tropical diseases pervaded cooler regions. He argued that the spread of malaria is more closely linked to deforestation, agricultural practices, human migration, poor public health services, civil war, and natural disasters. "Claims that malaria resurgence is due to climate change ignore these realities and disregard history," he wrote in an article about malaria's spread through England during the Little Ice Age, which began about 1450 and lasted for several hundred years, during which time the climate was cooler than it is today (McFarling 2002a, A-7).

S. I. Hay and colleagues investigated long-term meteorological trends in four high-altitude sites in East Africa where increases in malaria have been reported during the past two decades. "Here we show that temperature, rainfall, vapor pressure, and the number of months suitable for *P. falciparum* transmission have not changed significantly during the past century or during the period of reported malaria resurgence," said the researchers. Therefore, they found that associations between the resurgence of malaria and climate change at high altitudes in these areas "are overly simplistic" (Hay et al. 2002, 905).

FURTHER READING

Brown, Paul. "Global Warming Kills 150,000 a Year: Disease and Malnutrition the Biggest Threats, UN Organisations Warn at Talks on Kyoto." *London Guardian*, December 12, 2003d, 19.

Cifuentes, Luis, Victor H. Borja-Aburto, Nelson Gouveia, George Thurston, and Devra Lee Davis. "Hidden Health Benefits of Greenhouse Gas Mitigation." *Science* 252 (August 17, 2001):1257–1259.

Eilperin, Juliet. "U.S., China Got Climate Warnings Toned Down." *Washington Post*, April 7, 2007a, A-5. http://www.washingtonpost.com/wp-dyn/content/article/2007/04/06/AR2007040600291_pf.html.

Epstein, Paul R. "Climate, Ecology, and Human Health." December 18, 1998. http://www.iitap.iastate.edu/gccourse/issues/health/health.html.

"Findings." *Washington Post*, April 30, 2004, A-30.

Gelbspan, Ross. *The Heat Is On: The High Stakes Battle Over Earth's Threatened Climate.* Reading, MA: Addison-Wesley Publishing Co., 1997b.

Glick, Patricia. *Global Warming: The High Costs of Inaction.* San Francisco: Sierra Club, 1998. http://www.sierraclub.org/global-warming/inaction.html.

Hay, S. I., J. Cox, D. J. Rogers, S. E. Randolph, D. I. Stern, G. D. Shanks, M. F. Myers, and R. W. Snow. "Climate Change and the Resurgence of Malaria in the East African Highlands." *Nature* 425 (February 21, 2002):905–909.

Houghton, John. *Global Warming: The Complete Briefing.* Cambridge: Cambridge University Press, 1997.

Mann, Barbara. Personal communication. August 3, 1999.

Martens, Willem J. M., Theo H. Jetten, and Dana A. Focks. "Sensitivity of Malaria, Schistosomiasis, and Dengue to Global Warming." *Climatic Change* 35 (1997):145–156.

Martens, Pim. "How Will Climate Change Affect Human Health?" *American Scientist* 87, no. 6 (November/December 1999):534–541.

McFarling, Usha Lee. "Study Links Warming to Epidemics: The Survey Lists Species Hit by Outbreaks and Suggests That Humans Are Also in Peril." *Los Angeles Times*, June 21, 2002a, A-7.
"Melting Ice May Release Frozen Influenza Viruses." Environment News Service, November 27, 2006. http://www.ens-newswire.com/ens/nov2006/2006-11-27-09.asp#anchor3.
Sierra Club. "Global Warming: The High Costs of Inaction." 1999. http://www.sierraclub.org/globalwarming/resources/innactio.htm.
Surendran, Aparna. "Fossil Fuel Cuts Would Reduce Early Deaths, Illness, Study Says." *Los Angeles Times*, August 17, 2001, A-20.
Taubes, Gary. "Apocalypse Not." JunkScience.com, 1997. http://www.junkscience.com/news/taubes2.html.
Webb, Jason. "Mosquito Invasion as Argentina Warms." Reuters, 1998a. http://bonanza.lter.uaf.edu/~davev/nrm304/glbxnews.htm.

Human Influences as a Dominant Climate-Change Forcing

According to atmospheric scientists Thomas Karl and Kevin Trenberth, writing in *Science*, natural variations ceased being the main determinant of Earth's climate about 1950. Since then, they asserted, "Human influences have been the dominant detectable influence on climate change" (Karl and Trenberth 2003, 1720). The human overload of greenhouse gases in the atmosphere has become the main (although not the only) driver of the climate. Among human causes, land-use changes and urbanization are adding heat to the atmosphere, according to Karl and Trenberth. Models created by the two scientists anticipated a 90 percent probability that temperatures worldwide will rise between 3.1° and 8.9°F by the year 2100. They concluded that

> There is still considerable uncertainty about the rates of change that can be expected, but it is clear that these changes will be increasingly manifested in important and tangible ways, such as changes in extremes of temperature and precipitation, decreases in seasonal and perennial snow and ice extent, and sea-level rise. Anthropogenic climate change is now likely to continue for many centuries. We are venturing into the unknown with climate, and its associated impacts could be quite disruptive. (Karl and Trenberth 2003, 1719)

"Studying these annual temperature data, one gets the unmistakable feeling that temperature is rising and that the rise is gaining momentum," said Lester R. Brown, an economist and president of the Earth Policy Institute in Washington, D.C. (McFarling 2002d, A-5). A string of warmer years provides strong evidence that humans are in large part to blame for changing the climate, said Peter Frumhoff, an ecologist and senior scientist with the Union of Concerned Scientists in Cambridge, Massachusetts. "It's important [that] we pay attention to this drumbeat of evidence as the signal of human impact starts to emerge from the noise of natural climate patterns," he said (McFarling 2002d, A-5).

FURTHER READING

Karl, Thomas R., and Kevin E. Trenberth. "Modern Global Climate Change." *Science* 302 (December 5, 2003):1719–1723.
McFarling, Usha Lee. "NASA Finds 2002 Second Warmest Year on Record." *Los Angeles Times* in *Calgary Herald*, December 12, 2002d, A-5.

Human Rights, Global Warming, and the Arctic

The Inuit have assembled a human-rights case against the United States (specifically the George W. Bush administration) because global warming is threatening their way of life. They have invited the Washington-based Inter-American Commission on Human Rights to visit the Arctic to witness the devastation being caused by global warming. Sheila Watt-Cloutier, president until 2006 of the Inuit Circumpolar Conference, which represents 155,000 people inside the Arctic Circle, said, "We want to show that we are not powerless victims. These are drastic times for our people and require drastic measures" (Brown 2003b, 14).

The Inter-American Commission on Human Rights, based in Washington, D.C., was created in 1959 by the Organization of American States (OAS) to safeguard personal freedom and security among the citizens of member states. While the body cannot issue legally binding verdicts, a finding in favor of the Inuit against the United States could become useful evidence in future attempts to successfully sue the United States on climate-change grounds in international legal forums. Such a finding also might be cited in future suits against corporations in U.S. federal courts if and when legal protection is offered to the atmosphere and other resources held in common.

"People worry about the polar bear becoming extinct ... because there will be no ice from which they can hunt seals," said Watt-Cloutier.

> But the Inuit face extinction for the same reason and at the same time.... This is a David and Goliath story. Most people have lost contact with the natural world. They even think global warming has benefits, like wearing a T-shirt in November, but we know the planet is melting and with it our vibrant culture, our way of life. We are an endangered species, too. (Brown 2003b, 14)

"The ocean is too warm," said Watt-Cloutier. "Our elders, who instruct the young on the ways of the winter and what to expect, are at a loss. Last Christmas after the ice had formed the temperature rose to 4°C (39°F) and it rained. We'd never known it before" (Brown 2003b, 14).

Contents and Implications of the ICC Petition to the OAS

The Inuit Circumpolar Conference's (ICC) 167-page petition alleges violations of fundamental human rights among Arctic peoples who ring the Arctic Ocean from Nunavut (a semi-sovereign Inuit province of Canada), to Greenland, the Russian Federation, and Alaska. The petition, which was compiled in defense of Inuit rights as a people within the context of international human-rights law, seeks a declaration from the commission that emissions of greenhouse gases from the United States—the source of more than 25 percent of the world's cumulative greenhouse gases during the last century—is violating Inuit human rights as outlined in the 1948 American Declaration on the Rights and Duties of Man.

The petition asserts that practice of the Inuit's rights to culture, life, health, physical integrity and security, property, and subsistence have been imperiled by global warming. With accelerating loss of ice and snow, hunting, travel and other subsistence activities have become more dangerous, and in some cases impossible, the Inuit assert. In addition, drinking-water sources have been threatened. Some coastal communities may be forced to move to escape rising waters provoked by rising seas and increasing storminess.

The petition does not seek monetary damages. Instead, it seeks cessation of U.S. actions that violate Inuit rights to live in a cold environment—not a small task, since such action will involve a major restructuring of the U.S economic base to sharply curtail emissions of greenhouse gases. The petition anticipates the types of actions that will have to take place on a worldwide scale, including the rapidly expanding economies of China and India, to preserve the Inuit way of life, as well as a sustainable worldwide biosphere. The Inuit in this case are conscious of their role in the natural world that supports them, which will not survive climatic business as usual. The Inuit have asked for U.S. leadership in an international effort involving sharp reductions in greenhouse-gas emissions to avoid destruction of the order of nature that has sustained them for many thousands of years.

The petition begins by establishing that global warming is harming "every aspect of Inuit life and culture" (*Petition to the Inter-American Commission* 2005, 13–19). It next associates warming with human-provoked emissions of greenhouse gases (2005, 20–34). Details then follow regarding the specific perils faced by Inuit people, including dangers to hunters and others who need to travel and obtain food from the land, dangers of melting permafrost, harm to animals in the Arctic, coastal erosion and storm surges, and heat-related health problems (2005, 35–67). Next, the petition establishes that the United States historically has been the world's greatest national source of greenhouse gases (2005, 68–69). Because carbon dioxide, the major greenhouse gas, resides in the atmosphere for several centuries, the burden is cumulative, so while China during 2007 very likely passed the United States in present-day emissions, the United States remains by far the largest source of the historical burden, at about 27.5 percent of the worldwide total.

The petition documents violations of international law (2005, 95–97). Combining traditional knowledge of hunters and elders with wide-ranging peer-reviewed science, the petition alleges that Inuit human rights are being violated in several ways, including the following:

> 1. The right to life and physical security;
> 2. The right to personal property;
> 3. The right to health;
> 4. The right to practice traditional culture;
> 5. The right to use land traditionally used and occupied; and
> 6. The right to the means of subsistence. (*Petition to the Inter-American Commission* 2005, 70–102; 111–114)

According to Watt-Cloutier, The ICC petition illustrates three basic messages that arise from Inuit experience in the Arctic.

> 1. "Dangerous" climate change is already here;
> 2. Climate change in the Arctic is quickly going to get worse; and
> 3. Climate change in the Arctic is important globally. (Watt-Cloutier 2005)

The petition alleges that the United States has "consistently denied, distorted, and suppressed scientific evidence of the causes, rate, and magnitude of global warming" (*Petition to the Inter-American Commission* 2005, 109) as it reiterates that many of the dangers currently facing the Inuit, including retreat of protective sea ice, impaired access to vital resources, loss of homes, and other infrastructure, are a direct result of human-rights violations committed by the United States, which should be held accountable (*Petition to the Inter-American Commission* 2005, 103–110). The Petition concludes by documenting a lack of remedies within the U.S. legal system, as well as that country's lack of cooperation under international law in this venue (*Petition to the Inter-American Commission* 2005, 112–116).

The Inuit Petition as Global Precedent

The ICC and its co-petitioners acknowledge that the declaration they seek from the Commission will not be enforceable, but that it would have great moral value. The Petition is intended to educate and encourage the United States to join the community of nations in a global effort to combat climate change. The focus has been placed on the United States because of its refusal, until George W. Bush left office, to join in the Kyoto Protocol and other diplomatic efforts to reduce greenhouse-gas emissions.

In summary, the petition asks that the Commission make the following recommendation to the United States:

> 1. Adopt mandatory measures to limit its emissions of greenhouse gases in co-operation with the community of nations;
> 2. Take into account the impact of U.S. greenhouse-gas emissions on the Arctic and Inuit before approving all major government actions;
> 3. In consultation with the Inuit, develop a plan to protect Inuit culture and the Arctic environment and to mitigate any harm caused by U.S. greenhouse-gas emissions; and
> 4. In co-ordination with Inuit, develop a plan to help Inuit adapt to unavoidable climate change. (Watt-Cloutier 2005)

The petition also asks the Commission to declare that the United States has an obligation to take into account the impact of its emissions on the Arctic and Inuit people before approving all major government actions (*Petition to the Inter-American Commission* 2005, 109).

In addition to detailed documentation of the perils faced by Inuit people as a result of global warming, the ICC petition contains an extensive legal justification making the case that international law is a proper forum for this dispute, given that one of the fundamental norms of customary international law establishes every state's obligation not to knowingly allow its territory to be used for acts contrary to the rights of other states. Because the emission of greenhouse gases in one state causes harm to others, this norm provides the context for assessing states' human-rights obligations with respect to global warming. Under international law, a strong presumption exists compelling nations to work within the international framework to correct human-rights violations.

Inuit Hunters' Experiences

Simon Nattaq was one of 63 Inuit from Canada and Alaska whose testimonies are included in the petition. Nattaq lost both of his feet to frostbite when his snowmobile crashed through ice thinned by rising Arctic temperatures. All of his gear was lost in the water, leaving Nattaq stranded for two days. He now walks with prosthetic feet and believes that God kept him alive to warn the world about global warming. "Today I am here because the creator allowed it," said Nattaq, 61, a city counselor in Iqaluit (Duff-Brown 2007).

Thinning ice is of particular danger to hunters and other travelers across the Arctic. Ronald Brower of Barrow, Alaska, said:

> One of my sons ... was going to visit the next [whaling] crew ... and he fell right through the ice half-way to that camp. I've seen my fellow whalers trying to go whaling break through the ice, because it's melting from the bottom, and our snow machines have fallen through the ice. (*Petition to the Inter-American Commission* 2005, 40)

Inuit hunters' dogs have an ability to detect thin ice, which they may refuse to cross. Snowmobiles have no such sense.

Eugene Brower, also of Barrow, said that while pack ice was visible from the Alaskan north coast year-round until a few years ago, it has now retreated well over the horizon, forcing hunters to journey off shore as far as 90 miles to find bearded seals and walrus (*Petition to the Inter-American Commission* 2005, 41, 48). Some polar bears have drowned after swimming long distances, 100 miles or more, in search of ice floes.

Inuit life has changed in many ways because of the rapidly warming climate. Many Inuit once used ice cellars to store meat in frozen ground. The climate has warmed so much that the cellars are no longer safe in the summer, even as far north as Barrow, Alaska. Meat that once could be safely stored in permafrost has been spoiling (*Petition to the Inter-American Commission* 2005, 50). Other Inuit have reported painful heat rashes, a condition previously unknown (*Petition to the Inter-American Commission* 2005, 59). With increasing inability to hunt, Inuit are being forced to buy imported food at prices several times that paid for the same goods at lower latitudes. Diabetes and other problems associated with imported food have been increasing. Drinking water obtained by melting ice has become dirtier (*Petition to the Inter-American Commission* 2005, 62). A warmer climate increases risks for insect-vector diseases, even West Nile Virus (*Petition to the Inter-American Commission* 2005, 63).

Rationale for the Petition

Watt-Cloutier elaborated on the ICC's rationale for the petition:

> This petition is not about money, it is about encouraging the United States of America to join the world community to agree to deep cuts in greenhouse gas emissions needed to protect the Arctic environment and Inuit culture and, ultimately, the world. We submit this petition not in a spirit of confrontation—that is not the Inuit way—but as a means of inviting and promoting dialogue with the United States of America within the context of the climate change convention. Our purpose is to educate not criticize, and to inform, not condemn. I invite the United States of America to respond positively to our petition. As well, I invite governments and non-governmental organizations worldwide to support our petition and to never forget that, ultimately, climate change is a matter of human rights. (Watt-Cloutier 2005)

"What is happening affects virtually every facet of Inuit life," Watt-Cloutier told the Eleventh Conference of Parties (COP) to the United Nations Framework Convention on Climate Change, held in Montreal on December 7, 2005. "We are a people of the land, ice, snow, and animals. Our hunting culture thrives on the cold. We need it to be cold to maintain our culture and way of life. Climate change has become the ultimate threat to Inuit culture" (Watt-Cloutier 2005). In addition to effects mentioned above, the ICC petition also described several other widespread effects of warming on the Inuit world: melting permafrost, thinning and ablation of sea ice, receding glaciers, incursion of animal species previously unknown in the Arctic, increased coastal erosion, longer and warmer summers, and shorter winters.

Watt-Cloutier described how marine species that are dependent on sea ice, including polar bears, ice-living seals, walrus, and some marine birds, are likely to decline, with some species facing extinction. "For Inuit, warming is likely to disrupt or even destroy [our] hunting and food-sharing culture as reduced sea ice causes the animals on which they depend to decline, become less accessible, and possibly become extinct," said Watt-Cloutier (Watt-Cloutier 2005).

"Inuit are adaptable and resourceful," said Watt-Cloutier.

> We have to be to survive in the Arctic. But the ACIA [Arctic Climate Impact Assessment, 2004 report; www.acia.uaf.edu] foresees a time—within the lifetime of my eight-year-old grandson—when environmental change will be so great that Inuit will no longer be able to maintain their hunting culture. Already Inuit are struggling to adapt to the impacts of climate change.... How would you respond if an international assessment prepared by more than 300 scientists from 15 countries concluded that your age-old culture and economy was doomed, and that you were to become a footnote to globalization? (Watt-Cloutier 2005)

"The Arctic has gained broad recognition as the globe's 'barometer' of climate change," Watt-Cloutier said.

> I live in Iqaluit on Baffin Island. My back yard is the world's "sentinel" ecosystem for climate change.... An Inuk out on the land hunting for a seal with which to feed his family observes even minute changes to the environment. In a very real sense he is the sentinel—the first line of defense against climate change. That Inuk hunter illustrates something else—climate change is a human and

> family issue.... The Arctic can help us all look beyond narrowly defined national interests to create a global perspective. This is what's needed if we are to combat climate change. The "Voice from the North" can help determine the global level of greenhouse gas reductions required to achieve the goal of the convention. (Watt-Cloutier 2005)

"A Gift to the World"

The Inuit and their leaders are conscious that natural circumstances have compelled them to fill a role as world leaders in responding to toxic contamination by polychlorinated biphenyls, dioxins, and other chemicals, as well as global warming. This is not a free choice on their part; the alternative is an end of the world as they know it—and, if the effects of toxic chemicals are not addressed, their own demise.

Leaders among the Inuit see their Arctic home as a bridge between regions of the world, reflecting their mission to bring many peoples together in common pursuit of solutions to the climate crisis. The ICC petition to the OAS is phrased as an invitation to move toward an environmentally sustainable world. The ICC petition asserts that changes in their homeland today will follow in the rest of the world as multiple feedbacks cause warming to accelerate. Changes in the Arctic climate thus serve as "canary in the coal mine." The Inuit argue that their experiences should serve as a warning to everyone.

"In a very real sense our petition is a 'gift' from Inuit hunters and elders to the world," Watt-Cloutier told Global Green USA in Beverly Hills, California, April 1, 2006.

> It is an act of generosity from an ancient culture deeply tied to the natural environment and still in tune with its wisdom, to an urban, industrial, and 'modern' culture that has largely lost its sense of place and position in the natural world. Let us work together to protect the Arctic so that we may save the planet. (Watt-Cloutier 2006c)

FURTHER READING

Brown, Paul. "Global Warming is Killing Us Too, Say Inuit." *London Guardian*, December 11, 2003b, 14.

Duff-Brown, Beth. "Arctic Inuit Argue U.S. Pollution Devastates Centuries-old Hunting Traditions." Associated Press Worldstream, March 1, 2007. (LEXIS)

Petition to the Inter-American Commission on Human Rights Seeking Relief from Violations Resulting from Global Warming Caused by Acts and Omissions of the United States. Submitted by Sheila Watt-Cloutier with the Support of the Inuit Circumpolar Conference, and Behalf of All Inuit in the Arctic. Iqualuit, Nunavut, December 7, 2005. http://inuitcircumpolar.com/index.php?ID=316&Lang=En.

Watt-Cloutier, Sheila. "The Climate Change Petition by the Inuit Circumpolar Conference to the Inter-American Commission on Human Rights." Presentation at the Eleventh Conference of Parties to the UN Framework Convention on Climate Change, Montreal, December 7, 2005. http://inuitcircumpolar.com/index.php?ID=318&Lang=En.

Watt-Cloutier, Sheila. "The Arctic and the Global Environment: Making a Difference on Climate Change." Solutions for Communities Climate Summit, Hosted by Global Green USA. Beverly Hills, California, April 1, 2006c. http://inuitcircumpolar.com/index.php?ID=329&Lang=En.

Hurricanes, Intensity and Frequency of

Hurricanes thrive in warm water (and wither if sea-surface temperatures fall below 80°F). This fact, among others, has caused an intense debate to blow up regarding whether a warmer world will mean hurricanes will occur more often, and become more intense. While the temperature of the ocean undoubtedly plays a role in hurricane frequency and intensity, many other factors are at work, as well, making a simple linear association of hurricane intensity with global warming difficult to prove. As is the case with other phenomena, global warming's effect on hurricanes plays out in a broader climatic context.

An infrared satellite image provided by the National Oceanic and Atmospheric Administration shows the eye of Hurricane Rita coming ashore between Sabine Pass and Johnson's Bayou on the southwest Louisiana Gulf Coast on September 24, 2005. (AP/Wide World Photos)

Benjamin D. Santer of the Lawrence Livermore National Laboratory, U.S. Energy Department, has compiled data indicating that sea-surface temperatures are, indeed, rising, based on several climate models, ranging from 0.32°C to 0.67°C during the twentieth century. "For the period 1995–2005," Santer wrote in the *Proceedings of the National Academy of Sciences*, "We find an 84 percent chance that external forcing explains at least 67 percent of the observed SST [sea-surface temperature] increases in the two tropical cyclogenesis regions" (Santer et al. 2006, 13,905). "Even under modest scenarios for emissions, we're talking about sea surface temperature changes in these regions of a couple of degrees," Santer said. "That's much larger than anything we've already experienced, and that is worrying" (Revkin 2006h).

The relationship (or lack thereof) between hurricane intensity and warming atmospheric temperatures is complicated by the fact that water temperatures (like air temperatures) sometimes vary, over periods of several decades, with the long-term trend "signal" provoked by greenhouse-gas levels. For example, water temperatures in the Atlantic Ocean, which produces nearly all the hurricanes that have an impact on the United States, have been rising steadily since the 1970s, paralleling a general global rise in air temperatures. Frequency and intensity of hurricanes (as well as the number hitting U.S. coastlines and inflicting major damage) also have been rising during the same period.

Any study that takes the record back to the 1970s indicates a tight relationship between ocean warming, hurricane intensity, and air temperatures. During the 1950s and 1960s, air temperatures were generally cooler than during the 1970s, but hurricane intensity was higher—again, on an average. By 2005, this divergence was fueling a testy debate between some hurricane experts regarding whether, and to what degree, hurricane intensity and frequency was related to the overall warming trend. This debate often spilled over into the public realm as Florida and surrounding areas were smacked by four major hurricanes in 2004 (Hurricanes Charley, Frances, Ivan, and Jeanne), and as the 2005 season set records for the number of named hurricanes, including Katrina, Rita, and others that were exceptionally large and violent. Each of these hurricanes ranked in the top 10 of such storms to hit the United States in terms of insurance losses.

The Temperature-Intensity Argument

As early as 1986, Kerry Emanuel, a hurricane specialist at the Massachusetts Institute of Technology, published an article in the *Journal of the Atmospheric Sciences* (Emanuel 1986) in which he argued that hurricane intensity is governed, in part, by the degree of thermodynamic disequilibrium between the atmosphere and the underlying ocean. Therefore, Emanuel reasoned, warmer ocean waters would contribute to more intense tropical cyclones. By the year 2000, Emanuel's forecast of "hypercanes" had become gist for at least one best-selling book, *The Coming Superstorm* by Art Bell and Whitley Strieber, which rose to number 15 on the *New York Times* best-seller list for nonfiction hardbound books during January 2000. A major motion picture, *The Day after Tomorrow*, was loosely based on that book.

In 1987, Emanuel published an article in *Nature*, asserting that a 3°C increase in sea-surface temperature could increase the potential destructive power (as measured by the square of the wind speed) of storms by 40 to 50 percent (Emanuel 1987). Emanuel asserted in 1988 that a warming of the sea surface by 6° to 10°C (assuming no temperature change in the lower stratosphere) would make a supersize, ultrapowerful "hypercane" theoretically possible (Emanuel 1988a, 1988b).

Emanuel's theory has sparked intense debate among hurricane specialists. Critics of his theory responded that even the most pessimistic climate models do not project such a large amount of warming in the tropics. William Gray, perhaps the best-known hurricane specialist in the United States (due in part to his broadcasts over The Weather Channel) has assembled statistics indicating that Atlantic hurricane activity between 1970 and 1987 was less than half the level observed between 1947 and 1969, an indication that temperature rises already experienced have not caused more intense tropical storms.

Hurricanes increase in strength with the warmth of the water over which they travel. If warm surface water is shallow, the storm's own turbulence may mix enough underlying water to drive its surface temperature below the 80°F required to sustain the storm. Under present conditions, the top wind speed of a hurricane is probably about 200 miles per hour, and the lowest possible central pressure is probably about 885 millibars (McKibben 1989, 95).

Floodwaters from Hurricane Katrina fill streets near downtown New Orleans on August 30, 2005. (AP/Wide World Photos)

A study published in the August 4, 2005, edition of *Nature* indicated that the "dissipation of power" of Atlantic hurricanes had more than doubled during the previous 30 years, with a dramatic spike since 1995, with global warming and other variations in ocean temperatures working together (Emanuel 2005, 686–688). In this study, Emanuel was the first scientist to directly indicate a statistical relationship between warming and storm intensity (Merzer 2005). The trend reflects longer storm lifetimes and greater intensities, both of which Emanuel associates with increasing sea-surface temperatures. Emanuel wrote:

> The large upswing in the last decade is unprecedented and probably reflects the effect of global warming.... My results suggest that future warming may lead to an upward trend in tropical cyclone destructive potential and—taking into account an increasing coastal population—a substantial increase in hurricane-related losses in the 21st century. (Emanuel 2005, 686)

Researchers at the National Center for Atmospheric Research (NCAR) concluded that warmer sea-surface temperatures and altered wind patterns associated with global climate change are responsible for the increase in hurricane numbers and intensity ("Climate Change" 2007). Sea-surface temperatures have risen about 1.3°F during the last century, with the increases preceding rises in storm frequency, according to this study, by Greg Holland of NCAR, and Peter Webster of Georgia Institute of Technology. The study was published online by the Royal Society of London.

More Studies, More Debate

The Holland-Webster study's conclusions prompted more debate. "All signs that I've seen show that it's related to natural variability," said Eric Blake, a meteorologist at the National Hurricane Center, whose view is supported by some scientists. "There could be some impact of global warming, but its role is probably secondary or tertiary" ("Climate Change" 2007). Holland, however, said that "[e]ven a quiet year by today's standards would [have been] considered normal or slightly active compared to an average year in the early part of the twentieth century" ("Climate Change" 2007).

While tracing any specific storm to warming is a tenuous exercise, hurricanes may intensify generally as oceans warm. Hurricanes are essentially heat engines, so storms that approach their upper limits of intensity are expected to be slightly stronger—and produce more rainfall—in a warmer climate due to the higher sea-surface temperatures. According to a simulation study by a group of scientists at the National Oceanic and Atmospheric Administration's (NOAA) Geophysical Fluid Dynamics Laboratory (GFDL), a 5 to 12 percent increase in wind speeds for the strongest hurricanes (typhoons) in the northwest tropical Pacific is projected if tropical sea surfaces warm by a little over 2°C.

Although such an increase in the upper-limit intensity of hurricanes with global warming had been suggested on theoretical grounds by Emanuel more than a decade previously, Holland and Webster were the first to examine the question using a hurricane prediction model that was being used operationally to simulate realistic hurricane structures. Other models also indicate increases in precipitation of as much as 20 percent (Knutson, Tuleya, and Kurihara 1998, 1018; Knutson et al. 2001, 2458).

Another study reached similar conclusions. By the 2080s, warmer seas could cause the average hurricane to intensify about an extra half step on the Saffir-Simpson scale, according to a study conducted on supercomputers at the Commerce Department's GFDL in Princeton, New Jersey. The same study anticipated that rainfall up to 60 miles from hurricanes' cores could be nearly 20 percent heavier. This study is significant because it used half a dozen computer simulations of global climate devised by separate groups at institutions around the world.

Thomas R. Knutson and Robert E. Tuleya's models indicated that given sea-surface temperature increases of 0.8° to 2.4°C, hurricanes would become 14 percent more intense (based on central pressure), with a 6 percent increase in maximum wind speeds and an 18 percent rise in average precipitation rates within 100 kilometers of storm centers. At the time of this study, Tuleya, was a hurricane expert who had recently retired after 31 years at the fluid dynamics laboratory and taught at Old Dominion University in Norfolk, Virginia; Knutson worked at Princeton University. "One implication of the results," they wrote, "is that if the frequency of tropical cyclones remains the same over the coming century, a greenhouse-gas induced warming may lead to a gradually increasing risk in the occurrence of highly destructive category-5 storms" (Knutson and Tuleya 2004, 3477).

This study of hurricanes and warming was "by far and away the most comprehensive effort" to assess the question using powerful computer simulations, said Emanuel. "This clinches the issue" (Revkin 2004d, A-20). The study added that rising sea levels caused by global warming would lead to more flooding from hurricanes. With almost every combination of greenhouse-warmed oceans and atmosphere and formulas for storm dynamics, the results were the same: more powerful storms and more rainfall, said Tuleya (Revkin 2004d, A-20).

Mark A. Saunders and Adam S. Lea of the Benfield University College London, Hazard Research Centre, Department of Space and Climate Physics, isolated factors affecting hurricane strength and frequency in the Atlantic Ocean and Caribbean between 1965 and 2005, removing such influences as wind shear. Saunders and Lea found that the sensitivity of tropical Atlantic hurricane activity to August-September sea-surface temperature over the period considered was such that a 0.5°C increase in sea-surface temperature would be associated with an approximately 40 percent increase in hurricane frequency and activity. The results also indicated that local sea-surface warming was responsible for approximately 40 percent of the increase in hurricane activity relative to the 1950–2000 average between 1996 and 2005 (Saunders and Lea 2008, 557).

Other scientists lend support to Emanuel's case. S. T. O'Brien and colleagues asserted that doubling the level of atmospheric carbon dioxide (increasing tropical sea-surface temperatures between 1° and 4°C) will double the number of hurricanes, and increase their strength by 40 percent to 60 percent (O'Brien, Hayden, and

Shugart 1992). O'Brien and colleagues also believe that warmer average temperatures will extend the hurricane season, because the season effectively ends for any given area when water temperature falls below 26°C (80°F). R. J. Haarsma and colleagues estimated that a doubling of greenhouse-gases levels will increase the frequency of hurricanes by 50 percent and increase the average intensity of the storms by 20 percent (Haarsma, Mitchell, and Senior 1993).

Consistent Critics of the Climate-Hurricane Link

A report by Christopher W. Landsea with NOAA's Miami, Florida, office asserted that an increase in carbon-dioxide levels may raise the threshold temperature in which hurricanes thrive (presently above 80°F) in proportion to the rise in temperature (Landsea 1993, 1703–1713). This change might nullify the increase in strength that is implied by warmer water-surface temperatures. Landsea also speculates that increased frequency of El Niño–type weather patterns in the Pacific tend to dampen hurricane development in the Atlantic, although storm frequency and intensity often rises under El Niño conditions in the eastern Pacific.

Landsea contended that intensity of Atlantic hurricanes has decreased since the middle of the twentieth century. Landsea and colleagues asserted that hurricane frequency and intensity had not increased during the past half-century. They wrote that "a long-term (five decade) downward trend continues to be evident primarily in the frequency of intense hurricanes. In addition, the mean maximum intensity (i.e., averaged over all cyclones in a season) has decreased" (Landsea et al. 1996, 1700).

Landsea believes that any warming-induced change in hurricane frequency and intensity will probably be lost in the "noise" of year-to-year hurricane variability. He concluded, "Overall, these suggested changes are quite small compared to the observed large natural variability of hurricanes, typhoons and tropical cyclones. However, more study is needed to better understand the complex interaction between these storms and the tropical atmosphere and ocean" (Landsea 1999).

Three greenhouse skeptics (Sherwood Idso, Robert Balling, and R. S. Cerveny) have challenged Emanuel's theories (Idso, Balling, and Cerveny 1990). They assembled hurricane data for the central Atlantic, the east coast of the United States, the Gulf of Mexico, and the Caribbean Sea between 1947 and 1987, and compared their information with estimates of the sea-surface temperatures in the Northern Hemisphere. Idso, Balling, and Cerveny concluded that "[t]here is basically no trend of any sort in the number of hurricanes experienced in any of the four regions with respect to variations in temperature" (Idso, Balling, and Cerveny 1990, 261).

Doubts have been raised about the relationship of warming seas and intensity of tropical cyclones by scientists who have not been as publicly critical of global warming as Idso, Balling, and Cerveny. A. J. Broccoli and S. Manabe (1990) included cloud-related feedbacks in their models, and found a 15 percent reduction in the number of days with hurricanes, even assuming a temperature rise caused by a doubling of atmospheric carbon dioxide. Broccoli and Manabe noted that their results were easily skewed by (and highly dependent upon) how they decided to represent cloud processes within their models.

William M. Gray, professor emeritus of Atmospheric Sciences at Colorado State University, is a long-standing opponent of the idea that warming temperatures have anything to do with hurricanes. According to his tally, between 1957 and 2006, 83 hurricanes hit the United States, 34 of them major. Between 1900 and 1949, 101 hurricanes hit the same area, 39 of which were major, with wind speeds above 110 miles an hour. From 1966 to 2006, said Gray, only 22 major hurricanes hit the United States, whereas between 1925 and 1965, 39 such storms hit the same area. "Even though global mean temperatures have risen by an estimated 0.4°C and CO_2 by 20 percent, the number of major hurricanes hitting the United States declined," Gray wrote (2007, A-12). Since 1995, however, the number of major storms hitting the U.S. Atlantic and Gulf of Mexico coasts has risen sharply. Gray associated this increase with strengthening circulation in the Atlantic Ocean.

Several Variables Shape Hurricanes

While heat is important to hurricanes, they are very sensitive to other influences. Even dust storms blowing off the Sahara Desert in Africa can have an important effect. After the devastating hurricane seasons of 2004 and 2005, the next year was very quiet. A major reason why hurricane activity was so much lower in 2006 than in 2005 appears to have been an increase in dust levels over the ocean from the Sahara Desert. During the 2006 hurricane season, sea-surface temperatures remained relatively cool and only five hurricanes formed, one-third of the total in 2005.

William Lau and Kyu-Myong Kim at NASA's Goddard Space Flight Center, Greenbelt, Maryland, showed that airborne Saharan dust very probably caused 30 to 40 percent of the water-temperature drop in the Atlantic in areas where hurricanes usually form. The dust blocked sunlight in tropical regions between June 2005 and June 2006. The team's research was reported in December 2007 in the American Geophysical Union's *Geophysical Research Letters* ("Saharan Dust" 2007). Lau said:

> Previous studies have looked at how hot, dry air associated with a Saharan dust outbreak affects an individual storm, but our study is the first to focus on dust's radiative effect on sea surface temperatures, which may affect storms for the entire season. Nobody had suggested that link before. ("Saharan Dust" 2007)

Still other factors influence hurricane numbers and intensity, including El Niño conditions in the tropical eastern Pacific Ocean and the strength of the West African monsoon. If other conditions are right, a hurricane season can be intense even if surface waters are cooler than usual, and vice versa.

A study of the eastern Caribbean (near the Puerto Rican island of Vieques) covering 5,000 years, published in the May 24, 2007, edition of *Nature* revealed centuries-long periods in which all of these variables played important roles in the number and intensity of hurricanes. The study's authors, from the Woods Hole Oceanographic Institution, said that their findings did not conflict with other scientists' assertions that increasing hurricane activity can be linked to human-caused warming of the climate and seas.

Jeffrey P. Donnelly, the lead author of this study, said the findings pointed to the importance of figuring out an unresolved puzzle: whether global warming will affect the El Niño cycle one way or the other. More intense or longer Pacific warm-ups could stifle Atlantic and Caribbean hurricanes even with warmer seas, Dr. Donnelly said (Revkin 2007e).

"Warm sea-surface temperatures are clearly the fuel for intense hurricanes," he said. "What our work says is that without sea temperatures varying a lot, the climate system can flip back and forth between active and inactive regimes." He added that a disturbing possibility was a warming of waters while conditions in the Pacific and Africa are in their hurricane-nurturing mode. "If you flip that knob and also have warming seas," Dr. Donnelly said, "oh boy, who knows what could happen?" (Revkin 2007e).

Donnelly and Woodruff (2007) compiled a record of intense hurricane activity in the western North Atlantic Ocean for 5,000 years based on sediment cores from a Caribbean lagoon that contains coarse-grained deposits associated with intense hurricane landfalls. They wrote:

> The record indicates that the frequency of intense hurricane landfalls has varied on centennial to millennial scales over this interval. Comparison of the sediment record with paleoclimate records indicates that this variability was probably modulated by atmospheric dynamics associated with variations in the El Niño/Southern Oscillation and the strength of the West African monsoon, and suggests that sea surface temperatures as high as at present are not necessary to support intervals of frequent intense hurricanes. To accurately predict changes in intense hurricane activity, it is therefore important to understand how the El Niño/Southern Oscillation and the West African monsoon will respond to future climate change. (Donnelly and Woodruff 2007, 465)

Cyclones in Unusual Places

In addition to variations in frequency and strength, tropical cyclones have been appearing in places where they occur very rarely, if at all. During June 2007, for example, Tropical Cyclone Gonu, with sustained winds of more than 120 miles per hour, churned 35-foot-high waves and struck Oman, on the Arabian peninsula, causing at least 13 deaths. The storm was the strongest on record in the northwestern Arabian Sea. The cyclone hit the Omani coastal towns of Sur and Ras al Hadd with sustained winds more than 100 miles an hour. Judith Curry, a hurricane expert at Georgia Tech, said the cyclone's strength was "really rather amazing" for the region, and appeared to be amplified by sea temperatures hovering around 87°F. Even weakened, she said, the storm could prove disastrous in Oman or Iran. "Cyclones are very rare in this region and hence governments and people are unprepared," she said (Revkin 2007f).

The Debate Continues

Studies of increasing hurricane intensity released during September 2008 again swung the debate to global warming's causal role. Research published in the September 3, 2008, edition of *Nature* indicated that maximum wind speeds of the strongest tropical cyclones have increased significantly since 1981, driven by rising ocean temperatures. "It'll be pretty hard now for anyone to claim that cyclone activity has not increased," said Judith Curry of the Georgia Institute of Technology, Atlanta, who was not involved in this study, but whose work has reached similar conclusions (Schiermeier 2008b).

James Elsner, a climatologist at Florida State University in Tallahassee, and colleagues found that the strongest tropical storms have intensified, especially in the North Atlantic and northern Indian oceans. The scientists analyzed satellite data of cyclone wind speeds, finding only a slight increase in the average number or intensity of all storms. The strongest storms, however, were becoming more intense. The number of storms at category 4 and 5 on the Saffir–Simpson scale increased, and these are the ones most likely to cause catastrophic damage (Elsner et al. 2008, 92).

The team estimated that a 1°C increase in sea-surface temperature could contribute to a 31 percent increase in the global frequency of category 4 and 5 storms per year, from 13 to 17. Since 1970, the tropical oceans have warmed on average by around 0.5°C. Computer models suggest they may warm by a further 2°C by 2100 (Schiermeier 2008b).

Criticism of the Elsner study fell out along familiar lines, as Christopher Landsea, who has made similar arguments about similar studies in the past, said that the data rely on inaccurate information for storms in the Indian Ocean. According to Landsea, the data skew the record in the North Atlantic by beginning measurements in 1981, during a cyclical lull in hurricane activity, and ending measurements in 2006, during a relatively active period that began about 1995. The study's statistics are elegant, said Landsea, but they rely on suspect information. Knutson, also a frequent skeptic of studies linking ocean warming and hurricane intensity, told the *New York Times*, that the Elsner study covered too short a period to be reliable as a predictor of a trend. Thus, he said, "It's not a definite smoking gun for a greenhouse warming signal on hurricanes" (Chang 2008b, A-18).

J. Lighthill and colleagues (1994) argued that while global warming may exert some influence on cyclone formation in the tropics, natural variability is more important. This position drew a reply from Emanuel (1995). L. Bengtsson, a member of the German Max Planck Institut für Meteorologie, and colleagues contend that global warming will strengthen the upper-level westerlies in areas where hurricanes usually develop, inhibiting storm development and intensity, negating any boost the storms may get from warmer sea-surface temperatures (Bengtsson, Botzet, and Esch 1996). (This is also typical of the El Niño pattern that seems to be occurring more frequently as temperatures rise.)

In 1995 the Intergovernmental Panel on Climate Change (IPCC), in its *Second Assessment*, restated its earlier position that the state of science on the subject does not permit a conclusion regarding whether global warming will affect the number and intensity of tropical cyclones. Tom Karl and colleagues wrote two years later in *Scientific American* that "[o]verall, it seems unlikely that tropical cyclones will increase significantly on a global scale. In some regions, activity may escalate; in others, it will lessen" (Karl et al. 1997, 83).

In another study, Johan Nyberg and colleagues reconstructed hurricane activity in the North Atlantic Ocean for 270 years into the past, using proxy records for vertical wind shear and sea-surface temperature from corals and a marine sediment core. In an exercise of what scientists call "paleotempestology," samples are taken from lagoons into which storm tides wash, an event associated with strong winds and storm surges that occur only during very strong tropical storms. Like all proxies, these are far from perfect. They do not, for example, account for changes in dominant hurricanes' paths, since they sample only a very small fraction of the area over which the storms move (Elsner 2007, 648). Records would be required over a much larger area to give them value.

Nyberg and colleagues found that the average frequency of major hurricanes decreased gradually from the 1760s until the early 1990s, reaching a long-term low cycle during the 1970s and 1980s. After 1995, frequency increased to levels similar to other periods of high intensity in their record, "and thus appears to represent a recovery to normal hurricane activity, rather than a direct response to increasing sea-surface temperatures" (Nyberg et al. 2007, 698). The upshot of this and other research is that, while hurricanes are sustained by warm water, vertical wind shear (winds blowing from different directions at various heights that disturb hurricanes' circulation) can tear them apart, dispersing storm-sustaining heat. This research raises other questions: El Niño conditions may be fostered by warming oceans, but El Niño conditions in the Pacific tend to cause above-average wind sheer in the Atlantic, which seems to tear up hurricanes' circulation. The picture is not as simple, therefore, as equating warmer water with more frequent and intense hurricanes.

Wetter and Windier, but not More Frequent?

Global warming will not increase Atlantic hurricane intensity, even as it does feed storms that are wetter, according to research by Knutson, a meteorologist at NOAA's fluid dynamics lab and several colleagues. Knutson said that his computer model study, published electronically May 18, 2008, in *Nature Geoscience* argued "against the notion that we've already seen a really dramatic increase in Atlantic hurricane activity resulting from greenhouse warming" (Borenstein 2008).

"Our regional climate model of the Atlantic basin reproduces the observed rise in hurricane counts between 1980 and 2006, along with much of the interannual variability, when forced with observed sea surface temperatures and atmospheric conditions," wrote Knutson and colleagues.

> Here we assess, in our model system, the changes in large-scale climate that are projected to occur by the end of the twenty-first century by an ensemble of global climate models, and find that Atlantic hurricane and tropical storm frequencies are reduced. At the same time, near-storm rainfall rates increase substantially. Our results do not support the notion of large increasing trends in either tropical storm or hurricane frequency driven by increases in atmospheric greenhouse-gas concentrations. (Knutson et al. 2008)

Knutson's data indicate that the number of such storms will decline 18 percent by the end of the twenty-first century. The number hitting land will decline 30 percent, Knutson argued. However, according to Knutson's research, the greatest decrease in frequency will be among weaker storms. Rainfall within 30 miles of a hurricane should rise about 37 percent and wind strength should increase by about 2 percent, according to Knutson's models.

Knutson's study stoked debate over global warming and hurricanes. Emanuel said that his conclusions are "demonstrably wrong" because they are based on a computer model that does not examine storms correctly.

Kevin Trenberth, climate analysis chief at NCAR, said that Knutson's computer model is poor at assessing tropical weather and "fail[s] to replicate storms with any kind of fidelity." Trenberth said that it is not just the number of hurricanes "that matter, it is also the intensity, duration and size, and this study falls short on these issues" (Borenstein 2008). Knutson acknowledged that his model is coarse and lacks an accurate analysis of individual storms' strength.

NOAA hurricane meteorologist Landsea said Knutson's work was "very consistent with what's being said all along. I think global warming is a big concern, but when it comes to hurricanes the evidence for changes is pretty darn tiny" (Borenstein 2008).

Is Tropical Storm Activity Too Variable to Detect a Global-Warming Signal?

Short-term tropical storm activity varies so greatly in any one ocean basin that scientists often have trouble detecting long-term climate "signals" (or patterns) on a worldwide basis. When one area (the Atlantic Ocean, for example) is active another (the Northern Pacific, perhaps) may be quieter than usual. Year-to-year variation may be much greater than multiyear trends, confounding any attempt to generalize trends, whether with respect to global warming, or other trends. Thus, some scientists find a rise in storm numbers and intensity and attribute it at least in part to a warming climate, whereas others find the opposite. The debate brings to mind the old aphorism about quoting scripture—the outcome may say more about who is doing the quoting than the record being examined.

For example, into a context in which some scientists have found rising tropical storm intensity came a 2008 study by Florida State University researchers asserting that the 2007 and 2008 tropical storm seasons worldwide were among the quietest in the previous 30 years (even as the Atlantic ocean experienced an active season). The study, "Global Tropical Storm Activity," coauthored by Ryan Maue, agreed with research by Stan Goldenberg of NOAA that tropical storm activity is inherently variable and all but unconnected with a warming climate. Goldenberg called the idea of associating global warming with increasing tropical storm intensity "a simplistic notion" (Dorell 2008, 3-A).

"The Lemming-like March to the Sea"

Along beaches in many locales in the United States, ranks of condos have risen during the last few decades, heedless of the debate over hurricane intensity. At the same time, a panel of scientists warned of the "lemming-like march to the sea" (Revkin 2006e, D-2). The scientists, who disagree among themselves regarding whether global warming is intensifying hurricanes, said that the increasing concentration of housing on and near the coasts will cause increasing damage even if hurricanes do not intensify over time.

At a meeting of the National Association of Insurance Commissioners, Andrew Logan, insurance director of the Ceres investor coalition (representing $4 trillion in market capital), said that "insurance as we know it is threatened by a perfect storm of rising weather losses, rising global temperatures and more Americans living in harm's way" (Morrison and Sink 2007, A-25). Ceres estimated that losses related to catastrophic weather—the vast majority of insurance payouts—have increased 15-fold in the U.S. property-casualty industry in three decades.

A month after Hurricane Katrina hit the U.S. Gulf of Mexico in late summer 2005, many insurance executives expected the industry's cost to be about $35 billion, which equaled the inflation-adjusted private payout for the World Trade Center attacks in 2001. The previous U.S. industry record for a season of U.S. hurricanes had been about $27 billion in 2004, when four powerful hurricanes hit Florida. The pre-Katrina hurricane record payout (also inflation-adjusted) was $23 million for Hurricane Andrew in 1992 (Treaster 2005). Total insurance losses in the United States caused by damage from hurricanes Katrina, Rita, and Wilma amounted to $57.6 billion by the end of 2005, more than twice 2004's all-time record. During 2005, world insurance loses due to weather-related natural disasters estimated by the Munich Re Foundation, part of one of the world's largest reinsurance companies, reached an all-time record of more than U.S. $200 billion.

By the summer of 2006, the annual premium to insure some oil-and-gas platforms in the Gulf of Mexico had risen to $25 million, from $2 million (Pleven et al. 2006, A-1). Some businesses with exposure to hurricanes were canceling projects; others were gambling without insurance. Wal-Mart dropped its coverage for severe windstorms and decided to pay for damage internally. Omaha investor Warren Buffett's Berkshire Hathaway stepped up its investments in reinsurance even as others dropped out. "If you like to watch football, you probably enjoy the game a little more if you have a bet on it," Buffett told the *Wall Street Journal.* "I like to watch the Weather Channel" (Pleven et al. 2006, A-1).

During 2006, several major insurers (among them Allstate, the largest in the United States) were canceling or refusing to write new policies in some coastal areas of Florida, Louisiana, and New York deemed too risky because of anticipated hurricanes. Where insurance was available in coastal areas prone to hurricane landfalls, premiums had risen to several thousand dollars a month on some properties, rivaling the cost of a mortgage (Adams 2006, B-1). Insurance premiums on some 1,500-square-foot houses in Key West, Florida, doubled to $13,000 a year in 2006, $10,000 of which was being assessed for windstorm insurance (Waddell 2006).

By 2007, some insurance companies were canceling homeowners' policies en masse along parts of the U.S. east coast, citing potential hurricane threats. Between 2004 and 2007, more than three million homeowners' policies along the east coast were terminated for adverse storm risk. In many cases, exclusion from the insurance market was making many homes unsalable because coverage is required to obtain a mortgage. Some of the homes that had been refused private insurance were several miles inland. Some states (Florida was an example) were assuming the risk out of public funds. Florida's Citizens Property Insurance Corporation by 2007 became that state's largest homeowners' insurer, with 1.3 million policies (Vitello 2007). By 2007, Massachusetts also had a state high-risk pool with 200,000 policies, including 44 percent of homeowners on Cape Cod.

FURTHER READING

Adams, Marilyn, "Strapped Insurers Flee Coastal Areas." *USA Today*, April 26, 2006, B-1.

Bengtsson, L., M. Botzet, and M. Esch. "Will Greenhouse Gas-induced Warming over the Next 50 Years Lead to a Higher Frequency and Greater Intensity of Hurricanes? *Tellus* 48A (1996):57–73.

Borenstein, Seth. "Jump in Atlantic Hurricanes not Global Warming, says Study that Predicts Fewer Future Storms." Associated Press, May 19, 2008. (LEXIS)

Broccoli, A. J., and S. Manabe. "Can Existing Climate Models Be Used to Study Anthropogenic Changes in Tropical Cyclone Intensity?" *Geophysical Research Letters* 17 (1990):1917–1920.

Chan, Johnny. "Comment on 'Changes in Tropical Cyclone Number, Duration, and Intensity in a

Warming Environment.'" *Science* 311 (March 24, 2006):1713.

Chang, Kenneth. "Strongest Storms Grow Stronger Yet, Study Says." *New York Times*, October 4, 2008b, A-18.

"Climate Change Linked to Doubling of Atlantic Hurricanes." Environment News Service, July 30, 2007. http://www.ens-newswire.com/ens/jul2007/2007-07-30-01.asp.

Donnelly, Jeffrey P., and Jonathan D. Woodruff. "Intense Hurricane Activity over the Past 5,000 Years Controlled by El Niño and the West African Monsoon." *Nature* 447 (May 24, 2007):465–468.

Dorell, Oren. "Report: Tropical Storm Activity at 30-Year Low." *USA Today*, November 12, 2008, 3-A.

Elsner, James B. "Tempests in Time." *Nature* 447 (June 7, 2007):647–648.

Elsner, James B., James P. Kossin, and Thomas H. Jagger, "The Increasing Intensity of the Strongest Tropical Cyclones." *Nature* 455 (September 4, 2008):92–95.

Emanuel, Kerry A. "An Air-sea Interaction Theory for Tropical Cyclones. Part I: Steady-state Maintenance." *Journal of the Atmospheric Sciences* 43 (1986):585–604. See also http://www.wind.mit.edu/~emanuel/home.html.

Emanuel, Kerry A. "The Dependence of Hurricane Intensity on Climate." *Nature* 326, no. 2 (April 1987):483–485.

Emanuel, Kerry A. "The Maximum Intensity of Hurricanes." *Journal of the Atmospheric Sciences* 45 (1988a):1143–1156.

Emanuel, Kerry A. "Toward a General Theory of Hurricanes." *American Scientist* 76(1988b):370–379.

Emanuel, Kerry A. "Thermodynamic Control of Hurricane Intensity." *Nature* 401 (October 14, 1999):665–669.

Emanuel, Kerry A. "Increasing Destructiveness of Tropical Storms over the Past 30 Years." *Nature* 436 (August 4, 2005):686–688.

Gray, William M. "Hurricanes and Hot Air." *Wall Street Journal*, July 26, 2006, A-12.

Haarsma, R. J., J. F. B. Mitchell, and C. A. Senior. "Tropical Disturbances in a G[lobal] C[limate] M[odel]." *Climate Dynamics* 8 (1993):247–257.

Hoyos, C. D., P. A. Agudelo, P. J. Webster, and J. A. Curry. "Deconvolution of the Factors Contributing to the Increase in Global Hurricane Intensity" *Science* 312 (April 7, 2006):94–97.

Idso, S. B., R. C. Balling Jr., and R. S. Cerveny. "Carbon Dioxide and Hurricanes: Implications of Northern Hemispheric Warming for Atlantic/Caribbean Storms." *Meteorology and Atmospheric Physics* 42 (1990):259–263.

Karl, T. R., N. Nicholls, and J. Gregory. "The Coming Climate: Meteorological Records and Computer Models Permit Insights Into Some of the Broad Weather Patterns of a Warmer World." *Scientific American* 276 (1997):79–83.

Knutson, Thomas R., and Robert E. Tuleya. "Impact of CO_2-Induced Warming on Simulated Hurricane Intensity and Precipitation: Sensitivity to the Choice of Climate Model and Convective Parameterization." *Journal of Climate* 17, no. 18 (September 15, 2004):3477–3495.

Knutson, Thomas R., Robert E. Tuleya, and Yoshio Kurihara. "Simulated Increase of Hurricane Intensities in a CO_2-warmed Climate." *Science* 297 (February 13, 1998):1018–1020.

Knutson, Thomas R., Robert E. Tuleya, Weixing Shen, and Isaac Ginis. "Impact of CO_2-induced Warming on Hurricane Intensities as Simulated in a Hurricane Model with Ocean Coupling." *Journal of Climate* 14 (2001):2458–2469.

Knutson, Thomas R., Joseph J. Sirutis, Stephen T. Garner, Gabriel A. Vecchi, and Isaac M. Held. "Simulated Reduction in Atlantic Hurricane Frequency under Twenty-first-century Warming Conditions." *Nature Geoscience* (May 18, 2008), doi: 10.1038/ngeo202. http://www.nature.com/ngeo/journal/vaop/ncurrent/abs/ngeo202.html.

Landsea, C. W. "A Climatology of Intense (or Major) Atlantic Hurricanes." *Monthly Weather Review* 121 (1993):1703–1713.

Landsea, C. W. "NOAA: Report on Intensity of Tropical Cyclones." Miami, Florida, August 12, 1999. http://www.aoml.noaa.gov/hrd/tcfaq/tcfaqG.html#G3.

Landsea, C. W., N. Nicholls, W. M. Gray, and L. A. Avila. "Downward Trends Atlantic Hurricanes during the Past Five Decades." *Geophysical Research Letters* 23 (1996):1697–1700.

Lighthill, J., G. Holland, W. Gray, C. Landsea, G. Craig, J. Evans, Y. Kurihara, and C. Guard. "Global Climate Change and Tropical Cyclones." *Bulletin of the American Meteorological Society* 75 (1994):2147–2157.

McKibben, Bill. *The End of Nature*. New York: Random House, 1989.

Merzer, Martin. "Study: Global Warming Likely Making Hurricanes Stronger." *Miami Herald*, August 1, 2005, n.p. (LEXIS)

Morrison, John, and Alex Sink. "The Climate Change Peril That Insurers See." *Washington Post*, September 27, 2007, A-25. http://www.washingtonpost.com/wp-dyn/content/article/2007/09/26/AR2007092602070_pf.html.

Nyberg, Johan, Bjorn A. Malmgren, Amos Winter, Mark R. Jury, K. Halimeda Kilbourne, and Terrence M. Quinn. "Low Atlantic Hurricane Activity in the 1970s and 1980s Compared to the Past 270 Years." *Nature* 447 (June 7, 2007):698–701.

O'Brien, S. T., B. P. Hayden, and H. H. Shugart. "Global Climatic Change, Hurricanes, and a Tropical Forest." *Climatic Change* 22 (1992):175–90.

Pleven, Liam, Ian McDonald, and Karen Richardson. "As Hurricane Season Starts, Disaster Insurance Runs Short." *Wall Street Journal*, July 10, 2006, A-1, A-8.

Revkin, Andrew. "Global Warming Is Expected To Raise Hurricane Intensity." *New York Times*, September 30, 2004d, A-20.
Revkin, Andrew C. "Climate Experts Warn of More Coastal Building." *New York Times*, July 25, 2006e, D-2.
Revkin, Andrew. "Study Links Tropical Ocean Warming to Greenhouse Gases." *New York Times*, September 12, 2006h. http://www.nytimes.com/2006/09/12/science/12ocean.html.
Revkin, Andrew C. "Study Finds Hurricanes Frequent in Some Cooler Periods." *New York Times*, May 24, 2007e. http://www.nytimes.com/2007/05/24/science/earth/24storm.html.
Revkin, Andrew C. "Cyclone Nears Iran and Oman." *New York Times*, June 6, 2007f. http://www.nytimes.com/2007/06/06/world/middleeast/06storm.html.
"Saharan Dust Has Chilling Effect on North Atlantic." NASA Earth Observatory Press Release, December 14, 2007. http://earthobservatory.nasa.gov/Newsroom/NasaNews/2007/2007121425986.html.
Santer, B. D., T. M. L. Wigley, P. J. Gleckler, C. Bonfils, M. F. Wehner, K. AchutaRao, T. P. Barnett, J. S. Boyle, W. Bruggemann, M. Fiorino, N. Gillett, J. E. Hansen, P. D. Jones, S. A. Klein, G. A. Meehl, S. C. B. Raper, R. W. Reynolds, K. E. Taylor, and W. M. Washington. "Forced and Unforced Ocean Temperature Changes in Atlantic and Pacific Tropical Cyclogenesis Regions." *Proceedings of the National Academy of Sciences* 103, no. 38 (September 19, 2006):13,905–13,910.
Saunders, Mark A., and Adam S. Lea. "Large Contribution of Sea Surface Warming to Recent Increase in Atlantic Hurricane Activity." *Nature* 451 (January 31, 2008):557–560.
Schiermeier, Quirin. "Global Warming Blamed for Growth in Storm Intensity." *Nature* 455, September 3, 2008b. http://www.nature.com/news/2008/080903/full/news.2008.1079.html.
Treaster, Joseph B. "Gulf Coast Insurance Expected to Soar." *New York Times*, September 24, 2005. http://www.nytimes.com/2005/09/24/business/24insure.html.
Vitello, Paul. "Home Insurers Canceling in East." *New York Times*, October 16, 2007. http://www.nytimes.com/2007/10/16/nyregion/16insurance.html.
Waddell, Lynn. "Rising Insurance Rates Push Florida Homeowners to Brink." *New York Times*, June 29, 2006. http://www.nytimes.com/2006/06/29/us/29florida.html.

Hydrofluorocarbons (HFCs)

Solving One Problem, Aggravating Another

Following the ban of chlorofluorocarbon-laced refrigerant Freon, another chemical, hydrofluorocarbon (HFC) was introduced as an environmentally friendly replacement that would not further imperil stratospheric ozone. Shortly thereafter, however, HFC proved to be a potent greenhouse gas, with as much as 4,000 times the global-warming potential, molecule for molecule, of carbon dioxide. During 2000, the Coca-Cola company announced plans to phase out the use of HFCs in its cold-drink vending machines.

By the year 2000, HFCs were being widely used not only in air conditioning and refrigeration systems, but also in aerosol sprays that clean computers and cameras, among other devices. A 16-ounce spray can contains roughly the same amount of HFCs as an automobile's air-conditioning system.

Additionally, while air-conditioning and refrigeration systems use the same amount of HFC repeatedly in an enclosed loop, a spray can shoots HFCs into the atmosphere after one use. According to the Aerosol Industry Association of Japan, 1,850 tons of HFCs were distributed in about 4.5 million cans in 2003, up considerably from 1,050 tons in 1995. An estimated 80 percent of these HFC sprays are used to blow away dust. The gas specifically used for this purpose (HFC134a) has a global warming potential 1,300 times that of carbon dioxide ("Spray Cans" 2004). The Kiko Network, a Japanese organization involved in mitigating global warming, estimated that one aerosol can is as damaging, in terms of carbon dioxide released, as leaving on a 21-inch television set for four hours every day for 22 years ("Spray Cans" 2004).

FURTHER READING

"Spray Cans Warming the Planet, One Dust-busting Puff at a Time." *International Herald-Tribune* in *Herald Asahi* (Tokyo), May 27, 2004, n.p. (LEXIS)

Hydrogen Fuel Cells

Political correctness in relation to global warming in the automobile industry has become associated with the development of hydrogen fuel cells, especially after President George W. Bush used his State of the Union Address in January 2003 to propose $1.2 billion in research funding to develop hydrogen-fuel technologies. With those funds, Bush said that America could lead the world in developing clean, hydrogen-powered automobiles. Iceland, meanwhile, has made plans to become the world's first hydrogen economy (utilizing its geothermal resources).

Reykjavik's bus fleet has been retrofitted with fuel-cell engines, and hydrogen fueling stations have opened.

Jeremy Rifkin, a liberal social critic and author, published a book during September 2002 titled *The Hydrogen Economy: The Creation of the Worldwide Energy Web and the Redistribution of Power on Earth.* Rifkin believes that cheap hydrogen could make the twenty-first century more democratic and decentralized, much the way oil transformed the nineteenth and twentieth centuries by fueling the rise of powerful corporations and nation-states. With hydrogen, wrote Rifkin, "Every human being on Earth could be 'empowered'" (Coy 2002, 83).

Promise and Problems

As much as it has been touted as pollution-free, hydrogen fuel has been no free climatic lunch. Hydrogen, unlike oil or coal, does not exist in nature in a combustible form. Hydrogen is usually bonded with other chemical elements, and stripping them away to produce the pure hydrogen necessary to power a fuel cell requires large amounts of energy.

Unless an alternative source (such as Iceland's geothermal resource) becomes available, most hydrogen fuel will be produced from fossil fuels. Extraction of hydrogen from water via electrolysis and compression of the hydrogen to fit inside a tank that can be used in an automobile requires a great deal of electricity. Until electricity is routinely produced via solar, wind, and other renewable sources, the hydrogen car will require energy from conventional sources, including fossil fuels. Today, 97 percent of the hydrogen produced in the United States comes from processes that involve the burning of fossil fuels, including oil, natural gas, and coal.

In mid-2008, however, scientists at the Massachusetts Institute of Technology announced the discovery of a cobalt-phosphorus catalyst that can split hydrogen from oxygen atoms in water to create hydrogen gas. Before that discovery by MIT chemist Daniel Nocera and colleagues, reported in the *Journal of the American Chemical Society,* such a reaction was possible only using chemicals that worked under toxic conditions, or platinum, which is much too expensive for industrial-scale processes. "If we are going to use solar energy in a direct conversion process, we need to cover large areas. That makes a low-cost catalyst a must," said John Turner, an electrochemist at the National Renewable Energy Laboratory in Golden, Colorado ("Service" 2008, 620). "It's big-to-giant step," said Thomas Moore, a chemist at Arizona State University, Tempe. ""I'd say it's a breakthrough" ("Service" 2008, 620).

The chemistry of splitting the water molecule has been known for more than 200 years, but a process useful on industrial scale has eluded scientists. Nocera and colleagues' process still requires excess electricity to start the water-splitting reaction that is not recovered in the fuel it produces, and thus far, the catalyst can operate only at low levels of electrical current. Nocera said that work is ongoing to solve these problems. If these hurdles can be overcome, the next big challenge will be to determine whether the catalytic process can operate in saltwater. If so, sunlight could be used to generate hydrogen fuel from seawater for piping to storage on land. Once that step is mastered, humankind will be able to imitate plants in the ability to convert sunlight to energy.

Hydrogen Hype

Hydrogen fuel is such a fine fantasy that expectations have raced ahead of reality. On September 5, 2002, a coalition of companies, including DuPont and 3M asked Congress to spend $5.5 billion during the ensuing decade to advance fuel-cell development (Coy 2002, 83). A month later, the Ford Motor Co. announced that it would have a hydrogen car, the Focus, in limited service for commercial fleets in 2004. The car was to be an electric hybrid with a potential top speed of 115 miles an hour that was said to perform on a par with a gas-driven family car. By 2008, no such car existed.

In 2002, General Motors' (GM) Chief Executive Officer Richard Wagoner announced the company's launch of a $500 million hydrogen-car initiative for initial rollout in 2008. The cost could rise to billions of dollars annually if Wagoner chooses to mass-produce the new models. Wagoner defended the expense as a crucial investment in GM's future. "People say, 'How can you afford to spend so much on fuel cells?' and I say, 'How can you afford not to?'" Wagoner asked (Lippert 2002, E-3). At the Paris Motor Show in September 2002, GM unveiled the Hy-wire, a "concept car" that uses a hydrogen fuel cell, about six years behind Toyota and Honda. Ford Motor also purchased a sizable stake in British Columbia's Ballard Power, a world leader in the race to provide the first commercially available hydrogen fuel-celled automobile. GM also owns a stake in this company.

In Iceland, meanwhile, 85 percent of the country's 290,000 people use geothermal energy to heat their homes (Brown 2003, 166); Iceland's government, working with Shell and Daimler-Chrysler, in 2003 began to convert Reykjavik's city buses from internal combustion to fuel-cell engines, using hydroelectricity to electrolyze water and produce hydrogen. The next stage is to convert the country's automobiles, then its fishing fleet. These conversions are part of a systemic plan to divorce Iceland's economy from fossil fuels (Brown 2003, 168).

Other Problems with Hydrogen Power

Other challenges for hydrogen include reduction of its price to compete with other fuels, as well as construction of a distribution system analogous to gas stations, perhaps replacing them (Hansen 2008c).

Another problem with hydrogen fuel is storage in both vehicles and fueling stations. Hydrogen is flammable (far more prone to explode than gasoline), and must be stored at high pressure (up to 10,000 pounds per square inch). It is also far less dense than conventional fossil fuels, and so requires 50 times the storage space of gasoline. Liquid hydrogen avoids these problems, but it must be stored at 400°F below zero, not a practical solution for individual car owners or every neighborhood's fueling station. Nevertheless, the U.S. Energy Department by 2007 was pouring grant money ($170 million over five years) into developing a fleet of fuel-cell vehicles and fueling stations ("Hydrogen Car" 2007, 2-D).

Paul M. Grant, writing in *Nature*, provided an illustration:

> Let us assume that hydrogen is obtained by "splitting" water with electricity—electrolysis. Although this isn't the cheapest industrial approach to "make" hydrogen, it illustrates the tremendous production scale involved—about 400 gigawatts of continuously available electric power generation [would] have to be added to the grid, nearly doubling the present U.S. national average power capacity. (Grant 2003, 129–130)

That, calculated Grant, would represent the power-generating capacity of 200 Hoover Dams (Grant 2003, 129–130). At $1,000 per kilowatt, the cost of such new infrastructure would total about $400 billion. What about producing the 400 gigawatts with renewable energy? Grant estimated that, "with the wind blowing hardest, and the sun shining brightest," wind power generation would require a land area the size of New York State, or a layout of state-of-the-art photovoltaic solar cells half the size of Denmark (Grant 2003, 130). Grant's preferred solution to this problem is use of energy generated by nuclear fission.

Regardless of hydrogen fuel cells' present-day limitations, the European Union has advocated a transition to it from fossil fuels. The plan includes a $2 billion dollar E.U. commitment, over the course of several years, to bring industry, the research community, and government together in plans to make this transition. According to author Jeremy Rifkin,

> The E.U. decision to transform Europe into a hydrogen economy over the course of the next half century is likely to have as profound and far-reaching an impact on commerce and society as the changes that accompanied the harnessing of steam power and coal at the dawn of the industrial revolution and the introduction of the internal-combustion engine and the electrification of society in the 20th century. ("E.U. Plans" 2002)

Hydrogen Power and Stratospheric Ozone Depletion

Advocates of a hydrogen economy generally have ignored another potential problem. Some research indicates that leakage of hydrogen gas could cause problems in the Earth's stratospheric ozone layer. Writing in *Science*, researchers from the California Institute of Technology reported that the accumulation of leakage associated with a hydrogen economy could indirectly cause as much as a 10 percent decrease in stratospheric ozone. Michael Prather (2003, 581), however, has asserted that such damage is unlikely to be a problem with realistic hydrogen leakage rates.

Tracey K. Tromp and colleagues wrote:

> The widespread use of hydrogen fuel cells could have heretofore unknown environmental impacts due to unintended emissions of molecular hydrogen, including an increase in the abundance of water vapor in the stratosphere (plausibly as much as about 1 part per million by volume). This would cause stratospheric cooling, enhancement of the heterogeneous chemistry that destroys ozone, an increase in noctilucent clouds, and changes in tropospheric chemistry and atmosphere-biosphere interactions. (Tromp et al. 2003, 1740)

If hydrogen replaced all fossil fuels for transportation and to power buildings, Tromp and colleagues estimated that 60 to 120 trillion grams of hydrogen would be released into the atmosphere each year, four to eight times the amount released today from human sources. The scientists assumed a 10 to 20 percent loss rate due to leakage. Molecular hydrogen freely rises and mixes with stratospheric air, resulting in the creation of additional moisture at high altitudes and increased dampening of the stratosphere. Through a chain of chemical reactions, hydrogen cools the air and accelerates the destruction of ozone ("Hydrogen Leakage" 2003). The authors of this study estimated that hydrogen leakage could triple the amount of hydrogen in the stratosphere.

"We have an unprecedented opportunity this time to understand what we are getting into before we even switch to the new technology," said Tromp, the study's lead author. "It will not be like the case with the internal combustion engine, when we started learning the effects of carbon dioxide decades later" ("Hydrogen Leakage" 2003).

Refuting assertions that hydrogen leakage could deplete ozone, Martin G. Schultz and colleagues, writing in *Science*, asserted that "[a] possible rise in atmospheric hydrogen concentrations is unlikely to cause significant perturbations of the climate system" (Schultz et al. 2003, 624).

FURTHER READING

Brown, Lester R. *Plan B: Rescuing a Planet under Stress and a Civilization in Trouble.* New York: Earth Policy Institute/W.W. Norton, 2003.

Coy, Peter. "The Hydrogen Balm? Author Jeremy Rifkin Sees a Better, Post-Petroleum World." *Business Week*, September 30, 2002, 83.

"E.U. Plans to Become First Hydrogen Economy Superpower." *Industrial Environment* 12, no. 13 (December 2002):n.p. (LEXIS)

Grant, Paul M. "Hydrogen Lifts Off—with a Heavy Load: The Dream of Clean, Usable Energy Needs to Reflect Practical Reality." *Nature* 424 (July 10, 2003):129–130.

Hansen, James E. "Trip Report." August 5, 2008c. http://www.columbia.edu/~jeh1/mailings/20080804_TripReport.pdf.

"Hydrogen Car Has Far to Go." *Houston Chronicle* in *Omaha World-Herald*, September 9, 2007, 2-D.

"Hydrogen Leakage Could Expand Ozone Depletion." Environment News Service," June 13, 2003. http://ens-news.com/ens/jun2003/2003-06-13-09.asp.

Lippert, John. "General Motors Chief Weighs Future of Fuel Cells." *Toronto Star*, September 27, 2002, E-3.

Prather, Michael J. "An Environmental Experiment with H2?" *Science* 302 (October 24, 2003):581–582.

Rifkin, Jeremy. *The Hydrogen Economy: The Creation of the Worldwide Energy Web and the Redistribution of Power on Earth.* New York: Jeremy P. Tarcher/Putnam, 2002.

Schultz, Martin G., Thomas Diehl, Guy P. Brasseur, and Werner Zittel. "Air Pollution and Climate-Forcing Impacts of a Global Hydrogen Economy." *Science* 302 (October 24, 2003):624–627.

Service, Robert F. "New Catalyst Marks Major Step in the March toward Hydrogen Fuel." *Science* 321 (August 1, 2008):620.

Tromp, Tracey K. Run-Lie Shia, Mark Allen, John M. Eiler, and Y. L. Yung. "Potential Environmental Impact of a Hydrogen Economy on the Stratosphere." *Science* 300 (June 13, 2003):1740–1742.

Hydrological Cycle

Steady atmospheric warming contributes to increases in the moisture-holding capacity of the atmosphere, altering the hydrological cycle and the characteristics of precipitation. Such changes in the global rate and distribution of precipitation may have a greater direct effect on human well-being and ecosystem dynamics than changes in temperature itself.

While models of a warming climate generally agree that atmospheric moisture increases with temperature, theory as well as an increasing number of daily weather reports strongly indicate that changes in precipitation patterns may vary widely across time and space. Such changes will be highly uneven, episodic, and often nasty. Both droughts and deluges are likely to become more severe. They may even alternate in some regions. By 2000, the hydrological cycle seemed to be changing more rapidly than temperatures. With sustained warming, usually wet places often seemed to be receiving more rain than before; dry places often were experiencing less rain, subject to more persistent droughts. Some drought-stricken regions occasionally were doused with brief deluges that ran off cracking earth. In many places, the daily weather increasingly was becoming a question of drought or deluge.

In another regional study, Tim P. Barnett and colleagues combined a regional hydrologic and global climate model in a study implying that human-caused carbon-dioxide emissions have already greatly changed river flows and snow pack in the western United States. "Observations have shown that the hydrological cycle of the western United States changed significantly over the last half of the 20th century," the study said. It continued:

> We present a regional, multivariable climate change detection and attribution study, using a high-resolution hydrologic model forced by global climate models, focusing on the changes that have already affected this primarily arid region with a large and growing population. The results show that up to 60 percent of the climate-related trends of river flow, winter air temperature, and snow pack between 1950 and 1999 are human-induced. These results are robust to perturbation of study variables and methods. They portend, in conjunction with previous work, a coming crisis in water supply for the western United States. (Barnett et al. 2008, 1080)

Hydrological Changes Traced to the Advent of Industrialism

Despite these impacts, attention in climate models and paleoclimatic studies has concentrated mainly on temperature changes. Some studies have focused on changes in precipitation, however. One reconstruction of precipitation variability in the high mountains of northern Pakistan revealed dry conditions at the beginning of the past millennium and through the eighteenth and early nineteenth centuries, with precipitation increasing during the late nineteenth and the twentieth centuries to yield the wettest conditions of the past 1,000 years.

This study indicated a large-scale intensification of the hydrological cycle coincident with the onset of industrialization and global warming, with an unprecedented amplitude that argues for a human role. "We suggest," wrote the authors of this study, "that an unprecedented twentieth-century intensification of the hydrological cycle in western Central Asia has already occurred." This change is not restricted to the monsoon, but has been of "a broader seasonal and spatial extent," and "not simply a temporary fluctuation" (Treydte et al. 2006, 1181).

Andrew Revkin of the *New York Times* summarized the worldwide situation:

> A warmer world is more likely to be a wetter one, experts warn, with more evaporation resulting in more rain, in heavy and destructive downpours. But in a troublesome twist, that world may also include more intense droughts, as the increased evaporation parches soils between occasional storms. (Revkin 2002b, A-10)

"In a hotter climate, your chances of being caught with either too much or too little are higher," said John M. Wallace, a professor of atmospheric sciences at the University of Washington (Revkin 2002b, A-10).

Even as flooding rains have inundated some places, the percentage of Earth's land area affected by serious drought more than doubled between the 1970s and the early 2000s, according to an analysis by the National Center for Atmospheric Research (NCAR) in Boulder, Colorado. Increasing drought occurred over much of Europe and Asia, Canada, western and southern Africa, and eastern Australia. Rising global temperatures appear to be a major factor, said NCAR scientist Aiguo Dai (Dai, Trenberth, and Qian 2004, 1117; "U.S. N.S.F." 2005). Dai and colleagues found that the proportion of land areas experiencing very dry conditions increased from 10 to 15 percent during the early 1970s to about 30 percent by 2002. Almost half of that change was due to rising temperatures rather than decreases in rainfall or snowfall (Dai, Trenberth, and Qian 2004, 1117). "These results point to increased risk of droughts as human activity contributes to global warming," said Dai ("U.S. N.S.F." 2005).

Areas affected by severe drought have expanded worldwide even as the amount of water vapor in the atmosphere has increased. Worldwide average precipitation also has increased slightly. However, as Dai noted, "surface air temperatures over global land areas have increased sharply since the 1970s." Warming increases evaporation. "Droughts and floods are extreme climate events that are likely to change more rapidly than the average climate," said Dai ("U.S. N.S.F." 2005). "The warming-induced drying has occurred over most land areas since the 1970s," said Dai, "with the largest effects in northern mid- and high latitudes" ("U.S. N.S.F." 2005). Precipitation in the United States has run counter to that trend, however, increasing especially between the Rocky Mountains and Mississippi River.

Too Wet, Too Dry, Too Often

While warmer air holds more moisture, generally making rain (and sometimes snow) heavier, warmer air also increases evaporation, paradoxically intensifying drought at the same time. While models of a warming climate generally agree that atmospheric moisture will increase with temperature, theory as well as an increasing number of daily weather reports indicate strongly that increases in precipitation will not be evenly distributed across time and space. They will, in fact, be highly uneven, episodic, and sometimes damaging. Both drought and deluge are likely to become more severe.

By 2000, the hydrological cycle seemed to be changing more quickly than temperatures. In 2005, for example, India's annual monsoon brought a 37-inch rainfall in 24 hours to Mumbai (Bombay). India is accustomed to a drought-and-deluge cycle, but not like this. With sustained warming, usually wet places generally seemed to be receiving more rain and snow than before; dry places often received less precipitation, and became subject to more persistent drought.

Atmospheric moisture increases rapidly as temperatures rise; over the United States and Europe, atmospheric moisture increased 10 to 20 percent from 1980 to 2000. "That's why you see the impact of global warming mostly in intense storms and flooding like we have seen in Europe," Kevin Trenberth, an NCAR scientist, told London's *Financial Times* (Wolf 2000, 27; see also Trenberth 2003, 1205–1217). As if on cue to support climate models, the summer of 2002 featured a number of climatic extremes, especially regarding precipitation. Excessive rain deluged Europe and Asia, swamping cities and villages and killing at least 2,000 people, while drought and heat scorched much of the United States. Climate skeptics argued that weather is always variable, but other observers noted that extremes seemed to be more frequent than before.

Extreme Precipitation Events

Thomas Karl, director of the National Climatic Data Center in the United States government's National Oceanic and Atmospheric Administration, said, "It is likely that the frequency of heavy and extreme precipitation events has increased as global temperatures have risen." Karl testified, "This is particularly evident in areas where precipitation has increased, primarily in the mid-and-high latitudes of the Northern Hemisphere" (Hume 2003, A-13). Studies at the Goddard Institute for Space Studies indicate that the frequency of heavy downpours has indeed increased and suggest that this trend will intensify.

During 1998,

> [A]t least 56 countries suffered severe floods, while 45 baked in droughts that saw normally unburnable tropical forests go up in smoke from Mexico to Malaysia and from the Amazon to Florida.... Spring in the Northern Hemisphere is coming a week earlier [and] the altitude at which the atmosphere chills to freezing is rising by nearly 15 feet a year.

Storms Intensify

A powerful storm carrying hurricane-force winds battered Western Europe on Christmas Day, 1999, killing 80 people and leaving 3.4 million without electricity in France alone. Winds reached 120 miles an hour, Europe's strongest in at least 50 years. Ten days later, 300,000 people in France still had no electricity, as Francois Roussely, chairman of the electric utility EDF, estimated damage to that country's power infrastructure at $770 million. Total storm damage was being estimated at $5 billion across Western Europe. "It's a catastrophe without precedent in Europe," Roussely said. He said repairs to France's electrical system would take three years. The storm had been forecast, but the strength of its winds had not been anticipated. In many areas, wind speeds were the fastest ever recorded. "There's never been anything like this," said Jean-Claude Gayssot, France's transport minister (Daley 1999, A-10).

The worst storms experienced by England in several decades caused road and rail chaos across Western Europe during October and November 2000. Torrential rain and winds up to 90 mph uprooted trees, blocked roads, and cut electricity supplies across southern England and Wales. According to newspaper reports, a tornado ripped through a trailer park in Selsey in West Sussex less than 48 hours after a similar twister devastated parts of Bognor Regis.

Marilyn McKenzie Hedger, head of the United Kingdom Climate Impacts Program based at Oxford University, said, "These events should be a wake-up call to everyone to discover how we are going to cope with climate change" (Brown 2000b, 1). Michael Meacher, U.K. environment minister, said that while it would be foolish to blame global warming every time extreme weather conditions occur, "[t]he increasing frequency and intensity of extreme climate phenomena suggest that although global warming is certainly not the sole cause, it is very likely to be a major contributory factor" (Brown 2000b, 1). The storm included the lowest barometric pressure on record (951 millibars) during October in the United Kingdom.

Heavy storms dumped a foot of rain on swaths of New England during May 2006, the region's worst flooding in 70 years. The night of June 19, Houston was inundated by a foot of rain as well. A total of 5.19 inches of rain was recorded at Ronald Reagan Washington National Airport June 25, 2006, with storms continuing much of the following week. The heavy rains followed a dry period, and inundated the Mid-Atlantic states with the heaviest rainfalls on record outside of tropical storms. In four days (June 23 through 27), 12 inches of rain fell at Reagan National Airport. The roar of the falling rain was reported to have muffled even the sound of thunder (Barringer 2006a).

"The rain just came down with such intensity," said Wayne Robinson, director of emergency management for Dorchester County, on the Eastern Shore of Chesapeake Bay in Maryland. "We have reports of 11 inches Sunday, some as much as 15 inches over two days," said Robinson. "It just kept going across the roads, and the asphalt is just washed away. The culverts, the big pipes over little streams, those were just washed completely out" (Barringer 2006a). Later the same week, swollen rivers filled with several days of downpours wreaked havoc from Virginia to Vermont, causing the worst flooding on record except for the remains of hurricanes.

Global Warming May Aggravate Thunderstorms

As part of a general intensification of the hydrological cycle with warming, thunderstorms may become more intense and destructive, according to a study by NASA scientists at the Goddard Institute for Space Studies (GISS). "The strongest thunderstorms, the strongest severe storms and tornadoes are likely to happen more often and be stronger," said Tony Del Genio of GISS, co-author of the study, published in *Geophysical Research Letters* (Del Genio, Yao, and Jonas et al. 2007). However, the total number of storms may be fewer.

Individual thunderstorms are generally too small for most climate models to analyze, so some scientists caution that this study may be premature. However, Del Genio and colleagues looked at the forces that combine to create and sustain thunderstorms.

Del Genio's computer model indicates that global warming will produce more strong updrafts. "The consequences of stronger updrafts are more lightning and bigger hail," he said. Increasing carbon dioxide in the atmosphere also favors more frequent lightning that could ignite more fires in the drought-plagued Western United States. Warming may decrease wind shear in some (but not all) storms, moderating the effects of updrafts, Del Genio said. Harold Brooks, a scientist at NOAA's severe storms laboratory in Norman, Oklahoma, has published studies with results similar to the new NASA study, most notably regarding hail. Some of the severe hail that should be increasing could be baseball-size and come down at 100 mph, "falling like a major league fastball," he said (Borenstein 2007c).

A February Thunderstorm on Baffin Island

Unprecedented precipitation events have occurred in the Arctic, where temperatures have risen rapidly. The following account of a late-February thunderstorm was sent by Sheila Watt-Cloutier, who was serving at the time as president of the Inuit Circumpolar Conference. She lives in Iqaluit, on Baffin Island.

> Last night on February 26th on my daughter's 30th birthday so much rain fell that I woke up to several puddles and pools of water in my tundra backyard and since it was 6°C today the puddles/pools were not freezing. There was even lightning last night here in the Arctic on a February night. Much of the snow is melted on the back of my house and all the roads are already slushy and messy. All planes coming up from the south were cancelled because the runways were icy from the rain. I think Pangnirtung has been hit very hard with high winds and again the forecast for them tomorrow is 8 above. One would think we were April already!
>
> High winds are still gusting up to 90 km [kilometers per hour] as I write this and rain is forecasted tonight again. Unfortunately the predictions of the Arctic Climate Impact Assessment are unfolding before my very eyes.
>
> One of my friends said today the first thing she thought of were the caribou and how hard it is going to be for them to try and get to the lichen under the ice when it gets cold again and everything freezes. She said she was going to encourage her husband to go and get caribou soon while they are still healthy as come spring they will surely be

> skinny and not as healthy as they normally would had it not rained so much at this time of year and created that crust of ice separating them from their food source. (Watt-Cloutier 2006b)

Global Warming and Drought Cycles

Warming tends to accelerate the usual atmospheric processes by which air rises in the tropics (increasing rainfall) and subsides in the subtropics, decreasing rainfall. If one looks at a map of the world, major deserts usually concentrate in the subtropics, with the wettest areas usually nearest the equator. Warmer temperatures accelerate both tendencies. These are general tendencies; they are not always consistent based on latitude; the U.S. West (except for the northwest coast), for example, is markedly drier than the East because of access to moisture from the Gulf of Mexico. The foothills of the Himalayas also are very wet, due to orographic lift during the wettest months of the Indian monsoon.

Hoerling and Kumar asserted the combination of La Niña's cooler-than-average waters in the Eastern tropical Pacific and persistently warm water in the western tropical Pacific created the "perfect ocean" for a atmospheric circulation pattern that produced a globe-girdling drought across a wide swath of the Northern Hemisphere (Hoerling and Kumar 2003, 691). "An almost unbroken zonal belt of high pressure wrapped the middle latitudes," Hoerling said (Toner 2003b, 4-A). Some drought-stricken areas, which stretched from New England to Pakistan, received as little as half their average precipitation during the four-year period, according to the study.

Hoerling and Kumar studied actual climatic behavior and then tested it against three different models in three laboratories. Each time, they examined sea-surface temperatures between 1998 and 2002. The researchers were looking for conditions that would support intense, widespread drought (Kerr 2003a, 636). A total of 51 simulations produced strikingly similar results, anticipating less rain in the United States, southern Europe, and Southwest Asia. "The modeling results offer compelling evidence that the widespread mid-latitude drought was strongly determined by the tropical oceans," Hoerling and Kumar wrote. "It is thus more than figurative, although not definitive, to claim this ocean was 'perfect' for drought" (Radford, January 31, 2003, 18).

The researchers expressed concern that a generally warmer world could cause such patterns to appear more often and persist longer. Persistent warmth in the Indo-Western Pacific could become a persistent pattern. The area has warmed on an irregular basis since the 1970s, and may continue to do so. "There's a strong suspicion that the Indo-Western Pacific warming trend is related to the global-warming trend," said meteorologist Mathew Barlow of Atmospheric and Environmental Research, Inc. in Lexington, Massachusetts (Kerr 2003a, 636). *See also:* Drought and Deluge: Anecdotal Observations; Drought and Deluge: Scientific Issues; Hadley Cells

FURTHER READING

Barnett, Tim P., David W. Pierce, Hugo G. Hidalgo, Celine Bonfils, Benjamin D. Santer, Tapash Das, Govindasamy Bala, Andrew W. Wood, Toru Nozawa, Arthur A. Mirin, Daniel R. Cayan, and Michael D. Dettinger. "Human-Induced Changes in the Hydrology of the Western United States." *Science* 319 (February 22, 2008):1080–1083.

Barringer, Felicity. "Savage Storms Wreak Havoc Across the Washington Region." *New York Times*, June 27, 2006a. http://www.nytimes.com/2006/06/27/us/27rain.html?pagewanted=print.

Borenstein, Seth. "More Severe U.S. Storms Will Come with Global Warming, NASA Researchers Say." Associated Press, August 30, 2007c. (LEXIS)

Brown, Paul. "Global Warming: It's with Us Now." *London Guardian*, October 31, 2000b, 1.

Dai, Aiguo, Kevin E. Trenberth, and Taotao Qian. "A Global Dataset of Palmer Drought Severity Index for 1870–2002: Relationship with Soil Moisture and Effects of Surface Warming." *Journal of Hydrometeorology* 5, no. 6 (December 2004):1117–1130.

Daley, Suzanne. "Battered by Fierce Weekend Storm, Western Europe Begins an Enormous Cleanup Job." *New York Times*, Tuesday, December 28, 1999, A-10. [International Edition.]

Del Genio, Anthony D., Mao-Sung Yao, and Jeffrey Jonas. "Will Moist Convection Be Stronger in a Warmer Climate?" *Geophysical Research Letters* 34, no. 16 (August 17, 2007), L16703, doi: 10.1029/2007GL030525.

Hoerling, Martin, and Arun Kumar. "The Perfect Ocean for Drought." *Science* 299 (January 31, 2003): 691–694.

Hume, Stephen. "A Risk We Can't Afford: The Summer of Fire and the Winter of the Deluge Should Prove to the Nay-sayers That if We Wait too Long to React to Climate Change We'll Be in Grave Peril." *Vancouver Sun*, October 23, 2003, A-13.

Kerr, Richard A. "A Perfect Ocean for Four Years of Globe-Girdling Drought." *Science* 299 (January 31, 2003a):636.

Radford, Tim. "Scientists Discover the Harbinger of Drought: Subtle Temperature Changes in Tropical Seas May Trigger Northern Hemisphere's Long, Dry Spells." *London Guardian*, January 31, 2003, 18.

Revkin, Andrew C. "Forecast for a Warmer World: Deluge and Drought." *New York Times*, August 28, 2002b, A-10.

Toner, Mike. "Drought May Signal World Warming Trend." *Atlanta Journal-Constitution*, January 31, 2003b, 4-A.

Trenberth, Kevin E., Aiguo Dai, Roy M. Rassmussen, and David B. Parsons. "The Changing Character of Precipitation." *Bulletin of the American Meteorological Society* (September 2003):1205–1217.

Treydte, Kerstin S., Gerhard H. Schleser, Gerhard Helle, David C. Frank, Matthias Winiger, Gerald H. Haug, and Jan Esper. "The Twentieth Century Was the Wettest Period in Northern Pakistan over the Past Millennium." *Nature* 440 (April 27, 2006): 1179–1182.

"U.S. N.S.F.: Scientists Find Climate Change is Major Factor in Drought's Growing Reach." M2 Presswire, January 12, 2005. (LEXIS)

Watt-Cloutier, Sheila. Personal communication, March 1, 2006b.

Wolf, Martin. "Hot Air about Global Warming." *London Financial Times*, November 29, 2000, 27.